LIFE SCIENCES, ECOLOGY, AND HEALTH

MISCELLANEOUS

SOCIAL SCIENCES

APPLIED
CALCULUS

APPLIED CALCULUS

LAURENCE D. HOFFMANN

Claremont McKenna College

McGraw-Hill Book Company

New York   St. Louis   San Francisco   Auckland   Bogotá   Hamburg
Johannesburg   London   Madrid   Mexico   Montreal   New Delhi
Panama   Paris   São Paulo   Singapore   Sydney   Tokyo   Toronto

APPLIED CALCULUS

Copyright © 1983 by McGraw-Hill, Inc. All rights reserved.
This book includes Chapters 1 to 7 and Appendixes A and B of *Calculus for the Social, Managerial, and Life Sciences*, Second Edition by Laurence D. Hoffmann, copyright © 1980 by McGraw-Hill, Inc. All rights reserved.
Printed in the United States of America. Except as permitted under the United States Copyright Act of 1976, no part of this publication may be reproduced or distributed in any form or by any means, or stored in a data base or retrieval system, without the prior written permission of the publisher.

1 2 3 4 5 6 7 8 9 0   D O C D O C   8 9 8 7 6 5 4 3 2

ISBN 0-07-029319-8

This book was set in Aster by Progressive Typographers.
The editors were Peter R. Devine and James W. Bradley;
the production supervisor was Dennis J. Conroy.
New drawings were done by J & R Services, Inc.
The cover was designed by Anne Canevari Green.
R. R. Donnelley & Sons Company was printer and binder.

Library of Congress Cataloging in Publication Data

Hoffmann, Laurence D., date
    Applied calculus.

    "This book includes chapters 1 to 7 and appendixes A and B of Calculus for the social, managerial, and life sciences, second edition"—T.p. verso.
    Includes index.
    1. Calculus.   I. Hoffmann, Laurence D., date Calculus for the social, managerial, and life sciences.   II. Title.
QA303.H5688   1983   515   82-4702
ISBN 0-07-029319-8

CONTENTS

v

PREFACE

If you are preparing for a career in business, economics, psychology, sociology, architecture, or biology, and if you have taken high school algebra, then this book was written for you. Its primary goal is to teach you the techniques of differential and integral calculus that you are likely to encounter in undergraduate courses in your major and in your subsequent professional activities.

Applications The text is applications-oriented. Each new concept you learn is applied to a variety of practical situations. Special emphasis is placed on the techniques and strategies you will need to solve practical problems.

Problems You learn mathematics by doing it. Each section in this text is followed by an extensive set of problems. Many involve routine computation and are designed to help you master new techniques. Others ask you to apply the new techniques to practical situations. There is a proficiency test at the end of each chapter. At the back of the book you will find the answers to the odd-numbered problems and to all the proficiency test questions.

Level of rigor Theory for its own sake has been avoided. However, the main results are stated carefully and completely, and most of them are explained or justified. Whenever possible, explanations are informal and intuitive.

Duration The text is intended for use in a two-semester course. To allow for flexibility in the design of the course, the text actually contains more material than you will be able to complete in one year.

Relationship to the author's other calculus text

This text is an expansion of "Calculus for the Social, Managerial, and Life Sciences," second edition, McGraw-Hill, 1980. It consists of all seven chapters from the shorter text plus five new chapters on limits at infinity and improper integrals, multiple integrals, infinite series, numerical methods, and trigonometric functions. This expanded text does not replace the shorter one. Both will be available on a continuing basis.

Algebra review

If you need to brush up on your high school algebra, there is an algebra review in the appendix that includes worked examples and practice problems for you to do. You will be advised throughout the text when it might be appropriate to consult this material.

Numerical methods

Functions and data that arise in practical situations are often much more unruly than those in the simplified examples found in calculus books. To work with such functions and data you will need to use numerical methods of approximation. Several of these methods are discussed in Chapter 11. The discussion includes analyses of the accuracy of the various methods.

Computers and programming

A computer supplement by Professor Granville C. Henry and the author has been prepared to accompany this text. In the supplement you will be introduced to the use of library programs to solve calculus problems and will learn elementary programming in BASIC. In the process, you will develop an appreciation for the capabilities and limitations of both calculus and the computer.

Acknowledgments

Many people helped with the preparation of this book. My friend and former colleague, economist Jerry St. Dennis, offered generous advice and constant encouragement during the preparation of the first edition, and his influence is still evident in the current text.

Several reviewers read early versions of the manuscript. Especially valuable were the detailed comments of: George Feissner, State University of New York at Cortland; Charles Frady, Georgia State University; Alexander Hahn, University of Notre Dame; Charles Himmelberg, University of Kansas; Stanley Lukawecki, Clemson University; John G. Michaels, State University of New York at Brockport; and Robert Zink, Purdue University.

Reviewers of the shorter text upon which this one is based include: George Articolo, Rutgers University; Theodore J. Barth, University of

California at Riverside; Barbara Lee Bleau, Pennsylvania State University; Carl Eberhardt, University of Kentucky; Rodney T. Hansen, Montana State University; William Huebsch, Canisius College; Roger Johnes, De Paul University; V. J. Klaussen, California State University at Fullerton; Lowell Leake, University of Cincinnati; Robert A. Mills, Eastern Michigan University; Bill New, Cerritos College; J. A. Pfaltzgraff, University of North Carolina; Karen J. Schroeder, Bentley College; David Shea, University of Wisconsin; Paul Slepian, Howard University; George Springer, Indiana University; and Charles Votaw, Fort Hays State University.

Kathleen Dennison, Kenneth Drew, Mark Greaves, Craig Lytle, Paul Martinez, Tom McKay, Michael David Miller, Howard Newberg, Terry Phife, Jack Rann, Mansoora Rashid, Blaise Stoltenberg, Eric Swanson, Kenneth Wechsler, and Jeff Zeiger are current or former students at the Claremont Colleges who checked the manuscript for accuracy and who worked all the problems.

Carol B. Cole has been a friend and collaborator throughout the preparation of this book. I am especially grateful for her involvement.

My editors at McGraw-Hill have been particularly helpful, encouraging, and patient. Peter Devine, James W. Bradley, and Carol Napier are professional editors of the highest quality, and it has been a pleasure working with them.

Laurence D. Hoffmann

1 FUNCTIONS   In many practical situations, the value of one quantity may depend on the value of a second. For example, the consumer demand for beef may depend on its current market price; the amount of air pollution in a metropolitan area may depend on the number of cars on the road; the value of a bottle of wine may depend on its age. Such relationships can often be represented mathematically as **functions.**

Function   **A function is a rule that associates with each object in a set A, one and only one object in a set B.**

This definition is illustrated in Figure 1.1.

For most of the functions in this book, the sets A and B will be collections of real numbers. You can think of such a function as a rule

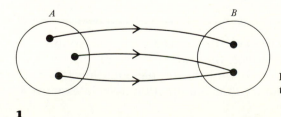

Figure 1.1 A visual representation of a function.

1

that associates "new" numbers with "old" numbers. To be called a function, the rule must have the property that it associates one and only one "new" number with each "old" number. Here is an example.

EXAMPLE 1.1

According to a certain function, the "new" number is obtained by adding 4 to the square of the "old" number. What number does this function associate with 3?

SOLUTION

The number associated with 3 is $3^2 + 4$, or 13.

Variables Often, you can write a function compactly by using a mathematical formula. It is traditional to let x denote the old number and y the new number, and write an equation relating x and y. For instance, you can express the function in Example 1.1 by the equation

$$y = x^2 + 4$$

The letters x and y that appear in such an equation are called **variables.** The numerical value of the variable y is determined by that of the variable x. For this reason, y is sometimes referred to as the **dependent variable** and x as the **independent variable.**

Functional notation There is an alternative notation for functions that is widely used and somewhat more versatile. A letter such as f is chosen to stand for the function itself, and the value that the function associates with x is denoted by $f(x)$ instead of y. The symbol $f(x)$ is read "f of x." Using this **functional notation,** you can rewrite Example 1.1 as follows.

EXAMPLE 1.2

Find $f(3)$ if $f(x) = x^2 + 4$.

SOLUTION

$$f(3) = 3^2 + 4 = 13$$

Observe the convenience and simplicity of this notation. In Example 1.2, the compact formula $f(x) = x^2 + 4$ completely defines the function, and the simple equation $f(3) = 13$ indicates that 13 is the number that the function associates with 3.

The next example illustrates how functional notation is used in a practical situation. Notice that in this example, as in many practical situations, letters other than f and x are used to represent the function and its independent variable.

EXAMPLE 1.3

Suppose the total cost in dollars of manufacturing q units of a certain commodity is given by the function $C(q) = q^3 - 30q^2 + 500q + 200$.

(a) Compute the cost of manufacturing 10 units of the commodity.
(b) Compute the cost of manufacturing the 10th unit of the commodity.

SOLUTION

(a) The cost of manufacturing 10 units is the value of the total cost function when $q = 10$. That is,

$$\begin{aligned} \text{Cost of 10 units} &= C(10) \\ &= (10)^3 - 30(10)^2 + 500(10) + 200 \\ &= \$3{,}200 \end{aligned}$$

(b) The cost of manufacturing the 10th unit is the difference between the cost of manufacturing 10 units and the cost of manufacturing 9 units. That is,

$$\begin{aligned} \text{Cost of 10th unit} &= C(10) - C(9) \\ &= 3{,}200 - 2{,}999 \\ &= \$201 \end{aligned}$$

The domain of a function The set of values of the independent variable for which a function can be evaluated is called the **domain** of the function. For instance, the function $f(x) = x^2 + 4$ in Example 1.2 can be evaluated for any real number x. Thus, the domain of this function is the set of all real numbers. The domain of the function $C(q) = q^3 - 30q^2 + 500q + 200$ in Example 1.3 is also the set of all real numbers (although $C(q)$ represents total cost only for nonnegative values of q). In the next example are two functions whose domains are restricted for algebraic reasons.

EXAMPLE 1.4

Find the domain of each of the following functions.

(a) $f(x) = \dfrac{1}{x - 3}$

(b) $g(x) = \sqrt{x - 2}$

SOLUTION

(a) Since division by any real number except zero is possible, the only value of x for which $f(x)$ cannot be evaluated is $x = 3$, the value that makes the denominator of f equal to zero. Hence the domain of f consists of all real numbers except 3.

(b) Since negative numbers do not have real square roots, the only values of x for which $g(x)$ can be evaluated are those for which $x - 2$ is nonnegative. It follows that the domain of g consists of all real numbers that are greater than or equal to 2.

Composition of functions

There are many situations in which a quantity is given as a function of one variable which, in turn, can be written as a function of a second variable. By combining the functions in an appropriate way, you can express the original quantity as a function of the second variable. This process is known as the **composition of functions.**

Composition of functions

> **The composite function $g[h(x)]$ is the function formed from the two functions $g(u)$ and $h(x)$ by substituting $h(x)$ for u in the formula for $g(u)$.**

The situation is illustrated in Figure 1.2.

EXAMPLE 1.5

Find the composite function $g[h(x)]$ if $g(u) = u^2 + 3u + 1$ and $h(x) = x + 1$.

SOLUTION

Replace u by $x + 1$ in the formula for g to get

$$g[h(x)] = (x + 1)^2 + 3(x + 1) + 1 = x^2 + 5x + 5$$

The problem in Example 1.5 could have been worded more compactly as follows: Find the composite function $g(x + 1)$ where $g(u) = u^2 + 3u + 1$. The use of this compact notation is illustrated further in the next example.

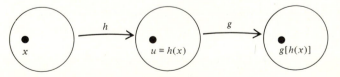

Figure 1.2 The composition of functions.

EXAMPLE 1.6

Find $f(x - 1)$ if $f(x) = 3x^2 + \dfrac{1}{x} + 5$.

SOLUTION

At first glance, this problem may look confusing because the letter x appears both as the independent variable in the formula defining f and as part of the expression $x - 1$. Because of this, you may wish to begin by writing the formula for f in more neutral terms, say as

$$f(\square) = 3(\square)^2 + \frac{1}{\square} + 5$$

To find $f(x - 1)$, you simply insert the expression $x - 1$ inside each box, getting

$$f(x - 1) = 3(x - 1)^2 + \frac{1}{x - 1} + 5$$

The next example illustrates how a composite function may arise in a practical problem.

EXAMPLE 1.7

An environmental study of a certain suburban community suggests that the average daily level of carbon monoxide in the air will be $c(p) = 0.5p + 1$ parts per million when the population is p thousand. It is estimated that t years from now the population of the community will be $p(t) = 10 + 0.1t^2$ thousand. Express the level of carbon monoxide in the air as a function of time.

SOLUTION

Since the level of carbon monoxide is related to the variable p by the equation

$$c(p) = 0.5p + 1$$

and the variable p is related to the variable t by the equation

$$p(t) = 10 + 0.1t^2$$

it follows that the composite function

$$c[p(t)] = 0.5(10 + 0.1t^2) + 1 = 6 + 0.05t^2$$

expresses the level of carbon monoxide in the air as a function of the variable t.

Problems In Problems 1 through 9, compute the indicated values of the given function.

1. $f(x) = 3x^2 + 5x - 2$; $f(1)$, $f(0)$, $f(-2)$

2. $h(t) = (2t + 1)^3$; $h(-1)$, $h(0)$, $h(1)$

3. $g(x) = x + \dfrac{1}{x}$; $g(-1)$, $g(1)$, $g(2)$

4. $f(x) = \dfrac{x}{x^2 + 1}$; $f(2)$, $f(0)$, $f(-1)$

5. $h(t) = \sqrt{t^2 + 2t + 4}$; $h(2)$, $h(0)$, $h(-4)$

6. $g(u) = (u + 1)^{3/2}$; $g(0)$, $g(-1)$, $g(8)$

7. $f(t) = (2t - 1)^{-3/2}$; $f(1)$, $f(5)$, $f(13)$

8. $g(x) = 4 + |x|$; $g(-2)$, $g(0)$, $g(2)$

9. $f(x) = x - |x - 2|$; $f(1)$, $f(2)$, $f(3)$

In Problems 10 through 22, specify the domain of the given function.

10. $f(x) = x^3 - 3x^2 + 2x + 5$

11. $g(x) = \dfrac{x^2 + 5}{x + 2}$

12. $f(t) = \dfrac{t + 1}{t^2 - t - 2}$

13. $y = \sqrt{x - 5}$

14. $f(x) = \sqrt{2x - 6}$

15. $g(t) = \sqrt{t^2 + 9}$

16. $h(u) = \sqrt{u^2 - 4}$

17. $f(t) = (2t - 4)^{3/2}$

18. $y = \dfrac{x - 1}{\sqrt{x^2 + 2}}$

19. $f(x) = (x^2 - 9)^{-1/2}$

20. $h(t) = \dfrac{\sqrt{t^2 - 4}}{\sqrt{t - 4}}$

21. $g(t) = \dfrac{1}{|t - 1|}$

22. $h(x) = \sqrt{|x - 3|}$

Manufacturing cost 23. Suppose the total cost in dollars of manufacturing q units of a certain commodity is given by the function $C(q) = q^3 - 30q^2 + 400q + 500$.
   (a) Compute the cost of manufacturing 20 units.
   (b) Compute the cost of manufacturing the 20th unit.

Worker efficiency 24. An efficiency study of the morning shift at a certain factory indicates that an average worker who arrives on the job at 8:00 A.M. will have assembled $f(x) = -x^3 + 6x^2 + 15x$ transistor radios x hours later.
   (a) How many radios will such a worker have assembled by 10:00 A.M.?

(b) How many radios will such a worker assemble between 9:00 A.M. and 10:00 A.M.?

Temperature change 25. Suppose that t hours past midnight, the temperature in Miami was $C(t) = -\frac{1}{6}t^2 + 4t + 10$ degrees Celsius.
  (a) What was the temperature at 2:00 P.M.?
  (b) By how much did the temperature increase or decrease between 6:00 P.M. and 9:00 P.M.?

Population growth 26. It is estimated that t years from now, the population of a certain suburban community will be $P(t) = 20 - \dfrac{6}{t + 1}$ thousand.
  (a) What will the population of the community be 9 years from now?
  (b) By how much will the population increase during the 9th year?
  (c) What will happen to the size of the population in the long run?

Experimental psychology 27. To study the rate at which animals learn, a group of psychology students performed an experiment in which a white rat was sent repeatedly through a laboratory maze. The students found that the time required for the rat to traverse the maze on the nth trial was approximately $f(n) = 3 + \dfrac{12}{n}$ minutes.
  (a) What is the domain of the function f?
  (b) For what values of n does $f(n)$ have meaning in the context of the psychology experiment?
  (c) How long did it take the rat to traverse the maze on the 3rd trial?
  (d) On which trial did the rat first traverse the maze in 4 minutes or less?
  (e) According to the function f, what will happen to the time required for the rat to traverse the maze as the number of trials increases? Will the rat ever be able to traverse the maze in less than 3 minutes?

Poiseuille's law 28. Biologists have found that the speed of blood in an artery is a function of the distance of the blood from the artery's central axis. According to *Poiseuille's law*, the speed (in centimeters per second) of blood that is r centimeters from the central axis of an artery is given by the function $S(r) = C(R^2 - r^2)$, where C is a constant and R is the radius of the artery. Suppose that for a certain artery, $C = 1.76 \times 10^5$ centimeters and $R = 1.2 \times 10^{-2}$ centimeters.
  (a) Compute the speed of the blood at the central axis of this artery.

(b) Compute the speed of the blood midway between the artery's wall and central axis.

Distribution cost 29. It is estimated that the number of worker-hours required to distribute new telephone books to x percent of the households in a certain rural community is given by the function $f(x) = \dfrac{600x}{300 - x}$.

(a) What is the domain of the function f?

(b) For what values of x does $f(x)$ have a practical interpretation in this context?

(c) How many worker-hours were required to distribute new telephone books to the first 50 percent of the households?

(d) How many worker-hours were required to distribute new telephone books to the entire community?

(e) What percentage of the households in the community had received new telephone books by the time 150 worker-hours had been expended?

Immunization 30. During a nationwide program to immunize the population against a virulent form of influenza, public health officials found that the cost of inoculating x percent of the population was approximately $f(x) = \dfrac{150x}{200 - x}$ million dollars.

(a) What is the domain of the function f?

(b) For what values of x does $f(x)$ have a practical interpretation in this context?

(c) What was the cost of inoculating the first 50 percent of the population?

(d) What was the cost of inoculating the second 50 percent of the population?

(e) What percentage of the population had been inoculated by the time 37.5 million dollars had been spent?

Speed of a moving object 31. A ball has been dropped from the top of a building. Its height (in feet) after t seconds is given by the function $H(t) = -16t^2 + 256$.

(a) How high will the ball be after 2 seconds?

(b) How far will the ball travel during the 3rd second?

(c) How tall is the building?

(d) When will the ball hit the ground?

In Problems 32 through 39, find the composite function $g[h(x)]$.

32. $g(u) = u^2 + 4$, $h(x) = x - 1$

33. $g(u) = 3u^2 + 2u - 6$, $h(x) = x + 2$

34. $g(u) = (2u + 10)^2$, $h(x) = x - 5$

35. $g(u) = (u - 1)^3 + 2u^2$, $h(x) = x + 1$

36. $g(u) = \dfrac{1}{u}$, $h(x) = x^2 + x - 2$

37. $g(u) = \dfrac{1}{u^2}$, $h(x) = x - 1$

38. $g(u) = u^2$, $h(x) = \dfrac{1}{x - 1}$

39. $g(u) = \sqrt{u + 1}$, $h(x) = x^2 - 1$

In Problems 40 through 47, find the indicated composite function.

40. $f(x + 1)$ where $f(x) = x^2 + 5$

41. $f(x - 2)$ where $f(x) = 2x^2 - 3x + 1$

42. $f(x + 3)$ where $f(x) = (2x - 6)^2$

43. $f(x - 1)$ where $f(x) = (x + 1)^5 - 3x^2$

44. $f\left(\dfrac{1}{x}\right)$ where $f(x) = 3x + \dfrac{2}{x}$

45. $f(x^2 + 3x - 1)$ where $f(x) = \sqrt{x}$

46. $f(x^2 - 2x + 9)$ where $f(x) = 2x - 20$

47. $f(x + 1)$ where $f(x) = \dfrac{x - 1}{x}$

In Problems 48 through 53, find functions $h(x)$ and $g(u)$ such that $f(x) = g[h(x)]$.

48. $f(x) = (x^5 - 3x^2 + 12)^3$

49. $f(x) = \sqrt{3x - 5}$

50. $f(x) = (x - 1)^2 + 2(x - 1) + 3$

51. $f(x) = \dfrac{1}{x^2 + 1}$

52. $f(x) = \sqrt{x + 4} - \dfrac{1}{(x + 4)^3}$

53. $f(x) = \sqrt{x + 3} - \dfrac{1}{(x + 4)^3}$

Air pollution 54. An environmental study of a certain suburban community suggests that the average daily level of carbon monoxide in the air will be $c(p) = 0.4p + 1$ parts per million when the population

is p thousand. It is estimated that t years from now the population of the community will be $p(t) = 8 + 0.2t^2$ thousand.
   (a) Express the future level of carbon monoxide in the community as a function of time.
   (b) What will the carbon monoxide level be 2 years from now?

Manufacturing cost 55. At a certain factory, the total cost of manufacturing q units during the daily production run is $C(q) = q^2 + q + 900$ dollars. On a typical workday, $q(t) = 25t$ units are manufactured during the first t hours of a production run.
   (a) Express the total manufacturing cost as a function of t.
   (b) How much will have been spent on production by the end of the 3rd hour?

Consumer demand 56. An importer of Brazilian coffee estimates that local consumers will buy approximately $\dfrac{4,374}{p^2}$ kilograms of the coffee per week when the price is p dollars per kilogram. It is estimated that t weeks from now the price of Brazilian coffee will be $0.04t^2 + 0.2t + 12$ dollars per kilogram.
   (a) Express the weekly consumer demand for the coffee as a function of t.
   (b) How many kilograms of the coffee will consumers be buying from the importer 10 weeks from now?

2 GRAPHS Graphs have visual impact. They also reveal information that may not be evident from verbal or algebraic descriptions. Two graphs depicting practical relationships are shown in Figure 2.1.

The graph in Figure 2.1a describes the effect that the market price of a commodity has on the manufacturer's total profit. According to the graph, profit will be small if the market price is either very low or

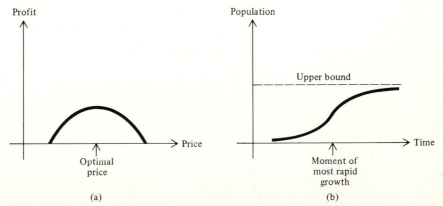

Figure 2.1 (a) A profit function. (b) Bounded population growth.

very high. (Can you explain why?) The fact that the graph has a peak suggests that there is an optimal selling price at which the manufacturer's profit will be greatest.

The graph in Figure 2.1b represents population growth when environmental factors impose an upper bound on the possible size of the population. It indicates that the rate of population growth increases at first and then decreases as the size of the population gets closer and closer to the upper bound.

The graph of a function

To represent a function $y = f(x)$ geometrically as a graph, it is traditional to use a rectangular coordinate system on which units for the independent variable x are marked on the horizontal axis and units for the dependent variable y on the vertical axis.

The graph of a function

> **The graph of a function f consists of all points whose coordinates (x, y) satisfy the equation $y = f(x)$.**

In Chapters 2 and 3 you will see efficient techniques you can use to draw accurate graphs of functions. For many functions, however, you can make a fairly good sketch by the elementary method of plotting points. This method is illustrated in the following examples.

EXAMPLE 2.1

Graph the function $f(x) = x^2$.

SOLUTION

Begin by computing $f(x)$ for several convenient values of x and summarize the results in a table.

x	-2	-1	$-\frac{1}{2}$	0	$\frac{1}{2}$	1	2
$f(x)$	4	1	$\frac{1}{4}$	0	$\frac{1}{4}$	1	4

Then plot the corresponding points $(x, f(x))$ and connect them by a smooth curve. The resulting graph is shown in Figure 2.2.

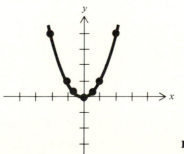

Figure 2.2 The graph of the function $y = x^2$.

EXAMPLE 2.2

Graph the function $f(x) = \dfrac{1}{x^2}$.

SOLUTION

As before, begin by computing $f(x)$ for several convenient values of x and summarize the results in a table. Because division by zero is impossible, you will not be able to calculate $f(0)$. This means that there will be no point on the graph whose x coordinate is zero, and consequently, there will be a break in the graph when $x = 0$. To find out what the graph looks like near $x = 0$, you should include in your table some values of x that are close to zero.

x	-3	-2	-1	$-\frac{1}{2}$	0	$\frac{1}{2}$	1	2	3
$f(x)$	$\frac{1}{9}$	$\frac{1}{4}$	1	4		4	1	$\frac{1}{4}$	$\frac{1}{9}$

Now plot the corresponding points $(x, f(x))$ and draw the graph as shown in Figure 2.3. Don't forget that there should be a break in the graph when $x = 0$.

Quadratic functions

The function $f(x) = x^2$ sketched in Figure 2.2 is an example of a **quadratic function.** In general, a quadratic function is a function of the form

$$f(x) = ax^2 + bx + c$$

where a, b, and c are constants and $a \neq 0$. The graph of a quadratic function is a **parabola,** which is the curve formed by the intersection of a circular cone and a plane as shown in Figure 2.4. The graph of a

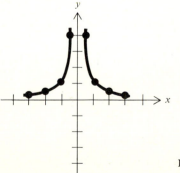

Figure 2.3 The graph of the function $y = \dfrac{1}{x^2}$.

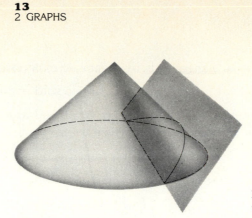

Figure 2.4 A parabola: the intersection of a cone and a plane.

quadratic function will be **concave upward** (as in Figure 2.5a) if the coefficient a is positive and **concave downward** (as in Figure 2.5b) if the coefficient a is negative. (You will prove this in Chapter 3 using calculus.)

Many profit functions in economics are quadratic. Here is an example.

EXAMPLE 2.3

A manufacturer can produce radios at a cost of $10 apiece and estimates that if they are sold for x dollars apiece, consumers will buy approximately $80 - x$ radios each month. Express the manufacturer's monthly profit as a function of the price at which the radios are sold, graph this function, and estimate the price at which the manufacturer's profit will be greatest.

SOLUTION

Begin by stating the desired relationship in words.

$$\text{Profit} = (\text{number of radios sold})(\text{profit per radio})$$

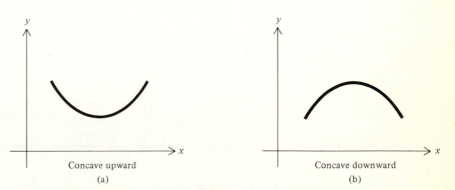

Concave upward
(a)

Concave downward
(b)

Figure 2.5 The graph of a quadratic function.

Now replace the words by algebraic expressions. You know that

$$\text{Number of radios sold} = 80 - x$$

Moreover, since the radios are produced at a cost of \$10 apiece and sold for x dollars apiece, it follows that

$$\text{Profit per radio} = x - 10$$

Let $P(x)$ denote the profit and conclude that

$$P(x) = (80 - x)(x - 10)$$

(Notice that the profit function is factored. Resist the temptation to multiply it out. It is already in its most convenient form.)

To sketch this function, compile a table of representative values and plot the corresponding points as shown in Figure 2.6.

x	10	20	30	40	50	60	70	80
$P(x)$	0	600	1,000	1,200	1,200	1,000	600	0

Notice that the graph crosses the x axis at (10, 0) and (80, 0). These points are known as the x **intercepts** of the graph. The location of the x intercepts is obvious from the factored form of the profit function $P(x) = (80 - x)(x - 10)$. Can you explain these x intercepts in economic terms?

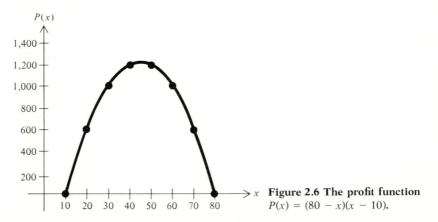

Figure 2.6 The profit function $P(x) = (80 - x)(x - 10)$.

The graph suggests that the price at which the manufacturer's profit will be greatest is approximately \$45. In Chapter 2 you will learn how to use calculus to find this optimal price exactly.

Polynomials   A **polynomial** is a function of the form

$$f(x) = a_0 + a_1 x + a_2 x^2 + \cdots + a_n x^n$$

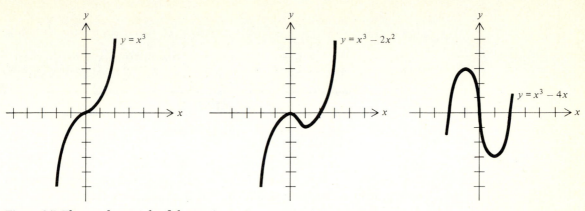

Figure 2.7 Three polynomials of degree 3.

where n is a nonnegative integer and $a_0, a_1, \ldots, a_n$ are constants. If $a_n \neq 0$, the integer n is said to be the **degree** of the polynomial. For example, the function $f(x) = 3x^5 - 6x^2 + 7$ is a polynomial of degree 5. Quadratic functions are polynomials of degree 2. It can be shown that the graph of a polynomial of degree n is an unbroken curve that crosses the x axis no more than n times. The graphs of three polynomials of degree 3 are shown in Figure 2.7.

Rational functions   The quotient of two polynomials is called a **rational function.** For example, the functions $f(x) = \dfrac{x^2 + 1}{x - 2}$ and $f(x) = \dfrac{1}{x^2}$ are both rational. So is the function $f(x) = 1 + \dfrac{1}{x}$ since it can be rewritten as $f(x) = \dfrac{x + 1}{x}$.

Since division by zero is impossible, there will be a break or **discontinuity** in the graph of a rational function at each value of x for which the denominator is equal to zero. You are already familiar with this

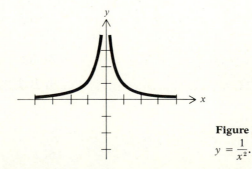

Figure 2.8 The graph of the rational function $y = \dfrac{1}{x^2}$.

phenomenon in the case of the rational function $f(x) = \frac{1}{x^2}$ whose graph (reproduced in Figure 2.8) has a discontinuity when $x = 0$.

Many cost functions in economics are rational. Here is an example.

EXAMPLE 2.4

For each shipment of raw materials, a manufacturer must pay an ordering fee to cover handling and transportation. After they are received, the raw materials must be stored until needed and storage costs result. If each shipment of raw materials is large, ordering costs will be low because few shipments are required, but storage costs will be high. If each shipment is small, ordering costs will be high because many shipments will be required, but storage costs will drop. A manufacturer estimates that if each shipment contains x units, the total cost of obtaining and storing the year's supply of raw materials will be $C(x) = x + \frac{160,000}{x}$ dollars. Sketch the relevant portion of this cost function and estimate the optimal shipment size.

SOLUTION

$C(x)$ is a rational function with a discontinuity at $x = 0$, and represents cost for nonnegative values of x. Compile a table for some representative nonnegative values of x and plot the corresponding points to get the graph shown in Figure 2.9.

x	100	200	300	400	500	600	700	800
$C(x)$	1,700	1,000	833	800	820	867	929	1,000

The graph indicates that total cost will be high if shipments are very small or very large and that the optimal shipment size is approximately 400 units.

Problems In Problems 1 through 18, sketch the graph of the given function. (If you wish, use a hand calculator to help with the computations.)

1. $f(x) = x$ 　　　　　　　　　　2. $f(x) = x^2$

3. $f(x) = x^3$ 　　　　　　　　　　4. $f(x) = x^4$

5. $f(x) = \frac{1}{x}$ 　　　　　　　　　　6. $f(x) = \frac{1}{x^2}$

7. $f(x) = \frac{1}{x^3}$ 　　　　　　　　　　8. $f(x) = \sqrt{x}$

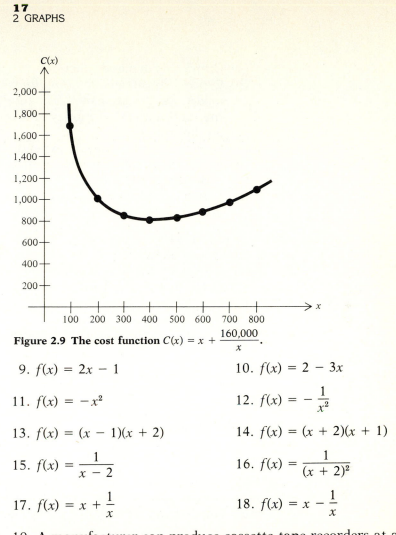

Figure 2.9 The cost function $C(x) = x + \dfrac{160,000}{x}$.

9. $f(x) = 2x - 1$

10. $f(x) = 2 - 3x$

11. $f(x) = -x^2$

12. $f(x) = -\dfrac{1}{x^2}$

13. $f(x) = (x - 1)(x + 2)$

14. $f(x) = (x + 2)(x + 1)$

15. $f(x) = \dfrac{1}{x - 2}$

16. $f(x) = \dfrac{1}{(x + 2)^2}$

17. $f(x) = x + \dfrac{1}{x}$

18. $f(x) = x - \dfrac{1}{x}$

Manufacturing cost 19. A manufacturer can produce cassette tape recorders at a cost of $20 apiece. It is estimated that if the tape recorders are sold for x dollars apiece, consumers will buy $120 - x$ of them a month. Express the manufacturer's monthly profit as a function of price, graph this function, and use the graph to estimate the optimal selling price.

Retail sales 20. A bookstore can obtain an atlas from the publisher at a cost of $5 per copy and estimates that if it sells the atlas for x dollars per copy, approximately $20(22 - x)$ copies will be sold each month. Express the bookstore's monthly profit from the sale of the atlas as a function of price, graph this function, and use the graph to estimate the optimal selling price.

Consumer expenditure 21. The consumer demand for a certain commodity is $D(p) = -200p + 12,000$ units per month when the market price is p dollars per unit.

(a) Graph this demand function.
(b) Express consumers' *total monthly expenditure* for the commodity as a function of p. (The total monthly expenditure is the total amount of money consumers spend each month on the commodity.)
(c) Graph the total monthly expenditure function.
(d) Discuss the economic significance of the p intercepts of the expenditure function.
(e) Use the graph in part (c) to estimate the market price that generates the greatest consumer expenditure.

Speed of a moving object 22. If an object is thrown vertically upward from the ground with an initial speed of 160 feet per second, its height (in feet) t seconds later is given by the function $H(t) = -16t^2 + 160t$.
(a) Graph the function $H(t)$.
(b) Use the graph in part (a) to determine when the object will hit the ground.
(c) Use the graph in part (a) to estimate how high the object will rise.

Distribution cost 23. It is estimated that the number of worker-hours required to distribute new telephone books to x percent of the households in a certain rural community is given by the function $f(x) = \dfrac{600x}{300 - x}$.
Sketch this function and specify what portion of the graph is relevant to the practical situation under consideration.

Immunization 24. Suppose that during a nationwide program to immunize the population against a virulent form of influenza, public health officials found that the cost of inoculating x percent of the population was approximately $f(x) = \dfrac{150x}{200 - x}$ million dollars. Sketch this function and specify what portion of the graph is relevant to the practical situation under consideration.

Experimental psychology 25. To study the rate at which animals learn, a group of psychology students performed an experiment in which a white rat was sent repeatedly through a laboratory maze. The students found that the time required for the rat to traverse the maze on the nth trial was approximately $f(n) = 3 + \dfrac{12}{n}$ minutes.
(a) Graph the function $f(n)$.
(b) What portion of the graph is relevant to the practical situation under consideration?
(c) What happens to the graph as n increases without bound? Interpret your answer in practical terms.

Inventory cost 26. A manufacturer estimates that if each shipment of raw materials contains x units, the total cost of obtaining and storing the year's supply of raw materials will be $C(x) = 2x + \dfrac{80,000}{x}$ dollars. Sketch the relevant portion of the graph of this cost function and estimate the optimal shipment size.

Production cost 27. A manufacturer estimates that if x machines are used, the cost of a production run will be $C(x) = 20x + \dfrac{2,000}{x}$ dollars. Sketch the relevant portion of this cost function and estimate how many machines the manufacturer should use to minimize cost.

Average cost 28. Suppose the total cost of manufacturing x units of a certain commodity is $C(x) = x^2 + 6x + 19$ dollars. Express the average cost per unit as a function of the number of units produced and, on the same set of axes, sketch the total cost and average cost functions. (*Hint:* Average cost is total cost divided by the number of units produced.)

Average cost 29. Suppose the total cost of manufacturing x units is given by the function $C(x) = x^2 + 4x + 16$. Express the average cost per unit as a function of the number of units produced and, on the same set of axes, sketch the total cost and average cost functions.

30. (a) Graph the functions $y = x^2$ and $y = x^2 + 3$. How are the graphs related?
    (b) Without further computation, graph the function $y = x^2 - 5$.
    (c) Suppose $g(x) = f(x) + c$, where c is a constant. How are the graphs of f and g related? Explain.

31. (a) Graph the functions $y = x^2$ and $y = -x^2$. How are the graphs related?
    (b) Suppose $g(x) = -f(x)$. How are the graphs of f and g related? Explain.

32. Graph the function $y = \dfrac{1}{x}$. Then, without further computation, graph the function $y = 3 - \dfrac{1}{x}$. (*Hint:* Combine the rules you discovered in Problems 30c and 31b.)

33. (a) Graph the functions $y = x^2$ and $y = (x - 2)^2$. How are the graphs related?
    (b) Without further computation, graph the function $y = (x + 1)^2$.
    (c) Suppose $g(x) = f(x - c)$, where c is a constant. How are the graphs of f and g related? Explain.

34. Graph the function $y = \dfrac{1}{x}$. Then, without further computation, graph the function $y = 3 - \dfrac{1}{x + 2}$.

Distance formula

35. Show that the distance d between the two points (x_1, y_1) and (x_2, y_2) is given by the formula

$$d = \sqrt{(x_2 - x_1)^2 + (y_2 - y_1)^2}$$

(*Hint:* Apply the pythagorean theorem to a right triangle whose hypotenuse is the line segment joining the two points.)

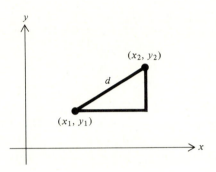

36. Compute the distance between the two given points using the formula in Problem 35.
    (a) $(1, 0)$ and $(0, 1)$
    (b) $(5, -1)$ and $(2, 3)$
    (c) $(2, 6)$ and $(2, -1)$

3 LINEAR FUNCTIONS

In many practical situations, the rate at which one quantity changes with respect to another is constant. Here is a simple example from economics.

EXAMPLE 3.1

A manufacturer's total cost consists of a fixed overhead of $200 plus production costs of $50 per unit. Express the total cost as a function of the number of units produced and draw the graph.

SOLUTION

Let x denote the number of units produced and $C(x)$ the corresponding total cost. Then,

Total cost = (cost per unit)(number of units) + overhead

where

$$\text{Cost per unit} = 50$$
$$\text{Number of units} = x$$
$$\text{Overhead} = 200$$

Hence,

$$C(x) = 50x + 200$$

The graph of this cost function is sketched in Figure 3.1.

The total cost in Example 3.1 increases at a constant rate of $50 per unit. As a result, its graph in Figure 3.1 is a straight line, increasing in height by 50 units for each 1-unit increase in x.

In general, a function whose value changes at a constant rate with respect to its independent variable is said to be a **linear function.** This is because the graph of such a function is a straight line. In algebraic terms, a linear function is a function of the form

$$f(x) = a_0 + a_1x$$

where a_0 and a_1 are constants. For example, the functions $f(x) = \frac{3}{2} + 2x$, $f(x) = -5x$, and $f(x) = 12$ are all linear. Linear functions are traditionally written in the form

$$y = mx + b$$

where m and b are constants. This standard notation will be used in the discussion that follows.

To work with linear functions, you will need to know the following things about straight lines.

The slope of a line

The **slope** of a line is the amount by which the y coordinate of a point on the line changes when the x coordinate is increased by 1. You can compute the slope of a nonvertical line if you know two of its points. Suppose (x_1, y_1) and (x_2, y_2) lie on a line as indicated in Figure 3.2.

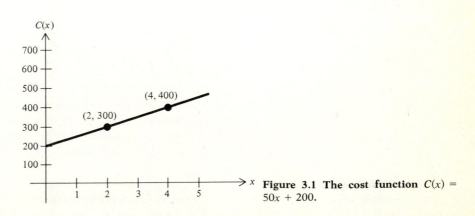

Figure 3.1 The cost function $C(x) = 50x + 200$.

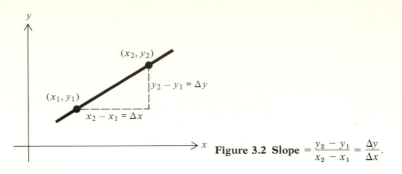

Figure 3.2 Slope $= \dfrac{y_2 - y_1}{x_2 - x_1} = \dfrac{\Delta y}{\Delta x}$.

Between these points, x changes by the amount $x_2 - x_1$ and y by the amount $y_2 - y_1$. The slope is the ratio

$$\text{Slope} = \frac{\text{change in } y}{\text{change in } x} = \frac{y_2 - y_1}{x_2 - x_1}$$

It is sometimes convenient to use the symbol Δy instead of $y_2 - y_1$ to denote the change in y. The symbol Δy is read "delta y." Similarly, the symbol Δx is used to denote $x_2 - x_1$.

The slope of a line

> **The slope of the nonvertical line passing through the points (x_1, y_1) and (x_2, y_2) is given by the formula**
>
> $$\text{Slope} = \frac{\Delta y}{\Delta x} = \frac{y_2 - y_1}{x_2 - x_1}$$

The use of this formula is illustrated in the following example.

EXAMPLE 3.2

Find the slope of the line joining the points $(3, -1)$ and $(-2, 5)$.

SOLUTION

$$\text{Slope} = \frac{\Delta y}{\Delta x} = \frac{5 - (-1)}{-2 - 3} = -\frac{6}{5}$$

The sign and magnitude of the slope of a line indicate the line's direction and steepness, respectively. The slope is positive if the height of the line increases as x increases and is negative if the height decreases as x increases. The absolute value of the slope is large if the slant of the line is severe and small if the slant of the line is gradual. The situation is illustrated in Figure 3.3.

The slope-intercept form of the equation of a line

The constants m and b in the equation $y = mx + b$ have geometric interpretations. The coefficient m is the slope of the corresponding line.

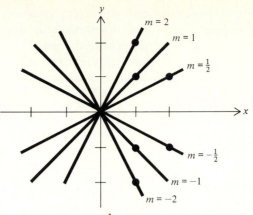

Figure 3.3 The direction and steepness of a line.

To see this, suppose that (x_1, y_1) and (x_2, y_2) are two points on the line $y = mx + b$. Then, $y_1 = mx_1 + b$ and $y_2 = mx_2 + b$ and so

$$\text{Slope} = \frac{y_2 - y_1}{x_2 - x_1} = \frac{(mx_2 + b) - (mx_1 + b)}{x_2 - x_1}$$

$$= \frac{mx_2 - mx_1}{x_2 - x_1} = \frac{m(x_2 - x_1)}{x_2 - x_1} = m$$

The constant b in the equation $y = mx + b$ is the value of y corresponding to $x = 0$. Hence, b is the height at which the line $y = mx + b$ crosses the y axis. The corresponding point $(0, b)$ is known as the y **intercept** of the line. The situation is illustrated in Figure 3.4.

Because the constants m and b in the equation $y = mx + b$ correspond to the slope and y intercept, respectively, this form of the equation of a line is known as the **slope-intercept form.**

The slope-intercept form of the equation of a line

The equation

$$y = mx + b$$

is the equation of the line whose slope is m and whose y intercept is $(0, b)$.

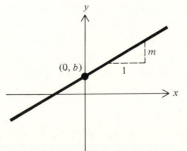

Figure 3.4 The slope and y intercept of the line $y = mx + b$.

The slope-intercept form of the equation of a line is particularly useful when geometric information about a line (such as its slope or y intercept) is to be determined from the line's algebraic representation. Here is a typical application.

EXAMPLE 3.3

Find the slope and y intercept of the line $3y + 2x = 6$ and draw the graph.

SOLUTION

The first step is to put the equation $3y + 2x = 6$ in slope-intercept form $y = mx + b$. To do this, solve for y to get

$$y = -\tfrac{2}{3}x + 2$$

and conclude that the slope is $-\tfrac{2}{3}$ and the y intercept is $(0, 2)$.

To graph a linear function, plot two of its points and draw a straight line through them. In this case, you already know one point, the y intercept $(0, 2)$. A convenient choice for the x coordinate of the second point is $x = 3$. The corresponding y coordinate is $y = -\tfrac{2}{3}(3) + 2 = 0$. Draw a line through the points $(0, 2)$ and $(3, 0)$ to obtain the graph shown in Figure 3.5.

Horizontal and vertical lines

Horizontal and vertical lines (Figure 3.6a and 3.6b) have particularly simple equations. The y coordinates of all the points on a horizontal line are the same. Hence, a horizontal line is the graph of a linear function of the form $y = b$, where b is a constant. The slope of a horizontal line is zero, since changes in x produce no changes in y.

The x coordinates of all the points on a vertical line are equal. Hence, vertical lines are characterized by equations of the form

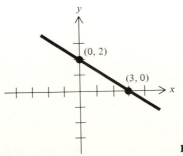

Figure 3.5 The line $3y + 2x = 6$.

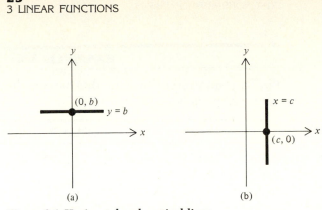

Figure 3.6 Horizontal and vertical lines.

$x = c$, where c is a constant. The slope of a vertical line is undefined. This is because only the y coordinates of points on a vertical line can change, and so the denominator of the quotient $\dfrac{\text{change in } y}{\text{change in } x}$ is zero.

The point-slope form of the equation of a line

Geometric information about a line can be obtained readily from the slope-intercept formula, $y = mx + b$. There is another form of the equation of a line, however, that is usually more efficient for problems in which a line's geometric properties are known and the goal is to find the equation of the line.

The point-slope form of the equation of a line

The equation

$$y - y_0 = m(x - x_0)$$

is an equation of the line that passes through the point (x_0, y_0) and that has slope equal to m.

The point-slope form of the equation of a line is simply the formula for slope in disguise. To see this, suppose that the point (x, y) lies on the line that passes through a given point (x_0, y_0) and that has slope m. Using the points (x, y) and (x_0, y_0) to compute the slope, you get

$$\frac{y - y_0}{x - x_0} = m$$

which you can put in point-slope form

$$y - y_0 = m(x - x_0)$$

by simply multiplying both sides by $x - x_0$.

The use of the point-slope form of the equation of a line is illustrated in the next two examples.

EXAMPLE 3.4

Find an equation of the line that passes through the point (5, 1) and whose slope is equal to $\frac{1}{2}$.

SOLUTION

Use the formula $y - y_0 = m(x - x_0)$ with $(x_0, y_0) = (5, 1)$ and $m = \frac{1}{2}$ to get

$$y - 1 = \tfrac{1}{2}(x - 5)$$

which you can rewrite as $y = \tfrac{1}{2}x - \tfrac{3}{2}$

Instead of the point-slope form, the slope-intercept form could have been used to solve the problem in Example 3.4. For practice, solve the problem this way. Notice that the solution based on the point-slope formula is much more efficient.

The next example illustrates how you can use the point-slope form to find an equation of a line that passes through two given points.

EXAMPLE 3.5

Find an equation of the line that passes through the points $(3, -2)$ and $(1, 6)$.

SOLUTION

First compute the slope

$$m = \frac{6 - (-2)}{1 - 3} = \frac{8}{-2} = -4$$

Then use the point-slope formula with $(1, 6)$ as the given point (x_0, y_0) to get

$$y - 6 = -4(x - 1) \qquad \text{or} \qquad y = -4x + 10$$

Convince yourself that the resulting equation would have been the same if you had chosen $(3, -2)$ to be the given point (x_0, y_0).

Practical applications   You can often use the techniques for translating geometric information into linear equations in practical problems that are not really geometric in nature. The following examples are typical.

EXAMPLE 3.6

Since the beginning of the year, the price of a loaf of whole-wheat bread at a local supermarket has been rising at a constant rate of 2 cents per month. By November first, the price had reached 64 cents per loaf. Express the price of the bread as a function of time and determine the price at the beginning of the year.

SOLUTION

Let x denote the number of months that have elapsed since the first of the year and y the price of a loaf of bread. Since y changes at a constant rate with respect to x, the function relating y to x must be linear and its graph a straight line. The fact that the price y increases by 2 each time x increases by 1 implies that the slope of the line is 2. The fact that the price was 64 cents on November first (10 months after the first of the year) implies that the line passes through the point (10, 64). To write an equation defining y as a function of x, use the point-slope formula $y - y_0 = m(x - x_0)$ with

$$m = 2 \quad \text{and} \quad (x_0, y_0) = (10, 64)$$

and get $\quad y - 64 = 2(x - 10) \quad$ or $\quad y = 2x + 44$

The corresponding line is shown in Figure 3.7. Notice that $y = 44$ when $x = 0$, which implies that the price of the bread at the beginning of the year was 44 cents per loaf.

EXAMPLE 3.7

The average SAT scores of incoming students at an eastern liberal arts college have been declining at a constant rate in recent years. In 1974, the average SAT score was 582 while in 1979 it was only 552. Express the average SAT score as a function of time. If the trend continues, what will the average SAT score of incoming students be in 1984?

SOLUTION

Let x denote the number of years since 1974 and y the average SAT score of incoming students. Since y changes at a constant rate with respect to x, the function relating y to x must be linear. Since $y = 582$

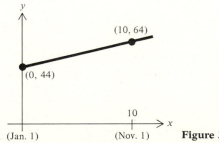

Figure 3.7 The rising price of bread: $y = 2x + 44.$

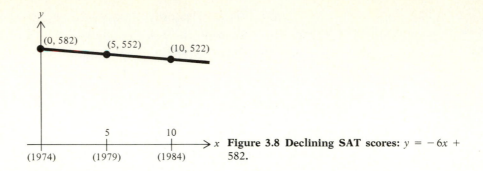

Figure 3.8 Declining SAT scores: $y = -6x + 582$.

when $x = 0$ and $y = 552$ when $x = 5$, the corresponding straight line must pass through the points (0, 582) and (5, 552). The slope of this line is

$$m = \frac{582 - 552}{0 - 5} = -6$$

Since one of the given points happens to be the y intercept (0, 582), use the slope-intercept form and conclude immediately that

$$y = -6x + 582$$

The corresponding line is shown in Figure 3.8.

To predict the average SAT score in 1984, compute y when $x = 10$ and get $y = -6(10) + 582 = 522$.

Problems

In Problems 1 through 5, find the slope (if possible) of the line that passes through the given pair of points.

1. (2, −3) and (0, 4)
2. (−1, 2) and (2, 5)
3. (2, 0) and (0, 2)
4. (5, −1) and (−2, −1)
5. (2, 6) and (2, −4)

In Problems 6 through 16, find the slope and y intercept (if they exist) of the given line and draw a graph.

6. $y = 3x$
7. $y = 5x + 2$
8. $y = 3x - 6$
9. $x + y = 2$
10. $3x + 2y = 6$
11. $2x - 4y = 12$
12. $5y - 3x = 4$
13. $4x = 2y + 6$
14. $\dfrac{x}{2} + \dfrac{y}{5} = 1$
15. $y = 2$
16. $x = -3$

In Problems 17 through 27, write an equation for the line with the given properties.

17. Through $(2, 0)$ with slope 1

18. Through $(-1, 2)$ with slope $\frac{2}{3}$

19. Through $(5, -2)$ with slope $-\frac{1}{2}$

20. Through $(0, 0)$ with slope 5

21. Through $(2, 5)$ and parallel to the x axis

22. Through $(2, 5)$ and parallel to the y axis

23. Through $(1, 0)$ and $(0, 1)$

24. Through $(2, 5)$ and $(1, -2)$

25. Through $(-2, 3)$ and $(0, 5)$

26. Through $(1, 5)$ and $(3, 5)$

27. Through $(1, 5)$ and $(1, -4)$

Manufacturing cost   28. A manufacturer's total cost consists of a fixed overhead of $5,000 plus production costs of $60 per unit. Express the total cost as a function of the number of units produced and draw the graph.

Manufacturing cost   29. During the summer, a group of students builds kayaks in a converted garage. The rental cost of the garage is $600 for the summer. The materials needed to build a kayak cost $25. Express the group's total cost as a function of the number of kayaks built and draw the graph.

Course registration   30. Students at a state college may preregister for their fall classes by mail during the summer. Those who do not preregister must register in person in September. The registrar can process 35 students per hour during the September registration period. After 4 hours in September, a total of 360 students have been registered.
   (a) Express the number of students registered as a function of time and draw the graph.
   (b) How many students were registered after 3 hours?
   (c) How many students preregistered during the summer?

Membership fees   31. Membership in a swimming club costs $150 for the 12-week summer season. If a member joins after the start of the season, the fee is prorated; that is, it is reduced linearly.
   (a) Express the membership fee as a function of the number of weeks that have elapsed by the time the membership is purchased and draw the graph.
   (b) Compute the cost of a membership that is purchased 5 weeks after the start of the season.

Linear depreciation 32. A doctor owns $1,500 worth of medical books which, for tax purposes, are assumed to depreciate linearly to zero over a 10-year period. That is, the value of the books decreases at a constant rate so that it is equal to zero at the end of 10 years. Express the value of the books as a function of time and draw the graph.

Linear depreciation 33. A manufacturer buys $20,000 worth of machinery that depreciates linearly so that its trade-in value after 10 years will be $1,000.
  (a) Express the value of the machinery as a function of its age and draw the graph.
  (b) Compute the value of the machinery after 4 years.

Water consumption 34. Since the beginning of the month, a local reservoir has been losing water at a constant rate. On the 12th of the month, the reservoir held 200 million gallons of water and on the 21st, it held only 164 million gallons.
  (a) Express the amount of water in the reservoir as a function of time and draw the graph.
  (b) How much water was in the reservoir on the 8th of the month?

Car pooling 35. To encourage motorists to form car pools, the transit authority in a major metropolitan area has been offering a special reduced rate at toll bridges for vehicles containing 4 or more persons. When the program began 30 days ago, 157 vehicles qualified for the reduced rate during the morning rush hour. Since then, the number of vehicles qualifying has been increasing at a constant rate and today, 247 vehicles qualified.
  (a) Express the number of vehicles qualifying each morning for the reduced rate as a function of time and draw the graph.
  (b) If the trend continues, how many vehicles will qualify during the morning rush hour 14 days from now?

Metric conversion 36. (a) Temperature measured in degrees Fahrenheit is a linear function of temperature measured in degrees Celsius. Use the facts that 0° Celsius is equal to 32° Fahrenheit and 100° Celsius is equal to 212° Fahrenheit to write an equation for this linear function.
  (b) Use the function you obtained in part (a) to convert 15° Celsius to Fahrenheit.
  (c) Convert 68° Fahrenheit to Celsius.

Appreciation of assets 37. The value of a certain rare book doubles every 10 years. The book was originally worth $3.
  (a) How much is the book worth when it is 30 years old? When it is 40 years old?

(b) Is the relationship between the value of the book and its age linear? Explain.

Parallel lines 38. What is the relationship between the slopes of parallel lines? Explain your answer. Using this relationship, write equations for the lines with the following properties:

(a) Through $(1, 3)$ and parallel to the line $4x + 2y = 7$.

(b) Through $(0, 2)$ and parallel to the line $2y - 3x = 5$.

(c) Through $(-2, 5)$ and parallel to the line through the points $(1, 2)$ and $(6, -1)$.

Perpendicular lines 39. Show that if a line L_1 with slope m_1 is perpendicular to a line L_2 with slope m_2, then $m_1 = -\dfrac{1}{m_2}$. (*Hint:* Find expressions for the slopes of the perpendicular lines L_1 and L_2 in the accompanying figure. Then apply the pythagorean theorem, together with the distance formula from Section 2, Problem 35, to the right triangle OAB to obtain the desired relationship between the slopes.)

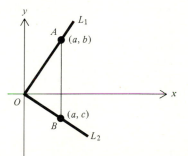

40. Using the result from Problem 39, write equations for the lines with the following properties:

(a) Through $(-1, 3)$ and perpendicular to the line $4x + 2y = 7$.

(b) Through $(0, 0)$ and perpendicular to the line $2y - 3x = 5$.

(c) Through $(2, 1)$ and perpendicular to the line joining $(0, 3)$ and $(2, -1)$.

4 INTERSECTIONS OF GRAPHS Sometimes it is necessary to determine when two functions are equal. This is the case, for example, when an economist wants to compute the market price at which the consumer demand for a commodity will be equal to its supply. It occurs when a manufacturer seeks to determine how many units must be sold before revenue exceeds cost. And it occurs when a political analyst attempts to predict how long it will take for the popularity of a certain challenger to reach that of the incumbent.

In geometric terms, the values of x for which two functions $f(x)$ and

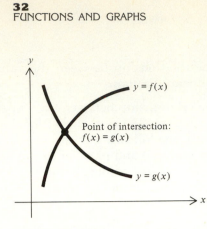

Figure 4.1 The intersection of two graphs.

$g(x)$ are equal are the x coordinates of the points at which their graphs intersect. The situation is illustrated in Figure 4.1.

To find the points of intersection algebraically, set $f(x)$ equal to $g(x)$ and solve for x. Here are three examples illustrating the algebraic techniques.

EXAMPLE 4.1

Where do the lines $y = 2x + 1$ and $y = -x + 4$ intersect?

SOLUTION

Solve the equation

$$2x + 1 = -x + 4$$

to get

$$3x = 3 \quad \text{or} \quad x = 1$$

To find the corresponding value of y, substitute $x = 1$ into either of the original equations $y = 2x + 1$ or $y = -x + 4$. You will get $y = 3$ from which you can conclude that $(1, 3)$ is the point of intersection. The situation is illustrated in Figure 4.2.

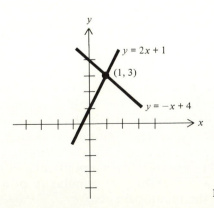

Figure 4.2 The lines $y = 2x + 1$ **and** $y = -x + 4$.

EXAMPLE 4.2

Find the points of intersection of the graphs of the functions $f(x) = 2x$ and $g(x) = x^2$.

SOLUTION

Set $f(x)$ equal to $g(x)$ and solve to get

$$2x = x^2$$
$$x^2 - 2x = 0$$
$$x(x - 2) = 0$$

Since the product of real numbers is equal to zero only when at least one of the factors equals zero, it follows that

$$x = 0 \quad \text{or} \quad x = 2$$

Now substitute these values of x into either equation, $y = f(x)$ or $y = g(x)$ and you will find that $y = 0$ when $x = 0$ and $y = 4$ when $x = 2$. It follows that the points of intersection are $(0, 0)$ and $(2, 4)$. The situation is illustrated in Figure 4.3.

In the next example, you will need the **quadratic formula** to find the points of intersection of the given graphs. (A review of the use of this formula can be found in Section A of the appendix at the back of the book.) You may also want to use a hand calculator to help with the computations.

EXAMPLE 4.3

Find the points of intersection of the line $y = 3x + 2$ and the parabola $y = x^2$.

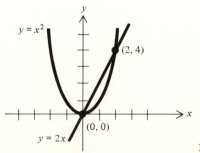

Figure 4.3 The curves $y = 2x$ and $y = x^2$.

SOLUTION

Rewrite the equation $\qquad x^2 = 3x + 2$

to get $\qquad x^2 - 3x - 2 = 0$

Since the expression $x^2 - 3x - 2$ has no obvious factors, use the quadratic formula to get

$$x = \frac{-(-3) \pm \sqrt{(-3)^2 - 4(1)(-2)}}{2(1)} = \frac{3 \pm \sqrt{17}}{2}$$

It follows that the x coordinates of the points of intersection are

$$x = \frac{3 + \sqrt{17}}{2} = 3.56 \qquad \text{and} \qquad x = \frac{3 - \sqrt{17}}{2} = -0.56$$

Computing the y coordinates from the equation $y = x^2$ you find that the points of intersection are $(3.56, 12.67)$ and $(-0.56, 0.31)$. (Due to round-off approximations, you will get slightly different values for the y coordinates if you use the equation $y = 3x + 2$.) The situation is illustrated in Figure 4.4.

Break-even analysis   Intersections of graphs arise in business in the context of **break-even analysis.** In a typical situation, a manufacturer wishes to determine how many units of a certain commodity have to be sold to make the total revenue equal to the total cost. Suppose that x denotes the number of units manufactured and sold, and let $C(x)$ and $R(x)$ be the corresponding total cost and total revenue, respectively. A pair of (linear) cost and revenue curves is sketched in Figure 4.5.

Because of fixed overhead costs, the total cost curve is initially higher than the total revenue curve. Hence, at low levels of production, the manufacturer suffers a loss. At higher levels of production, however, the total revenue curve is the higher one and the manufac-

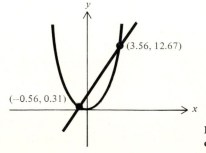

Figure 4.4 The line $y = 3x + 2$ and the parabola $y = x^2$.

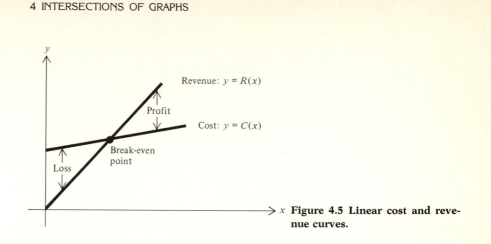

Figure 4.5 Linear cost and revenue curves.

turer realizes a profit. The point at which the two curves cross is called the **break-even point,** because when total revenue is equal to total cost, the manufacturer breaks even, experiencing neither a profit nor a loss. Here is an example.

EXAMPLE 4.4

A manufacturer can sell a certain product for $110 per unit. Total cost consists of a fixed overhead of $7,500 plus production costs of $60 per unit.

(a) How many units must the manufacturer sell to break even?
(b) What is the manufacturer's profit or loss if 100 units are sold?
(c) How many units must the manufacturer sell to realize a profit of $1,250?

SOLUTION

Let x denote the number of units manufactured and sold. Then the total revenue is given by the function

$$R(x) = 110x$$

and the total cost by the function

$$C(x) = 7,500 + 60x$$

(a) To find the break-even point, set $R(x)$ equal to $C(x)$ and solve, getting

$$110x = 7,500 + 60x$$
$$50x = 7,500$$
$$x = 150$$

It follows that the manufacturer will have to sell 150 units to break even. The situation is illustrated in Figure 4.6.

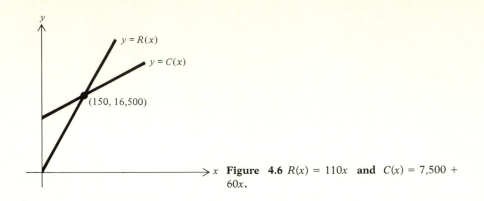

Figure **4.6** $R(x) = 110x$ and $C(x) = 7,500 + 60x$.

(b) The profit $P(x)$ is revenue minus cost. Hence,

$$P(x) = R(x) - C(x) = 110x - 7,500 - 60x = 50x - 7,500$$

The profit from the sale of 100 units is

$$P(100) = 5,000 - 7,500 = -2,500$$

The minus sign indicates a negative profit, or loss. It follows that the manufacturer will lose $2,500 if only 100 units are sold.

(c) To determine the number of units that must be sold to generate a profit of $1,250, set the formula for profit $P(x)$ equal to 1,250 and solve for x. You get

$$50x - 7,500 = 1,250$$
$$50x = 8,750$$
$$x = 175$$

from which you can conclude that 175 units must be sold to generate the desired profit.

The next example illustrates how break-even analysis can be used as a tool for decision-making.

EXAMPLE 4.5

A leading car rental agency charges $14 plus 15 cents per kilometer. A second agency charges $20 plus 5 cents per kilometer. Which agency offers the better deal?

SOLUTION

The answer depends on the number of kilometers the car is driven. For short trips, the first agency charges less than the second, but for long trips, the second charges less than the first. You can use

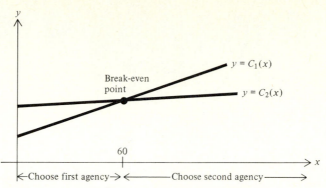

Figure 4.7 Car rental costs at competing agencies.

break-even analysis to determine the number of kilometers for which the two agencies charge the same amount.

Suppose a car is to be driven x kilometers. Then the first agency will charge

$$C_1(x) = 14 + 0.15x$$

dollars and the second will charge

$$C_2(x) = 20 + 0.05x$$

dollars. If you set these expressions equal to each other and solve, you get

$$14 + 0.15x = 20 + 0.05x$$
$$0.1x = 6$$
$$x = 60$$

This implies that the two agencies charge the same amount if the car is driven 60 kilometers. For shorter distances, the first agency offers the better deal and for longer distances, the second agency does. The situation is illustrated in Figure 4.7.

Market equilibrium An important economic application involving intersections of graphs arises in connection with the **law of supply and demand.** In this context, we think of the market price p of a commodity as determining the number of units of the commodity that manufacturers are willing to supply as well as the number of units that consumers are willing to buy. In most cases, manufacturers' supply $S(p)$ increases and consumers' demand $D(p)$ decreases as the market price p increases. A pair of supply and demand curves is sketched in Figure 4.8. (The letter q used to label the vertical axis stands for "quantity.")

(Actually, an economist's graph of these functions would not look

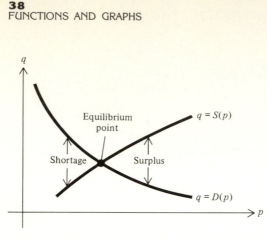

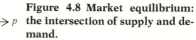

Figure 4.8 Market equilibrium: the intersection of supply and demand.

quite like the one in Figure 4.8. When dealing with supply and demand curves, economists usually depart from mathematical tradition and use the horizontal axis to represent the dependent variable q and the vertical axis for the independent variable p.)

The point of intersection of the supply and demand curves is called the point of **market equilibrium.** The p coordinate of this point (the **equilibrium price**) is the market price at which supply equals demand; that is, the market price at which there will be neither a surplus nor shortage of the commodity.

The law of supply and demand asserts that in a situation of pure competition, a commodity will tend to be sold at its equilibrium price. If the commodity is sold for more than the equilibrium price, there will be an unsold surplus on the market and retailers will tend to lower their prices. On the other hand, if the commodity is sold for less than the equilibrium price, the demand will exceed the supply and retailers will be inclined to raise their prices.

Here is an example.

EXAMPLE 4.6

Find the equilibrium price and the corresponding number of units supplied and demanded if the supply function for a certain commodity is $S(p) = p^2 + 3p - 70$ and the demand function is $D(p) = 410 - p$.

SOLUTION

Equate $S(p)$ and $D(p)$ and solve for p to get

$$p^2 + 3p - 70 = 410 - p$$
$$p^2 + 4p - 480 = 0$$
$$(p - 20)(p + 24) = 0$$
$$p = 20 \quad \text{or} \quad p = -24$$

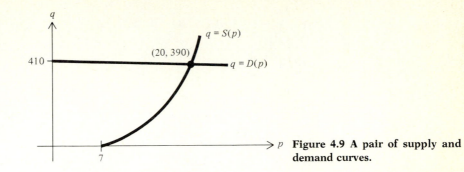

Figure 4.9 A pair of supply and demand curves.

Since only positive values of p are meaningful in this practical problem, you can conclude that the equilibrium price is \$20. Since the corresponding supply and demand are equal, use the simpler demand equation to compute this quantity. You get

$$D(20) = 410 - 20 = 390$$

and conclude that 390 units are supplied and demanded when the market is in equilibrium.

The supply and demand curves are sketched in Figure 4.9. Notice that the supply curve crosses the p axis when $p = 7$. (Verify this.) What is the economic interpretation of this fact?

Problems   In Problems 1 through 17, find the points of intersection (if any) of the given pair of curves and draw the graph.

1. $y = 3x + 5$ and $y = -x + 3$

2. $y = 5x - 14$ and $y = 4 - x$

3. $y = 3x + 8$ and $y = 3x - 2$

4. $y = x^2$ and $y = 6 - x$

5. $y = x^2 - x$ and $y = x - 1$

6. $y = x^3 - 6x^2$ and $y = -x^2$

7. $y = x^3$ and $y = x^2$

8. $y = x^3$ and $y = -x^3$

9. $y = x^2 + 2$ and $y = x$

10. $3y - 2x = 5$ and $y + 3x = 9$

11. $2x - 3y = -8$ and $3x - 5y = -13$

12. $y = \dfrac{1}{x}$ and $y = x^2$

13. $y = \dfrac{1}{x^2}$ and $y = 4$

14. $y = \dfrac{1}{x}$ and $y = \dfrac{1}{x^2}$

15. $y = \dfrac{1}{x^2}$ and $y = -x^2$

16. $y = x^2$ and $y = 2x + 2$

17. $y = x^2 - 2x$ and $y = x - 1$

Break-even analysis 18. A furniture manufacturer can sell dining-room tables for $70 apiece. The manufacturer's total cost consists of a fixed overhead of $8,000 plus production costs of $30 per table.
   (a) Determine how many tables the manufacturer must sell to break even.
   (b) Determine how many tables the manufacturer must sell to make a profit of $6,000.
   (c) Calculate the manufacturer's profit or loss if 150 tables are sold.
   (d) On the same set of axes, graph the manufacturer's total revenue and total cost functions. Explain how the overhead can be read off from the graph.

Break-even analysis 19. During the summer, a group of students builds kayaks in a converted garage. The rental for the garage is $600 for the summer, and the materials needed to build a kayak cost $25. The kayaks can be sold for $175 apiece.
   (a) How many kayaks must the students sell to break even?
   (b) How many kayaks must the students sell to make a profit of $450?

Checking accounts 20. The charge for maintaining a checking account at a certain bank is $2 per month plus 5 cents for each check that is written. A competing bank charges $1 per month plus 9 cents per check. Find a criterion for deciding which bank offers the better deal.

Membership fees 21. Membership in a private tennis club costs $500 per year and entitles the member to use the courts for a fee of $1 per hour. At a competing club, membership costs $440 per year and the charge for the use of the courts is $1.75 per hour. If only financial considerations are to be taken into account, how should a tennis player choose which club to join?

Property tax 22. Under the provisions of a proposed property tax bill, a homeowner will pay $100 plus 8 percent of the assessed value of the house. Under the provisions of a competing bill, the homeowner will pay $1,900 plus 2 percent of the assessed value. If only financial considerations are taken into account, how should a homeowner decide which bill to support?

Supply and demand 23. The supply and demand functions for a certain commodity are $S(p) = 4p + 200$ and $D(p) = -3p + 480$, respectively. Find the equilibrium price and the corresponding number of units supplied and demanded, and draw the supply and demand curves on the same set of axes.

Supply and demand 24. When electric blenders are sold for p dollars apiece, manufacturers are willing to supply $\frac{p^2}{10}$ blenders to local retailers while the local demand will be $60 - p$ blenders. At what market price will the manufacturers' supply of electric blenders be equal to the consumers' demand for the blenders? How many blenders will be sold at this price?

Supply and demand 25. The supply and demand functions for a certain commodity are $S(p) = p - 10$ and $D(p) = \frac{5,600}{p}$, respectively.

   (a) Find the equilibrium price and the corresponding number of units supplied and demanded.
   (b) Draw the supply and demand curves on the same set of axes.
   (c) Where does the supply function cross the p axis? What is the economic significance of this point?

Supply and demand 26. Suppose that the supply and demand functions for a certain commodity are $S(p) = ap + b$ and $D(p) = cp + d$, respectively.

   (a) What can you say about the signs of the coefficients a, b, c, and d if the supply and demand curves are oriented as shown in the following diagram?

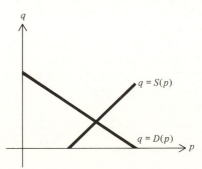

   (b) Express the equilibrium price in terms of the coefficients a, b, c, and d.

(c) Use your answer in part (b) to determine what happens to the equilibrium price as *a* increases.

(d) Use your answer in part (b) to determine what happens to the equilibrium price as *d* increases.

Spy story   27. The hero of a popular spy story has escaped from the headquarters of an international diamond smuggling ring in the tiny Mediterranean country of Azusa. Our hero, driving a stolen milk truck at 72 kilometers per hour, has a 40-minute head start on his pursuers who are chasing him in a Ferrari going 168 kilometers per hour. The distance from the smugglers' headquarters to the border, and freedom, is 83.8 kilometers. Will our hero make it?

Air travel   28. Two jets bound for Los Angeles leave New York 30 minutes apart. The first travels 880 kilometers per hour, while the second goes 1,040 kilometers per hour. At what time will the second plane pass the first?

5 FUNCTIONAL MODELS

A mathematical representation of a practical situation is called a **mathematical model.** In preceding sections, you saw models representing such quantities as manufacturing cost, air pollution levels, population size, supply, and demand. In this section, you will see examples illustrating some of the techniques you can use to build mathematical models of your own.

A profit function   In the following example, profit is expressed as a function of the price at which a product is sold.

EXAMPLE 5.1

A manufacturer can produce radios at a cost of $2 apiece. The radios have been selling for $5 apiece, and, at this price, consumers have been buying 4,000 radios a month. The manufacturer is planning to raise the price of the radios and estimates that for each $1 increase in the price, 400 fewer radios will be sold each month. Express the manufacturer's monthly profit as a function of the price at which the radios are sold.

SOLUTION

Begin by stating the desired relationship in words.

Profit = (number of radios sold)(profit per radio)

Since the goal is to express profit as a function of price, the independent variable is price and the dependent variable is profit. Let *x* denote the price at which the radios will be sold, and $P(x)$ the corresponding profit.

Next, express the number of radios sold in terms of the variable x. You know that 4,000 radios are sold each month when the price is $5 and that 400 fewer will be sold each month for each $1 increase in the price. Thus,

Number of radios sold = 4,000 − 400(number of $1 increases)

The number of $1 increases in the price is the difference $x - 5$ between the new and old selling prices. Hence,

$$
\begin{aligned}
\text{Number of radios sold} &= 4,000 - 400(x - 5) \\
&= 6,000 - 400x \\
&= 400(15 - x)
\end{aligned}
$$

The profit per radio is simply the difference between the selling price x and the cost $2. That is,

Profit per radio = $x - 2$

If you now substitute the algebraic expressions for the number of radios sold and the profit per radio into the verbal equation with which you began, you will get

$$P(x) = 400(15 - x)(x - 2)$$

The graph of this factored polynomial is sketched in Figure 5.1. (Actually, only the portion of the graph for $x \geq 5$ is relevant to the original problem as stated. Can you give a practical interpretation of the portion between $x = 2$ and $x = 5$?) Notice that the profit function reaches a maximum for some value of x near $x = 8$. In Chapter 2, you will learn how to use calculus to find this optimal selling price.

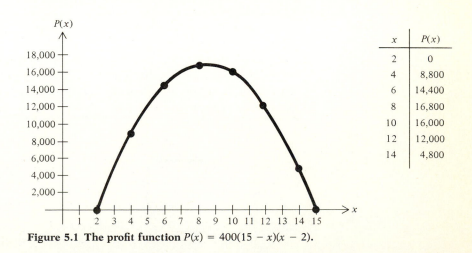

x	$P(x)$
2	0
4	8,800
6	14,400
8	16,800
10	16,000
12	12,000
14	4,800

Figure 5.1 The profit function $P(x) = 400(15 - x)(x - 2)$.

Elimination of variables

In the next example, the quantity you are seeking is expressed most naturally in terms of two variables. You will have to eliminate one of these before you can write the quantity as a function of a single variable.

EXAMPLE 5.2

The highway department is planning to build a picnic area for motorists along a major highway. It is to be rectangular with an area of 5,000 square meters and is to be fenced off on the three sides not adjacent to the highway. Express the number of meters of fencing required as a function of the length of the unfenced side.

SOLUTION

It is natural to start by introducing two variables, say x and y, to denote the length of the sides of the picnic area (Figure 5.2) and to express the number of meters F of fencing required in terms of these two variables:

$$F = x + 2y$$

Since the goal is to express the number of meters of fencing as a function of x alone, you must find a way to express y in terms of x. To do this, use the fact that the area is to be 5,000 square meters and write

$$xy = 5,000$$

Solve this equation for y

$$y = \frac{5,000}{x}$$

and substitute the resulting expression for y into the formula for F to get

$$F(x) = x + \frac{10,000}{x}$$

A graph of the relevant portion of this rational function is sketched in Figure 5.3. Notice that there is some length x for which the amount of required fencing is minimal. In Chapter 2, you will compute this optimal value of x using calculus.

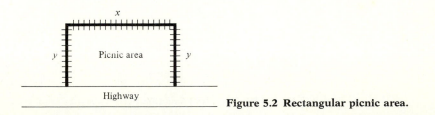

Figure 5.2 Rectangular picnic area.

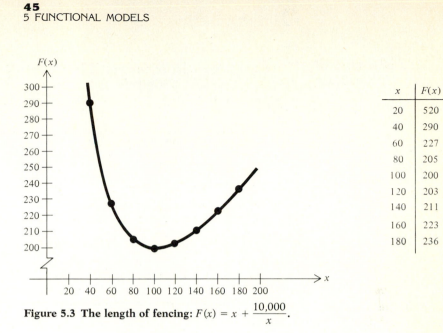

Figure 5.3 **The length of fencing:** $F(x) = x + \dfrac{10,000}{x}$.

x	$F(x)$
20	520
40	290
60	227
80	205
100	200
120	203
140	211
160	223
180	236

Functions involving multiple formulas

In the next example, you will need three formulas to define the desired function.

EXAMPLE 5.3

During the 1977 drought, residents of Marin County, California, were faced with a severe water shortage. To discourage excessive use of water, the County Water District initiated drastic rate increases. The monthly rate for a family of four was $1.22 per 100 cubic feet of water for the first 1,200 cubic feet, $10 per 100 cubic feet for the next 1,200 cubic feet, and $50 per 100 cubic feet thereafter. Express the monthly water bill for a family of four as a function of the amount of water used.

SOLUTION

Let x denote the number of hundred-cubic-feet units of water used by the family during the month and $C(x)$ the corresponding cost in dollars. If $0 \le x \le 12$, the cost is simply the cost per unit times the number of units used:

$$C(x) = 1.22x$$

If $12 < x \le 24$, each of the first 12 units costs $1.22, and so the total cost of these 12 units is $1.22(12) = 14.64$ dollars. Each of the remaining $x - 12$ units costs $10, and hence the total cost of these units is $10(x - 12)$ dollars. The cost of all x units is the sum

$$C(x) = 14.64 + 10(x - 12) = 10x - 105.36$$

If $x > 24$, the cost of the first 12 units is $1.22(12) = 14.64$ dollars, the cost of the next 12 units is $10(12) = 120$ dollars, and the cost of the remaining $x - 24$ units is $50(x - 24)$ dollars. The cost of all x units is the sum

$$C(x) = 14.64 + 120 + 50(x - 24) = 50x - 1,065.36$$

Combining these three formulas you get

$$C(x) = \begin{cases} 1.22x & \text{if } 0 \leq x \leq 12 \\ 10x - 105.36 & \text{if } 12 < x \leq 24 \\ 50x - 1,065.36 & \text{if } x > 24 \end{cases}$$

The graph of this function is shown in Figure 5.4. Notice that the graph consists of three line segments, each one steeper than the preceding one. What aspect of the practical situation is reflected by the increasing steepness of the lines?

Proportionality

The following concepts of proportionality are used frequently in the construction of mathematical models.

Direct proportionality

To say that Q is proportional (or directly proportional) to x means that there is a constant k for which

$$Q = kx$$

Inverse proportionality

To say that Q is inversely proportional to x means that there is a constant k for which

$$Q = \frac{k}{x}$$

Joint proportionality

To say that Q is jointly proportional to x and y means that Q is directly proportional to the product of x and y; i.e., there is a constant k for which

$$Q = kxy$$

Here is an example from biology.

EXAMPLE 5.4

When environmental factors impose an upper bound on its size, population grows at a rate that is jointly proportional to its current size and the difference between its current size and the upper bound. Express the rate of population growth as a function of the size of the population.

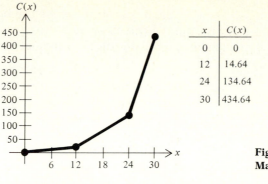

x	$C(x)$
0	0
12	14.64
24	134.64
30	434.64

Figure 5.4 The cost of water in Marin County.

SOLUTION

Let p denote the size of the population, $R(p)$ the corresponding rate of population growth, and b the upper bound placed on the population by the environment. Then,

$$\text{Difference between population and bound} = b - p$$

and

$$R(p) = kp(b - p)$$

where k is the constant of proportionality.

A graph of this factored polynomial is sketched in Figure 5.5. In Chapter 2, you will use calculus to compute the population size for which the rate of population growth is greatest.

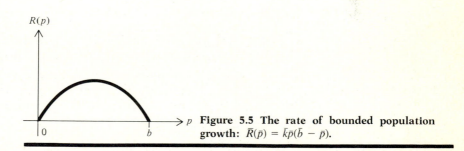

Figure 5.5 The rate of bounded population growth: $\bar{R}(\bar{p}) = \bar{k}\bar{p}(\bar{b} - \bar{p})$.

Problems

Retail sales

1. A college bookstore can obtain the book *Social Groupings of the American Dragonfly* from the publisher at a cost of $3 per book. The bookstore has been offering the book at a price of $15 per copy and, at this price, has been selling 200 copies a month. The bookstore is planning to lower its price to stimulate sales and estimates that for each $1 reduction in the price, 20 more books

will be sold each month. Express the bookstore's monthly profit from the sale of this book as a function of the selling price, draw the graph, and estimate the optimal selling price.

Retail sales

2. A manufacturer has been selling lamps at a price of $6 per lamp and, at this price, consumers have been buying 3,000 lamps per month. The manufacturer wishes to raise the price and estimates that for each $1 increase in the price, 1,000 fewer lamps will be sold each month. The manufacturer can produce the lamps at a cost of $4 per lamp. Express the manufacturer's monthly profit as a function of the price at which the lamps are sold, draw the graph, and estimate the optimal selling price.

Transportation costs

3. A bus company is willing to charter buses only to groups of 35 or more people. If a group contains exactly 35 people, each person pays $60. In larger groups, everybody's fare is reduced by 50 cents for each person in excess of 35. Express the bus company's revenue as a function of the size of the group, draw the graph, and estimate the size of the group that will maximize the revenue.

Agricultural yield

4. A Florida citrus grower estimates that if 60 orange trees are planted, the average yield per tree will be 400 oranges. The average yield will decrease by 4 oranges per tree for each additional tree planted on the same acreage. Express the grower's total yield as a function of the number of additional trees planted, draw the graph, and estimate the total number of trees the grower should plant to maximize yield.

Harvesting

5. Farmers can get $2 per bushel for their potatoes on July first, and after that, the price drops by 2 cents per bushel per day. On July first, a farmer has 80 bushels of potatoes in the field and estimates that the crop is increasing at a rate of 1 bushel per day. Express the farmer's revenue from the sale of the potatoes as a function of the time at which the crop is harvested, draw the graph, and estimate when the farmer should harvest the potatoes to maximize revenue.

Recycling

6. During the summer, members of a scout troop have been collecting used bottles that they plan to deliver to a glass company for recycling. So far, in 80 days, the scouts have collected 24,000 kilograms of glass for which the glass company currently offers 1 cent per kilogram. However, because bottles are accumulating faster than they can be recycled, the company plans to reduce by 1 cent each day the price it will pay for 100 kilograms of used glass. Assume that the scouts can continue to collect bottles at the same rate and that transportation costs make more than one trip to the glass company unfeasible. Express the revenue the

scouts will get from the recycled glass as a function of the number of additional days they continue to collect bottles. Draw the graph and estimate the most profitable time for the scouts to conclude their project and deliver the bottles.

Fencing

7. A city recreation department plans to build a rectangular playground 3,600 square meters in area. The playground is to be surrounded by a fence. Express the length of the fencing as a function of the length of one of the sides of the playground, draw the graph, and estimate the dimensions of the playground requiring the least amount of fencing.

Area

8. Express the area of a rectangular field whose perimeter is 320 meters as a function of the length of one of its sides. Draw the graph and estimate the dimensions of the field of maximum area.

Construction cost

9. A closed box with a square base is to have a volume of 250 cubic meters. The material for the top and bottom of the box costs $2 per square meter, and the material for the sides costs $1 per square meter. Express the construction cost of the box as a function of the length of its base.

Construction cost

10. An open box with a square base is to be built for $48. The sides of the box will cost $3 per square meter, and the base will cost $4 per square meter. Express the volume of the box as a function of the length of its base.

Volume

11. An open box is to be made from a square piece of cardboard, 18 inches by 18 inches, by removing a small square from each corner and folding up the flaps to form the sides. Express the volume of the resulting box as a function of the length x of a side of the removed squares. Draw the graph and estimate the value of x for which the volume of the resulting box is greatest.

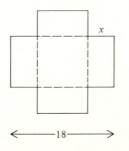

Packaging

12. A beer can can hold 12 fluid ounces, which is approximately 6.89π cubic inches. Express the surface area of the can as a function of its radius. (Recall that the volume of a cylinder of radius r and

height h is $\pi r^2 h$. The circumference of a circle of radius r is $2\pi r$ and its area is πr^2.)

Packaging 13. A cylindrical can is to hold 4π cubic inches of frozen orange juice. The cost per square inch of constructing the metal top and bottom is twice the cost per square inch of constructing the cardboard side. Express the cost of constructing the can as a function of its radius if the cost of the side is 0.02 cent per square inch.

Volume 14. A cylindrical can with no top has been made from 27π square inches of metal. Express the volume of the can as a function of its radius.

Admission fees 15. A local natural history museum charges admission to groups according to the following policy: Groups of fewer than 50 people are charged a rate of $1.50 per person, while groups of 50 people or more are charged a reduced rate of $1 per person.
    (a) Express the amount a group will be charged for admission to the museum as a function of its size and draw the graph.
    (b) How much money will a group of 49 people save in admission costs if it can recruit 1 additional member?

Discounts 16. A record club offers the following special sale: If 5 records are bought at the full price of $6 apiece, additional records can then be bought at half price. There is a limit of 9 records per customer. Express the cost of the records as a function of the number bought and draw the graph.

Postal rates 17. There was a time when the postal rate for letters weighing no more than 7 ounces was 13 cents for the first ounce or fraction thereof and 11 cents for each additional ounce or fraction thereof. Express the cost of sending a letter as a function of its weight and draw the graph.

Telegram rates 18. In 1977, the rate for interstate telegrams was $4.75 for 15 words or less plus 12 cents for each additional word. Express the cost of sending a telegram as a function of its length and draw the graph.

Income tax   19. The following table is taken from the 1972 federal income tax rate schedule for single taxpayers.

If the taxable income is:		The income tax is:	
Over . . .	but not over . . .		of the excess over . . .
$ 8,000	$10,000	$1,590 + 25%	$ 8,000
$10,000	$12,000	$2,090 + 27%	$10,000
$12,000	$14,000	$2,630 + 29%	$12,000
$14,000	$16,000	$3,210 + 31%	$14,000

(a) Express an individual's income tax as a function of the taxable income x for $8,000 < x \leq 16,000$ and draw the graph.

(b) Your graph in part (a) should consist of four line segments. Compute the slope of each segment. What happens to these slopes as the taxable income increases? Explain the behavior of the slopes in practical terms.

Transportation cost   20. A bus company has adopted the following pricing policy for groups wishing to charter its buses: Groups containing no more than 40 people will be charged a fixed amount of $2,400 (40 times $60). In groups containing between 40 and 80 people, everyone will pay $60 minus 50 cents for each person in excess of 40. The company's lowest fare of $40 per person will be offered to groups that have 80 members or more. Express the bus company's revenue as a function of the size of the group and draw the graph.

Population growth   21. In the absence of environmental constraints, population grows at a rate proportional to its size. Express the rate of population growth as a function of the size of the population.

Radioactive decay   22. A sample of radium decays at a rate proportional to the amount of radium remaining. Express the rate of decay of the sample as a function of the amount remaining.

Temperature change   23. The rate at which the temperature of an object changes is proportional to the difference between its own temperature and the temperature of the surrounding medium. Express this rate as a function of the temperature of the object.

The spread of an epidemic   24. The rate at which an epidemic spreads through a community is jointly proportional to the number of people who have caught the disease and the number who have not. Express this rate as a function of the number of people who have caught the disease.

Political corruption   25. The rate at which people are implicated in a government scandal is jointly proportional to the number of people already implicated and the number of people involved who have not yet been implicated. Express this rate as a function of the number of people who have been implicated.

Production cost
26. At a certain factory, setup cost is proportional to the number of machines used, and operating cost is inversely proportional to the number of machines used. Express the total cost as a function of the number of machines used.

Transportation cost
27. A truck is hired to transport goods from a factory to a warehouse. The driver's wages are figured by the hour and so are inversely proportional to the speed at which the truck is driven. The cost of gasoline is directly proportional to the speed. Express the total cost of operating the truck as a function of the speed at which it is driven.

The distance between moving objects
28. A car traveling east at 80 kilometers per hour and a truck traveling south at 60 kilometers per hour start at the same intersection. Express the distance between them as a function of time. (*Hint:* Use the pythagorean theorem.)

The distance between moving objects
29. A truck is 975 kilometers due east of a car and is traveling west at a constant speed of 60 kilometers per hour. Meanwhile, the car is going north at a constant speed of 90 kilometers per hour. Express the distance between the car and truck as a function of time.

Installation cost
30. A cable is to be run from a power plant on one side of a river 900 meters wide to a factory on the other side, 3,000 meters downstream. The cable will be run in a straight line from the power plant to some point *P* on the opposite bank, and then along the bank to the factory. The cost of running the cable across the water is $5 per meter, while the cost over land is $4 per meter.

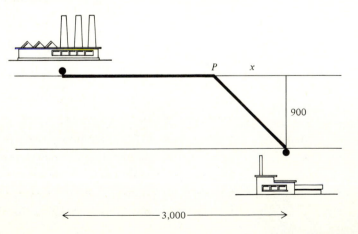

Let x be the distance from P to the point directly across the river from the power plant, and express the cost of installing the cable as a function of x.

Poster design   31. A rectangular poster contains 25 square centimeters of print surrounded by margins of 2 centimeters on each side and 4 centimeters on the top and bottom. Express the total area of the poster (printing plus margins) as a function of the width of the printed portion.

Production cost   32. A plastics firm has received an order from the city recreation department to manufacture 8,000 special Styrofoam kickboards for its summer swimming program. The firm owns several machines, each of which can produce 30 kickboards an hour. The cost of setting up the machines to produce these particular kickboards is $20 per machine. Once the machines have been set up, the operation is fully automated and can be overseen by a single production supervisor earning $4.80 per hour. Express the cost of producing the 8,000 kickboards as a function of the number of machines used, draw the graph, and estimate the number of machines the firm should use to minimize cost.

CHAPTER SUMMARY AND PROFICIENCY TEST

Important terms, symbols, and formulas

Function
Independent and dependent variables
Functional notation: $f(x)$
Domain of a function
Composition of functions: $g[h(x)]$
Graph of a function
Quadratic function: $f(x) = ax^2 + bx + c$
Polynomial
Rational function
Discontinuity
Linear function; constant rate of change

Slope: $m = \dfrac{\Delta y}{\Delta x} = \dfrac{y_2 - y_1}{x_2 - x_1}$

Slope-intercept formula: $y = mx + b$
Point-slope formula: $y - y_0 = m(x - x_0)$
Intersection of graphs; break-even analysis
Market equilibrium; law of supply and demand
Direct proportionality: $Q = kx$

Inverse proportionality: $Q = \dfrac{k}{x}$

Joint proportionality: $Q = kxy$

Proficiency test

1. Specify the domain of each of the following functions.
   (a) $f(x) = x^2 - 2x + 6$
   (b) $f(x) = \dfrac{x - 3}{x^2 + x - 2}$
   (c) $f(x) = \sqrt{x^2 - 9}$

2. As advances in technology result in the production of increasingly power-ful and compact calculators, the price of calculators currently on the market drops. It is estimated that x months from now, the price of a cer-tain model will be $P(x) = 40 + \dfrac{30}{x + 1}$ dollars.
   (a) What will the price be 5 months from now?
   (b) By how much will the price drop during the 5th month?
   (c) When will the price be \$43?
   (d) What will happen to the price in the long run?

3. Find the composite function $g[h(x)]$.
   (a) $g(u) = u^2 + 2u + 1$, $h(x) = 1 - x$
   (b) $g(u) = \dfrac{1}{2u + 1}$, $h(x) = x + 2$
   (c) $g(u) = \sqrt{1 - u}$, $h(x) = 2x + 4$

4. (a) Find $f(x - 2)$ where $f(x) = x^2 - x + 4$.
   (b) Find $f(x^2 + 1)$ where $f(x) = \sqrt{x} + \dfrac{2}{x - 1}$.
   (c) Find $f(x + 1) - f(x)$ where $f(x) = x^2$.

5. Find the value of c for which the curve $y = 3x^2 - 2x + c$ passes through the point $(2, 4)$.

6. Graph the following functions.
   (a) $f(x) = 4 - x^2$
   (b) $f(x) = \dfrac{-2}{x - 3}$
   (c) $f(x) = x + \dfrac{2}{x}$

7. The consumer demand for a certain commodity is $D(p) = -50p + 800$ units per month when the market price is p dollars per unit.
   (a) Graph this demand function.

    (b) Express consumers' total monthly expenditure for the commodity as a function of p and draw the graph.

    (c) Use the graph in part (b) to estimate the market price at which the total expenditure for the commodity is greatest.

8. A private college in the southwest has launched a fundraising campaign. College officials estimate that it will take $f(x) = \dfrac{10x}{150 - x}$ weeks to reach x percent of their goal.

    (a) Sketch the relevant portion of the graph of this function.

    (b) How long will it take to reach 50 percent of the campaign's goal?

    (c) How long will it take to reach 100 percent of the goal?

9. Find the slope and y intercept of the given line and draw the graph.

    (a) $y = 3x + 2$ (b) $5x - 4y = 20$

    (c) $2y + 3x = 0$ (d) $\dfrac{x}{3} + \dfrac{y}{2} = 4$

10. Find the equation of the line with slope 5 and y intercept $(0, -4)$.

11. Find the equation of the line that passes through $(1, 3)$ and has slope -2.

12. Find the equation of the line through the points $(2, 4)$ and $(1, -3)$.

13. Since the beginning of the year, the price of unleaded gasoline has been increasing at a constant rate of 2 cents per gallon per month. By June first, the price had reached 92 cents per gallon.

    (a) Express the price of unleaded gasoline as a function of time and draw the graph.

    (b) What was the price at the beginning of the year?

    (c) What will the price be on October first?

14. The circulation of a newspaper is increasing at a constant rate. Three months ago the circulation was 3,200. Today it is 4,400.

    (a) Express the circulation as a function of time and draw the graph.

    (b) What will the circulation be 2 months from now?

15. Find the points of intersection (if any) of the given pair of curves and draw the graph.

    (a) $y = -3x + 5$ and $y = 2x - 10$

    (b) $y = x + 7$ and $y = -2 + x$

    (c) $y = x^2 - 1$ and $y = 1 - x^2$

    (d) $y = x^2$ and $y = 15 - 2x$

    (e) $y = \dfrac{24}{x^2}$ and $y = 3x$

16. One plumber charges $25 plus $16 per half hour. A second charges $31 plus $14 per half hour. Find a criterion for deciding which plumber to call if only financial considerations are to be taken into account.

17. A manufacturer can sell a certain product for $80 per unit. Total cost consists of a fixed overhead of $4,500 plus production costs of $50 per unit.
    (a) How many units must the manufacturer sell to break even?
    (b) What is the manufacturer's profit or loss if 200 units are sold?
    (c) How many units must the manufacturer sell to realize a profit of $900?

18. A manufacturer can produce bookcases at a cost of $10 apiece. Sales figures indicate that if the bookcases are sold for x dollars apiece, approximately $50 - x$ will be sold each month. Express the manufacturer's monthly profit as a function of the selling price x, draw the graph, and estimate the optimal selling price.

19. A retailer can obtain cameras from the manufacturer at a cost of $50 apiece. The retailer has been selling the cameras at a price of $80 apiece, and, at this price, consumers have been buying 40 cameras a month. The retailer is planning to lower the price to stimulate sales and estimates that for each $5 reduction in the price, 10 more cameras will be sold each month. Express the retailer's monthly profit from the sale of the cameras as a function of the selling price. Draw the graph and estimate the optimal selling price.

20. A manufacturing firm has received an order to make 400,000 souvenir medals commemorating the 10th anniversary of the landing of Apollo 11 on the moon. The firm owns several machines, each of which can produce 200 medals per hour. The cost of setting up the machines to produce the medals is $80 per machine, and the total operating cost is $5.76 per hour. Express the cost of producing the 400,000 medals as a function of the number of machines used. Draw the graph and estimate the number of machines the firm should use to minimize cost.

21. Psychologists believe that when a person is asked to recall a set of facts, the rate at which the facts are recalled is proportional to the number of relevant facts in the subject's memory that have not yet been recalled. Express the recall rate as a function of the number of facts that have been recalled.

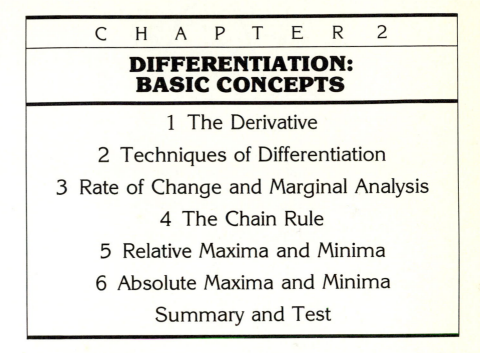

C H A P T E R 2

DIFFERENTIATION: BASIC CONCEPTS

1 The Derivative

2 Techniques of Differentiation

3 Rate of Change and Marginal Analysis

4 The Chain Rule

5 Relative Maxima and Minima

6 Absolute Maxima and Minima

Summary and Test

1 THE DERIVATIVE

Differentiation is a mathematical technique of exceptional power and versatility. It is one of the two central concepts in the branch of mathematics called **calculus** and has a variety of applications including curve sketching, the optimization of functions, and the analysis of rates of change.

A practical optimization problem

A typical problem to which calculus can be applied is the profit maximization problem you saw in Example 5.1 of Chapter 1. Recall that in that problem, a manufacturer's monthly profit from the sale of radios was $P(x) = 400(15 - x)(x - 2)$ dollars when the radios were sold for x dollars apiece. The graph of this profit function, which is reproduced in Figure 1.1, suggests that there is an optimal selling price x at which the manufacturer's profit will be greatest. In geometric terms, the optimal price is the x coordinate of the peak of the graph.

In this relatively simple example, the peak can be characterized in terms of lines that are **tangent** to the graph. In particular, the peak is

57

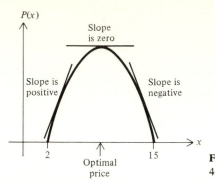

Figure 1.1 **The profit function** $P(x) =$
$400(15 - x)(x - 2)$.

the only point on the graph at which the tangent line is horizontal;
that is, at which the slope of the tangent is zero. To the left of the
peak, the slope of the tangent is positive. To the right of the peak, the
slope is negative. But just at the peak itself, the curve "levels off" and
the slope of its tangent is zero.

These observations suggest that you could solve the optimization
problem if you had a procedure for computing slopes of tangents.
Such a procedure shall now be developed. Throughout the development, you may rely on your intuitive understanding that the tangent
to a curve at a point is the line that indicates the direction of the
curve at that point.

The slope of
a tangent

The goal is to solve the following general problem: Given a point
$(x, f(x))$ on the graph of a function f, find the slope of the line that is
tangent to the graph at this point. The situation is illustrated in Figure 1.2.

In Chapter 1, Section 3, you learned that the slope of the line passing through two points (x_1, y_1) and (x_2, y_2) is given by the formula

$$\text{Slope} = \frac{\Delta y}{\Delta x} = \frac{y_2 - y_1}{x_2 - x_1}$$

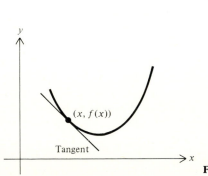

Figure 1.2 **A tangent to the curve** $y = f(x)$.

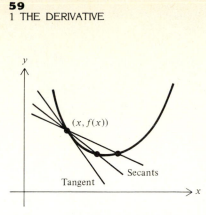

Figure 1.3 Secants approximating a tangent.

Unfortunately, in the present situation, you know only one point on the tangent line, namely the point of tangency $(x, f(x))$. Hence, direct computation of the slope is impossible and you are forced to adopt an indirect approach.

The strategy is to approximate the tangent by other lines whose slopes can be computed directly. In particular, consider lines joining the given point $(x, f(x))$ to neighboring points on the graph of f. These lines, shown in Figure 1.3, are called **secants** and are good approximations to the tangent provided the neighboring point is close to the given point $(x, f(x))$.

You can make the slope of the secant as close as you like to the slope of the tangent by choosing the neighboring point sufficiently close to the given point $(x, f(x))$. This suggests that you should be able to determine the slope of the tangent itself by first computing the slopes of related secants and then studying the behavior of these slopes as the neighboring points get closer and closer to the given point.

To compute the slope of a secant, first label the coordinates of the neighboring point as indicated in Figure 1.4. In particular, let Δx denote the change in the x coordinate between the given point $(x, f(x))$ and the neighboring point. The x coordinate of the neighboring point is $x + \Delta x$, and since the point lies on the graph of f, its y coordinate is $f(x + \Delta x)$.

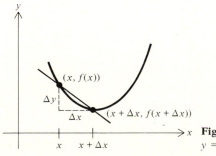

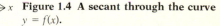

Figure 1.4 A secant through the curve $y = f(x)$.

Since the change in the y coordinate is $\Delta y = f(x + \Delta x) - f(x)$, it follows that

$$\text{Slope of secant} = \frac{\Delta y}{\Delta x} = \frac{f(x + \Delta x) - f(x)}{\Delta x}$$

Remember that this quotient is not the slope of the tangent but only an approximation of it. If Δx is small, however, the neighboring point $(x + \Delta x, f(x + \Delta x))$ is close to the given point $(x, f(x))$, and the approximation is a good one. In fact, the slope of the actual tangent is the number that this quotient approaches as Δx approaches zero.

A calculation based on these observations is performed in the following example.

EXAMPLE 1.1

Find the slope of the line that is tangent to the graph of the function $f(x) = x^2$ at the point $(2, 4)$.

SOLUTION

A sketch of f showing the given point $(2, 4)$ and a related secant is drawn in Figure 1.5.

Since the x coordinate of the given point is 2, it follows that the x coordinate of the neighboring point is $2 + \Delta x$, and the y coordinate of this point is $(2 + \Delta x)^2$. Hence,

$$\text{Slope of secant} = \frac{(2 + \Delta x)^2 - 4}{\Delta x}$$

Your goal is to find the number that this quotient approaches as Δx approaches zero. Before you can do this, you must rewrite the quotient in a simpler form. (Do you see what would happen if you let Δx approach zero in the numerator and denominator of the unsimplified quotient?) To simplify the quotient, expand the term $(2 + \Delta x)^2$,

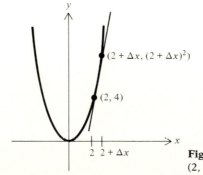

Figure 1.5 The curve $y = x^2$ and a secant through $(2, 4)$.

rewrite the numerator, and then divide numerator and denominator by Δx as follows.

$$\text{Slope of secant} = \frac{(2 + \Delta x)^2 - 4}{\Delta x}$$

$$= \frac{4 + 4\Delta x + (\Delta x)^2 - 4}{\Delta x}$$

$$= \frac{4\Delta x + (\Delta x)^2}{\Delta x}$$

$$= 4 + \Delta x$$

Now let Δx approach zero. Since $4 + \Delta x$ approaches 4 as Δx approaches zero, you can conclude that at the given point (2, 4), the slope of the tangent is 4.

The derivative   In the preceding example, you found the slope of the tangent to the curve $y = x^2$ at a particular point (2, 4). In the next example, you will perform the same calculation again, this time representing the given point algebraically as (x, x^2). The result will be a formula into which you can substitute any value of x to calculate the slope of the tangent to the curve at the point (x, x^2).

EXAMPLE 1.2

Derive a formula expressing the slope of the tangent to the curve $y = x^2$ as a function of the x coordinate of the point of tangency.

SOLUTION

Represent the point of tangency as (x, x^2) and the neighboring point as $(x + \Delta x, (x + \Delta x)^2)$ as shown in Figure 1.6.

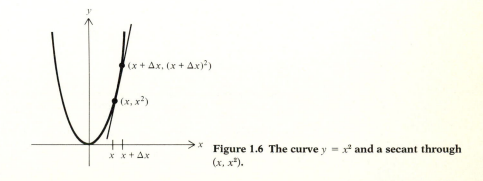

Figure 1.6 The curve $y = x^2$ **and a secant through** (x, x^2)**.**

Then,

$$\text{Slope of secant} = \frac{(x + \Delta x)^2 - x^2}{\Delta x}$$

$$= \frac{x^2 + 2x\,\Delta x + (\Delta x)^2 - x^2}{\Delta x}$$

$$= 2x + \Delta x$$

Since $2x + \Delta x$ approaches $2x$ as Δx approaches 0, you can conclude that at the point (x, x^2), the slope of the tangent is $2x$.

For example, at the point $(2, 4)$, $x = 2$ and so the slope of the tangent is $2(2) = 4$.

In Example 1.2, you started with a function f and derived a related function that expressed the slope of its tangent in terms of the x coordinate of the point of tangency. This derived function is known as the **derivative** of f and is frequently denoted by the symbol f', which is read "f prime." In Example 1.2 you discovered that the derivative of x^2 is $2x$; that is, you found that if $f(x) = x^2$, then $f'(x) = 2x$.

Here is a summary of the situation.

Geometric interpretation of the derivative

The derivative $f'(x)$ expresses the slope of the tangent to the curve $y = f(x)$ as a function of the x coordinate of the point of tangency.

How to compute the derivative of $f(x)$

Step 1. Form the difference quotient (the slope of a secant):

$$\frac{f(x + \Delta x) - f(x)}{\Delta x}$$

Step 2. Simplify the difference quotient algebraically.
Step 3. Let Δx approach zero in the simplified difference quotient. The resulting expression will be the derivative $f'(x)$ (the slope of the tangent).

That is, $\dfrac{f(x + \Delta x) - f(x)}{\Delta x} \to f'(x)$ **as** $\Delta x \to 0$

where the arrow, $\to$, means "approaches."

Here is another example.

EXAMPLE 1.3

Find the equation of the line that is tangent to the graph of the function $f(x) = \dfrac{1}{x}$ when $x = 2$.

SOLUTION

Since you will need to know the slope of the tangent, begin by finding the derivative. First form the difference quotient and simplify it algebraically as follows.

$$\frac{1/(x + \Delta x) - 1/x}{\Delta x} = \frac{1/(x + \Delta x) - 1/x}{\Delta x} \cdot \frac{x(x + \Delta x)}{x(x + \Delta x)}$$

$$= \frac{x - (x + \Delta x)}{x \, \Delta x(x + \Delta x)}$$

$$= \frac{-\Delta x}{x \, \Delta x(x + \Delta x)}$$

$$= \frac{-1}{x(x + \Delta x)}$$

Now, let Δx approach zero. Since

$$\frac{-1}{x(x + \Delta x)} \rightarrow -\frac{1}{x^2} \qquad \text{as} \qquad \Delta x \rightarrow 0$$

you can conclude that the derivative is

$$f'(x) = -\frac{1}{x^2}$$

To find the slope of the tangent when $x = 2$, compute $f'(2)$.

$$\text{Slope of tangent} = f'(2) = -\tfrac{1}{4}$$

To find the y coordinate of the point of tangency, compute $f(2)$.

$$y = f(2) = \tfrac{1}{2}$$

Now use the point-slope form of the equation of a line, $y - y_0 = m(x - x_0)$ with $m = -\tfrac{1}{4}$ and $(x_0, y_0) = (2, \tfrac{1}{2})$ and conclude that the equation of the tangent is

$$y - \tfrac{1}{2} = -\tfrac{1}{4}(x - 2) \qquad \text{or} \qquad y = -\tfrac{1}{4}x + 1$$

The situation is illustrated in Figure 1.7.

Notation   Symbols other than f' are sometimes used to denote the derivative. For example, if y rather than $f(x)$ is used to denote the function itself, the symbol $\dfrac{dy}{dx}$ $\left(\text{suggesting slope } \dfrac{\Delta y}{\Delta x}\right)$ is frequently used instead of $f'(x)$. Hence, instead of the statement

$$\text{If } f(x) = x^2, \text{ then } f'(x) = 2x$$

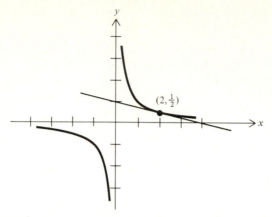

Figure 1.7 The curve $y = \dfrac{1}{x}$ **and the tangent when** $x = 2$.

you could write

$$\text{If } y = x^2, \text{ then } \frac{dy}{dx} = 2x$$

Sometimes the two notations are combined as in the statement

$$\text{If } f(x) = x^2, \text{ then } \frac{df}{dx} = 2x$$

By omitting reference to y and f altogether, you can condense these statements and write

$$\frac{d}{dx}(x^2) = 2x$$

to indicate that the derivative of x^2 is $2x$.

The maximization of profit

In the next example, you will see how to use the derivative to maximize the profit function that was discussed at the beginning of this section.

EXAMPLE 1.4

A manufacturer's profit from the sale of radios is given by the function $P(x) = 400(15 - x)(x - 2)$, where x is the price at which the radios are sold. Find the optimal selling price.

SOLUTION

For reference, the graph of this profit function is sketched once again in Figure 1.8.

Your goal is to find the value of x for which the profit $P(x)$ is greatest. This is the value of x for which the slope of the tangent is

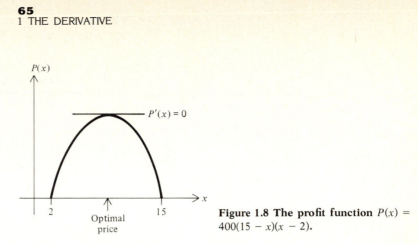

Figure 1.8 The profit function $P(x) = 400(15 - x)(x - 2)$.

zero. Since the slope of the tangent is given by the derivative, begin by computing $P'(x)$. In this case, it is easier to work with the unfactored form of the profit function

$$P(x) = -400x^2 + 6,800x - 12,000$$

First form the difference quotient and simplify it algebraically as follows.

$$\frac{P(x + \Delta x) - P(x)}{\Delta x}$$

$$= \frac{-400(x + \Delta x)^2 + 6,800(x + \Delta x) - 12,000 - (-400x^2 + 6,800x - 12,000)}{\Delta x}$$

$$= \frac{-400(\Delta x)^2 - 800x\,\Delta x + 6,800\,\Delta x}{\Delta x}$$

$$= -400\,\Delta x - 800x + 6,800$$

Now let Δx approach zero. Since

$$-400\,\Delta x - 800x + 6,800 \rightarrow -800x + 6,800 \qquad \text{as} \qquad \Delta x \rightarrow 0$$

it follows that the derivative is

$$P'(x) = -800x + 6,800$$

To find the value of x for which the slope of the tangent is zero, set the derivative equal to zero and solve the resulting equation for x as follows.

$$P'(x) = 0$$

$$-800x + 6,800 = 0$$

$$800x = 6,800$$

$$x = 8.5$$

It follows that $x = 8.5$ is the x coordinate of the peak of the graph and that the optimal selling price is $8.50 per radio.

Limits

As you have seen, the derivative is the value that a certain difference quotient approaches as the variable Δx approaches zero. In general, mathematicians use the word **limit** to denote the value that a function approaches as its variable approaches a specific number. Limits play a central role in modern mathematics and form the basis for a rigorous development of calculus. This important theoretical concept is discussed in more detail in Section B of the appendix.

Differentiability and continuity

Not all functions have a derivative for every value of x. Three functions that do not have derivatives when $x = 0$ are sketched in Figure 1.9.

At the point $(0, 0)$ in Figure 1.9a, the tangent line cannot be uniquely determined. As a result, the derivative, which gives the slope of the tangent, cannot be defined for $x = 0$. The function $f(x) = x^{2/3}$ in Figure 1.9b has a vertical tangent when $x = 0$. Since the slope of a vertical line is undefined, this function has no derivative when $x = 0$. The function $f(x) = \dfrac{1}{x}$ in Figure 1.9c has no derivative at $x = 0$ because the function itself is undefined for this value of x.

A function that has a derivative when $x = a$ is said to be **differentiable** at $x = a$. The graphs of differentiable functions must be "smooth." They cannot have corners or cusps as the graphs in Figure 1.9a and 1.9b do. Most of the functions you will encounter in this text will be differentiable at most points. It can be shown, for example, that polynomials are differentiable everywhere and that rational functions fail to have derivatives only at those values of x that make their denominators zero.

A function whose graph is an unbroken curve is said to be **continuous.** (A more formal definition is given in the appendix.) A function that is differentiable must be continuous, although, as the function in Figure 1.9a indicates, not every continuous function is differentiable.

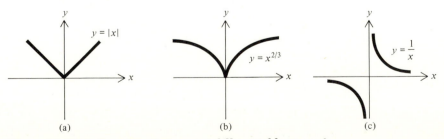

Figure 1.9 Three functions that are not differentiable at $x = 0$.

Problems In Problems 1 through 7, compute the derivative of the given function and find the slope of the line that is tangent to its graph for the specified value of x.

1. $f(x) = 5x - 3$; $x = 2$

2. $f(x) = x^2 - 1$; $x = -1$

3. $y = 2x^2 - 3x + 5$; $x = 0$

4. $y = x^3 - 1$; $x = 2$

5. $f(x) = \dfrac{2}{x}$; $x = \frac{1}{2}$

6. $f(x) = \dfrac{1}{x^2}$; $x = 2$

7. $y = \sqrt{x}$; $x = 9$

In Problems 8 through 11, find the equation of the line that is tangent to the graph of the given function for the specified value of x.

8. $f(x) = x^2 + x + 1$; $x = 2$

9. $f(x) = x^3 - x$; $x = -2$

10. $y = \dfrac{3}{x^2}$; $x = \frac{1}{2}$

11. $y = 2\sqrt{x}$; $x = 4$

12. Suppose $f(x) = x^2$.
    (a) Compute the slope of the secant joining the points on the graph of f whose x coordinates are $x = -2$ and $x = -1.9$.
    (b) Use calculus to compute the slope of the line that is tangent to the graph when $x = -2$ and compare this slope to your answer in part (a).

13. Suppose $f(x) = x^3$.
    (a) Compute the slope of the secant joining the points on the graph of f whose x coordinates are $x = 1$ and $x = 1.1$.
    (b) Use calculus to compute the slope of the line that is tangent to the graph when $x = 1$ and compare this slope with your answer in part (a).

14. (a) Find the derivative of the linear function $y = 3x - 2$.
    (b) Write an equation of the tangent to the graph of this function at the point $(-1, -5)$.
    (c) Explain how the answers to parts (a) and (b) could have been obtained from geometric considerations with no calculation whatsoever.

15. Sketch the graph of the function $y = x^2 - 3x$ and use calculus to find its lowest point.

16. Sketch the graph of the function $y = 1 - x^2$ and use calculus to find its highest point.

17. Sketch the graph of the function $y = x^3 - x^2$. Determine the values of x for which the derivative is zero. What happens to the graph at the corresponding points?

Maximization of profit

18. A manufacturer can produce tape recorders at a cost of \$20 apiece. It is estimated that if the tape recorders are sold for x

dollars apiece, consumers will buy $120 - x$ of them a month. Use calculus to determine the price at which the manufacturer's profit will be greatest.

19. What can you conclude about the graph of a function between $x = a$ and $x = b$ if its derivative is positive whenever $a \leq x \leq b$?

20. Sketch the graph of a function f whose derivative has all of the following properties.
    (a) $f'(x) > 0$ when $x < 1$ and when $x > 5$
    (b) $f'(x) < 0$ when $1 < x < 5$
    (c) $f'(1) = 0$ and $f'(5) = 0$

21. (a) Find the derivatives of the functions $y = x^2$ and $y = x^2 - 3$ and account geometrically for their similarity.
    (b) Without further computation, find the derivative of the function $y = x^2 + 5$.

22. (a) Find the derivative of the function $y = x^2 + 3x$.
    (b) Find the derivatives of the functions $y = x^2$ and $y = 3x$ separately.
    (c) How is the answer in part (a) related to the answers in part (b)?
    (d) In general, if $f(x) = g(x) + h(x)$, what would you guess is the relationship between the derivative of f and the derivatives of g and h?

23. (a) Compute the derivatives of the functions $y = x^2$ and $y = x^3$.
    (b) Examine your answers in part (a). Can you detect a pattern? What do you think is the derivative of $y = x^4$? How about the derivative of $y = x^{27}$?

2 TECHNIQUES OF DIFFERENTIATION

In Section 1, you learned how to find the derivative of a function by letting Δx approach zero in the expression for the slope of a secant. For even the simplest functions, this process is tedious and time-consuming. In this section, you will see some shortcuts. Justification of some of these shortcuts will be given at the end of the section, after you have had a chance to practice using them.

The derivative of a power function

A **power function** is a function of the form $f(x) = x^n$, where n is a real number. For example, $f(x) = x^2$, $f(x) = x^{-3}$, and $f(x) = x^{1/2}$ are all power functions. So are $f(x) = \dfrac{1}{x^2}$ and $f(x) = \sqrt[3]{x}$ since they can be rewritten as $f(x) = x^{-2}$ and $f(x) = x^{1/3}$, respectively. Here is a simple rule you can use to find the derivative of any power function. The proof of this rule will be given in Chapter 4.

The power rule

> **For any number n,**
>
> $$\frac{d}{dx}(x^n) = nx^{n-1}$$
>
> **That is, to find the derivative of x^n, reduce the power of x by 1 and multiply by the original power.**

According to this rule, the derivative of x^2 is $2x^1$ or $2x$, which agrees with the result you obtained in Example 1.2. Here are a few more calculations.

EXAMPLE 2.1

Differentiate (find the derivative of) each of the following functions.

(a) $y = x^{27}$

(b) $y = \dfrac{1}{x^{27}}$

(c) $y = \sqrt{x}$

(d) $y = \dfrac{1}{\sqrt{x}}$

SOLUTION

In each case, use exponential notation to write the function as a power function and then apply the general rule. (You can find a review of exponential notation in Section A of the appendix at the back of the book.)

(a) $\dfrac{d}{dx}(x^{27}) = 27x^{26}$

(b) $\dfrac{d}{dx}\left(\dfrac{1}{x^{27}}\right) = \dfrac{d}{dx}(x^{-27}) = -27x^{-27-1} = -27x^{-28} = -\dfrac{27}{x^{28}}$

(c) $\dfrac{d}{dx}(\sqrt{x}) = \dfrac{d}{dx}(x^{1/2}) = \tfrac{1}{2}x^{1/2-1} = \tfrac{1}{2}x^{-1/2} = \dfrac{1}{2\sqrt{x}}$

(d) $\dfrac{d}{dx}\left(\dfrac{1}{\sqrt{x}}\right) = \dfrac{d}{dx}(x^{-1/2}) = -\tfrac{1}{2}x^{-1/2-1} = -\tfrac{1}{2}x^{-3/2} = -\dfrac{1}{2\sqrt{x^3}}$

The derivative of a constant   The derivative of any constant function is zero. This is because the graph of a constant function $y = c$ is a horizontal line and its slope is zero.

The derivative of a constant

For any constant c,

$$\frac{d}{dx}(c) = 0$$

That is, the derivative of a constant is zero.

The derivative of a constant times a function

The next rule expresses the fact that the curve $v = cf(x)$ is c times as steep as the curve $y = f(x)$.

The constant multiple rule

For any constant c,

$$\frac{d}{dx}(cf) = c\frac{df}{dx}$$

That is, the derivative of a constant times a function is equal to the constant times the derivative of the function.

EXAMPLE 2.2

Differentiate the function $y = 3x^5$.

SOLUTION

You already know that $\frac{d}{dx}(x^5) = 5x^4$. Combining this with the constant multiple rule you get

$$\frac{d}{dx}(3x^5) = 3\frac{d}{dx}(x^5) = 3(5x^4) = 15x^4$$

The derivative of a sum

The next rule states that a sum can be differentiated term by term.

The sum rule

$$\frac{d}{dx}(f + g) = \frac{df}{dx} + \frac{dg}{dx}$$

That is, the derivative of a sum is the sum of the individual derivatives.

EXAMPLE 2.3

Differentiate the function $y = x^2 + 3x^5$.

SOLUTION

You know that $\frac{d}{dx}(x^2) = 2x$ and that $\frac{d}{dx}(3x^5) = 15x^4$. According to the sum rule, you simply add these derivatives to get the derivative of the sum $x^2 + 3x^5$. That is,

$$\frac{d}{dx}(x^2 + 3x^5) = \frac{d}{dx}(x^2) + \frac{d}{dx}(3x^5) = 2x + 15x^4$$

By combining the sum rule with the power and constant multiple rules, you can differentiate any polynomial. Here is an example.

EXAMPLE 2.4

Differentiate the polynomial $y = 5x^3 - 4x^2 + 12x - 8$.

SOLUTION

Differentiate this sum term by term.

$$\frac{dy}{dx} = \frac{d}{dx}(5x^3) + \frac{d}{dx}(-4x^2) + \frac{d}{dx}(12x) + \frac{d}{dx}(-8)$$
$$= 15x^2 - 8x + 12$$

The derivative of a product

Suppose you wanted to differentiate the product $y = x^2(3x + 1)$. You might be tempted to differentiate the factors x^2 and $3x + 1$ separately, and then multiply your answers. That is, since $\frac{d}{dx}(x^2) = 2x$ and $\frac{d}{dx}(3x + 1) = 3$, you might conclude that $\frac{dy}{dx} = 6x$. However, this answer is wrong. To see this, rewrite the function as $y = 3x^3 + x^2$ and observe that the derivative is $9x^2 + 2x$ and not $6x$. The derivative of a product is *not* the product of the individual derivatives. Here is the correct formula for the derivative of a product.

The product rule

$$\frac{d}{dx}(fg) = f\frac{dg}{dx} + g\frac{df}{dx}$$

That is, the derivative of a product is the first factor times the derivative of the second plus the second factor times the derivative of the first.

The use of this rule is illustrated in the next example.

EXAMPLE 2.5

Differentiate the function $y = x^2(3x + 1)$.

SOLUTION

According to the product rule,

$$\frac{d}{dx}[x^2(3x+1)] = x^2\frac{d}{dx}(3x+1) + (3x+1)\frac{d}{dx}(x^2)$$

$$= x^2(3) + (3x+1)(2x)$$

$$= 9x^2 + 2x$$

The derivative of a quotient

The derivative of a quotient is not the quotient of the individual derivatives. Here is the correct rule.

The quotient rule

$$\frac{d}{dx}\left(\frac{f}{g}\right) = \frac{g\frac{df}{dx} - f\frac{dg}{dx}}{g^2}$$

The quotient rule is probably the most complicated formula you have had to learn so far in this book. Here's one way to remember it. The numerator resembles the product rule except that it contains a minus sign, which makes the order in which the terms are written important. Begin by squaring the denominator (since this is easy to do) and then, while still thinking of the original denominator, copy it in the numerator. This gets you started with the proper term in the numerator, and you can easily write down the rest thinking of the product rule. Don't forget to insert the minus sign, without which the rule would not have been so hard to remember in the first place!

Using the quotient rule, you can now differentiate any rational function. Here is an example.

EXAMPLE 2.6

Differentiate the rational function $y = \frac{x^2 + 2x - 21}{x - 3}$.

SOLUTION

According to the quotient rule,

$$\frac{dy}{dx} = \frac{(x-3)\frac{d}{dx}(x^2+2x-21) - (x^2+2x-21)\frac{d}{dx}(x-3)}{(x-3)^2}$$

$$= \frac{(x-3)(2x+2) - (x^2+2x-21)(1)}{(x-3)^2}$$

$$= \frac{x^2 - 6x + 15}{(x-3)^2}$$

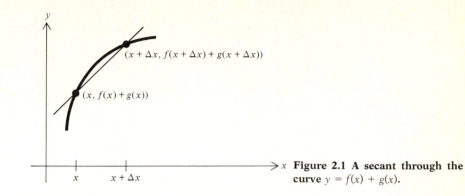

$(x + \Delta x, f(x + \Delta x) + g(x + \Delta x))$

$(x, f(x) + g(x))$

Figure 2.1 A secant through the curve $y = f(x) + g(x)$.

Suggestion: The quotient rule is somewhat cumbersome. Therefore, do not use it to differentiate a quotient like $y = \dfrac{1}{x^2}$ that can be rewritten as $y = x^{-2}$ and differentiated easily using the power rule.

Justification of the sum rule

To see why the sum rule is true, consider a secant through the graph of the function $f + g$ as shown in Figure 2.1.

Begin with the expression for the slope of this secant and rewrite it as the sum of two quotients, one involving f and the other involving g as follows.

$$\text{Slope of secant} = \frac{[f(x + \Delta x) + g(x + \Delta x)] - [f(x) + g(x)]}{\Delta x}$$

$$= \frac{f(x + \Delta x) - f(x)}{\Delta x} + \frac{g(x + \Delta x) - g(x)}{\Delta x}$$

To get the derivative of $f + g$, let Δx approach zero in this expression. Since

$$\frac{f(x + \Delta x) - f(x)}{\Delta x} \to \frac{df}{dx} \qquad \text{as} \qquad \Delta x \to 0$$

and

$$\frac{g(x + \Delta x) - g(x)}{\Delta x} \to \frac{dg}{dx} \qquad \text{as} \qquad \Delta x \to 0$$

it follows that

$$\frac{d}{dx}(f + g) = \frac{df}{dx} + \frac{dg}{dx}$$

and the sum rule is proved.

Justification of the product rule

To show that $\dfrac{d}{dx}(fg) = f\dfrac{dg}{dx} + g\dfrac{df}{dx}$, begin with the difference quotient and rewrite it using the algebraic "trick" of subtracting and then adding the quantity $f(x + \Delta x)g(x)$ in the numerator as follows.

$$\text{Slope of secant} = \frac{f(x + \Delta x)g(x + \Delta x) - f(x)g(x)}{\Delta x}$$

$$= \frac{f(x + \Delta x)g(x + \Delta x) - f(x + \Delta x)g(x)}{\Delta x}$$

$$+ \frac{f(x + \Delta x)g(x) - f(x)g(x)}{\Delta x}$$

$$= f(x + \Delta x)\frac{g(x + \Delta x) - g(x)}{\Delta x} + g(x)\frac{f(x + \Delta x) - f(x)}{\Delta x}$$

Now let Δx approach zero. Since $f(x + \Delta x)$ approaches $f(x)$, $\dfrac{g(x + \Delta x) - g(x)}{\Delta x}$ approaches $\dfrac{dg}{dx}$, and $\dfrac{f(x + \Delta x) - f(x)}{\Delta x}$ approaches $\dfrac{df}{dx}$ as Δx approaches zero, it follows that

$$\frac{d}{dx}(fg) = f\frac{dg}{dx} + g\frac{df}{dx}$$

and the product rule is proved.

Problems   In Problems 1 through 25, differentiate the given function. In each case, do as much of the computation as possible in your head.

1. $y = x^2 + 2x + 3$

2. $y = 3x^5 - 4x^3 + 9x - 6$

3. $f(x) = x^9 - 5x^8 + x + 12$

4. $f(x) = \frac{1}{4}x^8 - \frac{1}{2}x^6 - x + 2$

5. $y = \dfrac{1}{x} + \dfrac{1}{x^2}$

6. $y = \dfrac{3}{x} - \dfrac{2}{x^2} + 5$

7. $f(x) = \sqrt{x} + \dfrac{1}{\sqrt{x}}$

8. $f(x) = 2\sqrt{x} + \dfrac{4}{\sqrt{x}} - \sqrt{2}$

9. $y = -16x + \dfrac{2}{x} - x^{3/2} + \dfrac{1}{3x} + \dfrac{x}{3}$

10. $y = -\dfrac{2}{x^2} + x^{2/3} + \dfrac{1}{2\sqrt{x}} + \dfrac{x^2}{4} + \sqrt{5}$

11. $f(x) = (2x + 1)(3x - 2)$ (Use the product rule.)

12. $f(x) = (x - 5)(1 - 2x)$ (Use the product rule.)

13. $y = 10(3x + 1)(1 - 5x)$ (Use the product rule.)

14. $y = 400(15 - x)(x - 2)$ (Use the product rule.)

15. $f(x) = \frac{1}{3}(x^5 - 2x^3 + 1)$

16. $f(x) = -3(5x^3 - 2x + 5)$

17. $y = \dfrac{x + 1}{x - 2}$

18. $y = \dfrac{2x - 3}{5x + 4}$

19. $f(x) = \dfrac{x}{x^2 - 2}$

20. $f(x) = \dfrac{1}{x - 2}$

21. $y = \dfrac{3}{x + 5}$

22. $y = \dfrac{x^2 + 1}{1 - x^2}$

23. $f(x) = \dfrac{x^2 - 3x + 2}{2x^2 + 5x - 1}$

24. $f(x) = \dfrac{x^2 + 2x + 1}{3}$

25. $y = (2x + 1)(x - 3)(1 - 4x)$

In Problems 26 through 29, find the equation of the line that is tangent to the graph of the given function at the specified point.

26. $y = x^5 - 3x^3 - 5x + 2;\ (1, -5)$

27. $y = (x^2 + 1)(1 - x^3);\ (1, 0)$

28. $f(x) = \dfrac{x + 1}{x - 1};\ (0, -1)$

29. $f(x) = 1 - \dfrac{1}{x} + \dfrac{2}{\sqrt{x}};\ (4, \tfrac{7}{4})$

In Problems 30 through 33, find the equation of the line that is tangent to the graph of the given function at the point $(x, f(x))$ for the specified value of x.

30. $f(x) = x^4 - 3x^3 + 2x^2 - 6;\ x = 2$

31. $f(x) = x - \dfrac{1}{x^2};\ x = 1$

32. $f(x) = \dfrac{x^2 + 2}{x^2 - 2};\ x = -1$

33. $f(x) = (x^3 - 2x^2 + 3x - 1)(x^5 - 4x^2 + 2);\ x = 0$

34. (a) Differentiate the function $y = 2x^2 - 5x - 3$.
    (b) Now factor the function in part (a) as $y = (2x + 1)(x - 3)$ and differentiate using the product rule. Compare your answers.

35. (a) Use the quotient rule to differentiate the function $y = \dfrac{2x - 3}{x^3}$.
    (b) Now rewrite this function as $y = x^{-3}(2x - 3)$ and differentiate using the product rule.
    (c) Show that your answers to parts (a) and (b) are the same.

36. The product rule tells you how to differentiate the product of *any* two functions, while the constant multiple rule tells you how to differentiate products in which one of the factors is constant. Show that the two rules are consistent. That is, use the product rule to show that $\dfrac{d}{dx}(cf) = c\dfrac{df}{dx}$ if c is a constant.

37. Sketch the graph of the function $f(x) = x^2 - 4x - 5$ and use calculus to determine its lowest point.

38. Sketch the graph of the function $f(x) = 3 - 2x - x^2$ and use calculus to determine its highest point.

39. Find numbers a and b such that the lowest point on the graph of the function $f(x) = ax^2 + bx$ is $(3, -8)$.

40. Find numbers a, b, and c such that the graph of the function $f(x) = ax^2 + bx + c$ will have x intercepts at $(0, 0)$ and $(5, 0)$, and a tangent with slope 1 when $x = 2$.

41. Find the equations of all the tangents to the graph of the function $f(x) = x^2 - 4x + 25$ that pass through the origin $(0, 0)$.

42. Find all the points (x, y) on the graph of the function $y = 4x^2$ with the property that the tangent to the graph at (x, y) passes through the point $(2, 0)$.

Consumer expenditure

43. The consumer demand for a certain commodity is $D(p) = -200p + 12,000$ units per month when the market price is p dollars per unit.
    (a) Express consumers' total monthly expenditure for the commodity as a function of p and draw the graph.
    (b) Use calculus to determine the market price for which the consumer expenditure is greatest.

44. Show that $\quad \dfrac{d}{dx}(fgh) = fg\dfrac{dh}{dx} + fh\dfrac{dg}{dx} + gh\dfrac{df}{dx}$

   (*Hint:* Apply the product rule twice.)

45. Show that the quotient rule is true. (*Hint:* Show that the difference quotient is

$$\frac{1}{\Delta x}\left[\frac{f(x + \Delta x)}{g(x + \Delta x)} - \frac{f(x)}{g(x)}\right] = \frac{g(x)f(x + \Delta x) - f(x)g(x + \Delta x)}{g(x + \Delta x)g(x)\,\Delta x}$$

   Before letting Δx approach zero, rewrite this quotient using the trick of subtracting and adding $g(x)f(x)$ in the numerator.)

3 RATE OF CHANGE AND MARGINAL ANALYSIS

In this section, you will see how the derivative can be interpreted as a **rate of change.** Viewed in this way, a derivative may represent such quantities as the rate at which population grows, a manufacturer's marginal cost, the speed of a moving object, the rate of inflation, or the rate at which natural resources are being depleted.

You may have already sensed the connection between derivatives and rate of change. The derivative of a function is the slope of its tangent line, and the slope of any line is the rate at which it is rising or falling. The purpose of this section is to make this connection more precise. Here is a familiar practical situation that will serve as a model for the general discussion.

Average and instantaneous speed

Imagine that a car is moving along a straight road, and that $D(t)$ is its distance from its starting point after t hours. Suppose you want to determine the speed of the car at a particular time t but do not have access to the car's speedometer. Here's what you can do.

First record the position of the car at time t, and again at some later time $t + \Delta t$. That is, determine $D(t)$ and $D(t + \Delta t)$. Then compute the **average speed** of the car between the times t and $t + \Delta t$ as follows.

$$\text{Average speed} = \frac{\text{change in distance}}{\text{change in time}} = \frac{D(t + \Delta t) - D(t)}{\Delta t}$$

Since the speed of the car may fluctuate during the time interval from t to $t + \Delta t$, it is unlikely that this average speed is equal to the **instantaneous speed** (the speed shown on the speedometer) at time t. However, if Δt is small, the possibility of drastic changes in speed is small, and the average speed may be a fairly good approximation to the instantaneous speed. Indeed, you can find the instantaneous speed at time t by letting Δt approach zero in the expression for the average speed.

Notice that the expression $\dfrac{D(t + \Delta t) - D(t)}{\Delta t}$ for the average speed is exactly the difference quotient that appears in the definition of the derivative. As Δt approaches zero, this quotient approaches the derivative of D. It follows that the instantaneous speed at time t is just the derivative $D'(t)$ of the distance function.

Instantaneous speed

> **The instantaneous speed of a moving object is the derivative $D'(t)$ of its distance function. That is,**
>
> $$\text{Speed} = \text{derivative of distance}$$

Average and instantaneous rate of change

These ideas can be extended to more general situations. Suppose that y is a function of x, say $y = f(x)$. Corresponding to a change from x to $x + \Delta x$, the variable y changes by an amount $\Delta y = f(x + \Delta x) - f(x)$. Thus, the difference quotient

$$\frac{\text{Change in } y}{\text{Change in } x} = \frac{\Delta y}{\Delta x} = \frac{f(x + \Delta x) - f(x)}{\Delta x}$$

represents the resulting **average rate of change of y with respect to x**.

As the interval over which you are averaging becomes shorter (that is, as Δx approaches zero), the average rate of change approaches what you would intuitively call the **instantaneous rate of change of y with respect to** x, and the difference quotient approaches the derivative $\frac{dy}{dx} = f'(x)$. Hence, the instantaneous rate of change of y with respect to x is just the derivative $\frac{dy}{dx}$.

Instantaneous rate of change

> If $y = f(x)$, **the instantaneous rate of change of y with respect to x is given by the derivative of f. That is,**
>
> $$\text{Rate of change} = \frac{dy}{dx} = f'(x)$$

EXAMPLE 3.1

It is estimated that x months from now, the population of a certain community will be $P(x) = x^2 + 20x + 8{,}000$.

(a) At what rate will the population be changing 15 months from now?
(b) By how much will the population actually change during the 16th month?

SOLUTION

(a) The rate of change of the population is the derivative of the population function. That is,

$$\text{Rate of change} = P'(x) = 2x + 20$$

Since $\qquad\qquad P'(15) = 2(15) + 20 = 50$

it follows that 15 months from now, the population will be growing at a rate of 50 people per month.
(b) The actual change in the population during the 16th month is the difference between the population at the end of 16 months and the population at the end of 15 months. That is,

$$\text{Change in population} = P(16) - P(15) = 8{,}576 - 8{,}525$$
$$= 51 \quad \text{people}$$

The reason for the difference in the preceding example between the actual change in population during the 16th month in part (b) and the monthly rate of change at the beginning of that month in part (a) is that the rate of change of the population varied during the month. The instantaneous rate of change in part (a) can be thought of as the change in population that would occur during the 16th month if the rate of change of the population remained constant.

Marginal analysis in economics   In economics, the (instantaneous) rate of change of the total production cost with respect to the number of units produced is called the **marginal cost.** It is measured in dollars per unit and is often a good approximation to the cost of producing 1 additional unit.

Marginal cost

> **The marginal cost per unit is the (instantaneous) rate of change of total cost with respect to output. That is,**
>
> Marginal cost = derivative of total cost

EXAMPLE 3.2

Suppose the total cost in dollars of manufacturing q units of a certain commodity is $C(q) = 3q^2 + 5q + 10$.

(a) Derive a formula for the marginal cost.
(b) What is the marginal cost when 50 units have been produced?
(c) What is the actual cost of producing the 51st unit?

SOLUTION

(a) The marginal cost is the derivative $C'(q) = 6q + 5$.
(b) When 50 units have been produced, $q = 50$ and the marginal cost is $C'(50) = 305$ dollars per unit.
(c) The actual cost of producing the 51st unit is the difference between the cost of producing 51 units and the cost of producing 50 units. That is,

Cost of 51st unit $= C(51) - C(50) = 8,068 - 7,760 = \$308$

Notice that in the preceding example, the marginal cost in part (b) was close to, but not equal to the actual cost in part (c) of producing 1 additional unit. In geometric terms, the difference between these two quantities is the difference between the slope of a tangent to the cost curve and the slope of a nearby secant. The marginal cost $C'(50)$ in part (b) is the slope of the line that is tangent to the cost curve $C(q)$ when $q = 50$. The difference $C(51) - C(50)$ in part (c) is the slope

$$\frac{\text{Change in } C}{\text{Change in } q} = \frac{C(50 + 1) - C(50)}{1}$$

of the secant joining the two points on the cost curve whose q coordinates are 50 and 51. The situation is illustrated in Figure 3.1.

The answers to parts (b) and (c) of the preceding example were almost equal because the points $(50, C(50))$ and $(51, C(51))$ are close

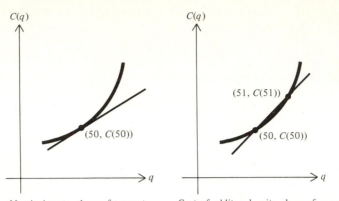

Marginal cost: slope of tangent   Cost of additonal unit: slope of secant

Figure 3.1 The relationship between marginal cost and the cost of 1 additional unit.

together and lie on a portion of the cost curve that is practically linear. For such points, the slope of the secant is a good approximation to the slope of the tangent. Because the similarity of the answers in parts (b) and (c) is typical, and because it is usually easier to compute the marginal cost for one value of q than the total cost for two values of q, economists often use the marginal cost to approximate the actual cost of producing 1 additional unit.

In general, the term **marginal analysis** in economics refers to the practice of using the derivative of a function to estimate the change in the dependent variable produced by a 1-unit increase in the size of the independent variable. In the next example, marginal analysis is used to estimate the effect of a 1-unit increase in the size of the labor force on the output of a factory.

EXAMPLE 3.3

It is estimated that the weekly output at a certain plant is $Q(x) = -x^3 + 60x^2 + 1{,}200x$ units, where x is the number of workers employed at the plant. Currently there are 30 workers employed at the plant. Use marginal analysis to estimate the change in the weekly output that will result from the addition of 1 more worker to the work force.

SOLUTION

The derivative

$$Q'(x) = -3x^2 + 120x + 1{,}200$$

is the rate of change of the output Q with respect to the number x of workers. For any value of x, this derivative is an approximation to the number of additional units that will be produced each week due to the hiring of the $(x + 1)$st worker. Hence, the change in the weekly output that will result if the number of workers is increased from 30 to 31 is approximately

$$Q'(30) = -3(30)^2 + 120(30) + 1,200 = 2,100 \qquad \text{units}$$

For practice, compute the change in output exactly and compare your answer to the approximation. Is the approximation a good one?

Percentage rate of change

In many practical situations, the rate of change of a quantity is not as significant as its **percentage rate of change.** For example, a yearly rate of change in population of 500 people in a city of 5 million would be negligible, while the same rate of change could have enormous impact in a town of 2,000. The percentage rate of change compares the rate of change of a quantity with the size of that quantity.

$$\text{Percentage rate of change of } Q = 100 \, \frac{\text{rate of change of } Q}{Q}$$

For example, a rate of change of 500 people per year in the population of a city of 5 million amounts to a percentage rate of change of only $\frac{100(500)}{5,000,000} = 0.01$ percent of the population per year. On the other hand, the same rate of change in a town of 2,000 is equal to a percentage rate of change of $\frac{100(500)}{2,000} = 25$ percent of the population per year.

Here is the formula for percentage rate of change written in terms of the derivative.

Percentage rate of change

> If $y = f(x)$, **the percentage rate of change of y with respect to x is given by the formula**
>
> $$\text{Percentage rate of change} = 100 \, \frac{f'(x)}{f(x)} = 100 \, \frac{dy/dx}{y}$$

EXAMPLE 3.4

The gross national product (GNP) of a certain country was $N(t) = t^2 + 5t + 100$ billion dollars t years after 1970.

(a) At what rate was the GNP changing in 1975?
(b) At what percentage rate was the GNP changing in 1975?

SOLUTION

(a) The rate of change of the GNP is the derivative $N'(t) = 2t + 5$. The rate of change in 1975 is $N'(5) = 2(5) + 5 = 15$ billion dollars per year.
(b) The percentage rate of change of the GNP in 1975 is

$$100 \, \frac{N'(5)}{N(5)} = 100 \, \frac{15}{150} = 10 \qquad \text{percent per year}$$

Problems

Newspaper circulation

1. It is estimated that t years from now, the circulation of a local newspaper will be $C(t) = 100t^2 + 400t + 5,000$.
   (a) Derive an expression for the rate at which the circulation will be changing t years from now.
   (b) At what rate will the circulation be changing 5 years from now? Will it be increasing or decreasing?
   (c) By how much will the circulation actually change during the 6th year?

Speed of a moving object

2. An object moves along a straight line so that after t minutes, its distance from a fixed reference point is $D(t) = 10t + \dfrac{5}{t+1}$ meters.
   (a) At what speed is the object moving at the end of 4 minutes?
   (b) How far does the object actually travel during the 5th minute?

Worker efficiency

3. An efficiency study of the morning shift at a certain factory indicates that an average worker who arrives on the job at 8:00 A.M. will have assembled $f(x) = -x^3 + 6x^2 + 15x$ transistor radios x hours later.
   (a) Derive a formula for the rate at which the worker will be assembling radios after x hours.
   (b) At what rate will the worker be assembling radios at 9:00 A.M.?
   (c) How many radios will the worker actually assemble between 9:00 A.M. and 10:00 A.M.?

Air pollution

4. An environmental study of a certain suburban community suggests that t years from now, the average level of carbon monoxide in the air will be $Q(t) = 0.05t^2 + 0.1t + 3.4$ parts per million.
   (a) At what rate will the carbon monoxide level be changing 1 year from now?
   (b) By how much will the carbon monoxide level change this year?
   (c) By how much will the carbon monoxide level change over the next 2 years?

Population growth

5. It is estimated that t years from now, the population of a certain suburban community will be $P(t) = 20 - \dfrac{6}{t+1}$ thousand.
   (a) Derive a formula for the rate at which the population will be changing with respect to time.
   (b) At what rate will the community's population be growing 1 year from now?

(c) By how much will the population actually increase during the 2nd year?

(d) At what rate will the population be growing 9 years from now?

(e) What will happen to the rate of population growth in the long run?

SAT scores

6. It is estimated that x years from now, the average SAT score of the incoming students at an eastern liberal arts college will be $f(x) = -6x + 582$.

(a) Derive an expression for the rate at which the average SAT score will be changing with respect to time.

(b) What is the significance of the fact that the expression in part (a) is a constant? What is the significance of the fact that the constant in part (a) is negative?

7. Use calculus to prove that if y is a linear function of x, the rate of change of y with respect to x is constant.

The distance between moving objects

8. Two cars leave an intersection at the same time. One travels east at a constant speed of 60 kilometers per hour, while the other goes north at a constant speed of 80 kilometers per hour. Find an expression for the rate at which the distance between the cars is changing with respect to time.

Marginal analysis

9. Suppose the total cost in dollars of manufacturing q units is $C(q) = 3q^2 + q + 500$.

(a) Use marginal analysis to estimate the cost of manufacturing the 41st unit.

(b) Compute the actual cost of manufacturing the 41st unit.

Marginal analysis

10. A manufacturer's total cost is $C(q) = 0.1q^3 - 0.5q^2 + 500q + 200$ dollars, where q is the number of units produced.

(a) Use marginal analysis to estimate the cost of manufacturing the 4th unit.

(b) Compute the actual cost of manufacturing the 4th unit.

Marginal analysis

11. A manufacturer's total monthly revenue is $R(q) = 240q + 0.05q^2$ dollars when q units are produced during the month. Currently, the manufacturer is producing 80 units a month and is planning to increase the monthly output by 1 unit.

(a) Use marginal analysis to estimate the additional revenue that will be generated by the production of the 81st unit.

(b) Use the revenue function to compute the actual additional revenue that will be generated by the production of the 81st unit.

Marginal analysis   12. It is estimated that the weekly output at a certain plant is $Q(x) = -x^2 + 2,100x$ units, where x is the number of workers employed at the plant. Currently there are 60 workers employed at the plant.
   (a) Use marginal analysis to estimate the effect that 1 additional worker will have on the weekly output.
   (b) Compute the actual change in the weekly output that will result if 1 additional worker is hired.

Marginal analysis   13. At a certain factory, the daily output is $600K^{1/2}$ units, where K denotes the capital investment measured in units of $1,000. The current capital investment is $900,000. Use marginal analysis to estimate the effect that an additional capital investment of $1,000 will have on the daily output.

Marginal analysis   14. At a certain factory, the daily output is $3,000K^{1/2}L^{1/3}$ units, where K denotes the firm's capital investment measured in units of $1,000, and L denotes the size of the labor force measured in worker-hours. Suppose that the current capital investment is $400,000 and that 1,331 worker-hours of labor are used each day. Use marginal analysis to estimate the effect that an additional capital investment of $1,000 will have on the daily output if the size of the labor force is not changed.

Population growth   15. It is projected that x months from now, the population of a certain town will be $P(x) = 2x + 4x^{3/2} + 5,000$.
   (a) At what rate will the population be changing 9 months from now?
   (b) At what percentage rate will the population be changing 9 months from now?

Annual earnings   16. The gross annual earnings of a certain company were $A(t) = 0.1t^2 + 10t + 20$ thousand dollars t years after its formation in 1975.
   (a) At what rate were the gross annual earnings of the company growing in 1979?
   (b) At what percentage rate were the gross annual earnings growing in 1979?

Property tax   17. Records indicate that x years after 1970, the average property tax on a three-bedroom home in a certain community was $T(x) = 20x^2 + 40x + 600$ dollars.
   (a) At what rate was the property tax increasing in 1976?
   (b) At what percentage rate was the property tax increasing in 1976?

Population growth   18. It is estimated that t years from now, the population of a certain town will be $P(t) = t^2 + 200t + 10,000$.

(a) Express the percentage rate of change of the population as a function of t, simplify this function algebraically, and draw its graph.

(b) What will happen to the percentage rate of change of the population in the long run?

Salary increases 19. Your starting salary will be \$12,000 and you will get a raise of \$1,000 each year.

(a) Express the percentage rate of change of your salary as a function of time and draw the graph.

(b) At what percentage rate will your salary be increasing after 1 year?

(c) What will happen to the percentage rate of change of your salary in the long run?

20. If y is a linear function of x, what will happen to the percentage rate of change of y with respect to x as x increases without bound? Explain.

Free-fall If an object is dropped or thrown vertically, its height (in feet) after t seconds is $H(t) = -16t^2 + S_0 t + H_0$, where S_0 is the initial speed of the object and H_0 is its initial height. Use this formula to solve Problems 21 through 24.

21. A stone is dropped (with initial speed zero) from the top of a building 144 feet high.

(a) When will the stone hit the ground? (That is, for what value of t is $H(t)$ equal to zero?)

(b) With what speed will the stone hit the ground?

22. A ball is thrown vertically upward from the ground ($H_0 = 0$) with an initial speed of 160 feet per second ($S_0 = 160$).

(a) When will the ball hit the ground?

(b) With what speed will the ball hit the ground?

(c) When will the ball reach its maximum height? (*Hint:* The speed of the ball will be zero when the ball reaches its maximum height.)

(d) How high will the ball rise?

23. A man standing on the top of a building throws a ball vertically upward. After 2 seconds, the ball passes him on the way down, and 2 seconds after that it hits the ground below.

(a) What was the initial speed of the ball?

(b) How high is the building?

(c) What is the speed of the ball when it passes the man on the way down?

(d) What is the speed of the ball as it hits the ground?

24. A ball is thrown vertically upward from the ground with a certain initial speed S_0.
    (a) Derive a formula for the time at which the ball hits the ground.
    (b) Use the result of part (a) to prove that the ball will be falling at a speed of S_0 feet per second when it hits the ground.

Manufacturing cost   25. Suppose the total manufacturing cost C at a certain factory is a function of the number q of units produced which, in turn, is a function of the number t of hours during which the factory has been operating.

(a) What quantity is represented by the derivative $\dfrac{dC}{dq}$? In what units is this quantity measured?

(b) What quantity is represented by the derivative $\dfrac{dq}{dt}$? In what units is this quantity measured?

(c) What quantity is represented by the product $\dfrac{dC}{dq}\dfrac{dq}{dt}$? In what units is this quantity measured?

4 THE CHAIN RULE

In many practical situations, a quantity of interest is given as a function of one variable which, in turn, can be thought of as a function of a second variable. In such cases, the rate of change of the quantity with respect to the second variable is equal to the rate of change of the quantity with respect to the first variable times the rate of change of the first variable with respect to the second.

For example, suppose the total manufacturing cost at a certain factory is a function of the number of units produced which, in turn, is a function of the number of hours during which the factory has been operating. Let C, q, and t denote the cost (in dollars), the number of units, and the number of hours, respectively. Then,

$$\frac{dC}{dq} = \begin{array}{l}\text{rate of change of cost} \\ \text{with respect to output}\end{array} \qquad \text{(dollars per unit)}$$

and

$$\frac{dq}{dt} = \begin{array}{l}\text{rate of change of output} \\ \text{with respect to time}\end{array} \qquad \text{(units per hour)}$$

The product of these two rates is the rate of change of cost with respect to time.

$$\frac{dC}{dq}\frac{dq}{dt} = \begin{array}{l}\text{rate of change of cost} \\ \text{with respect to time}\end{array} \qquad \text{(dollars per hour)}$$

Since the rate of change of cost with respect to time is also given by the derivative $\dfrac{dC}{dt}$, it follows that

$$\frac{dC}{dt} = \frac{dC}{dq}\frac{dq}{dt}$$

This formula is a special case of an important rule called the **chain rule.**

The chain rule

Suppose y is a function of u and u is a function of x. Then y can be regarded as a function of x and

$$\frac{dy}{dx} = \frac{dy}{du}\frac{du}{dx}$$

That is, the derivative of y with respect to x is the derivative of y with respect to u times the derivative of u with respect to x.

Notice that one way to remember the chain rule is to pretend that the derivatives $\dfrac{dy}{du}$ and $\dfrac{du}{dx}$ are quotients and cancel du, reducing the expression $\dfrac{dy}{du}\dfrac{du}{dx}$ on the right-hand side of the equation to the expression $\dfrac{dy}{dx}$ on the left.

Here are two examples illustrating the use of the chain rule.

EXAMPLE 4.1

Find $\dfrac{dy}{dx}$ if $y = u^3 - 3u^2 + 1$ and $u = x^2 + 2$.

SOLUTION

Since
$$\frac{dy}{du} = 3u^2 - 6u \qquad \text{and} \qquad \frac{du}{dx} = 2x$$

it follows that
$$\frac{dy}{dx} = \frac{dy}{du}\frac{du}{dx} = (3u^2 - 6u)(2x)$$

Notice that this derivative is expressed in terms of the variables x and u. Since you are thinking of y as a function of x, it is more natural to express $\dfrac{dy}{dx}$ in terms of x alone. To do this, substitute $x^2 + 2$ for u in the expression for $\dfrac{dy}{dx}$.

$$\frac{dy}{dx} = [3(x^2 + 2)^2 - 6(x^2 + 2)](2x) = 6x^3(x^2 + 2)$$

For practice, check this answer by first substituting $u = x^2 + 2$ into the original expression for y and then differentiating with respect to x.

In the next example, you will see how to use the chain rule to calculate a derivative for a particular value of the independent variable.

EXAMPLE 4.2

Find $\dfrac{dy}{dx}$ when $x = 1$ if $y = \dfrac{u}{u + 1}$ and $u = 3x^2 - 1$.

SOLUTION

First use the quotient rule to get

$$\frac{dy}{du} = \frac{(u + 1)(1) - u(1)}{(u + 1)^2} = \frac{1}{(u + 1)^2}$$

To evaluate this derivative when $x = 1$, put $x = 1$ in the formula $u = 3x^2 - 1$ to get $u = 2$ and then substitute $u = 2$ into the expression for $\dfrac{dy}{du}$. This gives

$$\frac{dy}{du} = \frac{1}{(2 + 1)^2} = \frac{1}{9}$$

Next, compute the derivative of u with respect to x

$$\frac{du}{dx} = 6x$$

and evaluate it when $x = 1$ to get

$$\frac{du}{dx} = 6(1) = 6$$

Finally, multiply the values of $\dfrac{dy}{du}$ and $\dfrac{du}{dx}$ to conclude that when $x = 1$,

$$\frac{dy}{dx} = \frac{dy}{du}\frac{du}{dx} = \frac{1}{9}(6) = \frac{2}{3}$$

The chain rule for powers   In Section 2, you learned the rule

$$\frac{d}{dx}(x^n) = nx^{n-1}$$

for differentiating power functions. There is a closely related rule (which is actually a special case of the chain rule in disguise) that you can use to differentiate functions of the form $[h(x)]^n$, that is, functions that are powers of other functions. According to this rule, you begin by computing $n[h(x)]^{n-1}$ and then multiply this expression by the derivative of $h(x)$.

The chain rule for powers

> **For any real number n and any differentiable function h,**
>
> $$\frac{d}{dx}[h(x)]^n = n[h(x)]^{n-1}\frac{d}{dx}[h(x)]$$

EXAMPLE 4.3

Differentiate the function $f(x) = (2x^4 - x)^3$.

SOLUTION

One way to do this problem is to expand the function and rewrite it as

$$f(x) = 8x^{12} - 12x^9 + 6x^6 - x^3$$

and then differentiate this polynomial term by term to get

$$f'(x) = 96x^{11} - 108x^8 + 36x^5 - 3x^2$$

But see how much easier it is to use the chain rule for powers. According to this rule,

$$f'(x) = 3(2x^4 - x)^2\frac{d}{dx}(2x^4 - x)$$
$$= 3(2x^4 - x)^2(8x^3 - 1)$$

Not only is this method easier, but the answer even comes out in factored form!

EXAMPLE 4.4

Differentiate the function $f(x) = \dfrac{1}{(2x + 3)^5}$.

SOLUTION

Rewrite the function as

$$f(x) = (2x + 3)^{-5}$$

and apply the chain rule to get

$$f'(x) = -5(2x + 3)^{-6}\frac{d}{dx}(2x + 3)$$
$$= -5(2x + 3)^{-6}(2)$$
$$= -\frac{10}{(2x + 3)^6}$$

For practice, compute $f'(x)$ again, this time using the quotient rule. Which method do you prefer?

EXAMPLE 4.5

Differentiate the function $f(x) = \sqrt{\dfrac{x+1}{x-1}}$.

SOLUTION

First rewrite the function as

$$f(x) = \left(\frac{x+1}{x-1}\right)^{1/2}$$

Then apply the chain rule to get

$$f'(x) = \frac{1}{2}\left(\frac{x+1}{x-1}\right)^{-1/2}\frac{d}{dx}\left(\frac{x+1}{x-1}\right)$$

Now use the quotient rule to get

$$\frac{d}{dx}\left(\frac{x+1}{x-1}\right) = \frac{(x-1)(1)-(x+1)(1)}{(x-1)^2} = -\frac{2}{(x-1)^2}$$

and substitute the result into the equation for $f'(x)$ to get

$$f'(x) = \frac{1}{2}\left(\frac{x+1}{x-1}\right)^{-1/2}\left[-\frac{2}{(x-1)^2}\right]$$

$$= -\frac{(x+1)^{-1/2}}{(x-1)^{3/2}}$$

$$= -\frac{1}{(x+1)^{1/2}(x-1)^{3/2}}$$

To see that the chain rule for powers is really nothing more than a special case of the chain rule, think of the function $y = [h(x)]^n$ as the composite function formed from $y = u^n$ and $u = h(x)$. Then,

$$\frac{dy}{du} = nu^{n-1} = n[h(x)]^{n-1}$$

and the chain rule

$$\frac{dy}{dx} = \frac{dy}{du}\frac{du}{dx}$$

can be rewritten as

$$\frac{d}{dx}[h(x)]^n = n[h(x)]^{n-1}\frac{d}{dx}[h(x)]$$

which is precisely the chain rule for powers.

Related rates   In many problems, a quantity is given as a function of one variable which, in turn, can be written as a function of a second variable, and the goal is to find the rate of change of the original quantity with respect to the second variable. Such problems are sometimes called **related rates problems** and can be solved by means of the chain rule. Here is an example.

EXAMPLE 4.6

An environmental study of a certain suburban community suggests that the average daily level of carbon monoxide in the air will be $c(p) = \sqrt{0.5p^2 + 17}$ parts per million when the population is p thousand. It is estimated that t years from now, the population of the community will be $p(t) = 3.1 + 0.1t^2$ thousand. At what rate will the carbon monoxide level be changing with respect to time 3 years from now?

SOLUTION

The goal is to find $\dfrac{dc}{dt}$ when $t = 3$. First compute the derivatives

$$\frac{dc}{dp} = \frac{1}{2}p(0.5p^2 + 17)^{-1/2} \quad \text{and} \quad \frac{dp}{dt} = 0.2t$$

When $t = 3$, $\qquad p = p(3) = 3.1 + 0.1(3)^2 = 4$

so $\qquad \dfrac{dc}{dp} = \dfrac{1}{2}(4)[0.5(16) + 17]^{-1/2} = \dfrac{4}{2\sqrt{25}} = 0.4$

and $\qquad \dfrac{dp}{dt} = 0.2(3) = 0.6$

Use the chain rule to conclude that

$$\frac{dc}{dt} = \frac{dc}{dp}\frac{dp}{dt} = 0.4(0.6) = 0.24 \qquad \text{parts per million per year}$$

In some related rates problems, you are given information about the rate of change of some of the variables instead of explicit formulas relating the variables. You will learn how to solve problems of this type in Chapter 3, Section 3.

Problems   In Problems 1 through 10, use the chain rule to compute the derivative $\dfrac{dy}{dx}$.

1. $y = u^2 + 1$, $u = 3x - 2$ 2. $y = 2u^2 - u + 5$, $u = 1 - x^2$

3. $y = \sqrt{u}$, $u = x^2 + 2x - 3$ 4. $y = u^2$, $u = \sqrt{x}$

5. $y = \dfrac{1}{u^2}, u = x^2 + 1$ 6. $y = \dfrac{1}{u}, u = 3x^2 + 5$

7. $y = \dfrac{1}{\sqrt{u}}, u = x^2 - 9$ 8. $y = u^2 + u - 2, u = \dfrac{1}{x}$

9. $y = \dfrac{1}{u - 1}, u = x^2$ 10. $y = u^2, u = \dfrac{1}{x - 1}$

In Problems 11 through 16, use the chain rule to compute the derivative $\dfrac{dy}{dx}$ for the given value of x.

11. $y = 3u^4 - 4u + 5, u = x^3 - 2x - 5; x = 2$

12. $y = u^5 - 3u^2 + 6u - 5, u = x^2 - 1; x = 1$

13. $y = \sqrt{u}, u = x^2 - 2x + 6; x = 3$

14. $y = 3u^2 - 6u + 2, u = \dfrac{1}{x^2}; x = \tfrac{1}{3}$

15. $y = \dfrac{1}{u}, u = 3 - \dfrac{1}{x^2}; x = \tfrac{1}{2}$

16. $y = \dfrac{1}{u + 1}, u = x^3 - 2x + 5; x = 0$

In Problems 17 through 31, differentiate the given function.

17. $f(x) = (2x + 1)^4$ 18. $f(x) = \sqrt{6x - 3}$

19. $f(x) = \dfrac{1}{5x - 6}$ 20. $f(x) = \dfrac{2}{(6x + 2)^2}$

21. $f(x) = \dfrac{1}{\sqrt{4x - 1}}$ 22. $f(x) = (3x^4 - 7x^2 + 9)^5$

23. $f(x) = (x^5 - 4x^3 - 7)^8$ 24. $f(x) = \sqrt{5x^6 - 12}$

25. $f(x) = \dfrac{3}{(1 - x^2)^4}$ 26. $f(x) = \dfrac{1}{\sqrt{5x^3 + 2}}$

27. $f(x) = \sqrt{\dfrac{3x + 1}{2x - 1}}$ 28. $f(x) = \left(\dfrac{x + 2}{2 - x}\right)^3$

29. $f(x) = (x + 2)^3(2x - 1)^5$ 30. $f(x) = 2(3x + 1)^4(5x - 3)^2$

31. $f(x) = \dfrac{(x + 1)^5}{(1 - x)^4}$

In Problems 32 through 35, find an equation of the line that is tangent to the graph of f for the given value of x.

32. $f(x) = (3x^2 + 1)^2$; $x = -1$

33. $f(x) = (x^2 - 3)^5(2x - 1)^3$; $x = 2$

34. $f(x) = \dfrac{1}{(2x - 1)^6}$; $x = 1$

35. $f(x) = \left(\dfrac{x + 1}{x - 1}\right)^3$; $x = 3$

36. Differentiate the function $f(x) = (3x + 5)^2$ by two different methods, first using the chain rule and then the product rule. Show that the two answers are really the same.

37. Prove the chain rule for powers for $n = 2$ by using the product rule to compute $\dfrac{dy}{dx}$ if $y = [h(x)]^2$.

38. Prove the chain rule for powers for $n = 3$ by using the product rule and the result of Problem 37 to compute $\dfrac{dy}{dx}$ if $y = [h(x)]^3$. (*Hint:* Begin by writing y as $h(x)[h(x)]^2$.)

Annual earnings 39. The gross annual earnings of a certain company were $f(t) = \sqrt{10t^2 + t + 236}$ thousand dollars t years after its formation in January 1975.
   (a) At what rate were the gross annual earnings of the company growing in January 1979?
   (b) At what percentage rate were the gross annual earnings growing in January 1979?

Manufacturing cost 40. At a certain factory, the total cost of manufacturing q units during the daily production run is $C(q) = 0.2q^2 + q + 900$ dollars. From experience it has been determined that approximately $q(t) = t^2 + 100t$ units are manufactured during the first t hours of a production run. Compute the rate at which the total manufacturing cost is changing with respect to time 1 hour after production commences.

Air pollution 41. It is estimated that t years from now, the population of a certain suburban community will be $p(t) = 20 - \dfrac{6}{t + 1}$ thousand. An environmental study indicates that the average daily level of carbon monoxide in the air will be $c(p) = 0.5\sqrt{p^2 + p + 58}$ parts per million when the population is p thousand. Find the rate at which the level of carbon monoxide will be changing with respect to time 2 years from now.

Consumer demand 42. When electric blenders are sold for p dollars apiece, local consumers will buy $D(p) = \dfrac{8,000}{p}$ blenders a month. It is estimated that t months from now, the price of the blenders will be $p(t) = 0.04t^{3/2} + 15$ dollars. Compute the rate at which the monthly demand for the blenders will be changing with respect to time 25 months from now. Will the demand be increasing or decreasing?

Consumer demand 43. An importer of Brazilian coffee estimates that local consumers will buy approximately $D(p) = \dfrac{4,374}{p^2}$ pounds of the coffee per week when the price is p dollars per pound. It is estimated that t weeks from now, the price of Brazilian coffee will be $p(t) = 0.02t^2 + 0.1t + 6$ dollars per pound. At what rate will the weekly demand for the coffee be changing 10 weeks from now? Will the demand be increasing or decreasing?

5 RELATIVE MAXIMA AND MINIMA

In Example 1.4, you used calculus to maximize a profit function like the one shown in Figure 5.1. In particular, you observed that the maximum value corresponded to the unique point on the graph at which the slope of the tangent was zero, and you set the derivative equal to zero and solved for x.

The simplicity of this example is misleading. In general, not every point at which the derivative of a function is zero is a peak of its graph. Two functions whose derivatives are zero when $x = 0$ are sketched in Figure 5.2. Both have horizontal tangents at $(0, 0)$, but the function $y = x^2$ in Figure 5.2a reaches its *lowest* point at $(0, 0)$, while the function $y = x^3$ in Figure 5.2b has neither a maximum nor a minimum at this point.

The situation is further complicated by the existence of functions that have maxima or minima at points at which the derivative is not even defined. Two such functions are sketched in Figure 5.3.

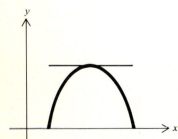

Figure 5.1 A profit function.

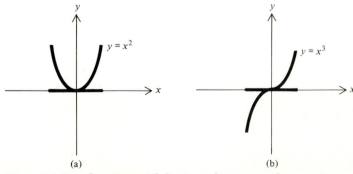

(a) (b)

Figure 5.2 Two functions with horizontal tangents when $x = 0$.

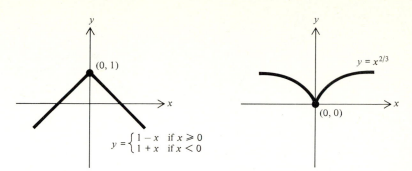

Figure 5.3 Two functions with extrema where the derivative is undefined.

In this section, you will learn a systematic procedure you can use to locate and identify maxima and minima of differentiable functions. In the process, you will also see how to use derivatives to help you sketch the graphs of functions.

Relative maxima and minima

A **relative maximum** of a function is a peak, a point on the graph of the function that is higher than any neighboring point on the graph. A **relative minimum** is the bottom of a valley, a point on the graph that is lower than any neighboring point. The function sketched in Figure 5.4 has a relative maximum at $x = b$ and relative minima at $x = a$ and $x = c$. Notice that a relative maximum need not be the highest point on a graph. It is maximal only *relative to* neighboring points. Similarly, a relative minimum need not be the lowest point on the graph.

Increase and decrease of functions

A function is said to be **increasing** if its graph is rising as x increases, and **decreasing** if its graph is falling as x increases. The function in Figure 5.5 is increasing for $a < x < b$ and for $x > c$. It is decreasing for $x < a$ and for $b < x < c$.

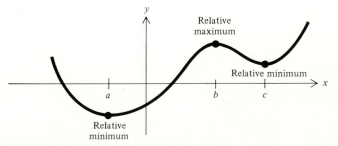

Figure 5.4 Relative maxima and minima.

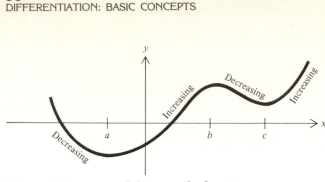

Figure 5.5 Increase and decrease of a function.

If you know the intervals on which a function is increasing and decreasing, you can easily identify its relative maxima and minima. A relative maximum occurs when the function stops increasing and starts decreasing. In Figure 5.5, this happens when $x = b$. A relative minimum occurs when the function stops decreasing and starts increasing. In Figure 5.5, this happens when $x = a$ and $x = c$.

The sign of the derivative

You can find out where a differentiable function is increasing or decreasing by checking the sign of its derivative. This is because the derivative is the slope of the tangent. When the derivative is positive, the slope of the tangent is positive and the function is increasing. When the derivative is negative, the slope of the tangent is negative and the function is decreasing. The situation is illustrated in Figure 5.6.

Here is a more precise statement of the situation.

The geometric significance of the sign of the derivative

If $f'(x) > 0$ whenever $a < x < b$, then f is increasing for $a < x < b$.

If $f'(x) < 0$ whenever $a < x < b$, then f is decreasing for $a < x < b$.

Critical points

Since a function is increasing when its derivative is positive and decreasing when its derivative is negative, the only points at which it

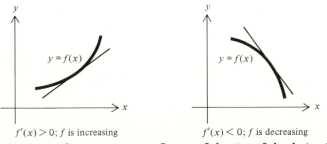

$f'(x) > 0$; f is increasing $f'(x) < 0$; f is decreasing

Figure 5.6 The geometric significance of the sign of the derivative.

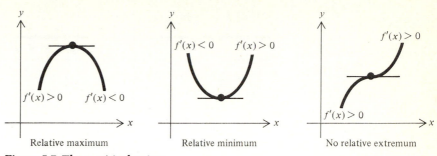

Figure 5.7 Three critical points.

can have a relative maximum or minimum are those at which its derivative is either zero or undefined. A point at which the derivative of a function is zero or undefined is said to be a **critical point** of the function. Every relative extremum is a critical point. However, as you saw in Figure 5.2b, not every critical point is necessarily a relative extremum.

If the sign of the derivative is positive to the left of a critical point and negative to the right of it, the critical point is a relative maximum. If the sign of the derivative is negative to the left of a critical point and positive to the right of it, the critical point is a relative minimum. If the sign of the derivative is the same on both sides of the critical point, the point is neither a relative maximum nor a relative minimum. The situation is illustrated in Figure 5.7.

Curve sketching The preceding observations suggest the following general procedure you can use to sketch functions and find their relative extrema.

How to use the derivative to graph a function

Step 1. Compute the derivative $f'(x)$.

Step 2. Find the x coordinates of the critical points by setting $f'(x)$ equal to zero and solving for x. Also include any values of x for which the derivative is undefined. Substitute these values of x into the function $f(x)$ to get the y coordinates of the critical points.

Step 3. Plot the critical points on the graph. These are the only points at which relative extrema can possibly occur.

Step 4. Determine where the function is increasing or decreasing by checking the sign of the derivative on the intervals whose endpoints are the x coordinates of the critical points.

Step 5. Draw the graph so that it increases on the intervals on which the derivative is positive, decreases on the intervals on which the derivative is negative, and levels off where $f'(x) = 0$.

Here are some examples.

EXAMPLE 5.1

Determine where the function $f(x) = 2x^3 + 3x^2 - 12x - 7$ is increasing and where it is decreasing, find its relative extrema, and draw the graph.

SOLUTION

Begin by computing and factoring the derivative.

$$f'(x) = 6x^2 + 6x - 12 = 6(x + 2)(x - 1)$$

From the factored form of the derivative you can see that $f'(x) = 0$ when $x = -2$ and when $x = 1$. Since $f(-2) = 13$ and $f(1) = -14$, it follows that the critical points are $(-2, 13)$ and $(1, -14)$. Begin the sketch (Figure 5.8a) by plotting these critical points. (To help you remember that the graph should have horizontal tangents at these points, you can draw a short horizontal line segment through each.)

To determine where the function is increasing and where it is decreasing, check the sign of the derivative for $x < -2$, for $-2 < x < 1$, and for $x > 1$.

When $x < -2$, both $x + 2$ and $x - 1$ are negative, so the derivative $f'(x) = 6(x + 2)(x - 1)$ is positive. Hence, f is increasing when $x < -2$.

When $-2 < x < 1$, the term $x + 2$ is positive while the term $x - 1$ is still negative. Hence the derivative is negative and f is decreasing when $-2 < x < 1$.

Finally, when $x > 1$, both $x + 2$ and $x - 1$ are positive. Hence the derivative is positive and f is increasing when $x > 1$.

These observations are summarized in the following table.

Interval	Sign of $f'(x)$	Increasing or decreasing
$x < -2$	+	increasing
$-2 < x < 1$	−	decreasing
$x > 1$	+	increasing

Complete the sketch in Figure 5.8b by drawing the graph so that it increases for $x < -2$ and $x > 1$, decreases for $-2 < x < 1$, and levels off at the critical points. Notice that the function has a relative maximum at the critical point $(-2, 13)$ and a relative minimum at the critical point $(1, -14)$.

EXAMPLE 5.2

Determine where the function $f(x) = 2 + (x - 1)^3$ is increasing and where it is decreasing, find its relative extrema, and draw the graph.

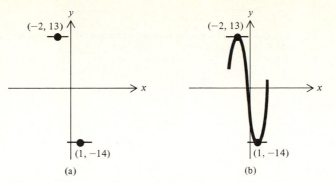

Figure 5.8 Steps leading to the graph of $y = 2x^3 + 3x^2 - 12x - 7$.

SOLUTION

To find the critical points, compute the derivative

$$f'(x) = 3(x - 1)^2$$

which is zero when $x = 1$. The corresponding critical point is $(1, 2)$.

To determine where the function is increasing and where it is decreasing, check the sign of the derivative for $x < 1$ and for $x > 1$.

Interval	Sign of $f'(x)$	Increasing or decreasing
$x < 1$	+	increasing
$x > 1$	+	increasing

Draw the graph using this information as shown in Figure 5.9. Notice that since f is increasing on both sides of the critical point $(1, 2)$, this point is neither a relative maximum nor a relative minimum.

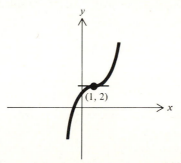

Figure 5.9 The graph of $y = 2 + (x - 1)^3$.

EXAMPLE 5.3

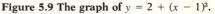

Determine where the rational function $f(x) = \dfrac{x^2}{x - 2}$ is increasing and where it is decreasing, find its relative extrema, and draw the graph.

SOLUTION

Use the quotient rule to get

$$f'(x) = \frac{(x - 2)(2x) - x^2(1)}{(x - 2)^2} = \frac{x(x - 4)}{(x - 2)^2}$$

This derivative is zero when $x = 0$ and $x = 4$ and the corresponding critical points are $(0, 0)$ and $(4, 8)$. It is also undefined at $x = 2$, but this does not correspond to a critical point because the function itself is also undefined when $x = 2$.

To find where the function is increasing and where it is decreasing, check the sign of the derivative on the intervals determined by the critical points. Since $f'(x)$ is undefined at $x = 2$, its sign for $0 < x < 2$ could be different from its sign for $2 < x < 4$, and so you will have to check these two intervals separately.

Interval	Sign of $f'(x)$	Increasing or decreasing
$x < .0$	+	increasing
$0 < x < 2$	−	decreasing
$2 < x < 4$	−	decreasing
$x > 4$	+	increasing

Now draw the graph using this information as shown in Figure 5.10. Notice that f has a relative maximum at the critical point $(0, 0)$ and a relative minimum at the critical point $(4, 8)$. Notice also that there is a break in the graph when $x = 2$.

In the next example, the relative extremum occurs at a critical point at which the derivative is not defined.

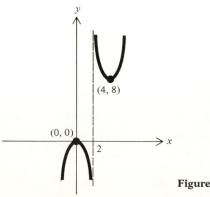

Figure 5.10 The graph of $y = \dfrac{x^2}{x - 2}$.

EXAMPLE 5.4

Determine where the function $f(x) = x^{2/3}$ is increasing and where it is decreasing, find its relative extrema, and draw the graph.

SOLUTION

The derivative

$$f'(x) = \tfrac{2}{3}x^{-1/3} = \frac{2}{3\sqrt[3]{x}}$$

is never zero, but is undefined when $x = 0$. Hence the corresponding point $(0, 0)$ is the only critical point. Check the sign of the derivative for $x < 0$ and for $x > 0$.

Interval	Sign of $f'(x)$	Increasing or decreasing
$x < 0$	$-$	decreasing
$x > 0$	$+$	increasing

The graph is shown in Figure 5.11. Notice that the tangent to the graph at the relative minimum $(0, 0)$ is vertical and its slope is undefined. This corresponds to the fact that the derivative is undefined when $x = 0$.

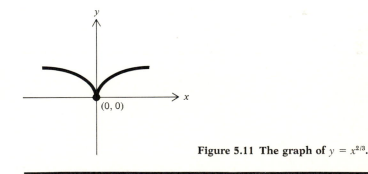

Figure 5.11 The graph of $y = x^{2/3}$.

In the next example, you will need the quadratic formula to find the critical points of the given function. You may want to use a calculator to help with the computations.

EXAMPLE 5.5

Determine where the function $f(x) = x^3 + 2x^2 - x + 1$ is increasing and where it is decreasing, find its relative extrema, and draw the graph.

SOLUTION

The derivative of f is

$$f'(x) = 3x^2 + 4x - 1$$

which has no obvious factors. According to the quadratic formula, $3x^2 + 4x - 1 = 0$ when

$$x = \frac{-4 \pm \sqrt{4^2 - 4(3)(-1)}}{2(3)} = \frac{-4 \pm \sqrt{28}}{6}$$

Hence, the x coordinates of the critical points are

$$x = \frac{-4 + \sqrt{28}}{6} = 0.22 \quad \text{and} \quad x = \frac{-4 - \sqrt{28}}{6} = -1.55$$

and the corresponding y coordinates are

$$y = f(0.22) = 0.89 \quad \text{and} \quad y = f(-1.55) = 3.63$$

For each of the intervals $x < -1.55$, $-1.55 < x < 0.22$, and $x > 0.22$, use any convenient value of x in the interval to check the sign of the derivative. The values $x = -2$, $x = 0$, and $x = 1$ would be reasonable choices.

Interval	Sign of $f'(x)$	Increasing or decreasing
$x < -1.55$	$+$	increasing
$-1.55 < x < 0.22$	$-$	decreasing
$x > 0.22$	$+$	increasing

The graph is shown in Figure 5.12.

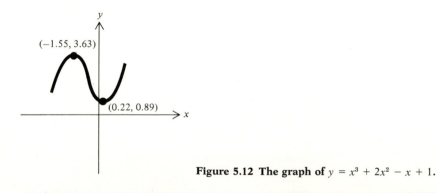

Figure 5.12 The graph of $y = x^3 + 2x^2 - x + 1$.

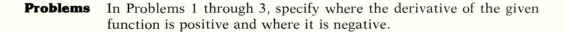

Problems   In Problems 1 through 3, specify where the derivative of the given function is positive and where it is negative.

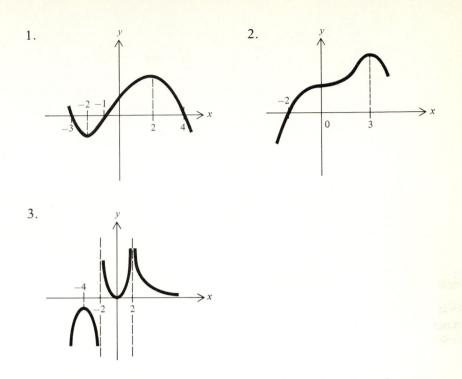

In Problems 4 through 30, determine where the given function is increasing and where it is decreasing, find its relative extrema, and draw the graph.

4. $f(x) = x^2 - 4x + 5$

5. $f(x) = x^3 + 3x^2 + 1$

6. $f(x) = x^3 - 3x - 4$

7. $f(x) = \frac{1}{3}x^3 - 9x + 2$

8. $f(x) = x^5 - 5x^4 + 100$

9. $f(x) = 3x^5 - 5x^3$

10. $f(x) = 3x^4 - 8x^3 + 6x^2 + 2$

11. $f(x) = 324x - 72x^2 + 4x^3$

12. $f(x) = 2x^3 + 6x^2 + 6x + 5$

13. $f(x) = 10x^6 + 24x^5 + 15x^4 + 3$

14. $f(x) = (x - 1)^5$

15. $f(x) = 3 - (x + 1)^3$

16. $f(x) = (x^2 - 1)^5$

17. $f(x) = (x^2 - 1)^4$

18. $f(x) = (x^3 - 1)^4$

19. $f(x) = \dfrac{x^2}{x - 1}$

20. $f(x) = \dfrac{x^2}{x + 2}$

21. $f(x) = \dfrac{x^2 - 3x}{x + 1}$

22. $f(x) = \dfrac{1}{x^2 - 9}$

23. $f(x) = x + \dfrac{1}{x}$

24. $f(x) = 2x + \dfrac{18}{x} + 1$

25. $f(x) = 6x^2 + \dfrac{12{,}000}{x}$

26. $f(x) = 1 + x^{1/3}$ 27. $f(x) = x^{3/5}$

28. $f(x) = 2 + (x - 1)^{2/3}$ 29. $f(x) = x^3 - 2x^2 - 3x + 2$

30. $f(x) = x^3 - 3x^2 + 2x + 1$

31. Sketch a graph of a function that has all of the following properties.
    (a) $f'(x) > 0$ when $x < -5$ and when $x > 1$
    (b) $f'(x) < 0$ when $-5 < x < 1$
    (c) $f(-5) = 4$ and $f(1) = -1$

32. Sketch a graph of a function that has all of the following properties.
    (a) $f'(x) < 0$ when $x < -1$
    (b) $f'(x) > 0$ when $-1 < x < 3$ and when $x > 3$
    (c) $f'(-1) = 0$ and $f'(3) = 0$

33. Sketch a graph of a function that has all of the following properties.
    (a) $f'(x) > 0$ when $x > 2$
    (b) $f'(x) < 0$ when $x < 0$ and when $0 < x < 2$
    (c) f is undefined when $x = 0$

34. Sketch a graph of a function that has all of the following properties.
    (a) $f'(x) > 0$ when $-1 < x < 3$ and when $x > 6$
    (b) $f'(x) < 0$ when $x < -1$ and when $3 < x < 6$
    (c) $f'(-1) = 0$ and $f'(6) = 0$
    (d) $f'(x)$ is undefined when $x = 3$

35. Find constants a, b, and c such that the graph of the function $f(x) = ax^2 + bx + c$ has a relative maximum at (5, 12) and crosses the y axis at (0, 3).

36. Use calculus to prove that the relative extremum of the quadratic function $y = ax^2 + bx + c$ occurs when $x = -\dfrac{b}{2a}$.

37. Use calculus to prove that the relative extremum of the quadratic function $y = (x - p)(x - q)$ occurs midway between its x intercepts.

38. Find the largest and smallest values of the function $f(x) = 2x^3 + 3x^2 - 12x - 7$ on the interval $-3 \le x \le 0$.

6 ABSOLUTE MAXIMA AND MINIMA

In most practical optimization problems, the goal is to find the **absolute maximum** or **absolute minimum** of a particular function on some interval, rather than a relative maximum or minimum. The ab-

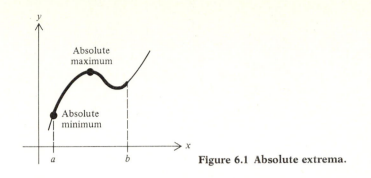

Figure 6.1 Absolute extrema.

solute maximum of a function on an interval is the largest value of the function on the interval. The absolute minimum is the smallest value. Absolute extrema often coincide with relative extrema, but not always. For example, on the interval $a \leq x \leq b$, the absolute maximum and relative maximum of the function in Figure 6.1 are the same. The absolute minimum, on the other hand, occurs at the endpoint $x = a$, which is not a relative minimum.

Absolute extrema on closed intervals

A **closed interval** is an interval of the form $a \leq x \leq b$, that is, an interval that contains both of its endpoints. A function that is continuous on a closed interval attains an absolute maximum and an absolute minimum on that interval. An absolute extremum can occur either at a relative extremum in the interval or at an endpoint $x = a$ or $x = b$. The possibilities are illustrated in Figure 6.2.

These observations suggest the following simple procedure for locating and identifying absolute extrema of continuous functions on closed intervals.

How to find the absolute extrema of a continuous function f on a closed interval $a \leq x \leq b$

Step 1. Find the x coordinates of all the critical points of f in the interval $a \leq x \leq b$.

Step 2. Compute $f(x)$ at these critical points and at the endpoints $x = a$ and $x = b$.

Step 3. Select the largest and smallest values of $f(x)$ obtained in Step 2. These are the absolute maximum and absolute minimum, respectively.

The procedure is illustrated in the following example.

EXAMPLE 6.1

Find the absolute maximum and absolute minimum of the function $f(x) = 2x^3 + 3x^2 - 12x - 7$ on the interval $-3 \leq x \leq 0$.

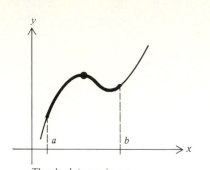

The absolute maximum
coincides with a relative maximum

The absolute maximum
occurs at an endpoint

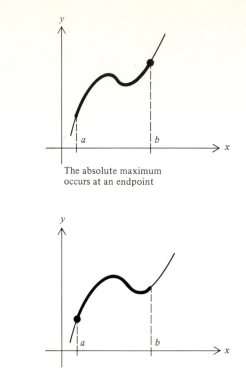

The absolute minimum coincides
with a relative minimum

The absolute minimum
occurs at an endpoint

Figure 6.2 Absolute extrema of a continuous function on a closed interval.

SOLUTION

From the derivative

$$f'(x) = 6x^2 + 6x - 12 = 6(x + 2)(x - 1)$$

you see that the critical points occur when $x = -2$ and $x = 1$. Of these, only $x = -2$ lies in the interval $-3 \le x \le 0$. Compute $f(x)$ at $x = -2$ and at the endpoints $x = -3$ and $x = 0$.

$$f(-2) = 13 \qquad f(-3) = 2 \qquad f(0) = -7$$

Comparing these three values, you can conclude that the absolute maximum of f on the interval $-3 \le x \le 0$ is $f(-2) = 13$ and the absolute minimum is $f(0) = -7$.

Notice that you did not have to classify the critical points or draw the graph to locate the absolute extrema. The sketch in Figure 6.3 is presented only for the sake of illustration.

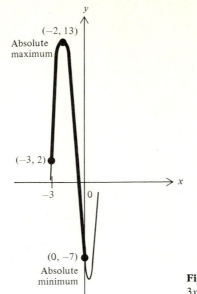

Figure 6.3 The absolute extrema of $y = 2x^3 + 3x^2 - 12x - 7$ on $-3 \leq x \leq 0$.

Applications   Here are two practical applications of this technique.

EXAMPLE 6.2

For several weeks, the highway department has been recording the speed of freeway traffic flowing past a certain downtown exit. The data suggest that between the hours of 1:00 P.M. and 6:00 P.M. on a normal weekday, the speed of the traffic at the exit is approximately $S(t) = 2t^3 - 21t^2 + 60t + 40$ kilometers per hour, where t is the number of hours past noon. At what time between 1:00 P.M. and 6:00 P.M. is the traffic moving the fastest, and at what time is it moving the slowest?

SOLUTION

The goal is to find the absolute maximum and absolute minimum of the function $S(t)$ on the interval $1 \leq t \leq 6$. From the derivative

$$S'(t) = 6t^2 - 42t + 60 = 6(t - 2)(t - 5)$$

you get the t coordinates of the critical points

$$t = 2 \quad \text{and} \quad t = 5$$

both of which lie in the interval $1 \leq t \leq 6$.

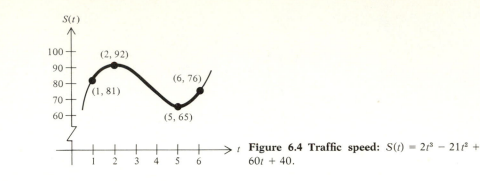

Figure 6.4 Traffic speed: $S(t) = 2t^3 - 21t^2 + 60t + 40.$

Now compute $S(t)$ for these values of t and at the endpoints $t = 1$ and $t = 6$ to get

$$S(1) = 81 \qquad S(2) = 92 \qquad S(5) = 65 \qquad S(6) = 76$$

Since the largest of these values is $S(2) = 92$ and the smallest is $S(5) = 65$, you can conclude that the traffic is moving fastest at 2:00 P.M. when its speed is 92 kilometers per hour and slowest at 5:00 P.M. when its speed is 65 kilometers per hour.

For reference, the graph of S is sketched in Figure 6.4.

The next example comes from biology.

EXAMPLE 6.3

When you cough, the radius of your trachea (windpipe) decreases, affecting the speed of the air in the trachea. If r_0 is the normal radius of the trachea, the relationship between the speed S of the air and the radius r of the trachea during a cough is given by a function of the form $S(r) = ar^2(r_0 - r)$, where a is a positive constant. Find the radius r for which the speed of the air is greatest.

SOLUTION

The radius r of the contracted trachea cannot be greater than the normal radius r_0 nor less than zero. Hence, the goal is to find the absolute maximum of $S(r)$ on the interval $0 \leq r \leq r_0$.

First differentiate $S(r)$ with respect to r using the product rule and factor the derivative as follows. (Note that a and r_0 are constants.)

$$\begin{aligned} S'(r) &= -ar^2 + (r_0 - r)(2ar) \\ &= ar[-r + 2(r_0 - r)] \\ &= ar(2r_0 - 3r) \end{aligned}$$

Then set the factored derivative equal to zero and solve to get the r coordinates of the critical points.

$$ar(2r_0 - 3r) = 0$$
$$r = 0 \quad \text{or} \quad r = \tfrac{2}{3}r_0$$

Both of these values of r lie in the interval $0 \leq r \leq r_0$, and one is actually an endpoint of the interval. Compute $S(r)$ for these two values of r and at the other endpoint $r = r_0$ to get

$$S(0) = 0 \qquad S(\tfrac{2}{3}r_0) = \frac{4a}{27}\, r_0{}^3 \qquad S(r_0) = 0$$

Comparing these values, you can conclude that the speed of the air is greatest when the radius of the contracted trachea is $\tfrac{2}{3}r_0$, that is, when it is two-thirds the radius of the uncontracted trachea.

A graph of the function S is sketched in Figure 6.5. Notice that the r intercepts of the graph are obvious from the factored function $S(r) = ar^2(r_0 - r)$. Notice also that the graph was drawn so that it has a horizontal tangent when $r = 0$, reflecting the fact that $S'(0) = 0$.

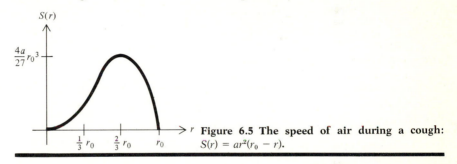

Figure 6.5 The speed of air during a cough: $S(r) = ar^2(r_0 - r)$.

Absolute extrema on intervals that are not closed

When the interval on which you wish to maximize or minimize a function is not of the form $a \leq x \leq b$, you will have to modify the procedure illustrated in the preceding examples. This is because there is no longer any guarantee that the function actually has an absolute maximum or minimum on the interval in question. On the other hand, if an absolute extremum does exist and the function is continuous, the absolute extremum will still occur at a relative extremum or at an endpoint that is contained in the interval. Some of the possibilities are illustrated in Figure 6.6.

To find the absolute extrema of a continuous function on an interval that is not of the form $a \leq x \leq b$, you still evaluate the function at the critical points and endpoints that are contained in the interval in question. But before you can draw any final conclusions, you must find out if the function actually has relative extrema on this interval. One good way to do this is to use the derivative to determine where the function is increasing and where it is decreasing, and then sketch the graph. The technique is illustrated in the following two examples.

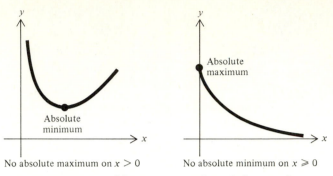

No absolute maximum on $x > 0$ No absolute minimum on $x \geq 0$

Figure 6.6 Extrema of functions on unbounded intervals.

EXAMPLE 6.4

When q units of a certain commodity are produced, the total manufacturing cost is $C(q) = 3q^2 + 5q + 75$ dollars. At what level of production will the average cost per unit be smallest?

SOLUTION

The average cost per unit is the total cost divided by the number of units produced. Thus, if q units are produced, the average cost is

$$A(q) = \frac{C(q)}{q} = \frac{3q^2 + 5q + 75}{q} = 3q + 5 + \frac{75}{q}$$

dollars per unit. Since only positive values of q are meaningful in this context, the goal is to find the absolute minimum of the function $A(q)$ on the unbounded interval $q > 0$.

The derivative of A is

$$A'(q) = 3 - \frac{75}{q^2}$$

If you set this equal to zero and solve, you get

$$3 - \frac{75}{q^2} = 0$$
$$q^2 = 25$$
$$q = \pm 5$$

Since -5 is not in the interval $q > 0$, and since this interval does not contain its only endpoint $q = 0$, you can conclude that the absolute minimum of $A(q)$, if one exists, is $A(5) = 35$.

To verify that this is actually the absolute minimum, observe that $A'(q)$ is negative for $0 < q < 5$ and positive for $q > 5$. It follows that A

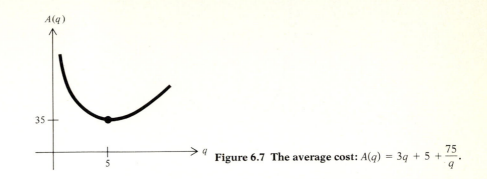

Figure 6.7 The average cost: $A(q) = 3q + 5 + \dfrac{75}{q}$.

is decreasing for $0 < q < 5$ and increasing for $q > 5$ as indicated in the sketch in Figure 6.7, and that $A(5) = 35$ is indeed the absolute minimum of A on the interval $q > 0$. That is, the average cost per unit is minimal when 5 units are produced.

EXAMPLE 6.5

At a certain factory, setup cost is proportional to the number of machines used, and operating cost is inversely proportional to the number of machines used. Show that the total cost is minimal when the setup cost equals the operating cost.

SOLUTION

Let x be the number of machines used and $C(x)$ the corresponding total cost. Since

$$\text{Total cost} = \text{setup cost} + \text{operating cost}$$

and

$$\text{Setup cost} = k_1 x \qquad \text{and} \qquad \text{operating cost} = \frac{k_2}{x}$$

it follows that

$$C(x) = k_1 x + \frac{k_2}{x}$$

Since only positive values of x are meaningful in this context, the goal is to minimize $C(x)$ on the interval $x > 0$.

The derivative of C is

$$C'(x) = k_1 - \frac{k_2}{x^2}$$

If you set this derivative equal to zero and solve, you get

$$k_1 - \frac{k_2}{x^2} = 0 \qquad \text{or} \qquad k_1 = \frac{k_2}{x^2}$$

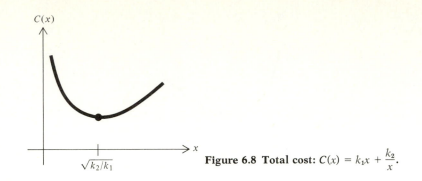

Figure 6.8 Total cost: $C(x) = k_1x + \dfrac{k_2}{x}$.

At this point, notice that if you multiply both sides of the equation by x, you will get

$$k_1x = \frac{k_2}{x}$$

But $\qquad k_1x =$ setup cost $\qquad$ and $\qquad \dfrac{k_2}{x} =$ operating cost

and so this equation says

$$\text{Setup cost} = \text{Operating cost}$$

which is exactly what you were trying to show!

All that remains is to verify that the function $C(x)$ actually has an absolute minimum where its derivative is equal to zero. To do this, continue to solve the equation $C'(x) = 0$ for x to get $x = \sqrt{\dfrac{k_2}{k_1}}$ and observe that $C'(x)$ is negative if $0 < x < \sqrt{\dfrac{k_2}{k_1}}$ and positive if $x > \sqrt{\dfrac{k_2}{k_1}}$. It follows that C is decreasing for $0 < x < \sqrt{\dfrac{k_2}{k_1}}$ and increasing for $x > \sqrt{\dfrac{k_2}{k_1}}$ as shown in Figure 6.8, and hence the point at which the derivative is zero is indeed the absolute minimum you were seeking.

Problems

In Problems 1 through 16, find the absolute maximum and absolute minimum (if any) of the given function on the specified interval.

1. $f(x) = x^2 + 4x + 5; \; -3 \le x \le 1$

2. $f(x) = x^3 + 3x^2 + 1; \; -3 \le x \le 2$

3. $f(x) = \frac{1}{3}x^3 - 9x + 2; \; 0 \le x \le 2$

4. $f(x) = x^5 - 5x^4 + 1; \; 0 \le x \le 5$

5. $f(x) = 3x^5 - 5x^3;\ -2 \le x \le 0$

6. $f(x) = 10x^6 + 24x^5 + 15x^4 + 3;\ -1 \le x \le 1$

7. $f(x) = (x^2 - 4)^5;\ -3 \le x \le 2$ 8. $f(x) = \dfrac{x^2}{x - 1};\ -2 \le x \le -\frac{1}{2}$

9. $f(x) = x + \dfrac{1}{x};\ \frac{1}{2} \le x \le 3$ 10. $f(x) = \dfrac{1}{x^2 - 9};\ 0 \le x \le 2$

11. $f(x) = x + \dfrac{1}{x};\ x > 0$ 12. $f(x) = 2x + \dfrac{32}{x};\ x > 0$

13. $f(x) = \dfrac{1}{x};\ x > 0$ 14. $f(x) = \dfrac{1}{x^2};\ x > 0$

15. $f(x) = \dfrac{1}{x + 1};\ x \ge 0$ 16. $f(x) = \dfrac{1}{(x + 1)^2};\ x \ge 0$

Profit   17. A manufacturer can produce radios at a cost of $5 apiece and esti-mates that if they are sold for x dollars apiece, consumers will buy $20 - x$ radios a day. At what price should the manufacturer sell the radios to maximize profit?

Consumer expenditure   18. The demand function for a certain commodity is $D(p) = 160 - 2p$, where p is the price at which the commodity is sold. At what price is the total consumer expenditure for the commodity greatest?

Social action   19. Suppose that x years after its founding in 1960, a certain civil rights organization had a membership of $f(x) = 100(2x^3 - 45x^2 + 264x)$.
     (a) At what time between 1960 and 1974 was the membership of the organization largest? What was the membership at that time?
     (b) At what time between 1961 and 1974 was the membership the smallest? What was the membership at that time?

Broadcasting   20. An all-news radio station has made a survey of the listening habits of local residents between the hours of 5:00 P.M. and mid-night. The survey indicates that the percentage of the local adult population that is tuned in to the station x hours after 5:00 P.M. is $f(x) = \frac{1}{8}(-2x^3 + 27x^2 - 108x + 240)$.
     (a) At what time between 5:00 P.M. and midnight are the most people listening to the station?
     (b) At what time between 5:00 P.M. and midnight are the fewest people listening?

Circulation   21. Poiseuille's law asserts that the speed of blood that is r centime-ters from the central axis of an artery of radius R is $S(r) =$

$c(R^2 - r^2)$, where c is a positive constant. Where is the speed of the blood greatest?

Respiration

22. Biologists define the flow F of air in the trachea by the formula $F = SA$, where S is the speed of the air and A the area of a cross section of the trachea.

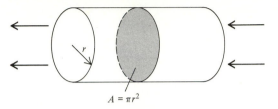

$A = \pi r^2$

(a) Assume that the cross section of the trachea is circular and use the formula from Example 6.3 for the speed of the air in the trachea during a cough to express the flow as a function of the radius r.
(b) Find the radius r for which the flow is greatest.

Population growth

23. When environmental factors impose an upper bound on its possible size, population grows at a rate that is jointly proportional to its current size and the difference between its current size and the upper bound. Show that the rate of population growth will be greatest when the population has climbed to 50 percent of its upper bound.

The spread of an epidemic

24. The rate at which an epidemic spreads through a community is jointly proportional to the number of people who have caught the disease and the number who have not. Show that the epidemic is spreading most rapidly when half the people have caught the disease.

Transportation cost

25. A truck is hired to transport goods from a factory to a warehouse. The driver's wages are figured by the hour and so are inversely proportional to the speed at which the truck is driven. The amount of gasoline used is directly proportional to the speed at which the truck is driven, and the price of gasoline remains constant during the trip. Show that the total cost is smallest at the speed for which the driver's wages are equal to the cost of the gasoline used.

Microeconomics

26. An economic law states that profit is maximized when marginal revenue equals marginal cost. (Marginal revenue and marginal cost are the derivatives of total revenue and total cost, respectively.)
(a) Use the theory of extrema to explain why this law is true.
(b) What assumptions about the shape of the profit curve are implicit in this law?

Average cost

27. Suppose the total cost in dollars of manufacturing q units is given by the function $C(q) = 3q^2 + q + 48$.
    (a) Express the average manufacturing cost per unit as a function of q.
    (b) For what value of q is the average cost the smallest?
    (c) For what value of q is the average cost equal to the marginal cost? Compare this value with your answer in part (b).
    (d) On the same set of axes, graph the total cost, marginal cost, and average cost functions.

Microeconomics

28. An economic law states that average cost is smallest when it equals marginal cost.
    (a) Prove this law. (*Hint:* If $C(q)$ represents the total cost of manufacturing q units, then average cost is $A(q) = \dfrac{C(q)}{q}$. Average cost will be smallest when $A'(q) = 0$. Use the quotient rule to compute $A'(q)$, then set $A'(q)$ equal to zero, and the economic law will emerge.)
    (b) What assumptions about the shape of the average cost graph are implicit in this economic law?

CHAPTER SUMMARY AND PROFICIENCY TEST

Important terms, symbols, and formulas

Secant line; tangent line

Derivative: $\dfrac{f(x + \Delta x) - f(x)}{\Delta x} \to f'(x)$ as $\Delta x \to 0$

Notation: $f'(x)$; $\dfrac{dy}{dx}$

Slope of tangent = derivative

Continuous function; differentiable function

Power rule: $\dfrac{d}{dx}(x^n) = nx^{n-1}$

Derivative of a constant: $\dfrac{d}{dx}(c) = 0$

Constant multiple rule: $\dfrac{d}{dx}(cf) = c\dfrac{df}{dx}$

Sum rule: $\dfrac{d}{dx}(f + g) = \dfrac{df}{dx} + \dfrac{dg}{dx}$

Product rule: $\dfrac{d}{dx}(fg) = f\dfrac{dg}{dx} + g\dfrac{df}{dx}$

Quotient rule: $\dfrac{d}{dx}\left(\dfrac{f}{g}\right) = \dfrac{g\dfrac{df}{dx} - f\dfrac{dg}{dx}}{g^2}$

Instantaneous speed = derivative of distance
Instantaneous rate of change = derivative
Marginal cost = derivative of total cost
Marginal analysis

Percentage rate of change of y with respect to $x = 100\,\dfrac{dy/dx}{y}$

Chain rule: $\dfrac{dy}{dx} = \dfrac{dy}{du}\dfrac{du}{dx}$

Chain rule for powers: $\dfrac{d}{dx}[h(x)]^n = n[h(x)]^{n-1}\dfrac{d}{dx}[h(x)]$

Relative maxima and minima

f is increasing; $f'(x) > 0$

f is decreasing; $f'(x) < 0$

Critical point: $f'(x) = 0$ or $f'(x)$ is undefined
Absolute maxima and minima
Closed interval: $a \leq x \leq b$
Absolute extrema

Proficiency test

1. Use the definition to find the derivative of the given function.

    (a) $f(x) = x^2 - 3x + 1$

    (b) $f(x) = \dfrac{1}{x - 2}$

2. Differentiate the following functions.

    (a) $f(x) = 6x^4 - 7x^3 + 2x + \sqrt{2}$

    (b) $f(x) = x^3 - \dfrac{1}{x^2} + 2\sqrt{x} - \dfrac{3}{x}$

    (c) $y = \dfrac{x + 3}{2x + 1}$

    (d) $y = (2x + 5)^3(x + 1)^2$

    (e) $f(x) = (5x^4 - 3x^2 + 2x + 1)^{10}$

    (f) $f(x) = \sqrt{x^2 + 1}$

    (g) $y = \left(\dfrac{x + 1}{1 - x}\right)^2$

    (h) $y = (3x + 1)\sqrt{6x + 5}$

3. Find the equation of the line that is tangent to the graph of f at the point $(x, f(x))$ for the given value of x.

    (a) $f(x) = x^2 - 3x + 2$; $x = 1$

    (b) $f(x) = \dfrac{4}{x - 3}$; $x = 1$

    (c) $f(x) = \dfrac{x}{x^2 + 1}$; $x = 0$

    (d) $f(x) = \sqrt{x^2 + 5}$; $x = -2$

4. After x weeks, the number of people using a new rapid transit system was approximately $N(x) = 6x^3 + 500x + 8{,}000$.

    (a) At what rate was the use of the system changing after 8 weeks?

    (b) By how much did the use of the system change during the 8th week?

5. It is estimated that the weekly output at a certain plant is $Q(x) = 50x^2 + 9{,}000x$ units, where x is the number of workers employed at the plant. Currently there are 30 workers employed at the plant.

    (a) Use calculus to estimate the change in the weekly output that will result from the addition of 1 worker to the force.

    (b) Compute the actual change in output that will result from the addition of 1 worker.

6. It is projected that t months from now, the population of a certain town will be $P(t) = 3t + 5t^{3/2} + 6{,}000$. At what percentage rate will the population be changing 4 months from now?

7. Find $\dfrac{dy}{dx}$.

    (a) $y = 5u^2 + u - 1$, $u = 3x + 1$

    (b) $y = \dfrac{1}{u^2}$, $u = 2x + 3$

8. Find $\dfrac{dy}{dx}$ for the given value of x.

    (a) $y = u^3 - 4u^2 + 5u + 2$, $u = x^2 + 1$; $x = 1$

    (b) $y = \sqrt{u}$, $u = x^2 + 2x - 4$; $x = 2$

9. At a certain factory, approximately $q(t) = t^2 + 50t$ units are manufactured during the first t hours of a production run, and the total cost of manufacturing q units is $C(q) = 0.1q^2 + 10q + 400$ dollars. Find the rate at which the manufacturing cost is changing with respect to time 2 hours after production commences.

10. Determine where the given function is increasing and where it is decreasing, find its relative extrema, and draw the graph.

(a) $f(x) = -2x^3 + 3x^2 + 12x - 5$

(b) $f(x) = 3x^5 - 20x^3$

(c) $f(x) = \dfrac{x^2}{x + 1}$

(d) $f(x) = 2x + \dfrac{8}{x} + 2$

11. Find the absolute maximum and absolute minimum (if any) of the given function on the specified interval.

(a) $f(x) = -2x^3 + 3x^2 + 12x - 5;\ -3 \le x \le 3$

(b) $f(x) = -3x^4 + 8x^3 - 10;\ 0 \le x \le 3$

(c) $f(x) = \dfrac{x^2}{x + 1};\ -\frac{1}{2} \le x \le 1$

(d) $f(x) = 2x + \dfrac{8}{x} + 2;\ x > 0$

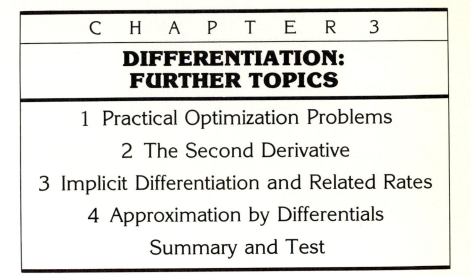

C H A P T E R 3

DIFFERENTIATION: FURTHER TOPICS

1 Practical Optimization Problems

2 The Second Derivative

3 Implicit Differentiation and Related Rates

4 Approximation by Differentials

Summary and Test

1 PRACTICAL OPTIMIZATION PROBLEMS

In this section, you will learn how to combine the techniques of model-building from Chapter 1, Section 5, with the optimization techniques of Chapter 2, Section 6, to solve practical optimization problems.

The first step in solving such a problem is to decide precisely what you are to optimize. Once you have identified this quantity, choose a letter to represent it. Some people are most comfortable using the standard letter f for this purpose. Others find it helpful to choose a letter more closely related to the quantity, such as R for revenue or A for area.

Your goal is to represent the quantity to be optimized as a function of some other variable so that you can apply calculus. It is usually a good idea to express the desired function in words before trying to represent it mathematically.

Once the function has been expressed in words, the next step is to choose an appropriate variable. Often, the choice is obvious. Sometimes you will be faced with a choice among several natural variables. When this happens, think ahead and try to choose the variable that leads to the simplest functional representation. In some problems, the quantity to be optimized is expressed most naturally in terms of two variables. If so, you will have to find a way to write one of these variables in terms of the other.

119

The next step is to express the quantity to be optimized as a function of the variable you have chosen. In most problems, the function has a practical interpretation only when the variable lies in a certain interval. Once you have written the function and identified the appropriate interval, the difficult part is done and the rest is routine. To complete the problem, simply apply the techniques you learned in Chapter 2, Section 6, to optimize your function on the specified interval.

Here are some examples to illustrate this procedure.

EXAMPLE 1.1

A cable is to be run from a power plant on one side of a river 900 meters wide to a factory on the other side, 3,000 meters downstream. The cost of running the cable under the water is $5 per meter while the cost over land is $4 per meter. What is the most economical route over which to run the cable?

SOLUTION

To help you visualize the situation, begin by drawing a diagram as shown in Figure 1.1.

(Notice that in drawing the diagram in Figure 1.1, you have assumed that the cable should be run in a *straight line* from the power plant to some point P on the opposite bank. Do you see why this assumption is justified?)

Your goal is to minimize the cost of installing the cable. Let C denote this cost and represent C as follows.

$C = 5$(number of meters of cable under water)
$$+ 4(\text{number of meters of cable over land})$$

Since you wish to describe the optimal route over which to run the cable, it will be convenient to choose a variable in terms of which you can easily locate the point P. Two reasonable choices for the variable x are illustrated in Figure 1.2.

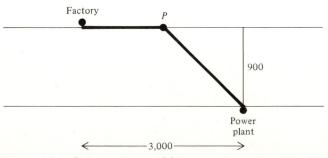

Figure 1.1 Relative positions of factory, river, and power plant.

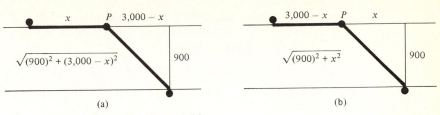

Figure 1.2 Two choices for the variable x.

Before plunging into the calculations, take a minute to decide which choice of variables is more advantageous. In Figure 1.2a, the distance across the water from the power plant to the point P is (by the pythagorean theorem) $\sqrt{(900)^2 + (3,000 - x)^2}$, and the corresponding total cost function is

$$C(x) = 5\sqrt{(900)^2 + (3,000 - x)^2} + 4x$$

In Figure 1.2b, the distance across the water is $\sqrt{(900)^2 + x^2}$, and the total cost function is

$$C(x) = 5\sqrt{(900)^2 + x^2} + 4(3,000 - x)$$

The second of these functions is the more attractive since the term $3,000 - x$ is merely multiplied by 4, while in the first function it is squared and appears under the radical. Hence, you should choose x as in Figure 1.2b and work with the total cost function

$$C(x) = 5\sqrt{(900)^2 + x^2} + 4(3,000 - x)$$

Since the distances x and $3,000 - x$ cannot be negative, your goal is to find the absolute minimum of the function C on the interval $0 \leq x \leq 3,000$.

To find the critical points, compute the derivative

$$C'(x) = \frac{5x}{\sqrt{(900)^2 + x^2}} - 4$$

and set it equal to zero to get

$$\frac{5x}{\sqrt{(900)^2 + x^2}} - 4 = 0$$

$$\sqrt{(900)^2 + x^2} = \tfrac{5}{4}x$$

Square both sides of this equation and solve for x to get

$$(900)^2 + x^2 = \tfrac{25}{16}x^2$$

$$x^2 = \tfrac{16}{9}(900)^2$$

$$x = \pm 1,200$$

Since only the positive value $x = 1,200$ is in the interval $0 \leq x \leq 3,000$, compute $C(x)$ at this value of x and at the endpoints $x = 0$ and $x = 3,000$. Since

$$C(0) = 16,500 \qquad C(1,200) = 14,700$$

$$C(3,000) = 1,500\sqrt{109} > 15,000$$

you can conclude that the installation cost is minimal if the cable reaches the opposite bank 1,200 meters downstream from the power plant.

In the next example, the quantity to be optimized is expressed most naturally in terms of *two* variables. Fortunately, additional information in the problem allows you to write one of these variables in terms of the other so that ultimately you have a function of just one variable.

EXAMPLE 1.2

The highway department is planning to build a picnic area for motorists along a major highway. It is to be rectangular with an area of 5,000 square meters and is to be fenced off on the three sides not adjacent to the highway. What is the least amount of fencing that will be needed to complete the job?

SOLUTION

Label the sides of the picnic area as indicated in Figure 1.3 and let F denote the amount of fencing required. Then,

$$F = x + 2y$$

The fact that the area is to be 5,000 tells you that

$$xy = 5,000 \qquad \text{or} \qquad y = \frac{5,000}{x}$$

To rewrite F in terms of the single variable x, substitute $y = \dfrac{5,000}{x}$ into the equation $F = x + 2y$, getting

$$F(x) = x + \frac{10,000}{x}$$

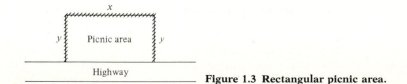

Figure 1.3 Rectangular picnic area.

Since $F(x)$ has a practical interpretation for any positive value of x, your goal is to find the absolute minimum of $F(x)$ on the interval $x > 0$.

To find the critical points, set the derivative

$$F'(x) = 1 - \frac{10,000}{x^2}$$

equal to zero and solve for x, getting

$$1 - \frac{10,000}{x^2} = 0$$

$$x^2 = 10,000$$

$$x = \pm 100$$

Only the positive value $x = 100$ lies in the interval $x > 0$. To find the absolute minimum of $F(x)$ on this interval, observe that $F'(x)$ is negative if $0 < x < 100$ and positive if $x > 100$. It follows that F is decreasing for $0 < x < 100$ and increasing for $x > 100$, as indicated in Figure 1.4, and that $x = 100$ does indeed correspond to the absolute minimum of $F(x)$ on the interval $x > 0$.

The smallest amount of fencing that will be needed to complete the job is $F(100) = 200$ meters.

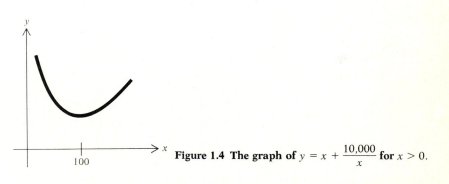

Figure 1.4 The graph of $y = x + \dfrac{10,000}{x}$ for $x > 0$.

Optimization of discrete functions   In the next example, the application of calculus to a practical problem gives an answer that makes no sense in the practical context. Additional analysis is needed to obtain a meaningful solution.

EXAMPLE 1.3

A bus company will charter a bus that holds 50 people to groups of 35 or more. If a group contains exactly 35 people, each person pays $60. In larger groups, everybody's fare is reduced by $1 for each person in excess of 35. Determine the size of the group for which the bus company's revenue will be greatest.

SOLUTION

Let R denote the bus company's revenue. Then,

$$R = \text{(number of people in the group)(fare per person)}$$

You could let x denote the number of people in the group, but it is slightly more convenient to let x denote the number of people in excess of 35. Then,

$$\text{Number of people in the group} = 35 + x$$
$$\text{Fare per person} = 60 - x$$

and the revenue function is

$$R(x) = (35 + x)(60 - x)$$

Since x represents the number of people in excess of 35, and since the total size of the group must be between 35 and 50, it follows that x must be in the interval $0 \leq x \leq 15$. The goal, therefore, is to find the absolute maximum of $R(x)$ on this closed interval.

The derivative

$$R'(x) = (35 + x)(-1) + (60 - x)(1) = 25 - 2x$$

is zero when $x = 12.5$. Since

$$R(12.5) = 2{,}256.25 \qquad R(0) = 2{,}100 \qquad R(15) = 2{,}250$$

it follows that the absolute maximum of $R(x)$ on the interval $0 \leq x \leq 15$ occurs when $x = 12.5$.

But x represents a certain number of people and must be a whole number. Hence, $x = 12.5$ cannot be the solution to the bus company's optimization problem. To find the optimal *integer* value of x, observe that R is increasing for $0 < x < 12.5$ and decreasing for $x > 12.5$, as indicated in Figure 1.5.

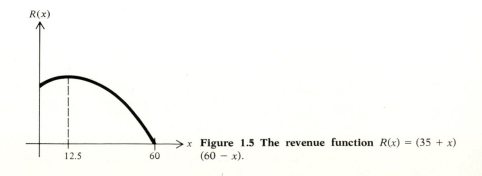

Figure 1.5 The revenue function $R(x) = (35 + x)(60 - x)$.

It follows that either $x = 12$ or $x = 13$ is the optimal integer value of x you are seeking. Since

$$R(12) = 2,256 \quad \text{and} \quad R(13) = 2,256$$

you can conclude that the bus company's revenue will be greatest when the group contains either 12 or 13 people in excess of 35; that is, for groups of 47 or 48.

In the preceding example, the graph of revenue as a function of x was actually a collection of discrete points corresponding to the integer values of x as indicated in Figure 1.6a. Since calculus cannot be used to study such a function, you worked with the differentiable function $R(x) = (35 + x)(60 - x)$ that was defined for all values of x and whose graph (Figure 1.6b) "connected" the points in Figure 1.6a. After applying calculus to this continuous model, you obtained a mathematical solution that was not the solution of the discrete practical problem, but that did suggest where to look for the practical solution.

An inventory problem For each shipment of raw materials, a manufacturer must pay an ordering fee to cover handling and transportation. When the raw materials arrive, they must be stored until needed and storage costs result. If each shipment of raw materials is large, few shipments will be needed and ordering costs will be low. Storage costs, however, will be high. If each shipment is small, ordering costs will be high because many shipments will be needed, but storage costs will be low. A manufacturer would like to determine the shipment size that will minimize total cost. The problem can be solved using calculus. Here is an example.

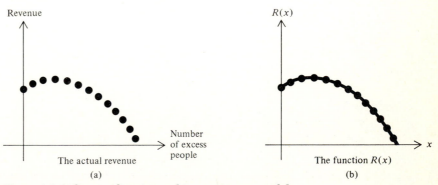

Figure 1.6 A discrete function and its continuous model.

EXAMPLE 1.4

A bicycle manufacturer buys 6,000 tires a year from a distributor and is trying to decide how often to order the tires. The ordering fee is $20 per shipment, the storage cost is 96 cents per tire per year, and each tire costs 25 cents. Suppose that the tires are used at a constant rate throughout the year, and that each shipment arrives just as the preceding shipment has been used up. How many tires should the manufacturer order each time to minimize cost?

SOLUTION

$$\text{Total cost} = \text{storage cost} + \text{ordering cost} + \text{cost of the tires}$$

Let x denote the number of tires in each shipment and $C(x)$ the corresponding total cost.

When a shipment arrives, all x tires are placed in storage and then withdrawn for use at a constant rate. The inventory decreases linearly until there are no tires left, at which time the next shipment arrives. The situation is illustrated in Figure 1.7a.

The average number of tires in storage during the year is $\frac{x}{2}$, and the total yearly storage cost is the same as if $\frac{x}{2}$ tires were kept in storage for the entire year (Figure 1.7b). (This assertion, although reasonable, is not really obvious and you have every right to be unconvinced. In Chapter 6, you will learn how to prove this fact using integral calculus.) It follows that

$$\text{Storage cost} = \frac{x}{2} \text{ (cost of storing 1 tire 1 year)}$$

$$= \frac{x}{2} (0.96)$$

$$= 0.48x$$

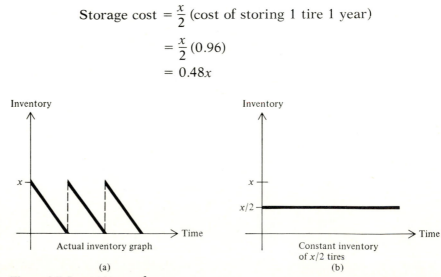

Figure 1.7 Inventory graphs.

The other two components of the total cost are easier to analyze.

Ordering cost = (ordering cost per shipment)(number of shipments)

Since 6,000 tires are ordered during the year and each shipment contains x tires, the number of shipments is $\dfrac{6,000}{x}$ and so

$$\text{Ordering cost} = 20 \left(\frac{6,000}{x}\right) = \frac{120,000}{x}$$

Moreover,

$$\begin{aligned}
\text{Cost of tires} &= (\text{total number of tires ordered})(\text{cost per tire}) \\
&= 6,000(0.25) \\
&= 1,500
\end{aligned}$$

Hence,
$$C(x) = 0.48x + \frac{120,000}{x} + 1,500$$

The goal is to minimize $C(x)$ on the interval $0 < x \leq 6,000$. Compute the derivative

$$C'(x) = 0.48 - \frac{120,000}{x^2}$$

and set it equal to zero to get

$$x^2 = \frac{120,000}{0.48} = 250,000$$

or
$$x = \pm 500$$

Take the positive value $x = 500$. It is easy to check that C is decreasing for $0 < x < 500$ and increasing for $x > 500$ as indicated in Figure 1.8. Hence the absolute minimum of $C(x)$ on the interval $0 < x \leq 6,000$ occurs when $x = 500$, and you can conclude that to minimize cost, the manufacturer should order the tires in lots of 500.

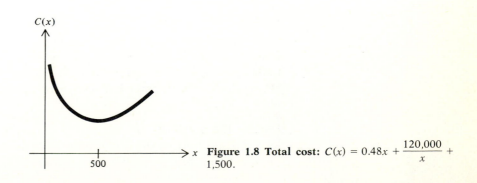

$C(x)$

500

Figure 1.8 Total cost: $C(x) = 0.48x + \dfrac{120,000}{x} + 1,500$.

Problems

Fencing

1. There are 320 meters of fencing available to enclose a rectangular field. How should this fencing be used so that the enclosed area is as large as possible?

Geometry

2. Prove that of all rectangles with a given perimeter, the square has the largest area.

Fencing

3. A city recreation department plans to build a rectangular playground having an area of 3,600 square meters and surround it by a fence. How can this be done using the least amount of fencing?

Geometry

4. Prove that of all rectangles with a given area, the square has the smallest perimeter.

Retail sales

5. A college bookstore can obtain the book *Social Groupings of the American Dragonfly* from the publisher at a cost of $3 per book. The bookstore has been offering the book at a price of $15 per copy, and at this price, has been selling 200 copies a month. The bookstore is planning to lower its price to stimulate sales and estimates that for each $1 reduction in the price, 20 more books will be sold each month. At what price should the bookstore sell the book to generate the greatest possible profit?

Manufacturing

6. A manufacturer has been selling lamps for $6 apiece, and at this price, consumers have been buying 3,000 lamps per month. The manufacturer wishes to raise the price and estimates that for each $1 increase in the price, 1,000 fewer lamps will be sold each month. The manufacturer can produce the lamps at a cost of $4 per lamp. At what price should the manufacturer sell the lamps to generate the greatest possible profit?

Agricultural yield

7. A Florida citrus grower estimates that if 60 orange trees are planted, the average yield per tree will be 400 oranges. The average yield will decrease by 4 oranges per tree for each additional tree planted on the same acreage. How many trees should the grower plant to maximize the total yield?

Harvesting

8. Farmers can get $2 per bushel for their potatoes on July first, and after that, the price drops by 2 cents per bushel per day. On July first, a farmer has 80 bushels of potatoes in the field and estimates that the crop is increasing at a rate of 1 bushel per day. When should the farmer harvest the potatoes to maximize revenue?

Recycling

9. During the summer, members of a scout troop have been collecting used bottles that they plan to deliver to a glass company for recycling. So far, in 80 days, the scouts have collected 24,000

kilograms of glass for which the glass company currently offers 1 cent per kilogram. However, because bottles are accumulating faster than they can be recycled, the company plans to reduce by 1 cent each day the price it will pay for 100 kilograms of used glass. Assume that the scouts can continue to collect bottles at the same rate and that transportation costs make more than one trip to the glass company unfeasible. What is the most profitable time for the scouts to conclude their summer project and deliver the bottles?

Construction cost

10. A closed box with a square base is to have a volume of 250 cubic meters. The material for the top and bottom of the box costs $2 per square meter, and the material for the sides costs $1 per square meter. Can the box be constructed for less than $300?

Construction cost

11. A carpenter has been asked to build an open box with a square base. The sides of the box will cost $3 per square meter, and the base will cost $4 per square meter. What are the dimensions of the box of greatest volume that can be constructed for $48?

Postal regulations

12. According to postal regulations, the girth plus length of parcels sent by fourth-class mail may not exceed 72 inches. What is the largest possible volume of a rectangular parcel with two square sides that can be sent by fourth-class mail?

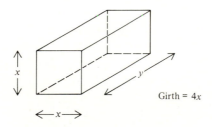

Construction

13. An open box is to be made from a square piece of cardboard, 18 inches by 18 inches, by removing a small square from each corner and folding up the flaps to form the sides. What are the dimensions of the box of greatest volume that can be constructed in this way?

The distance between moving objects

14. A truck is 975 kilometers due east of a car and is traveling west at a constant speed of 60 kilometers per hour. Meanwhile, the car is going north at a constant speed of 90 kilometers per hour. At what time will the car and truck be closest to each other? (*Hint:* You will simplify the calculation if you minimize the *square* of the distance between the car and truck rather than the distance itself. Can you explain why this simplification is justified?)

Installation cost 15. For the summer, the company that is installing the cable in Example 1.1 has hired a temporary employee with a Ph.D. in mathematics. The mathematician, recalling a problem from first-year calculus, asserts that no matter how far downstream the factory is located (beyond 1,200 meters), it would be most economical to have the cable reach the opposite bank 1,200 meters downstream from the power plant. The supervisor, amused by the naïveté of the over-educated employee, replies: "Any fool can see that if the factory is further away, the cable should reach the opposite bank further downstream. It's just common sense." Who is right? And why?

Spy story 16. It is noon, and the hero of a popular spy story (the same fellow who escaped from the diamond smugglers in Chapter 1, Section 4, Problem 27) is driving a jeep through the sandy desert in the tiny principality of Alta Loma. He is 32 kilometers from the nearest point on a straight, paved road. Down the road 16 kilometers is a power plant in which a band of international saboteurs has placed a time bomb set to explode at 12:50 P.M. The jeep can travel 48 kilometers per hour in the sand and 80 kilometers per hour on the paved road. If he arrives at the power plant in the shortest possible time, how long will our hero have to defuse the bomb?

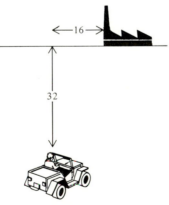

Poster design 17. A printer receives an order to produce a rectangular poster containing 25 square centimeters of print surrounded by margins of 2 centimeters on each side and 4 centimeters on the top and bottom. What are the dimensions of the smallest piece of paper that can be used to make the poster? (*Hint:* An unwise choice of variables will make the calculations unnecessarily complicated.)

Packaging 18. Use the fact that 12 fluid ounces is (approximately) 6.89π cubic inches to find the dimensions of the 12-ounce beer can that can be constructed using the least amount of metal. Compare these dimensions with those of one of the beer cans in your refrigerator. What do you think accounts for the difference? (Recall that the volume of a cylinder of radius r and height h is $\pi r^2 h$. The circumference of a circle of radius r is $2\pi r$ and its area is πr^2.)

Packaging 19. A cylindrical can is to hold 4π cubic inches of frozen orange juice. The cost per square inch of constructing the metal top and bottom is twice the cost per square inch of constructing the cardboard side. What are the dimensions of the least expensive can?

Volume 20. What is the maximum possible volume of a cylindrical can with no top that can be made from 27π square inches of metal?

Production cost 21. A plastics firm has received an order from the city recreation department to manufacture 8,000 special Styrofoam kickboards for its summer swimming program. The firm owns 10 machines, each of which can produce 30 kickboards an hour. The cost of setting up the machines to produce the kickboards is $20 per machine. Once the machines have been set up, the operation is fully automated and can be overseen by a single production supervisor earning $4.80 per hour.
   (a) How many of the machines should be used to minimize the cost of production?
   (b) How much will the supervisor earn during the production run if the optimal number of machines is used?
   (c) How much will it cost to set up the optimal number of machines?

Transportation cost 22. For speeds between 40 and 65 miles per hour, a truck gets $\dfrac{480}{x}$ miles per gallon when driven at a constant speed of x miles per hour. Gasoline costs 70 cents per gallon and the driver is paid $5.25 per hour. What is the most economical constant speed between 40 and 65 miles per hour at which to drive the truck?

Worker efficiency 23. An efficiency study of the morning shift at a certain factory indicates that an average worker who arrives on the job at 8:00 A.M. will have assembled $f(x) = -x^3 + 6x^2 + 15x$ transistor radios x hours later. The study indicates further that after a 15-minute coffee break, the average worker can assemble $g(x) = -\frac{1}{3}x^3 + x^2 + 23x$ radios in x hours. Determine the time between 8:00 A.M. and noon at which a 15-minute coffee break should be scheduled so that the average worker will assemble the maximum number of radios by lunchtime at 12:15 P.M.

Retail sales 24. A retailer has bought several cases of a certain imported wine. As the wine ages, its value initially increases, but eventually the wine will pass its prime and its value will decrease. Suppose that x years from now, the value of a case will be changing at the rate of $53 - 10x$ dollars per year. Suppose, in addition, that storage rates will remain fixed at $3 per case per year. When should the retailer sell the wine to obtain the greatest possible profit?

Construction cost 25. It is estimated that the cost of constructing an office building that is n floors high is $C(n) = 2n^2 + 500n + 600$ thousand dollars. How many floors should the building have to minimize the average cost per floor? (Remember that your answer should be a whole number.)

Production cost 26. Each machine at a certain factory can produce 50 units per hour. The setup cost is $80 per machine, and the operating cost is $5 per hour. How many machines should be used to produce 8,000 units at the least possible cost? (Remember that the answer should be a whole number.)

Inventory 27. An electronics firm uses 600 cases of transistors each year. The cost of storing one case for a year is 90 cents, and the ordering fee is $30 per shipment. How frequently should the transistors be ordered to keep total cost at a minimum? (Assume that the transistors are used at a constant rate throughout the year and that each shipment arrives just as the previous shipment has been used up.)

Inventory 28. A local tavern expects to use 800 bottles of bourbon this year. The bourbon costs $4 per bottle, the ordering fee is $10 per shipment, and the cost of storing the bourbon is 40 cents per bottle per year. The bourbon is consumed at a constant rate throughout the year, and each shipment arrives just as the previous shipment has been used up.
(a) How many bottles should the tavern order in each shipment to minimize cost?
(b) How often should the tavern order the bourbon?

(c) How will the answers to parts (a) and (b) change if the cost of the bourbon is increased to $4.30 per bottle?

Inventory 29. Through its franchised stations, an oil company gives out 16,000 road maps per year. The cost of setting up a press to print the maps is $100. In addition, production costs are 6 cents per map and storage costs are 20 cents per map per year. The maps are distributed at a uniform rate throughout the year and are printed in equal batches timed so that each arrives just as the preceding batch has been used up. How many maps should the oil company print in each batch to minimize cost?

Inventory 30. A manufacturer receives raw materials in equal shipments arriving at regular intervals throughout the year. The cost of storing the raw materials is directly proportional to the size of each shipment, while the total yearly ordering cost is inversely proportional to the shipment size. Show that total cost will be lowest if the size of the shipments is such that the storage cost and total ordering cost are equal.

Production cost 31. A manufacturing firm receives an order for Q items. Each of the firm's machines can produce n items per hour. The setup cost is s dollars per machine and the operating cost is p dollars per hour.
(a) Derive a formula for the number of machines that should be used to keep total cost as low as possible.
(b) Show that the total cost is minimal when the cost of setting up the machines is equal to the cost of operating the machines.

2 THE SECOND DERIVATIVE

In many practical problems, one seeks to determine when the rate of change of a given quantity is greatest or smallest. For example, a factory owner may wish to determine when an employee is working at maximum efficiency; that is, when the employee's rate of production is greatest. A traffic engineer may wish to find out when freeway traffic is moving most slowly. An economist may wish to predict when the rate of inflation will peak.

To find out when the rate of change of a function is greatest or smallest, you first take the derivative of the function to get an expression for its rate of change. Then you maximize or minimize this rate using the optimization techniques you learned in Sections 5 and 6 of Chapter 2. To do this, you have to differentiate again and work with the derivative of the derivative of the original function. This derivative of the derivative is known as the **second derivative** of the function.

The second derivative

> **The second derivative of f is the derivative of its derivative f' and is denoted by the symbol f''.**

The derivative f' is sometimes called the **first derivative** to distinguish it from the second derivative f''. If the function is denoted by y instead of f, the symbol $\dfrac{d^2y}{dx^2}$ is often used instead of f'' to denote the second derivative.

You don't have to use any new rules to find the second derivative of a function. Just find the first derivative and then differentiate again.

EXAMPLE 2.1

Compute the second derivative of the function $f(x) = 5x^4 - 3x^2 - 3x + 7$.

SOLUTION

Compute the first derivative

$$f'(x) = 20x^3 - 6x - 3$$

and then differentiate again to get

$$f''(x) = 60x^2 - 6$$

EXAMPLE 2.2

Compute the second derivative of the function $y = (x^2 + 1)^5$.

SOLUTION

Compute the first derivative using the chain rule.

$$\frac{dy}{dx} = 5(x^2 + 1)^4(2x) = 10x(x^2 + 1)^4$$

Then differentiate again using the product rule to get

$$\frac{d^2y}{dx^2} = 10x[4(x^2 + 1)^3(2x)] + 10(x^2 + 1)^4$$

$$= 80x^2(x^2 + 1)^3 + 10(x^2 + 1)^4$$
$$= 10(x^2 + 1)^3(9x^2 + 1)$$

The maximum efficiency of a worker

Here is a description of an industrial situation that can be analyzed with the aid of the second derivative.

The number of units a factory worker can produce in x hours is

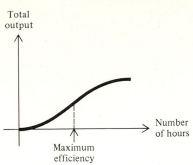

Figure 2.1 **The output of a factory worker.**

usually given by a function like the one whose graph is shown in Figure 2.1.

The graph reflects the fact that at first, the rate of production is low but increases as the worker settles into a routine. There comes a time at which the worker is performing at maximum efficiency, after which fatigue sets in and the rate of production decreases.

The moment of maximum efficiency (sometimes called the **point of diminishing returns**) is the time at which the worker's rate of production is greatest. In geometric terms, it is the point at which the graph of the output function is steepest. The next example illustrates how you can find the point of maximum efficiency using the second derivative.

EXAMPLE 2.3

An efficiency study of the morning shift at a factory indicates that an average worker who arrives on the job at 8:00 A.M. will have turned out $Q(t) = -t^3 + 9t^2 + 12t$ units t hours later. At what time during the morning is the worker performing most efficiently?

SOLUTION

The worker's rate of production is the derivative

$$R(t) = Q'(t) = -3t^2 + 18t + 12$$

Assuming that the morning shift runs from 8:00 A.M. until noon, the goal is to maximize the function $R(t)$ on the interval $0 \leq t \leq 4$. The derivative of R is

$$R'(t) = Q''(t) = -6t + 18$$

which is zero when $t = 3$. Comparing

$$R(0) = 12 \qquad R(3) = 39 \qquad R(4) = 36$$

you can conclude that the rate of production will be greatest and the worker performing most efficiently when $t = 3$; that is, at 11:00 A.M.

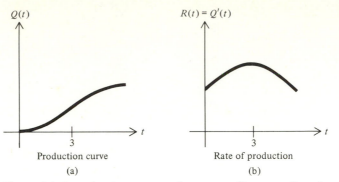

Production curve

(a)

Rate of production

(b)

Figure 2.2 A production curve and corresponding rate of production.

The graphs of the output $Q(t)$ and its derivative, the rate of production $R(t)$, are sketched in Figure 2.2. Notice that the production curve is steepest and the rate of production greatest when $t = 3$.

Concavity The point of diminishing returns for the production curve in Figure 2.2a occurs when $t = 3$. Before this point, the worker's rate of production is increasing, and after this point, it is decreasing. In geometric terms, the production curve is turning in a counterclockwise direction for $t < 3$ and in a clockwise direction for $t > 3$. It is customary to use the following notions of **concavity** to describe the direction in which a curve turns.

Concavity

A curve is said to be concave downward if its tangent turns in a clockwise direction as it moves along the curve from left to right.

A curve is said to be concave upward if its tangent turns in a counterclockwise direction as it moves along the curve from left to right.

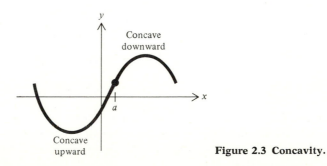

Figure 2.3 Concavity.

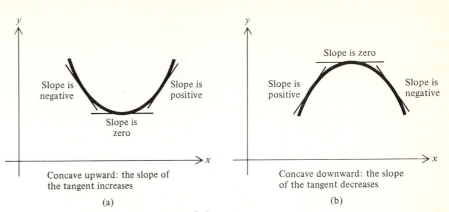

Figure 2.4 Concavity and the slope of the tangent.

For example, the curve in Figure 2.3 is concave upward for $x < a$ and concave downward for $x > a$.

When a curve is concave upward as in Figure 2.4a, the slope of its tangent *increases* as x increases. When a curve is concave downward as in Figure 2.4b, the slope of its tangent *decreases* as x increases.

The sign of the second derivative

The relationship between concavity and the slope of the tangent leads to a simple characterization of concavity in terms of the sign of the second derivative. Here is the argument.

Suppose the second derivative f'' is positive on an interval. This implies that the first derivative f' must be increasing on the interval. But f' is the slope of the tangent. Hence the slope of the tangent is increasing, and so the graph of f is concave upward on the interval. On the other hand, if f'' is negative on an interval, then f' is decreasing. This implies that the slope of the tangent is decreasing, and so the graph of f is concave downward on the interval.

Here is a summary of these important observations.

The geometric significance of the sign of the second derivative

If $f''(x) > 0$ whenever $a < x < b$, then f is concave upward for $a < x < b$.

If $f''(x) < 0$ whenever $a < x < b$, then f is concave downward for $a < x < b$.

Inflection points

A point at which the concavity of a function changes is called an **inflection point**. The function in Figure 2.3 had an inflection point at $x = a$. If the second derivative of a function is defined at an inflection point, its value there must be zero. Inflection points can also occur where the second derivative is undefined. Points at which the second derivative of a function is zero or undefined are called **second-order critical points.** (Points at which the first derivative of a function is zero or undefined are sometimes called **first-order critical points.**) Second-order critical points are to inflection points as first-order critical points are to relative extrema. In particular, every inflection point is a second-order critical point, but not every second-order critical point is necessarily an inflection point.

Curve sketching using the first and second derivatives

The preceding observations can be combined with the techniques developed in Chapter 2, Section 5, to get the following procedure you can use to obtain detailed graphs of functions using calculus.

How to use calculus to graph a function $f(x)$

Step 1. Compute the derivative $f'(x)$, find the x coordinates of the first-order critical points, and plot the critical points on the graph.

Step 2. Compute the second derivative $f''(x)$, find the x coordinates of the second-order critical points, and plot these critical points on the graph.

Step 3. Use the x coordinates of the first- and second-order critical points to divide the x axis into a collection of intervals. Check the signs of the first and second derivatives on each of these intervals.

Step 4. Draw the graph on each interval according to the following table:

Sign of f'	Sign of f''	Increasing or decreasing	Concavity	Shape
+	+	increasing	up	⌣
−	+	decreasing	up	⌣
+	−	increasing	down	⌢
−	−	decreasing	down	⌢

Here is an example.

EXAMPLE 2.4

Determine where the function $f(x) = x^4 + 8x^3 + 18x^2 - 8$ is increasing, decreasing, concave upward, and concave downward. Find the relative extrema and inflection points and draw the graph.

SOLUTION

The first derivative

$$f'(x) = 4x^3 + 24x^2 + 36x = 4x(x + 3)^2$$

is zero when $x = 0$ and $x = -3$, and the corresponding first-order critical points are $(0, -8)$ and $(-3, 19)$.

The second derivative

$$f''(x) = 12x^2 + 48x + 36 = 12(x + 3)(x + 1)$$

is zero when $x = -3$ and $x = -1$, and the corresponding second-order critical points are $(-3, 19)$ and $(-1, 3)$.

Plot these critical points and check the signs of $f'(x)$ and $f''(x)$ on each of the intervals defined by their x coordinates.

Interval	$f'(x)$	$f''(x)$	Increasing or decreasing	Concavity	Shape
$x < -3$	−	+	decreasing	up	
$-3 < x < -1$	−	−	decreasing	down	
$-1 < x < 0$	−	+	decreasing	up	
$x > 0$	+	+	increasing	up	

Draw the graph as shown in Figure 2.5 by connecting the critical points with a curve of appropriate shape on each interval.

Notice that the first-order critical point $(0, -8)$ is a relative minimum while the first-order critical point $(-3, 19)$ is neither a relative minimum nor a relative maximum, and that both of the second-order critical points $(-3, 19)$ and $(-1, 3)$ are inflection points.

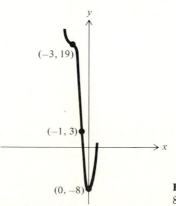

Figure 2.5 The graph of the function $y = x^4 + 8x^3 + 18x^2 - 8$.

The second derivative test

Here is a simple test involving the sign of the second derivative that you can use to classify first-order critical points.

The second derivative test

> **Suppose $f'(a) = 0$.**
>
> **If $f''(a) > 0$, then f has a relative minimum at $x = a$.**
>
> **If $f''(a) < 0$, then f has a relative maximum at $x = a$.**
>
> **However, if $f''(a) = 0$, the test is inconclusive and f may have a relative maximum, a relative minimum, or no relative extremum at all at $x = a$.**

To see why the second derivative test works, look at Figure 2.6 that shows the four possibilities that can occur when $f'(a) = 0$.

Figure 2.6a suggests that at a relative maximum, f must be concave downward and so $f''(a) \leq 0$. Figure 2.6b suggests that at a relative minimum, f must be concave upward and so $f''(a) \geq 0$. On the other hand, Figures 2.6c and 2.6d suggest that if a point at which $f'(a) = 0$ is not a relative extremum it must be an inflection point and so $f''(a)$, if it is defined, must be zero. It follows that if $f'(a) = 0$ and $f''(a) < 0$, the corresponding critical point must be a relative maximum, while

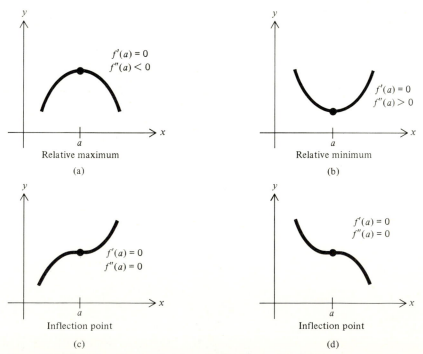

Figure 2.6 Behavior of a graph when the first derivative is zero.

if $f'(a) = 0$ and $f''(a) > 0$, the corresponding critical point must be a relative minimum.

The use of the second derivative test is illustrated in the following example.

EXAMPLE 2.5

Use the second derivative test to find the relative maxima and minima of the function $f(x) = 2x^3 + 3x^2 - 12x - 7$.

SOLUTION

Since the derivative

$$f'(x) = 6x^2 + 6x - 12 = 6(x + 2)(x - 1)$$

is zero when $x = -2$ and $x = 1$, the corresponding points $(-2, 13)$ and $(1, -14)$ are the first-order critical points of f. To test these points, compute the second derivative

$$f''(x) = 12x + 6$$

and evaluate it at $x = -2$ and $x = 1$. Since

$$f''(-2) = -18 < 0$$

it follows that $(-2, 13)$ is a relative maximum, and since

$$f''(1) = 18 > 0$$

it follows that $(1, -14)$ is a relative minimum.

For reference, the graph of f is sketched in Figure 2.7.

The function in the preceding example is the same one you analyzed in Example 5.1 of Chapter 2 using only the first derivative. No-

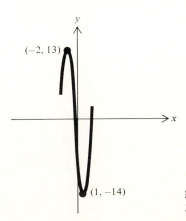

Figure 2.7 The graph of the function $y = 2x^3 + 3x^2 - 12x - 7$.

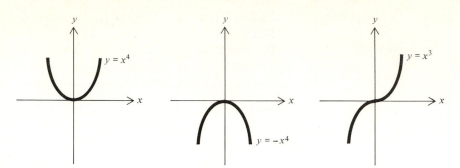

Figure 2.8 Three functions whose first and second derivatives are zero at $x = 0$.

tice the relative ease with which you can now identify the extrema. Using the second derivative test, you compute $f''(x)$ only at the critical points themselves. Using the first derivative, you had to investigate the sign of $f'(x)$ over entire intervals.

There are, however, some disadvantages to the second derivative test. For many functions (such as rational functions) the work involved in computing the second derivative is time-consuming and may diminish the efficiency of the test. Moreover, if both $f'(a)$ and $f''(a)$ are zero, the second derivative test tells you nothing whatsoever about the nature of the critical point. This is illustrated in Figure 2.8 which shows the graphs of three functions whose first and second derivatives are both zero when $x = 0$.

Remember that the second derivative test is a test for *relative extrema* only and tells you nothing about the *absolute* extrema of a function. In most practical optimization problems, you will have to use something more than the second derivative test to verify that a particular critical point is actually the desired absolute extremum.

Problems   In Problems 1 through 8, find the second derivative of the given function.

1. $f(x) = 5x^{10} - 6x^5 - 27x + 4$ 2. $y = \frac{2}{5}x^5 - 4x^3 + 9x - 6$

3. $y = x^2 - \dfrac{1}{x^2}$ 4. $f(x) = 5\sqrt{x} + \dfrac{3}{x^2} + 5x^{-1/3}$

5. $f(x) = (x^2 + 1)^5$ 6. $y = \sqrt{1 - x^2}$

7. $y = \dfrac{1}{x - 2}$ 8. $f(x) = \dfrac{x + 1}{x - 1}$

Efficiency   9. An efficiency study of the morning shift at a certain factory indicates that an average worker who arrives on the job at 8:00 A.M. will have assembled $Q(t) = -t^3 + 6t^2 + 15t$ transistor radios t hours later. At what time during the morning is the worker performing at maximum efficiency?

Efficiency   10. A college student spent 2 months doing the research for a history term paper and 4 hours actually writing it. After t hours of writing, the student had completed $P(t) = -\frac{1}{3}t^3 + 2t^2 + 2t$ pages. At what time was the student writing most efficiently?

11. At what point does the tangent to the curve $y = 2x^3 - 3x^2 + 6x$ have the smallest slope? What is the slope of the tangent at this point?

12. For what value of x in the interval $-1 \le x \le 4$ is the graph of the function $f(x) = 2x^2 - \frac{1}{3}x^3$ steepest? What is the slope of the tangent at this point?

Acceleration   The **acceleration** of a moving object is the rate of change of its speed with respect to time. Use this concept in Problems 13 through 15.

13. An object moves along a straight line so that after t seconds, its distance from a fixed reference point is $D(t) = t^3 - 12t^2 + 100t + 12$ meters. Find the acceleration of the object after 3 seconds. Is the object slowing down or speeding up at this time?

14. If an object is dropped or thrown vertically, its height (in feet) after t seconds is $H(t) = -16t^2 + S_0 t + H_0$, where S_0 is the initial speed of the object and H_0 is its initial height.
   (a) Derive an expression for the acceleration of the object.
   (b) How does the acceleration vary with time?
   (c) What is the significance of the fact that the answer to part (a) is negative?

15. After t hours of an 8-hour trip, a car traveling on a California freeway has gone $D(t) = 64t + \frac{10}{3}t^2 - \frac{2}{9}t^3$ kilometers.
   (a) Derive a formula expressing the acceleration of the car as a function of time.
   (b) At what rate is the speed of the car changing at the end of 6 hours? Is the speed increasing or decreasing at this time?
   (c) By how much does the speed of the car actually change during the 7th hour?
   (d) At what time during the 8-hour trip is the car traveling the fastest?

In Problems 16 and 17, determine where the second derivative of the function is positive, where it is negative, and where it is zero.

16.

17.

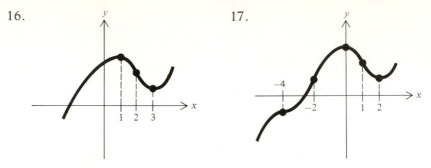

In Problems 18 through 35, determine where the given function is increasing, decreasing, concave upward, and concave downward. Find the relative extrema and inflection points and draw the graph.

18. $f(x) = x^3 + 3x^2 + 1$

19. $f(x) = \frac{1}{3}x^3 - 9x + 2$

20. $f(x) = x^3 - 3x^2 + 3x + 1$

21. $f(x) = x^4 - 4x^3 + 10$

22. $f(x) = x^5 - 5x$

23. $f(x) = (x - 2)^3$

24. $f(x) = (x - 2)^4$

25. $f(x) = (x^2 - 5)^3$

26. $f(x) = (x^2 - 3)^2$

27. $f(x) = x + \dfrac{1}{x}$

28. $f(x) = 1 + 2x + \dfrac{18}{x}$

29. $f(x) = \dfrac{x^2}{x - 3}$

30. $f(x) = \dfrac{x^2 - 3x}{x + 1}$

31. $f(x) = (x + 1)^{1/3}$

32. $f(x) = (x + 1)^{2/3}$

33. $f(x) = (x + 1)^{4/3}$

34. $f(x) = (x + 1)^{5/3}$

35. $f(x) = \sqrt{x^2 + 1}$

36. Sketch the graph of a function that has all of the following properties.
    (a) $f'(x) > 0$ when $x < -1$ and when $x > 3$
    (b) $f'(x) < 0$ when $-1 < x < 3$
    (c) $f''(x) < 0$ when $x < 2$
    (d) $f''(x) > 0$ when $x > 2$

37. Sketch the graph of a function that has all of the following properties.
    (a) $f'(x) > 0$ when $x < 2$ and when $2 < x < 5$
    (b) $f'(x) < 0$ when $x > 5$
    (c) $f'(2) = 0$

(d) $f''(x) < 0$ when $x < 2$ and when $4 < x < 7$

(e) $f''(x) > 0$ when $2 < x < 4$ and when $x > 7$

38. Sketch the graph of a function that has all of the following properties.

  (a) $f'(x) > 0$ when $x < 1$

  (b) $f'(x) < 0$ when $x > 1$

  (c) $f''(x) > 0$ when $x < 1$ and when $x > 1$

  What can you say about the derivative of f when $x = 1$?

39. The derivative of a certain function is $f'(x) = x^2 - 4x$.

  (a) On what intervals is f increasing? Decreasing?

  (b) On what intervals is f concave upward? Concave downward?

  (c) Find the x coordinates of the relative extrema and inflection points of f.

40. The derivative of a certain function is $f'(x) = x^2 - 2x - 8$.

  (a) On what intervals is f increasing? Decreasing?

  (b) On what intervals is f concave upward? Concave downward?

  (c) Find the x coordinates of the relative extrema and inflection points of f.

In Problems 41 through 43, the first derivatives of certain functions are sketched. In each case, determine where the function itself is increasing, decreasing, concave upward, and concave downward, and find the x coordinates of its relative extrema and inflection points.

41.   42.

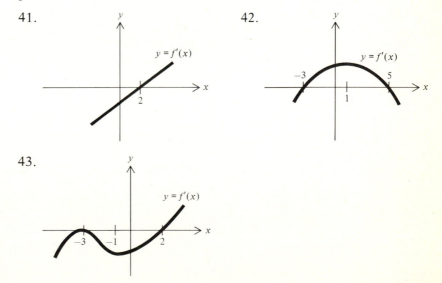

43.

44. Use calculus to show that the graph of the quadratic function $y = ax^2 + bx + c$ is concave upward if a is positive and concave downward if a is negative.

In Problems 45 through 50, use the second derivative test to find the relative maxima and minima of the given function.

45. $f(x) = x^3 + 3x^2 + 1$
46. $f(x) = x^4 - 2x^2 + 3$

47. $f(x) = (x^2 - 9)^2$
48. $f(x) = x + \dfrac{1}{x}$

49. $f(x) = 2x + 1 + \dfrac{18}{x}$
50. $f(x) = \dfrac{x^2}{x - 2}$

3 IMPLICIT DIFFERENTIATION AND RELATED RATES

In Chapter 2, Section 4, you learned how to use the chain rule to solve certain related rates problems. In these problems, one variable was given as a function of a second variable which, in turn, could be written as a function of a third. In this section you will learn a slightly different technique that you can use to solve related rates problems in which you are given information about the *rate of change* of some of the variables instead of explicit formulas relating all the variables. The technique is illustrated in the following example.

EXAMPLE 3.1

An environmental study of a certain community indicates that there will be $Q(p) = p^2 + 3p + 1,200$ units of a harmful pollutant in the air when the population is p thousand. The population is currently 30,000 and is increasing at a rate of 2,000 per year. At what rate is the level of air pollution increasing?

SOLUTION

If t denotes time (measured in years), the rate of change of the pollution level with respect to time is $\dfrac{dQ}{dt}$ and the rate of change of the population with respect to time is $\dfrac{dp}{dt}$. In this problem, you know that $\dfrac{dp}{dt} = 2$ and your goal is to find $\dfrac{dQ}{dt}$ when $p = 30$. You can do this by differentiating both sides of the equation

$$Q = p^2 + 3p + 1,200$$

with respect to t.

So that you won't forget that p is really a function of t, temporarily replace p by the expression $p(t)$ and rewrite the equation as

$$Q = [p(t)]^2 + 3p(t) + 1,200$$

Now differentiate both sides with respect to t, using the chain rule for powers to differentiate $[p(t)]^2$ and the constant multiple rule to differentiate $3p(t)$. You will get

$$\frac{dQ}{dt} = 2p(t)\frac{dp}{dt} + 3\frac{dp}{dt}$$

or, more simply,
$$\frac{dQ}{dt} = 2p\frac{dp}{dt} + 3\frac{dp}{dt}$$

Now substitute the given information $p = 30$ and $\dfrac{dp}{dt} = 2$ into this equation to get

$$\frac{dQ}{dt} = 2(30)(2) + 3(2) = 126$$

which tells you that the level of air pollution is currently increasing at a rate of 126 units per year.

In the preceding example, you started with an equation relating certain variables and differentiated both sides of it with respect to a new variable. This process is known as **implicit differentiation** because the original variables are not expressed *explicitly* as functions of the new variable.

Related rates problems   In many practical applications of implicit differentiation, the variable with respect to which you differentiate will denote time and the derivatives will represent rates of change with respect to time. Here are two more examples of this type. In each case, you will have to use geometry to obtain the initial equation.

EXAMPLE 3.2

A man 6 feet tall is walking away from a streetlight 20 feet high at the rate of 7 feet per second. At what rate is the length of his shadow increasing?

SOLUTION

Let x denote the length (in feet) of the man's shadow and y the distance between the man and the streetlight as shown in Figure 3.1, and let t denote time (measured in seconds).

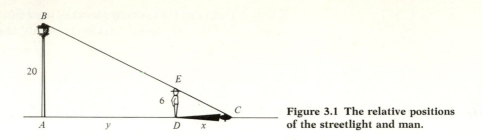

Figure 3.1 The relative positions of the streetlight and man.

You know that $\dfrac{dy}{dt} = 7$, and your goal is to find $\dfrac{dx}{dt}$. From the similar triangles ABC and DEC in Figure 3.1, you get the proportion

$$\frac{x + y}{20} = \frac{x}{6}$$

which you can rewrite as

$$x = \frac{3}{7} y$$

Differentiating both sides of this equation with respect to t, you get

$$\frac{dx}{dt} = \frac{3}{7} \frac{dy}{dt}$$

Use the fact that $\dfrac{dy}{dt} = 7$ to conclude that

$$\frac{dx}{dt} = \frac{3}{7} (7) = 3$$

That is, the man's shadow is increasing at the rate of 3 feet per second.

EXAMPLE 3.3

A water tank is in the shape of an inverted cone 20 feet high with a circular base 5 feet in radius. Water is running out of the bottom of the tank at the constant rate of 2 cubic feet per minute. How fast is the water level falling when the water is 8 feet deep?

SOLUTION

Let V denote the volume of the water in the tank after t minutes, let h be the corresponding water level, and let r be the radius of the surface of the water as shown in Figure 3.2.

You know that $\dfrac{dV}{dt} = -2$ (the minus sign indicating that the volume is decreasing), and your goal is to find $\dfrac{dh}{dt}$ when $h = 8$.

Start with the formula $\qquad V = \dfrac{1}{3}\pi r^2 h$

for the volume of a cone. Use the proportion

$$\frac{5}{20} = \frac{r}{h}$$

obtained from the similar triangles in Figure 3.2 to write r in terms of h as

$$r = \frac{h}{4}$$

and substitute this expression into the formula for volume to get

$$V = \frac{1}{48}\pi h^3$$

Differentiate both sides of this equation with respect to t. Don't forget to use the chain rule for powers when you differentiate h^3. (To help you remember, you can temporarily write h^3 as $[h(t)]^3$.) You will get

$$\frac{dV}{dt} = \frac{1}{16}\pi h^2 \frac{dh}{dt}$$

Substitute $h = 8$ and $\dfrac{dV}{dt} = -2$ into this equation and solve for $\dfrac{dh}{dt}$ to get

$$-2 = \frac{1}{16}\pi(64)\frac{dh}{dt}$$

$$\frac{dh}{dt} = -\frac{1}{2\pi}$$

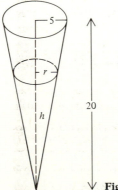

Figure 3.2 Conical water tank.

from which you can conclude that when the water is 8 feet deep, the water level is falling at a rate of $\dfrac{1}{2\pi}$ feet per minute.

Suggestion: So that you would not forget to use the chain rule for powers when you were first learning implicit differentiation, it was suggested that you temporarily replace p by $p(t)$ in Example 3.1 and h by $h(t)$ in Example 3.3. As soon as you feel comfortable differentiating implicitly, try to leave out these unnecessary steps. Your work will look much more professional as a result.

Implicit functions   In each of the examples in this section, you started with an equation relating two variables and differentiated both sides of it with respect to a third variable. Sometimes, you will encounter problems involving implicit differentiation in which the variable with respect to which you wish to differentiate is the same as one of the original variables. This occurs, for example, when you want to compute the derivative $\dfrac{dy}{dx}$ of a function whose equation is not easily solved for y in terms of x. A function expressed by such an equation is said to be an **implicit function.** Here is an example.

EXAMPLE 3.4

Find $\dfrac{dy}{dx}$ if $x^2y + 2y^3 = 3x + 2y$.

SOLUTION

Since there is no obvious way to solve for y, you will have to use implicit differentiation. That is, you will have to differentiate both sides of the equation, as it stands, with respect to x. Remember that y is really a function of x and that you will have to use the chain rule to differentiate powers of y. From the product rule, you get

$$\frac{d}{dx}(x^2y) = x^2\frac{dy}{dx} + 2xy$$

from the chain rule for powers, you get

$$\frac{d}{dx}(2y^3) = 6y^2\frac{dy}{dx}$$

and from the constant multiple rule, you get

$$\frac{d}{dx}(3x) = 3 \quad \text{and} \quad \frac{d}{dx}(2y) = 2\frac{dy}{dx}$$

Hence,
$$x^2 \frac{dy}{dx} + 2xy + 6y^2 \frac{dy}{dx} = 3 + 2 \frac{dy}{dx}$$

Solving this equation for $\frac{dy}{dx}$, you conclude that

$$\frac{dy}{dx} = \frac{3 - 2xy}{x^2 + 6y^2 - 2}$$

Notice that the formula for $\frac{dy}{dx}$ contains both the independent variable x and the dependent variable y. This is usual when derivatives are computed implicitly.

In the next example, you will see how to use implicit differentiation to find the slope of a tangent.

EXAMPLE 3.5

Find the slope of the line that is tangent to the curve $x^2y^3 - 6 = 5y^3 + x$ when $x = 2$.

SOLUTION

Differentiate both sides of the equation with respect to x to get

$$3x^2y^2 \frac{dy}{dx} + 2xy^3 = 15y^2 \frac{dy}{dx} + 1$$

and solve for $\frac{dy}{dx}$ to get
$$\frac{dy}{dx} = \frac{1 - 2xy^3}{3x^2y^2 - 15y^2}$$

The desired slope is the value of this derivative when $x = 2$. Before you can compute this value, you have to find the value of y that corresponds to $x = 2$. To do this, substitute $x = 2$ into the original equation and solve, getting
$$4y^3 - 6 = 5y^3 + 2$$
$$y^3 = -8$$
$$y = -2$$

Now substitute $x = 2$ and $y = -2$ into the formula for $\frac{dy}{dx}$ to get

$$\text{Slope of tangent} = \frac{1 - 2(2)(-2)^3}{3(2)^2(-2)^2 - 15(-2)^2} = -\frac{11}{4}$$

Problems   1. It is estimated that the annual advertising revenue received by a certain newspaper will be $R(x) = 0.5x^2 + 3x + 160$ thousand dollars when its circulation is x thousand. The circulation of the paper is currently 10,000 and is increasing at a rate of 2,000 per year. At what rate is the annual advertising revenue increasing?

2. Hospital officials estimate that approximately $N(p) = p^2 + 5p + 900$ people will seek treatment in the emergency room each year if the population of the community is p thousand. The population is currently 20,000 and is growing at the rate of 1,200 per year. At what rate is the number of people seeking treatment in the emergency room increasing?

3. A pebble is dropped into a lake and an expanding circular ripple results. When the radius of the ripple is 8 inches, the radius is increasing at a rate of 3 inches per second. At what rate is the area of the ripple changing at this time?

4. The volume of a spherical balloon is increasing at the rate of 3 cubic inches per second. At what rate is the radius of the balloon increasing when the radius is 2 inches? (*Hint:* The volume of a sphere of radius r is $\frac{4}{3}\pi r^3$.)

5. A man 6 feet tall is walking away from a streetlight 20 feet high at a rate of 7 feet per second. How fast is the shadow of the man's head moving along the ground?

6. A 20-foot ladder is leaning against a wall. The foot of the ladder is slipping away from the wall at a rate of 2 feet per second. At what rate is the top of the ladder moving down the wall when the bottom is 12 feet from the base of the wall?

7. A car, traveling north at 40 miles per hour, and a truck, traveling east at 30 miles per hour, leave an intersection at the same time. At what rate will the distance between them be changing 3 hours later?

8. A man is standing at the end of a pier 12 feet above the water and is pulling in a rope attached to a rowboat at the rate of 6 feet of rope per minute. How fast is the boat moving in the water when it is 16 feet from the pier?

9. A water tank is in the shape of an inverted cone 40 feet high with a circular base of radius 20 feet. Water is flowing into the tank at a constant rate of 80 cubic feet per minute. How fast is the water level rising when the water is 12 feet deep?

10. Suppose that in Problem 9, water is also flowing out of the bottom of the tank. At what rate should the water be allowed to flow out so that the water level will be rising at a rate of only 0.5 feet per minute when the water is 12 feet deep?

11. A ball is dropped from a height of 160 feet. A light is located at the same level, 10 feet away from the initial position of the ball. The height of the ball after t seconds is $H(t) = -16t^2 + 160$ feet. How fast is the ball's shadow moving along the ground 1 second after the ball is dropped?

In Problems 12 through 20, find $\dfrac{dy}{dx}$ by implicit differentiation.

12. $3x + 4y = 8$

13. $x^2 + y^2 = 25$

14. $x^2 + y = x^3 + y^2$

15. $x^3 + y^3 = xy$

16. $xy = 1$

17. $y^2 + 2xy^2 - 3x + 1 = 0$

18. $\dfrac{1}{x} + \dfrac{1}{y} = 1$

19. $(2x + y)^3 = x$

20. $(x - 2y)^2 = y$

In Problems 21 through 25, find the slope of the line that is tangent to the given curve at the specified value of x.

21. $x^2 = y^3$; $x = 8$

22. $\dfrac{1}{x} - \dfrac{1}{y} = 2$; $x = \frac{1}{4}$

23. $xy = 2$; $x = 2$

24. $x^2y^3 - 2xy = 6x + y + 1$; $x = 0$

25. $(1 - x + y)^3 = x + 7$; $x = 1$

In Problems 26 through 29, find $\dfrac{dy}{dx}$ in two ways: by implicit differentiation of the given equation, and by differentiation of an explicit formula for y. In each case, show that the two answers are really the same.

26. $x^2 + y^3 = 12$

27. $xy + 2y = x^2$

28. $x + \dfrac{1}{y} = 5$

29. $xy - x = y + 2$

In Problems 30 through 32, find the second derivative $\dfrac{d^2y}{dx^2}$ by implicit differentiation.

30. $xy = 1$

31. $x^2 = y^3$

32. $xy = y^2 + 1$

**4 APPROXI-
MATION BY
DIFFERENTIALS**

As you saw in Chapter 2, Section 3, the derivative of a function is its instantaneous rate of change and is often used as an estimate of the change in the value of the function produced by a 1-unit increase in the size of its independent variable. For example, economists frequently use marginal cost (the derivative of total cost) as an estimate of the cost of producing 1 additional unit. In this section, you will learn how to use calculus to estimate how *any* small change (not necessarily an increase of 1) in the size of the independent variable will affect the value of a function.

The technique is based on the fact that

$$\text{Change in } y \approx \left(\begin{array}{c}\text{rate of change of } y \\ \text{with respect to } x\end{array}\right) (\text{change in } x)$$

where the symbol $\approx$ is an approximation sign. If the rate of change of y with respect to x happens to be constant, the approximation sign can be replaced by an equals sign. If the rate of change is not constant, the approximation will be good provided the change in x is small.

Letting Δx denote the change in x and Δy the change in y, and, representing the rate of change of y with respect to x by the derivative $\dfrac{dy}{dx}$, you can rewrite the approximation formula compactly as follows.

Approximation formula   **If Δx is small,** $\Delta y \approx \dfrac{dy}{dx} \Delta x$

You can also write the formula using functional notation as follows.

**Approximation formula
(functional notation)**   **If Δx is small,** $f(x + \Delta x) - f(x) \approx f'(x) \Delta x$

Notice that if you divide both sides of the approximation formula by Δx you get

$$\frac{f(x + \Delta x) - f(x)}{\Delta x} \approx f'(x)$$

This shows that the approximation formula is simply a restatement of the fact that the difference quotient is close to the derivative when Δx is small.

The use of the approximation formula is illustrated in the following examples.

EXAMPLE 4.1

The total cost in dollars of manufacturing x units of a certain commodity is $f(x) = 3x^2 + 5x + 10$. The current level of production is 40 units. Estimate how the total cost will change if 40.5 units are produced.

SOLUTION

In this problem, $x = 40$, the change in x is $\Delta x = 0.5$, and the change in cost is $f(40.5) - f(40)$. Hence, by the approximation formula

$$\text{Change in cost} = f(40.5) - f(40) \approx f'(40)(0.5)$$

The derivative of f is

$$f'(x) = 6x + 5$$

and its value when $x = 40$ is

$$f'(40) = 6(40) + 5 = 245$$

Hence,

$$\text{Change in cost} \approx f'(40)(0.5) = 245(0.5) = \$122.50$$

For practice, compute the actual change in cost due to the increase in the level of production from 40 to 40.5 and compare your answer with the approximation. Is the approximation a good one?

In the next example, the approximation formula is used to estimate the maximum error in a calculation that is based on figures obtained by imperfect measurement.

EXAMPLE 4.2

You measure the side of a cube, find it to be 10 centimeters long, and conclude that the volume of the cube is $10^3 = 1,000$ cubic centimeters. If your measurement of the side is accurate to within 2 percent, approximately how accurate is your calculation of the volume?

SOLUTION

The volume of the cube is $V(x) = x^3$, where x is the length of a side. The error you make in computing the volume if you take the length of the side to be 10 when it is really $10 + \Delta x$ is

$$\text{Error in volume} = V(10 + \Delta x) - V(10) \approx V'(10) \, \Delta x$$

Your measurement of the side can be off by as much as 2 percent; that is, by as much as $0.02(10) = 0.2$ centimeters in either direction. Hence the maximum error in your measurement of the side is $\Delta x = \pm 0.2$, and the corresponding maximum error in your calculation of the volume is

$$\text{Maximum error in volume} = V(10 \pm 0.2) - V(10) \approx V'(10)(\pm 0.2)$$

Since $V'(x) = 3x^2$ and $V'(10) = 300$

it follows that

Maximum error in volume $\approx 300(\pm 0.2) = \pm 60$

This says that, at worst, your calculation of the volume as 1,000 cubic centimeters is off by approximately 60 cubic centimeters or 6 percent.

Differentials The expression $f'(x)\,\Delta x$ on the right-hand side of the approximation formula $f(x + \Delta x) - f(x) \approx f'(x)\,\Delta x$ is sometimes called the **differential** of f and is denoted by df. Similarly, the expression $\dfrac{dy}{dx}\,\Delta x$ on the right-hand side of the other form of the approximation formula $\Delta y \approx \dfrac{dy}{dx}\,\Delta x$ is known as the differential of y and is denoted by dy. Hence, if Δx is small,

$$\Delta y \approx dy \qquad \text{where} \qquad dy = \frac{dy}{dx}\,\Delta x$$

Geometric interpretation The approximation of Δy by its differential dy has a simple geometric interpretation that is illustrated in Figure 4.1.

Since the slope of the tangent is $\dfrac{dy}{dx}$, the differential $dy = \dfrac{dy}{dx}\,\Delta x$ must be the change in the height of the tangent corresponding to a change from x to $x + \Delta x$. On the other hand, Δy is the change in the height of the curve corresponding to this change in x. Hence, approximating Δy by the differential dy is the same as approximating the change in height of a curve by the change in height of its tangent. If Δx is small, this approximation is a good one.

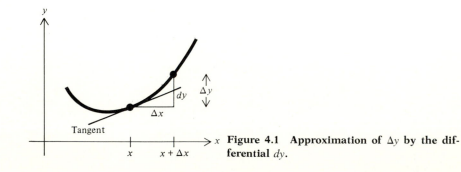

Figure 4.1 Approximation of Δy by the differential dy.

Problems   In each of the following problems, use calculus to obtain the required estimate.

Newspaper circulation
1. It is projected that t years from now the circulation of a local newspaper will be $C(t) = 100t^2 + 400t + 5{,}000$. Estimate the amount by which the circulation will increase during the next 6 months.

Population growth
2. It is projected that t years from now, the population of a certain suburban community will be $P(t) = 20 - \dfrac{6}{t + 1}$ thousand. By approximately how much will the population increase during the next quarter-year?

Air pollution
3. An environmental study of a certain community suggests that t years from now, the average level of carbon monoxide in the air will be $Q(t) = 0.05t^2 + 0.1t + 3.4$ parts per million. By approximately how much will the carbon monoxide level change during the coming 6 months?

Manufacturing
4. A manufacturer's total monthly revenue is $R(q) = 240q + 0.05q^2$ dollars when q units are produced during the month. Currently the manufacturer is producing 80 units a month and is planning to increase the monthly output by 0.65 units. Estimate how the total monthly revenue will change as a result.

Manufacturing
5. A manufacturer's total cost is $C(q) = 0.1q^3 - 0.5q^2 + 500q + 200$ dollars when the level of production is q units. The current level of production is 4 units, and the manufacturer is planning to decrease this to 3.9 units. Estimate how the total cost will change as a result.

Efficiency
6. An efficiency study of the morning shift at a certain factory indicates that an average worker who arrives on the job at 8:00 A.M. will have assembled $f(x) = -x^3 + 6x^2 + 15x$ transistor radios x hours later. Approximately how many radios will the worker assemble between 9:00 A.M. and 9:15 A.M.?

Production
7. At a certain factory, the daily output is $Q(K) = 600K^{1/2}$ units, where K denotes the capital investment measured in units of $1,000. The current capital investment is $900,000. Estimate the effect that an additional capital investment of $800 will have on the daily output.

Production
8. At a certain factory, the daily output is $Q(L) = 60{,}000L^{1/3}$ units, where L denotes the size of the labor force measured in worker-hours. Currently 1,000 worker-hours of labor are used each day. Estimate the effect on output that will be produced if the labor force is cut to 940 worker-hours.

Circulation of blood 9. The speed of blood flowing along the central axis of a certain artery is $S(R) = 1.8 \times 10^5 R^2$ centimeters per second, where R is the radius of the artery. A medical researcher measures the radius of the artery to be 1.2×10^{-2} centimeters and makes an error of 5×10^{-4} centimeters. Estimate the amount by which the calculated value of the speed of the blood will differ from the true speed if the incorrect value of the radius is used in the formula.

Area 10. You measure the radius of a circle to be 12 centimeters and use the formula $A = \pi r^2$ to calculate the area. If your measurement of the radius is accurate to within 3 percent, approximately how accurate is your calculation of the area?

Volume 11. You measure the radius of a sphere to be 6 inches and use the formula $V = \frac{4}{3}\pi r^3$ to calculate the volume. If your measurement of the radius is accurate to within 1 percent, approximately how accurate is your calculation of the volume?

Volume 12. Estimate what will happen to the volume of a cube if the length of each side is decreased by 1 percent.

Area 13. Estimate what will happen to the area of a circle if the radius is increased by 1 percent.

Circulation of blood 14. According to Poiseuille's law, the speed of blood flowing along the central axis of an artery of radius R is $S(R) = cR^2$, where c is a constant. What percentage error will you make in the calculation of $S(R)$ from this formula if you make a 1 percent error in the measurement of R?

Volume 15. A soccer ball made of leather $\frac{1}{8}$ inch thick has an inner diameter of $8\frac{1}{2}$ inches. Estimate the volume of its leather shell.

CHAPTER SUMMARY AND PROFICIENCY TEST

Important terms, symbols, and formulas

Optimization

Second derivative: $f''(x)$; $\dfrac{d^2y}{dx^2}$

Concave upward: $f''(x) > 0$
Concave downward: $f''(x) < 0$
Inflection point: concavity changes
Second-order critical point: $f''(x) = 0$ or $f''(x)$ is undefined
Second derivative test when $f'(a) = 0$:
    If $f''(a) > 0$, f has a relative minimum at $x = a$
    If $f''(a) < 0$, f has a relative maximum at $x = a$

Implicit differentiation; related rates

Approximation formulas: $\Delta y \approx \dfrac{dy}{dx} \Delta x$; $f(x + \Delta x) - f(x) \approx f'(x) \Delta x$

Differential: $dy = \dfrac{dy}{dx} \Delta x$

Proficiency test

1. A retailer can obtain cameras from the manufacturer at a cost of $50 apiece. The retailer has been selling the cameras at a price of $80 apiece, and at this price, consumers have been buying 40 cameras a month. The retailer is planning to lower the price to stimulate sales and estimates that for each $5 reduction in the price, 10 more cameras will be sold each month. At what price should the retailer sell the cameras to maximize profit?

2. You wish to use 300 meters of fencing to surround two identical adjacent rectangular plots as shown in the accompanying figure. How should you do this to make the combined area of the plots as large as possible?

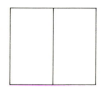

3. A manufacturing firm has received an order to make 400,000 souvenir medals commemorating the 10th anniversary of the landing of Apollo 11 on the moon. The firm owns 20 machines, each of which can produce 200 medals per hour. The cost of setting up the machines to produce the medals is $80 per machine, and the total operating cost is $5.76 per hour. How many machines should be used to minimize the cost of producing the 400,000 medals?

4. A citrus grower estimates that if 60 lemon trees are planted in a grove, the average yield per tree will be 475 lemons. The average yield will decrease by 5 lemons per tree for each additional tree planted in the grove. How many trees should the grower plant to maximize the total yield? (Remember that the answer should be a whole number.)

5. Suppose the consumer demand for a certain commodity is $D(p) = mp + b$ units per month when the market price is p dollars per unit.
   (a) Assume that $m < 0$ and $b > 0$ and sketch this demand function, labeling the points at which the graph intersects the coordinate axes. Explain in economic terms why the assumptions about the signs of m and b are reasonable.
   (b) Express consumers' total monthly expenditure for the commodity

as a function of p and sketch the graph of this function. Where does the graph cross the p axis?

(c) Use calculus to show that the market price at which consumer expenditure will be greatest is the value of p that is midway between the origin and the p intercept of the demand curve.

6. A postal clerk spends 4 hours each morning sorting mail. During that time, the clerk can sort approximately $f(t) = -t^3 + 7t^2 + 200t$ letters in t hours. At what time during this period is the clerk performing at peak efficiency?

7. Determine where the given function is increasing, decreasing, concave upward, and concave downward. Find the relative extrema and inflection points and draw the graph.

   (a) $f(x) = x^2 - 6x + 1$

   (b) $f(x) = x^3 - 3x^2 + 2$

   (c) $f(x) = \dfrac{x^2 + 3}{x - 1}$

8. Sketch the graph of a function that has all of the following properties.

   (a) $f'(x) > 0$ when $x < 0$ and when $x > 5$

   (b) $f'(x) < 0$ when $0 < x < 5$

   (c) $f''(x) > 0$ when $-6 < x < -3$ and when $x > 2$

   (d) $f''(x) < 0$ when $x < -6$ and when $-3 < x < 2$

9. Use the second derivative test to find the relative maxima and minima of the given function.

   (a) $f(x) = -2x^3 + 3x^2 + 12x - 5$

   (b) $f(x) = \dfrac{x^2}{x + 1}$

   (c) $f(x) = 2x + \dfrac{8}{x} + 2$

10. Water is flowing into a cylindrical tank 10 feet in diameter at a constant rate of 25 cubic feet per minute. How fast is the water level rising?

11. A truck is 360 kilometers due east of a car and is traveling west at a constant speed of 60 kilometers per hour. Meanwhile, the car is going north at a constant speed of 90 kilometers per hour. At what rate is the distance between the car and truck changing 2 hours later? Is this distance increasing or decreasing?

12. Find $\dfrac{dy}{dx}$ by implicit differentiation.

    (a) $5x + 3y = 12$

    (b) $x^2y = 1$

    (c) $(2x + 3y)^5 = x + 1$

13. Use implicit differentiation to find the slope of the line that is tangent to the given curve for the specified value of x.
    (a) $xy^3 = 8$; $x = 1$
    (b) $x^2y - 2xy^3 + 6 = 2x + 2y$; $x = 0$

14. At a certain factory, the daily output is $Q(L) = 20,000L^{1/2}$ units, where L denotes the size of the labor force measured in worker-hours. Currently 900 worker-hours of labor are used each day. Use differentials to estimate the effect on output that will be produced if the labor force is cut to 885 worker-hours.

15. The level of air pollution in a certain city is proportional to the square of the population. Use differentials to estimate the percentage by which the air-pollution level will increase if the population increases by 5 percent.

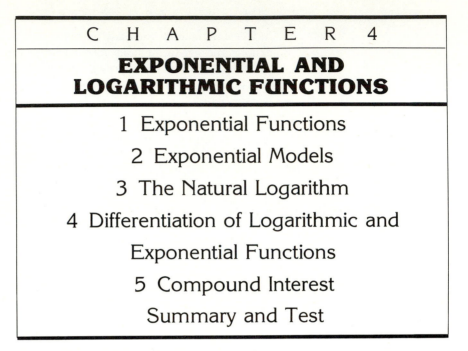

C H A P T E R 4

EXPONENTIAL AND LOGARITHMIC FUNCTIONS

1 Exponential Functions

2 Exponential Models

3 The Natural Logarithm

4 Differentiation of Logarithmic and

Exponential Functions

5 Compound Interest

Summary and Test

1 EXPONENTIAL FUNCTIONS

First impressions can be deceiving. Take the algebraic expression $\left(1 + \dfrac{1}{n}\right)^n$ for instance. At first glance this expression may look no more interesting than many other algebraic expressions. Yet the number it approaches as n increases without bound turns out to be one of the most important and useful numbers in mathematics.

To get a feel for what happens to the expression $\left(1 + \dfrac{1}{n}\right)^n$ as n increases, use your calculator to complete the following table. (Round off your answers to three decimal places.)

n	1	2	5	10	100	1,000	10,000	100,000
$\left(1 + \dfrac{1}{n}\right)^n$	2.000	2.250	2.488	2.594	2.705			

It turns out that as n increases without bound, the expression $\left(1 + \dfrac{1}{n}\right)^n$ approaches an irrational number, traditionally denoted by

162

the letter e, whose value is approximately 2.718. (A rigorous proof of this fact requires techniques beyond the scope of this book.)

The number e

$$\left(1 + \frac{1}{n}\right)^n \to e \approx 2.718 \text{ as } n \text{ increases without bound}$$

Functions involving powers of e play a central role in applied mathematics. They are used in demography to forecast population size, in finance to calculate the value of investments, in archaeology to date ancient artifacts, in psychology to study learning phenomena, in public health to analyze the spread of epidemics, and in industry to estimate the reliability of products. You will see some of these applications in Section 2. To illustrate how the number e might arise in practice, here is a brief discussion of compound interest.

Compound interest

Suppose a sum of money is invested and the interest is compounded only once. If P is the initial investment (the principal) and r is the interest rate (expressed as a decimal), the balance B after the interest is added will be

$$B = P + Pr = P(1 + r) \qquad \text{dollars}$$

This says that to compute the balance at the end of an interest period, you multiply the balance at the beginning of the period by the expression $1 + r$, where r is the interest rate per period.

At most banks, interest is compounded more than once a year. The interest that is added to the account during one period will itself earn interest during the subsequent periods. If the annual interest rate is r and interest is compounded k times, the year is divided into k equal interest periods and the interest rate during each is $\frac{r}{k}$. To compute the balance at the end of any period, you multiply the balance at the beginning of that period by the expression $1 + \frac{r}{k}$. Hence, the balance at the end of the first period is

$$P\left(1 + \frac{r}{k}\right)$$

The balance at the end of the second period is

$$P\left(1 + \frac{r}{k}\right)\left(1 + \frac{r}{k}\right) = P\left(1 + \frac{r}{k}\right)^2$$

The balance at the end of the third period is

$$P\left(1 + \frac{r}{k}\right)^2\left(1 + \frac{r}{k}\right) = P\left(1 + \frac{r}{k}\right)^3$$

At the end of 1 year, the interest has been compounded k times, and the balance is

$$P \left(1 + \frac{r}{k}\right)^k$$

and at the end of t years, the interest has been compounded kt times, and the balance is given by the function

$$B(t) = P \left(1 + \frac{r}{k}\right)^{kt}$$

Compound interest

> **If P dollars is invested at an annual interest rate r and interest is compounded k times per year, the balance $B(t)$ after t years will be**
>
> $$B(t) = P \left(1 + \frac{r}{k}\right)^{kt} \qquad \text{dollars}$$

As the frequency with which the interest is compounded increases, the corresponding balance $B(t)$ also increases. Hence a bank that compounds interest frequently may attract more customers than one that offers the same interest rate but that compounds interest less often. The question arises: What happens to the balance at the end of t years as the frequency with which the interest is compounded increases without bound? That is, what will the balance be at the end of t years if interest is compounded not quarterly, not monthly, not daily, but continuously? In mathematical terms: What happens to the expression $P \left(1 + \frac{r}{k}\right)^{kt}$ as k increases without bound? The answer turns out to involve the number e. Here is the argument.

To simplify the calculation, let $n = \frac{k}{r}$. Then, $k = nr$ and so

$$P \left(1 + \frac{r}{k}\right)^{kt} = P \left(1 + \frac{1}{n}\right)^{nrt} = P \left[\left(1 + \frac{1}{n}\right)^{n}\right]^{rt}$$

Since n increases without bound as k does, and since $\left(1 + \frac{1}{n}\right)^{n}$ approaches e as n increases without bound, it follows that $P \left(1 + \frac{r}{k}\right)^{kt}$ approaches Pe^{rt} as k increases without bound.

Here is a summary of the situation.

Continuously compounded interest

> **If P dollars is invested at an annual interest rate r and interest is compounded continuously, the balance $B(t)$ after t years will be**
>
> $$B(t) = Pe^{rt} \qquad \text{dollars}$$

The following numerical example illustrates what happens to a bank balance as the interest is compounded with increasing frequency. The calculations were done on a pocket calculator.

EXAMPLE 1.1

Suppose $1,000 is invested at an annual interest rate of 6 percent. Compute the balance after 10 years if the interest is compounded

(a) quarterly   (b) monthly   (c) continuously

SOLUTION

(a) To compute the balance after 10 years if the interest is compounded quarterly, use the formula $B(t) = P \left(1 + \dfrac{r}{k}\right)^{kt}$, with $t = 10$, $P = 1,000$, $r = 0.06$, and $k = 4$. You will get

$$B(10) = 1,000(1 + \tfrac{0.06}{4})^{40} = \$1,814.02$$

(b) This time, take $t = 10$, $P = 1,000$, $r = 0.06$, and $k = 12$ to get

$$B(10) = 1,000(1 + \tfrac{0.06}{12})^{120} = \$1,819.40$$

(c) Now use the formula $B(t) = Pe^{rt}$, with $t = 10$, $P = 1,000$, and $r = 0.06$ and conclude that

$$B(10) = 1,000e^{0.6} = \$1,822.12$$

This value, $1,822.12, is an upper bound for the possible balance. No matter how often interest is compounded, $1,000 invested at an annual interest rate of 6 percent cannot grow to more than $1,822.12 in 10 years.

Exponential functions

The function $B(t) = Pe^{rt}$ is closely related to a class of functions called **exponential functions.** In general, an exponential function is a function of the form $f(x) = a^x$, where a is a positive constant. In an exponential function, the independent variable x is the **exponent** of a positive constant a known as the **base** of the function. Thus an exponential function is different from a power function $f(x) = x^n$ in which the base is the variable and the exponent is the constant.

You are probably already familiar with the following four rules that define the expression a^x for all *rational* values of x. (These definitions, as well as the properties of exponents, are discussed in more detail in the algebra review at the back of the book.)

Definition of a^x for rational values of x (and $a > 0$)

Integer powers: If n is a positive integer,

$$a^n = a \cdot a \cdots a$$

where the product $a \cdot a \cdots a$ contains n factors.

Fractional powers: If n and m are positive integers,

$$a^{n/m} = (\sqrt[m]{a})^n$$

where $\sqrt[m]{}$ denotes the positive mth root.

Negative powers:

$$a^{-x} = \frac{1}{a^x}$$

Zero power: $a^0 = 1$

For example,

$$3^4 = 3 \cdot 3 \cdot 3 \cdot 3 = 81$$

$$4^{1/2} = \sqrt{4} = 2$$

$$4^{-3/2} = \frac{1}{4^{3/2}} = \frac{1}{8}$$

$$3^{-4} = \frac{1}{3^4} = \frac{1}{81}$$

$$4^{3/2} = (\sqrt{4})^3 = 8$$

$$27^{-2/3} = \frac{1}{(\sqrt[3]{27})^2} = \frac{1}{9}$$

The graphs of exponential functions

The graphs of four exponential functions are shown in Figure 1.1.

One way to obtain a rough sketch of an exponential function very quickly is to find its y intercept and determine its behavior as x increases without bound and as x decreases without bound. The technique is illustrated in the next example.

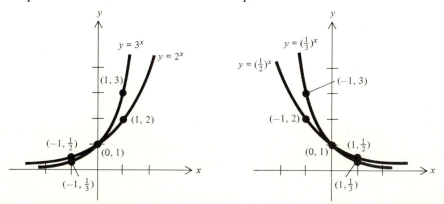

Figure 1.1 The graphs of four exponential functions.

EXAMPLE 1.2

Sketch the function $f(x) = a^x$ if $0 < a < 1$ and if $a > 1$.

SOLUTION

In both cases, the y intercept is $(0, 1)$ since $f(0) = a^0 = 1$. There are no x intercepts since, for positive a, a^x is always positive. To determine the behavior of the graph as x increases or decreases without bound, consider the two cases $0 < a < 1$ and $a > 1$ separately.

If $0 < a < 1$, the value of the product $a^n = a \cdot a \cdots a$ approaches zero as the number n of factors increases. This suggests that a^x approaches zero as x increases without bound. On the other hand, the value of the product $a^{-n} = \dfrac{1}{a} \cdot \dfrac{1}{a} \cdots \dfrac{1}{a}$ increases without bound $\left(\text{since } \dfrac{1}{a} > 1\right)$ as the number n of factors increases. This suggests that a^x increases without bound as x decreases without bound. A graph with these features is sketched in Figure 1.2a.

If $a > 1$, a^x increases without bound as x increases without bound, and a^x approaches zero as x decreases without bound. The corresponding graph is sketched in Figure 1.2b.

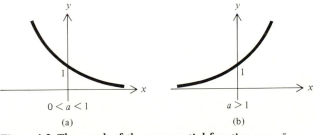

Figure 1.2 The graph of the exponential function $y = a^x$.

Since $e > 1$, the graph of the special exponential function $y = e^x$ resembles the sketch in Figure 1.2b.

Irrational exponents   The graphs in Figure 1.2 should not really have been drawn as unbroken curves since a^x has been defined only for *rational* values of x. However, it can be shown (using techniques beyond the scope of this book) that because there are "so many" rational numbers, there can be only *one* unbroken curve that passes through all the points (x, a^x) for which x is rational. That is, there exists a unique continuous function $f(x)$ that is defined for all real numbers x and that is equal to a^x

when x is rational. When x is irrational, one *defines* a^x to be the value $f(x)$ of this function. Fortunately, in practical work you will rarely, if ever, have to deal explicitly with a^x for an irrational value of x.

The laws of exponents

You may recall that for rational values of the independent variable, exponential functions obey certain laws of exponents. It can be shown that these useful laws remain valid when the variable is allowed to assume arbitrary real values.

Laws of exponents

The product law: $a^r a^s = a^{r+s}$

The quotient law: $\dfrac{a^r}{a^s} = a^{r-s}$

The power law: $(a^r)^s = a^{rs}$

The use of these laws is illustrated in the next two examples.

EXAMPLE 1.3

Evaluate the following expressions.

(a) $(2^{-3})^2$ (b) $\dfrac{7^{1/2}(7^{3/2})}{7^3}$ (c) $4^{2/5}(64^{1/5})$

SOLUTION

(a) $(2^{-3})^2 = 2^{-3(2)} = 2^{-6} = \frac{1}{64}$

(b) $\dfrac{7^{1/2}(7^{3/2})}{7^3} = \dfrac{7^{(1/2+3/2)}}{7^3} = 7^{(1/2+3/2-3)} = 7^{-1} = \dfrac{1}{7}$

(c) $4^{2/5}(64^{1/5}) = 4^{2/5}(4^3)^{1/5} = 4^{2/5}(4^{3/5}) = 4^{(2/5+3/5)} = 4$

EXAMPLE 1.4

Solve each of the following equations for n.

(a) $\dfrac{a^3}{a^5} = a^n$ (b) $a^5 a^n = a^2$

(c) $(a^n)^2 = a^{12}$ (d) $2^3 n = 2^5$

SOLUTION

(a) Since $\dfrac{a^3}{a^5} = a^{3-5} = a^{-2}$, it follows that $n = -2$.

(b) Since $a^5 a^n = a^{5+n}$, it follows that $5 + n = 2$, or $n = -3$.

(c) Since $(a^n)^2 = a^{2n}$, it follows that $2n = 12$, or $n = 6$.

(d) Multiply both sides of the equation by 2^{-3} to get

$$2^{-3}(2^3 n) = 2^{-3}(2^5) \quad \text{or} \quad n = 2^2 = 4$$

The next example illustrates a useful computational trick that is based on the power law for exponents.

EXAMPLE 1.5

Find $f(6)$ if $f(x) = e^{kx}$ and $f(2) = 5$.

SOLUTION

You do not have to know the value of k or of e to solve this problem! The fact that $f(2) = 5$ tells you that

$$e^{2k} = 5$$

and using the power law for exponents, you can rewrite the expression for $f(6)$ in terms of this quantity to get

$$f(6) = e^{6k} = (e^{2k})^3 = 5^3 = 125$$

Problems

1. Program a computer or a programmable calculator to evaluate $\left(1 + \dfrac{1}{n}\right)^n$ for $n = 1,000, 2,000, \ldots , 50,000$.

2. Program a computer or a programmable calculator to evaluate $\left(1 + \dfrac{1}{n}\right)^n$ for $n = -1,000, -2,000, \ldots , -50,000$. On the basis of these calculations, what can you conjecture about the behavior of $\left(1 + \dfrac{1}{n}\right)^n$ as n *decreases* without bound?

3. Learn how to use your calculator to find powers of e. In particular, find e^2, e^{-2}, $e^{0.05}$, $e^{-0.05}$, e^0, e, $\sqrt{e}$, and $\dfrac{1}{\sqrt{e}}$. (Round off your answers to three decimal places.)

4. Sketch the curves $y = 3^x$ and $y = 4^x$ on the same set of axes.

5. Sketch the curves $y = (\frac{1}{3})^x$ and $y = (\frac{1}{4})^x$ on the same set of axes.

In Problems 6 through 13, sketch the given function.

6. $f(x) = e^x$ 7. $f(x) = e^{-x}$

8. $f(x) = 2 + e^x$ 9. $f(x) = 3 + e^{-x}$

10. $f(x) = 2 - 3e^x$ 11. $f(x) = 3 - 2e^x$

12. $f(x) = 5 - 3e^{-x}$ 13. $f(x) = 3 - 5e^{-x}$

In Problems 14 through 29, evaluate the given expression without using a calculator.

14. 2^5 15. 2^{-5}

16. 2^0

17. $9^{1/2}$

18. $9^{-1/2}$

19. $27^{1/3}$

20. $27^{-1/3}$

21. $(\frac{1}{4})^{-2}$

22. $(\frac{1}{4})^{1/2}$

23. $(\frac{1}{4})^{-3/2}$

24. $\dfrac{2^3(2^4)}{(2^2)^3}$

25. $\dfrac{5^{2/3}(5^{7/3})}{5^4}$

26. $\dfrac{3^{-4}(3^2)}{(3^3)^{-2}}$

27. $\dfrac{2(32^{3/5})}{2^5}$

28. $\dfrac{\sqrt{8}(\sqrt{2})^3}{4}$

29. $[\sqrt{27}(3^{5/2})]^{1/2}$

In Problems 30 through 43, solve the given equation for n. (Assume $a > 0$ and $a \neq 1$.)

30. $a^3 a^5 = a^n$

31. $\dfrac{a^2}{a^6} = a^n$

32. $\dfrac{a^2}{a^{-2}} = a^n$

33. $(a^2)^3 = a^n$

34. $a^3 a^{-2} = a^n$

35. $a^3 a^n = \dfrac{1}{a^2}$

36. $(a^2)^n = a^6$

37. $(a^n)^2 = \dfrac{1}{a^3}$

38. $a^{2/5} a^n = \dfrac{1}{a}$

39. $(a^n)^3 = \sqrt{a}$

40. $\sqrt[n]{a}(a^{5/3}) = a^2$

41. $\dfrac{a^5(a^{1/3})}{a^n} = a$

42. $3^5 n = 3^8$

43. $2^{-1/2} n = 2^{3/2}$

44. Find $f(2)$ if $f(x) = e^{kx}$ and $f(1) = 20$.

45. Find $f(9)$ if $f(x) = e^{kx}$ and $f(3) = 2$.

46. Find $f(8)$ if $f(x) = Ae^{kx}$, $f(0) = 20$, and $f(2) = 40$.

47. Find $f(4)$ if $f(x) = 50 - Ae^{-kx}$, $f(0) = 20$, and $f(2) = 30$.

48. Find $f(2)$ if $f(x) = 50 - Ae^{kx}$, $f(0) = 30$, and $f(4) = 5$.

Compound interest   49. Suppose $1,000 is invested at an annual interest rate of 7 percent. Compute the balance after 10 years if the interest is compounded
    (a) annually   (b) quarterly
    (c) monthly   (d) continuously

Compound interest 50. A sum of money is invested at a certain fixed interest rate, and the interest is compounded continuously. After 10 years, the money has doubled. How will the balance at the end of 20 years compare with the initial investment?

Compound interest 51. (a) Solve the equation $B = Pe^{rt}$ for P.
(b) How much money should be invested today at an annual interest rate of 6 percent compounded continuously so that 10 years from now it will be worth $10,000? [*Hint:* Use the result from part (a).]

Effective interest rate 52. When a bank offers an annual interest rate of $100r$ percent and compounds the interest more than once a year, the total interest earned during a year is greater than $100r$ percent of the balance at the beginning of that year. The actual percentage by which the balance grows during a year is sometimes called the **effective interest rate,** while the advertised rate of $100r$ percent is called the **nominal interest rate.** Find the effective interest rate if the nominal rate is 6 percent and interest is compounded
(a) quarterly (b) continuously

2 EXPONENTIAL MODELS

Functions involving powers of e play a central role in applied mathematics. Here is a sampling of practical situations from the social, managerial, and natural sciences that can be described mathematically in terms of such functions.

Exponential growth

A quantity $Q(t)$ that increases according to a law of the form $Q(t) = Q_0e^{kt}$, where Q_0 and k are positive constants, is said to experience **exponential growth.** For example, if interest is compounded continuously, the resulting bank balance $B(t) = Pe^{rt}$ grows exponentially. Also, in the absence of environmental constraints, population increases exponentially. As you will see later in this book, quantities that grow exponentially are characterized by the fact that their rate of growth is proportional to their size.

To sketch the function $Q(t) = Q_0e^{kt}$, observe that $Q(t)$ is always positive, that $Q(0) = Q_0$, and that $Q(t)$ increases without bound as t increases without bound and approaches zero as t decreases without bound. A sketch is drawn in Figure 2.1.

Here is an example from biology.

EXAMPLE 2.1

Biologists have determined that under ideal conditions, the number of bacteria in a culture grows exponentially. Suppose that 2,000 bac-

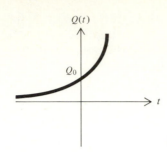

Figure 2.1 Exponential growth: $Q(t) = Q_0e^{kt}$.

teria are initially present in a certain culture and that 6,000 are present 20 minutes later. How many bacteria will be present at the end of 1 hour?

SOLUTION

Let $Q(t)$ denote the number of bacteria present after t minutes. Since the number of bacteria grows exponentially, and since 2,000 bacteria were initially present, you know that Q is a function of the form

$$Q(t) = 2,000e^{kt}$$

Since 6,000 bacteria are present after 20 minutes, it follows that

$$6,000 = 2,000e^{20k} \quad \text{or} \quad e^{20k} = 3$$

To find the number of bacteria present at the end of 1 hour, compute $Q(60)$ using the power law for exponents as follows.

$$Q(60) = 2,000e^{60k} = 2,000(e^{20k})^3 = 2,000(3)^3 = 54,000$$

Exponential decay A quantity $Q(t)$ that decreases according to a law of the form $Q(t) = Q_0e^{-kt}$, where Q_0 and k are positive constants, is said to experience **exponential decay** or, equivalently, to **decrease exponentially.** Radioactive substances decay exponentially. Sales of many products decrease exponentially when advertising is discontinued. Quantities that decrease exponentially are characterized by the fact that their rate of decrease is proportional to their size. A sketch of the function $Q(t) = Q_0e^{-kt}$ is shown in Figure 2.2.

EXAMPLE 2.2

A certain industrial machine depreciates so that its value after t years is given by a function of the form $Q(t) = Q_0e^{-0.04t}$. After 20 years, the machine is worth $8,986.58. What was its original value?

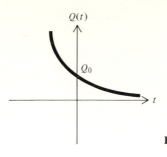

Figure 2.2 Exponential decay: $Q(t) = Q_0 e^{-kt}$.

SOLUTION

Your goal is to find Q_0. Since $Q(20) = 8,986.58$, you have

$$Q_0 e^{-0.8} = 8,986.58$$

Multiplying both sides of this equation by $e^{0.8}$, you get

$$Q_0 = 8,986.58 e^{0.8} = \$20,000$$

Learning curves   The graph of a function of the form $Q(t) = B - Ae^{-kt}$, where B, A, and k are positive constants, is sometimes called a **learning curve.** The name arose when psychologists discovered that functions of this form often describe the relationship between the efficiency with which an individual performs a task and the amount of training or experience the individual has had.

To sketch the function $Q(t) = B - Ae^{-kt}$, observe that $Q(0) = B - A$, that $Q(t)$ approaches B as t increases without bound (since Ae^{-kt} approaches zero), and that $Q(t)$ decreases without bound as t does. A sketch is drawn in Figure 2.3. The behavior of the graph as t increases without bound reflects the fact that eventually an individual will approach peak efficiency, and additional training will have little effect on performance.

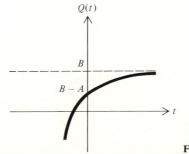

Figure 2.3 A learning curve: $Q(t) = B - Ae^{-kt}$.

EXAMPLE 2.3

The rate at which a postal clerk can sort mail is a function of the clerk's experience. The postmaster of a large city estimates that after t months on the job, the average clerk can sort $Q(t) = 700 - 400e^{-0.5t}$ letters per hour.

(a) How many letters can a new employee sort per hour?
(b) How many letters can a clerk with 6 months' experience sort per hour?
(c) Approximately how many letters will the average clerk ultimately be able to sort per hour?

SOLUTION

(a) The number of letters a new employee can sort per hour is

$$Q(0) = 700 - 400 = 300$$

(b) After 6 months, the average clerk can sort

$$Q(6) = 700 - 400e^{-0.5(6)} = 700 - 400e^{-3} = 680$$

letters per hour.

(c) As t increases without bound, $Q(t)$ approaches 700. Hence, the average clerk will ultimately be able to sort approximately 700 letters per hour. The situation is illustrated in Figure 2.4.

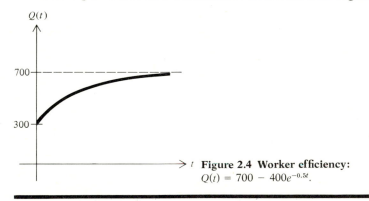

Figure 2.4 Worker efficiency: $Q(t) = 700 - 400e^{-0.5t}$.

Logistic curves   The graph of a function of the form $Q(t) = \dfrac{B}{1 + Ae^{-Bkt}}$, where B, A, and k are positive constants, is an S-shaped or **sigmoidal curve.** The term **logistic curve** is also used to refer to such a curve. A sketch of the function $Q(t) = \dfrac{B}{1 + Ae^{-Bkt}}$ is shown in Figure 2.5. Notice that the curve

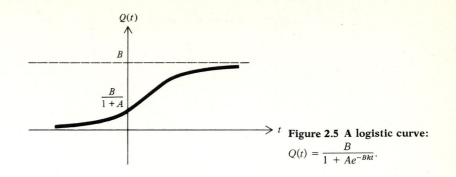

Figure 2.5 A logistic curve:
$$Q(t) = \frac{B}{1 + Ae^{-Bkt}}.$$

crosses the vertical axis at a height of $\dfrac{B}{1 + A}$ and that $Q(t)$ approaches B as t increases without bound.

Logistic curves are rather accurate models of population growth when environmental factors impose an upper bound on the possible size of the population. They also describe the spread of epidemics and rumors in a community. Here is a typical example.

EXAMPLE 2.4

Public health records indicate that t weeks after the outbreak of a rare form of influenza, approximately $Q(t) = \dfrac{20}{1 + 19e^{-1.2t}}$ thousand people had caught the disease.

(a) How many people had the disease when it first broke out?
(b) How many had caught the disease by the end of the 2nd week?
(c) If the trend continues, approximately how many people in all will contract the disease?

SOLUTION

(a) Since
$$Q(0) = \frac{20}{1 + 19} = 1$$

it follows that 1,000 people had the disease initially.

(b) Since
$$Q(2) = \frac{20}{1 + 19e^{-1.2(2)}} = 7.343$$

it follows that 7,343 people had caught the disease by the end of the 2nd week.

(c) Since $Q(t)$ approaches 20 as t increases without bound, it follows that approximately 20,000 people will eventually contract the disease. The situation is illustrated in Figure 2.6.

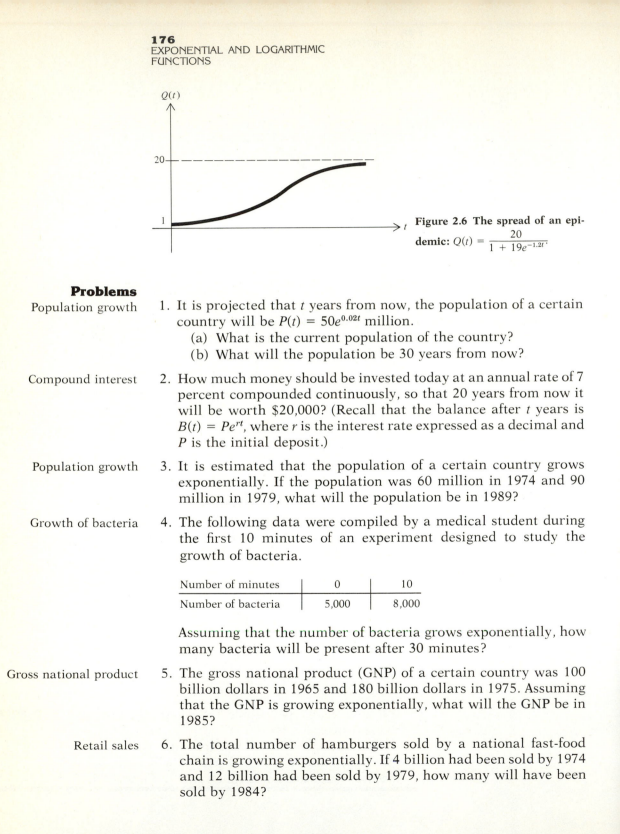

Figure 2.6 The spread of an epidemic: $Q(t) = \dfrac{20}{1 + 19e^{-1.2t}}$.

Problems

Population growth

1. It is projected that t years from now, the population of a certain country will be $P(t) = 50e^{0.02t}$ million.
    (a) What is the current population of the country?
    (b) What will the population be 30 years from now?

Compound interest

2. How much money should be invested today at an annual rate of 7 percent compounded continuously, so that 20 years from now it will be worth $20,000? (Recall that the balance after t years is $B(t) = Pe^{rt}$, where r is the interest rate expressed as a decimal and P is the initial deposit.)

Population growth

3. It is estimated that the population of a certain country grows exponentially. If the population was 60 million in 1974 and 90 million in 1979, what will the population be in 1989?

Growth of bacteria

4. The following data were compiled by a medical student during the first 10 minutes of an experiment designed to study the growth of bacteria.

Number of minutes	0	10
Number of bacteria	5,000	8,000

Assuming that the number of bacteria grows exponentially, how many bacteria will be present after 30 minutes?

Gross national product

5. The gross national product (GNP) of a certain country was 100 billion dollars in 1965 and 180 billion dollars in 1975. Assuming that the GNP is growing exponentially, what will the GNP be in 1985?

Retail sales

6. The total number of hamburgers sold by a national fast-food chain is growing exponentially. If 4 billion had been sold by 1974 and 12 billion had been sold by 1979, how many will have been sold by 1984?

Population density

7. The population density x miles from the center of a certain city is $D(x) = 12e^{-0.07x}$ thousand people per square mile.
   (a) What is the population density at the center of the city?
   (b) What is the population density 10 miles from the center of the city?

Radioactive decay

8. The amount of a sample of a radioactive substance remaining after t years is given by a function of the form $Q(t) = Q_0 e^{-0.0001t}$. At the end of 5,000 years, 2,000 grams of the substance remain. How many grams were present initially?

Radioactive decay

9. A radioactive substance decays exponentially. If 500 grams of the substance were present initially and 400 grams are present 50 years later, how many grams will be present after 200 years?

Product reliability

10. A statistical study indicates that the fraction of the electric toasters manufactured by a certain company that are still in working condition after t years of use is approximately $f(t) = e^{-0.2t}$.
    (a) What fraction of the toasters can be expected to work for at least 3 years?
    (b) What fraction of the toasters can be expected to fail during their 3rd year of use?
    (c) What fraction of the toasters can be expected to fail before 1 year of use?

Product reliability

11. A manufacturer of toys has found that the fraction of its plastic battery-operated toy oil tankers that sink in less than t days is approximately $f(t) = 1 - e^{-0.03t}$.
    (a) Sketch this reliability function. What happens to the graph as t increases without bound?
    (b) What fraction of the tankers can be expected to float for at least 10 days?
    (c) What fraction of the tankers can be expected to sink between the 15th and 20th days?

Sales

12. Once the initial publicity surrounding the release of a new book is over, sales of the hardcover edition tend to decrease exponentially. At the time publicity was discontinued, a certain book was experiencing sales of 25,000 copies per month. One month later, sales of the book had dropped to 10,000 copies per month. What will the sales be after 1 more month?

Recall from memory

13. Psychologists believe that when a person is asked to recall a set of facts, the number of facts recalled after t minutes is given by a function of the form $Q(t) = A(1 - e^{-kt})$, where k is a positive con-

stant and A is the total number of relevant facts in the person's memory.

(a) Sketch the function Q.

(b) What happens to the graph of Q as t increases without bound? Explain this behavior in practical terms.

Advertising

14. When professors select texts for their courses, they usually choose from among the books already on their shelves. For this reason, most publishers send complimentary copies of new texts to professors teaching related courses. The mathematics editor at a major publishing house estimates that if x thousand complimentary copies are distributed, the first-year sales of a certain new mathematics text will be approximately $f(x) = 20 - 15e^{-0.2x}$ thousand copies.

(a) Sketch this sales function.

(b) How many copies can the editor expect to sell in the first year if no complimentary copies are sent out?

(c) How many copies can the editor expect to sell in the first year if 10,000 complimentary copies are sent out?

(d) If the editor's estimate is correct, what is the most optimistic projection for the first-year sales of the text?

Depreciation

15. When a certain industrial machine is t years old, its resale value will be $V(t) = 4{,}800e^{-t/5} + 400$ dollars.

(a) Sketch the function V. What happens to the value of the machine as t increases without bound?

(b) How much was the machine worth when it was new?

(c) How much will the machine be worth after 10 years?

Efficiency

16. The daily output of a worker who has been on the job for t weeks is given by a function of the form $Q(t) = 40 - Ae^{-kt}$. Initially the worker could produce 20 units a day, and after 1 week the worker can produce 30 units a day. How many units will the worker produce per day after 3 weeks?

Newton's law of heating

17. A cool drink is removed from a refrigerator on a hot summer day and placed in a room whose temperature is 30° Celsius. According to a law of physics, the temperature of the drink t minutes later is given by a function of the form $f(t) = 30 - Ae^{-kt}$. If the temperature of the drink was 10° Celsius when it left the refrigerator and 15° Celsius after 20 minutes, what will the temperature of the drink be after 40 minutes?

The spread of an epidemic

18. Public health records indicate that t weeks after the outbreak of a rare form of influenza, approximately $f(t) = \dfrac{6}{3 + 9e^{-0.8t}}$ thousand people had caught the disease.

(a) How many people had the disease initially?

(b) How many had caught the disease by the end of 3 weeks?

(c) If the trend continues, approximately how many people in all will contract the disease?

Population growth

19. It is estimated that t years from now, the population of a certain country will be $P(t) = \dfrac{80}{8 + 12e^{-0.06t}}$ million.

(a) What is the current population?

(b) What will the population be 50 years from now?

(c) What will happen to the population in the long run?

The spread of an epidemic

20. An epidemic spreads through a community so that t weeks after its outbreak, the number of people who have been infected is given by a function of the form $f(t) = \dfrac{B}{1 + Ce^{-kt}}$, where B is the number of residents in the community who are susceptible to the disease. If $\frac{1}{5}$ of the susceptible residents were infected initially and $\frac{1}{2}$ had been infected by the end of the 4th week, what fraction of the susceptible residents will have been infected by the end of the 8th week?

The spread of a rumor

21. A traffic accident was witnessed by $\frac{1}{10}$ of the residents of a small town. The number of residents who had heard about the accident t hours later is given by a function of the form $f(t) = \dfrac{B}{1 + Ce^{-kt}}$, where B is the population of the town. If $\frac{1}{4}$ of the residents had heard about the accident after 2 hours, how long did it take for $\frac{1}{2}$ of the residents to hear the news?

3 THE NATURAL LOGARITHM

In many practical problems, a number x is known and the goal is to find the corresponding number y such that $x = e^y$. This number y is called the **natural logarithm** of x and is denoted by the symbol ln x. The letter l in the symbol ln stands for "logarithm" and the letter n for "natural." The word "ln" is virtually unpronounceable. It can be read as "el en," "log," or "Lynn."

The natural logarithm

> **Corresponding to each positive number x there is a unique power y such that $x = e^y$. This power y is called the natural logarithm of x and is denoted by ln x. Thus,**
>
> $$y = \ln x \quad \text{if and only if} \quad x = e^y$$

EXAMPLE 3.1

Find:

(a) $\ln e$ (b) $\ln 1$

SOLUTION

(a) According to the definition, $\ln e$ is the unique number y such that $e = e^y$. Clearly this number is $y = 1$. Hence, $\ln e = 1$.
(b) Similarly, $\ln 1$ is the unique number y such that $1 = e^y$. Since $e^0 = 1$, it follows that $\ln 1 = 0$.

The relationship between e^x and $\ln x$

The next example establishes two important identities that show that logarithmic and exponential functions have a certain "neutralizing" effect on each other.

EXAMPLE 3.2

Simplify the following expressions.

(a) $e^{\ln x}$ (for $x > 0$)   (b) $\ln e^x$

SOLUTION

(a) According to the definition, $\ln x$ is the unique number y for which $x = e^y$. Hence, $e^{\ln x} = e^y = x$.
(b) Similarly, $\ln e^x$ is the unique number y for which $e^x = e^y$. Clearly this number y is x itself. Hence, $\ln e^x = x$.

The two identities derived in Example 3.2 show that the composite functions $\ln e^x$ and $e^{\ln x}$ leave the variable x unchanged. In general, two functions f and g for which $f[g(x)] = x$ and $g[f(x)] = x$ are said to be **inverses** of one another. Thus the exponential function $y = e^x$ and the logarithmic function $y = \ln x$ are inverses.

The inverse relationship between e^x and $\ln x$

$$e^{\ln x} = x \quad (\text{if } x > 0) \quad \text{and} \quad \ln e^x = x$$

The next example illustrates how you can use the inverse relationship between e^x and $\ln x$ to solve equations.

EXAMPLE 3.3

Solve each of the following equations for x.

(a) $3 = e^{20x}$ (b) $2 \ln x = 1$

SOLUTION

(a) Take the natural logarithm of each side of the equation to get

$$\ln 3 = \ln e^{20x} \qquad \text{or} \qquad \ln 3 = 20x$$

Solve for x, using a calculator (or the natural logarithm table in the back of the book) to find $\ln 3$.

$$x = \frac{\ln 3}{20} = \frac{1.0986}{20} = 0.0549$$

(b) To isolate $\ln x$ on the left-hand side of the equation, divide both sides by 2, getting

$$\ln x = \frac{1}{2}$$

Then apply the exponential function to both sides of the equation and conclude that

$$e^{\ln x} = e^{1/2} \qquad \text{or} \qquad x = \sqrt{e}$$

The graph of $\ln x$ There is an easy way to obtain the graph of the logarithmic function $y = \ln x$ from that of the exponential function $y = e^x$. The method is based on the geometric fact that the point (b, a) is the reflection across the line $y = x$ of the point (a, b), as illustrated in Figure 3.1.

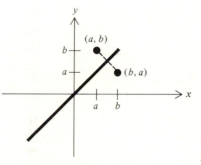

Figure 3.1 **Reflection of points across the line** $y = x$.

   The graph of $y = \ln x$ is the reflection across the line $y = x$ of the graph of $y = e^x$. To see this, suppose that (a, b) is a point on the curve $y = \ln x$. Then $b = \ln a$, or, equivalently, $a = e^b$. Hence the reflected point (b, a) can be written as (b, e^b), which is clearly a point on the curve $y = e^x$. Conversely, if (a, b) is on the curve $y = e^x$, it follows that $b = e^a$ and so $a = \ln b$. Hence the reflected point is $(b, a) = (b, \ln b)$ which lies on the curve $y = \ln x$.

   The graph of the exponential function $y = e^x$, the line $y = x$, and the graph of the logarithmic function $y = \ln x$ are sketched in Figure 3.2. Notice that $\ln x$ is defined only for positive values of x, that $\ln 1 = 0$,

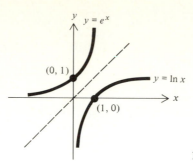

Figure 3.2 The graphs of $y = e^x$ and $y = \ln x$.

that $\ln x$ increases without bound as x increases without bound, and that $\ln x$ decreases without bound as x approaches zero from the right.

Properties of the natural logarithm

The laws of exponents can be used to derive the following important properties of logarithms.

Properties of logarithms

> **The logarithm of a product:** $\ln uv = \ln u + \ln v$
>
> **The logarithm of a quotient:** $\ln \dfrac{u}{v} = \ln u - \ln v$
>
> **The logarithm of a power:** $\ln u^v = v \ln u$

The first of these properties states that logarithms transform multiplication into the simpler operation of addition. The derivation of this property will be given in Example 3.5. The second property states that logarithms transform division into subtraction, and the third states that logarithms reduce exponentiation to multiplication. The derivations of these two properties are left as problems for you to do.

The next example illustrates the use of these algebraic properties of logarithms.

EXAMPLE 3.4

(a) Find $\ln \sqrt{ab}$ if $\ln a = 3$ and $\ln b = 7$.

(b) Show that $\ln \dfrac{1}{x} = -\ln x$.

(c) Find x if $2^x = e^3$.

SOLUTION

(a)
$$\ln \sqrt{ab} = \ln (ab)^{1/2} = \tfrac{1}{2} \ln ab$$
$$= \tfrac{1}{2}(\ln a + \ln b) = \tfrac{1}{2}(3 + 7) = 5$$

(b) $$\ln \frac{1}{x} = \ln 1 - \ln x = 0 - \ln x = -\ln x$$

(c) Take the natural logarithm of each side of the equation $2^x = e^3$ to get

$$x \ln 2 = 3 \qquad \text{or} \qquad x = \frac{3}{\ln 2}$$

EXAMPLE 3.5

Derive the formula $\ln uv = \ln u + \ln v$.

SOLUTION

Use the fact that e^x and $\ln x$ are inverses to get

$$e^{\ln uv} = uv \qquad e^{\ln u} = u \qquad e^{\ln v} = v$$

Now use these identities together with the product law for exponents to get

$$e^{\ln uv} = uv = e^{\ln u}e^{\ln v} = e^{\ln u + \ln v}$$

Finally, compare the powers of e at each end of this string of equalities to conclude that

$$\ln uv = \ln u + \ln v$$

Logarithms with other bases

You may be wondering how natural logarithms are related to the logarithms you studied in high school. The following discussion will show you the connection.

The graph of the exponential function $y = a^x$ (shown in Figure 3.3 for $a > 1$) suggests that for each positive number y there corresponds a unique number x such that $y = a^x$. This power x is called the **logarithm of y to the base a** and is denoted by $\log_a y$. Thus,

$$x = \log_a y \qquad \text{if and only if} \qquad y = a^x$$

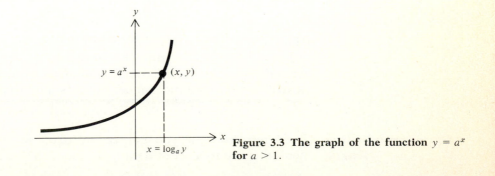

Figure 3.3 The graph of the function $y = a^x$ for $a > 1$.

The logarithms you used in high school algebra to simplify numerical calculations were logarithms to the base 10. For numerical work, 10 is a particularly convenient base for logarithms because the standard decimal representation of numbers is based on powers of 10. Natural logarithms are logarithms to the base e. Because of the importance of the special exponential function e^x, natural logarithms arise frequently in applied work.

Population growth

In the next example, you will see how to use logarithms to determine how much time it takes a quantity that grows exponentially to double in size.

EXAMPLE 3.6

The world's population is growing at a rate of about 2 percent per year. As you will see in Chapter 5, this implies that if the growth rate remains constant, the population t years from now will be given by a function of the form $P(t) = P_0 e^{0.02t}$, where P_0 is the current population. If this model of population growth is correct, how long will it take for the world's population to double?

SOLUTION

Your goal is to find the value of t for which $P(t) = 2P_0$. That is, you want to solve the equation

$$2P_0 = P_0 e^{0.02t}$$

for t. To do this, divide each side by P_0 to get

$$2 = e^{0.02t}$$

and then take the logarithm of each side to get

$$\ln 2 = 0.02t$$

or
$$t = \frac{\ln 2}{0.02} = 34.66$$

That is, the world's population will double in about $34\frac{2}{3}$ years.

Notice that the constant P_0 representing the current population was eliminated from the equation at an early stage and that the final answer is therefore independent of this quantity.

Exponential curve fitting

In the next example, you will see how to use logarithms to fit an exponential curve to a set of data.

EXAMPLE 3.7

The population density x miles from the center of a city is given by a function of the form $Q(x) = Ae^{-kx}$. Find this function if the population density at the center of the city is 15 thousand people per square mile and the density 10 miles from the center is 9 thousand per square mile.

SOLUTION

For simplicity, express the density in units of 1,000 people per square mile. The fact that $Q(0) = 15$ tells you that $A = 15$. The fact that $Q(10) = 9$ tells you that

$$9 = 15e^{-10k} \qquad \text{or} \qquad \tfrac{3}{5} = e^{-10k}$$

Taking the logarithm of each side of this equation, you get

$$\ln \tfrac{3}{5} = -10k$$

or

$$k = -\frac{\ln \tfrac{3}{5}}{10} = 0.051$$

It follows that the desired exponential function is $Q(x) = 15e^{-0.051x}$.

Carbon dating Carbon 14 (^{14}C) is a radioactive isotope of carbon that is widely used to date ancient fossils and artifacts. Here is an outline of the technique.

The carbon dioxide in the air contains ^{14}C as well as carbon 12 (^{12}C), a nonradioactive isotope. Scientists have found that the ratio of ^{14}C to ^{12}C in the air has remained approximately constant throughout history. Living plants absorb carbon dioxide from the air, and so the ratio of ^{14}C to ^{12}C in a living plant is the same as that in the air itself. When a plant dies, the absorption of carbon dioxide ceases. The ^{12}C already in the plant remains while the ^{14}C decays, and the ratio of ^{14}C to ^{12}C decreases exponentially. The ratio of ^{14}C to ^{12}C in a fossil t years after it was alive is approximately $R(t) = R_0 e^{-kt}$, where $k = \dfrac{\ln 2}{5,730}$ and R_0 is the ratio of ^{14}C to ^{12}C in the atmosphere. By comparing $R(t)$ with R_0, scientists can estimate the age of the fossil.

EXAMPLE 3.8

An archaeologist has found a fossil in which the ratio of ^{14}C to ^{12}C is $\tfrac{1}{5}$ the ratio found in the atmosphere. Approximately how old is the fossil?

SOLUTION

The age of the fossil is the value of t for which $R(t) = \frac{1}{5}R_0$, that is, for which

$$\tfrac{1}{5}R_0 = R_0 e^{-kt}$$

Dividing by R_0 and taking logarithms, you get

$$\ln \tfrac{1}{5} = -kt \qquad \text{or} \qquad t = -\frac{\ln \tfrac{1}{5}}{k} = \frac{\ln 5}{k}$$

Since $k = \dfrac{\ln 2}{5{,}730}$, you find that

$$t = \frac{5{,}730 \ln 5}{\ln 2} = 13{,}304.65$$

and you can conclude that the fossil is approximately 13,305 years old.

Problems
1. Learn how to use your calculator to find natural logarithms. In particular, find $\ln 1$, $\ln 2$, $\ln e$, $\ln 5$, $\ln \tfrac{4}{5}$, and $\ln e^2$. What happens if you try to find $\ln 0$ or $\ln -2$? Why?

In Problems 2 through 7, evaluate the given expression without using tables or a calculator.

2. $\ln e^3$

3. $\ln \sqrt{e}$

4. $e^{\ln 5}$

5. $e^{2 \ln 3}$

6. $e^{3 \ln 2 - 2 \ln 5}$

7. $\ln \dfrac{e^3 \sqrt{e}}{e^{1/3}}$

In Problems 8 through 19, solve the given equation for x.

8. $2 = e^{0.06x}$

9. $\dfrac{Q_0}{2} = Q_0 e^{-1.2x}$

10. $3 = 2 + 5e^{-4x}$

11. $-2 \ln x = b$

12. $-\ln x = \dfrac{t}{50} + C$

13. $5 = 3 \ln x - \tfrac{1}{2} \ln x$

14. $\ln x = \tfrac{1}{3}(\ln 16 + 2 \ln 2)$

15. $\ln x = 2(\ln 3 - \ln 5)$

16. $3^x = e^2$

17. $a^k = e^{kx}$

18. $a^{x+1} = b$

19. $x^{\ln x} = e$

20. Find $\ln \dfrac{1}{\sqrt{ab^3}}$ if $\ln a = 2$ and $\ln b = 3$.

21. Find $\frac{1}{a} \ln \left(\frac{\sqrt{b}}{c}\right)^a$ if $\ln b = 6$ and $\ln c = -2$.

Compound interest 22. How quickly will money double if it is invested at an annual interest rate of 6 percent compounded continuously?

Compound interest 23. Money deposited in a certain bank doubles every 13 years. The bank compounds interest continuously. What annual interest rate does the bank offer?

Compound interest 24. How quickly will money double if it is invested at an annual interest rate of 7 percent and interest is compounded
(a) continuously   (b) annually   (c) quarterly

Population growth 25. Based on the estimate that there are 10 billion acres of arable land on the earth and that each acre can produce enough food to feed 4 people, some demographers believe that the earth can support a population of no more than 40 billion people. The population of the earth was approximately 3 billion in 1960 and 4 billion in 1975. If the population of the earth were growing exponentially, when would the population reach the theoretical limit of 40 billion?

The **half-life** of a radioactive substance is the time it takes for 50 percent of a sample of the substance to decay. Problems 26 through 29 deal with this concept.

Radioactive decay 26. The amount of a certain radioactive substance remaining after t years is given by a function of the form $Q(t) = Q_0 e^{-0.003t}$. Find the half-life of the substance.

Radioactive decay 27. Radium decays exponentially. Its half-life is 1,690 years. How long will it take for a 50-gram sample of radium to be reduced to 5 grams?

Radioactive decay 28. A radioactive substance decays exponentially. Show that the amount $Q(t)$ of the substance remaining after t years is $Q(t) = Q_0 e^{-(\ln 2/\lambda)t}$, where Q_0 is the amount present initially and λ is the half-life of the substance.

Radioactive decay 29. A radioactive substance decays exponentially. Show that the amount $Q(t)$ of the substance remaining after t years is $Q(t) = Q_0(\frac{1}{2})^{t/\lambda}$, where Q_0 is the amount present initially and λ is the half-life of the substance. (*Hint:* Use the result of Problem 28.)

Advertising 30. The mathematics editor at a major publishing house estimates that if x thousand complimentary copies are distributed to instructors, the first-year sales of a new mathematics text will be approximately $f(x) = 20 - 15e^{-0.2x}$ thousand copies. According to this estimate, approximately how many complimentary copies should the editor send out to generate first-year sales of 12,000 copies?

Growth of bacteria 31. A medical student studying the growth of bacteria in a certain culture has compiled the following data.

Number of minutes	0	20
Number of bacteria	6,000	9,000

Use these data to find an exponential function of the form $Q(t) = Q_0 e^{kt}$ expressing the number of bacteria in the culture as a function of time.

Gross national product 32. An economist has compiled the following data on the gross national product (GNP) of a certain country.

Year	1965	1975
GNP (in billions)	100	180

Use these data to predict the GNP in 1985 if the GNP is growing
(a) linearly   (b) exponentially

Worker efficiency 33. An efficiency expert hired by a manufacturing firm has compiled the following data relating workers' output to their experience.

Experience (months)	0	6
Output (units per hour)	300	410

The expert believes that the output Q is related to experience t by a function of the form $Q(t) = 500 - Ae^{-kt}$. Find the function of this form that fits the data.

Population growth 34. According to a logistic model based on the assumption that the earth can support no more than 40 billion people, the world's population (in billions) t years after 1960 is given by a function of the form $P(t) = \dfrac{40}{1 + Ce^{-kt}}$, where C and k are positive constants. Find the function of this form that is consistent with the facts that the world's population was approximately 3 billion in 1960 and 4 billion in 1975.

Carbon dating 35. An archaeologist has found a fossil in which the ratio of ^{14}C to ^{12}C is $\frac{1}{3}$ the ratio found in the atmosphere. Approximately how old is the fossil?

Carbon dating 36. How old is a fossil in which the ratio of ^{14}C to ^{12}C is $\frac{1}{2}$ the ratio found in the atmosphere?

37. Use one of the laws of exponents to prove that $\ln \dfrac{u}{v} = \ln u - \ln v$.

38. Use one of the laws of exponents to prove that $\ln u^v = v \ln u$.

39. Express $\log_a x$ in terms of the natural logarithm.

4 DIFFERENTIATION OF LOGARITHMIC AND EXPONENTIAL FUNCTIONS

Both the logarithmic function $y = \ln x$ and the exponential function $y = e^x$ have simple derivatives.

Logarithmic functions

The derivative of $\ln x$

The derivative of $\ln x$ is simply $\dfrac{1}{x}$.

$$\frac{d}{dx}(\ln x) = \frac{1}{x}$$

You will see an outline of the proof of this formula a little later in this section. First, here are some examples illustrating its use.

EXAMPLE 4.1

Differentiate the function $f(x) = x \ln x$.

SOLUTION

Combine the product rule with the formula for the derivative of $\ln x$ to get

$$f'(x) = x\left(\frac{1}{x}\right) + \ln x = 1 + \ln x$$

EXAMPLE 4.2

Differentiate the function $y = \ln(2x^3 + 1)$.

SOLUTION

Let $u = 2x^3 + 1$ and $y = \ln u$ and apply the chain rule to get

$$\frac{dy}{dx} = \frac{dy}{du}\frac{du}{dx} = \frac{1}{u}(6x^2) = \frac{6x^2}{2x^3 + 1}$$

The preceding example illustrates the following general rule for differentiating the logarithm of a differentiable function.

The chain rule for logarithmic functions

$$\frac{d}{dx}[\ln h(x)] = \frac{h'(x)}{h(x)}$$

That is, to differentiate $\ln h$, simply divide the derivative of h by h itself.

Here is another example.

EXAMPLE 4.3

Differentiate the function $f(x) = \ln (x^2 + 1)^3$.

SOLUTION

As a preliminary step, simplify $f(x)$ using a property of logarithms to get

$$f(x) = 3 \ln (x^2 + 1)$$

Now apply the chain rule for logarithms and conclude that

$$f'(x) = 3 \frac{2x}{x^2 + 1} = \frac{6x}{x^2 + 1}$$

Convince yourself that the final answer would have been the same if you had not made the initial simplification of $f(x)$.

The proof that
$$\frac{d}{dx}(\ln x) = \frac{1}{x}$$

The proof that $\frac{d}{dx}(\ln x) = \frac{1}{x}$ is based on the fact that $\left(1 + \frac{1}{n}\right)^n$ approaches e as n increases (or decreases) without bound. To derive the formula for the derivative of $f(x) = \ln x$, form the difference quotient $\frac{f(x + \Delta x) - f(x)}{\Delta x}$ and rewrite it using the properties of logarithms as follows.

$$\frac{f(x + \Delta x) - f(x)}{\Delta x} = \frac{\ln (x + \Delta x) - \ln x}{\Delta x}$$

$$= \frac{1}{\Delta x} \ln \left(\frac{x + \Delta x}{x}\right)$$

$$= \ln \left(\frac{x + \Delta x}{x}\right)^{1/\Delta x}$$

$$= \ln \left(1 + \frac{\Delta x}{x}\right)^{1/\Delta x}$$

To find the derivative of $\ln x$, let Δx approach zero in the simplified form of the difference quotient. This will be easier to do if you first let $n = \frac{x}{\Delta x}$. Then,

$$\frac{\Delta x}{x} = \frac{1}{n} \quad \text{and} \quad \frac{1}{\Delta x} = \frac{n}{x}$$

and so $\quad \frac{f(x + \Delta x) - f(x)}{\Delta x} = \ln \left(1 + \frac{1}{n}\right)^{n/x} = \ln \left[\left(1 + \frac{1}{n}\right)^n\right]^{1/x}$

As Δx approaches zero, $n = \dfrac{x}{\Delta x}$ increases or decreases without bound, depending on the sign of Δx. Since

$$\left(1 + \frac{1}{n}\right)^n \to e$$

as n increases or decreases without bound, it follows that

$$\ln\left[\left(1 + \frac{1}{n}\right)^n\right]^{1/x} \to \ln e^{1/x} = \frac{1}{x}$$

as n increases or decreases without bound. Hence,

$$\frac{d}{dx}(\ln x) = \frac{1}{x}$$

and the proof is complete.

Exponential functions To find the formula for the derivative of e^x, differentiate both sides of the equation

$$\ln e^x = x$$

with respect to x, using the chain rule for logarithms to differentiate $\ln e^x$. You get

$$\frac{d/dx\,(e^x)}{e^x} = 1 \qquad \text{or} \qquad \frac{d}{dx}(e^x) = e^x$$

That is, e^x is its own derivative!

The derivative of e^x

$$\frac{d}{dx}(e^x) = e^x$$

You should have no trouble combining the formula for the derivative of e^x with the chain rule to get the following formula for the derivative of $e^{h(x)}$ where h is a differentiable function of x.

Chain rule for exponential functions

$$\frac{d}{dx}[e^{h(x)}] = h'(x)\, e^{h(x)}$$

That is, to compute the derivative of $e^{h(x)}$, simply multiply $e^{h(x)}$ by the derivative of the exponent $h(x)$.

Here are two examples that illustrate the use of these rules.

EXAMPLE 4.4

Differentiate the function $f(x) = e^{x^2+1}$.

SOLUTION

By the chain rule,

$$f'(x) = \left[\frac{d}{dx}(x^2 + 1)\right] e^{x^2+1} = 2xe^{x^2+1}$$

EXAMPLE 4.5

Differentiate the function $f(x) = xe^{2x}$.

SOLUTION

By the product rule,

$$f'(x) = x\frac{d}{dx}(e^{2x}) + e^{2x}\frac{d}{dx}(x)$$

$$= 2xe^{2x} + e^{2x} = (2x + 1)e^{2x}$$

Exponential growth In Section 2 you saw that a quantity $Q(t)$ that increases according to a law of the form $Q(t) = Q_0e^{kt}$, where Q_0 and k are positive constants, is said to experience exponential growth. The next example shows that if a quantity grows exponentially, its rate of growth is proportional to its size. The converse of this fact will be established in Chapter 5.

EXAMPLE 4.6

If $Q(t) = Q_0e^{kt}$, find expressions for the (instantaneous) rate of change of Q with respect to t and the percentage rate of change of Q with respect to t.

SOLUTION

The (instantaneous) rate of change of Q with respect to t is the derivative

$$Q'(t) = kQ_0e^{kt} = kQ(t)$$

This says that the rate of change, $Q'(t)$, is proportional to $Q(t)$ itself and that the constant k that appears in the exponent of $Q(t)$ is the constant of proportionality.

The percentage rate of change of Q with respect to t is

$$100\frac{Q'(t)}{Q(t)} = 100\frac{kQ(t)}{Q(t)} = 100k$$

This says that the constant k that appears in the exponent of the function $Q(t) = Q_0e^{kt}$ is the percentage rate of change of Q expressed as a decimal. This is consistent with what you already know about compound interest; namely that if interest is compounded continuously, the balance after t years is $B(t) = Pe^{rt}$, where r is the interest rate expressed as a decimal.

Optimization   Now that you know how to differentiate exponential functions, you can use calculus to find maximum and minimum values of functions involving powers of e. Here is an example.

EXAMPLE 4.7

The consumer demand for a certain commodity is $D(p) = 5{,}000e^{-0.02p}$ units per month when the market price is p dollars per unit. Determine the market price that will result in the greatest consumer expenditure.

SOLUTION

The consumer expenditure for the commodity is the price per unit times the number of units sold. That is,

$$E(p) = pD(p) = 5{,}000pe^{-0.02p}$$

Since only nonnegative values of p are meaningful in this context, the goal is to maximize the function $E(p)$ for $p \geq 0$.

The derivative of E is

$$E'(p) = 5{,}000[-0.02pe^{-0.02p} + e^{-0.02p}]$$
$$= 5{,}000e^{-0.02p}(1 - 0.02p)$$

Since $e^{-0.02p}$ is never zero, $E'(p) = 0$ if and only if

$$1 - 0.02p = 0 \qquad \text{or} \qquad p = \tfrac{1}{0.02} = 50$$

To verify that $p = 50$ is really the optimal price you are seeking, check the sign of $E'(p)$ for $0 < p < 50$ and for $p > 50$. Since $5{,}000e^{-0.02p}$ is always positive, it follows that $E'(p)$ is positive if $0 < p < 50$ and negative if $p > 50$. Hence E is increasing for $0 < p < 50$ and decreasing for $p > 50$ as shown in Figure 4.1, and you can conclude that consumer expenditure will be greatest when the market price is $50 per unit.

Curve sketching   In the next example, calculus is used to obtain a detailed sketch of the function $f(x) = \dfrac{1}{\sqrt{2\pi}}\, e^{-x^2/2}$. This function is known as the **standard**

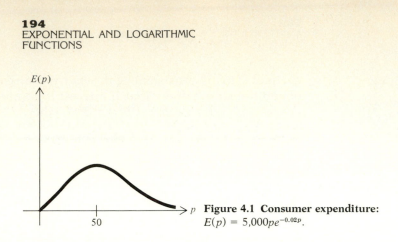

Figure 4.1 Consumer expenditure:
$E(p) = 5{,}000pe^{-0.02p}$.

normal probability density function and is one of the most important functions in probability and statistics. As you shall see, its graph is the famous "bell-shaped curve" that is used by social scientists to describe the distributions of many quantities including people's heights and IQ scores.

EXAMPLE 4.8

Determine where the function

$$f(x) = \frac{1}{\sqrt{2\pi}} e^{-x^2/2}$$

is increasing, decreasing, concave upward, and concave downward. Find the relative extrema and inflection points and draw the graph.

SOLUTION

The first derivative is

$$f'(x) = \frac{-x}{\sqrt{2\pi}} e^{-x^2/2}$$

Since $e^{-x^2/2}$ is always positive, $f'(x)$ is zero if and only if $x = 0$. Hence the corresponding point $\left(0, \dfrac{1}{\sqrt{2\pi}}\right)$ is the only first-order critical point.

Using the product rule, you find that the second derivative is

$$f''(x) = \frac{x^2}{\sqrt{2\pi}} e^{-x^2/2} - \frac{1}{\sqrt{2\pi}} e^{-x^2/2} = \frac{1}{\sqrt{2\pi}} (x^2 - 1)e^{-x^2/2}$$

which is zero if $x = \pm 1$. Hence the corresponding points

$$\left(1, \frac{e^{-1/2}}{\sqrt{2\pi}}\right) \quad \text{and} \quad \left(-1, \frac{e^{-1/2}}{\sqrt{2\pi}}\right)$$

are the second-order critical points.

Use your calculator to get

$$\frac{1}{\sqrt{2\pi}} = 0.40 \qquad \text{and} \qquad \frac{e^{-1/2}}{\sqrt{2\pi}} = 0.24$$

and plot the critical points on the graph.

Now check the signs of the first and second derivatives on each of the intervals defined by the x coordinates of the critical points.

Interval	$f'(x)$	$f''(x)$	Increasing or decreasing	Concavity	Shape
$x < -1$	+	+	increasing	up	⌣
$-1 < x < 0$	+	−	increasing	down	⌢
$0 < x < 1$	−	−	decreasing	down	⌢
$x > 1$	−	+	decreasing	up	⌣

Complete the graph as shown in Figure 4.2 by connecting the critical points with a curve of appropriate shape on each interval. Notice that the graph gets closer and closer to the x axis as $|x|$ increases without bound, reflecting the fact that $e^{-x^2/2}$ approaches zero as $|x|$ increases without bound.

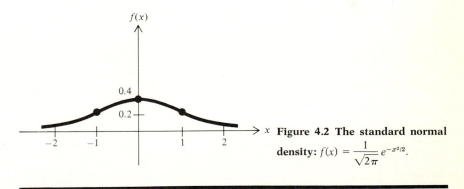

Figure 4.2 The standard normal density: $f(x) = \dfrac{1}{\sqrt{2\pi}} e^{-x^2/2}$.

Logarithmic differentiation Sometimes you can simplify the work involved in differentiating a function if you first take its logarithm. This technique, called **logarithmic differentiation,** is illustrated in the following example.

EXAMPLE 4.9

Differentiate the function $f(x) = \dfrac{\sqrt[3]{x + 1}}{x - 1}$.

SOLUTION

You could do this problem using the quotient rule, but the resulting computation would be somewhat tedious. (Try it.)

A more efficient approach is to work with the natural logarithm of f:

$$\ln f(x) = \tfrac{1}{3} \ln (x + 1) - \ln (x - 1)$$

(Notice that by introducing the logarithm, you eliminate the quotient and the cube root.)

Now use the chain rule for logarithms to differentiate both sides of this equation to get

$$\frac{f'(x)}{f(x)} = \frac{1}{3(x + 1)} - \frac{1}{x - 1} = \frac{-2(x + 2)}{3(x + 1)(x - 1)}$$

and solve for $f'(x)$ by multiplying both sides of this equation by $f(x) = \dfrac{\sqrt[3]{x + 1}}{x - 1}$ to get

$$f'(x) = \frac{-2(x + 2)}{3(x + 1)(x - 1)} \frac{\sqrt[3]{x + 1}}{x - 1} = \frac{-2(x + 2)}{3(x + 1)^{2/3}(x - 1)^2}$$

Problems In Problems 1 through 24, differentiate the given function.

1. $f(x) = e^{5x}$

2. $f(x) = 3e^{4x+1}$

3. $f(x) = e^{x^2+2x-1}$

4. $f(x) = e^{1/x}$

5. $f(x) = 30 + 10e^{-0.05x}$

6. $f(x) = x^2 e^x$

7. $f(x) = (x^2 + 3x + 5)e^{6x}$

8. $f(x) = xe^{-x^2}$

9. $f(x) = \dfrac{x}{e^x}$

10. $f(x) = \dfrac{x^2}{e^x}$

11. $f(x) = (1 - 3e^x)^2$

12. $f(x) = \sqrt{1 + e^x}$

13. $f(x) = \ln x^3$

14. $f(x) = \ln 2x$

15. $f(x) = \ln (x^2 + 5x - 2)$

16. $f(x) = \ln \sqrt{x^2 + 1}$

17. $f(x) = x^2 \ln x$

18. $f(x) = x \ln \sqrt{x}$

19. $f(x) = \dfrac{\ln x}{x}$

20. $f(x) = \dfrac{x}{\ln x}$

21. $f(x) = \ln \left(\dfrac{x + 1}{x - 1}\right)$

22. $f(x) = e^x \ln x$

23. $f(x) = \ln e^{2x}$

24. $f(x) = e^{\ln x + 2 \ln 3x}$

Population growth 25. It is projected that t years from now, the population of a certain country will be $P(t) = 50e^{0.02t}$ million.

    (a) At what rate will the population be changing 10 years from now?

    (b) At what percentage rate will the population be changing t years from now? Does this percentage rate depend on t or is it constant?

Compound interest 26. Money is deposited in a bank offering interest at an annual rate of 6 percent compounded continuously. Find the percentage rate of change of the balance with respect to time.

Depreciation 27. A certain industrial machine depreciates so that its value after t years is $Q(t) = 20{,}000e^{-0.4t}$ dollars.

    (a) At what rate is the machine depreciating after 5 years?

    (b) At what percentage rate is the value of the machine changing after t years? Does this percentage rate depend on t or is it constant?

Exponential decay 28. Show that if a quantity decreases exponentially, its rate of decrease is proportional to its size.

Exponential decay 29. Show that if a quantity decreases exponentially, its percentage rate of change is constant.

The spread of an epidemic 30. Public health records indicate that t weeks after the outbreak of a rare form of influenza, approximately $Q(t) = \dfrac{80}{4 + 76e^{-1.2t}}$ thousand people had caught the disease. At what rate was the disease spreading at the end of the 2nd week?

Population growth 31. According to a logistic model based on the assumption that the earth can support no more than 40 billion people, the world's population (in billions) t years after 1960 will be approximately

$$P(t) = \dfrac{40}{1 + 12e^{-0.08t}}$$

    (a) If this model is correct, at what rate will the world's population be increasing in 1985?

    (b) If this model is correct, at what percentage rate will the world's population be increasing in 1985?

Newton's law of heating **32.** A cool drink is removed from a refrigerator on a hot summer day and placed in a room whose temperature is 30° Celsius. According to a law of physics, the temperature of the drink t minutes later is given by a function of the form $f(t) = 30 - Ae^{-kt}$. Show that the rate of change of the temperature of the object with respect to time is proportional to the difference between the temperature of the room and that of the object.

Marginal analysis **33.** The mathematics editor at a major publishing house estimates that if x thousand complimentary copies are distributed to professors, the first-year sales of a certain new text will be $f(x) = 20 - 15e^{-0.2x}$ thousand copies. Currently, the editor is planning to distribute 10,000 complimentary copies.

(a) Use marginal analysis to estimate the increase in first-year sales that will result if 1,000 additional complimentary copies are distributed.

(b) Calculate the actual increase in first-year sales that will result from the distribution of the additional 1,000 complimentary copies. Is the estimate in part (a) a good one?

Consumer expenditure **34.** The consumer demand for a certain commodity is $D(p) = 3{,}000e^{-0.01p}$ units per month when the market price is p dollars per unit. Express consumers' total monthly expenditure for the commodity as a function of p, and determine the market price that will result in the greatest consumer expenditure.

Profit maximization **35.** A manufacturer can produce radios at a cost of $5 apiece and estimates that if they are sold for x dollars apiece, consumers will buy approximately $1{,}000e^{-0.1x}$ radios per week. At what price should the manufacturer sell the radios to maximize profit?

In Problems 36 through 43, determine where the given function is increasing, decreasing, concave upward, and concave downward. Find the relative extrema and inflection points and draw the graph.

36. $f(x) = xe^x$ **37.** $f(x) = xe^{-x}$

38. $f(x) = xe^{2-x}$ **39.** $f(x) = e^{-x^2}$

40. $f(x) = x^2e^{-x}$ **41.** $f(x) = e^x + e^{-x}$

42. $f(x) = \dfrac{6}{1 + e^{-x}}$ **43.** $f(x) = x - \ln x$ (for $x > 0$)

In Problems 44 through 49, use logarithmic differentiation to find $f'(x)$.

44. $f(x) = \sqrt{\dfrac{x + 1}{x - 1}}$ **45.** $f(x) = \dfrac{(x + 2)^5}{(3x - 5)^6}$

46. $f(x) = (x + 1)^3(x - 6)^2(2x + 1)$ 47. $f(x) = 2^x$

48. $f(x) = a^x$ (for $a > 0$)   49. $f(x) = x^x$ (for $x > 0$)

The power rule
50. Use logarithmic differentiation to prove the power rule:
$$\frac{d}{dx}(x^n) = nx^{n-1}.$$

Population growth
51. It is estimated that t years from now, the population of a certain country will be $P(t) = \dfrac{160}{1 + 8e^{-0.01t}}$ million. When will the population be growing most rapidly?

The spread of an epidemic
52. An epidemic spreads through a community so that t weeks after its outbreak, the number of residents who have been infected is given by a function of the form $f(t) = \dfrac{A}{1 + Ce^{-kt}}$, where A is the total number of susceptible residents. Show that the epidemic is spreading most rapidly when half of the susceptible residents have been infected.

Percentage rate of change
53. Show that the percentage rate of change of f with respect to x is
$$100 \frac{d}{dx}[\ln f(x)].$$

In Problems 54 and 55, use the formula from Problem 53 to find the specified percentage rate of change.

Exponential growth
54. A quantity grows exponentially according to the law $Q(t) = Q_0e^{kt}$. Find the percentage rate of change of Q with respect to t.

Population growth
55. It is projected that x years from now the population of a certain town will be approximately $P(x) = 5,000\sqrt{x^2 + 4x + 19}$. At what percentage rate will the population be changing 3 years from now?

5 COMPOUND INTEREST
In this section you will learn more about the concept of compound interest, which was introduced in Section 1. To begin the discussion, here is a summary of what you should already know.

When money is invested it (usually) earns **interest.** The amount of money invested is called the **principal.** Interest that is computed on the principal alone is called **simple interest,** and the balance after t years is given by the following formula.

Simple interest
> **If P dollars is invested at an annual simple interest rate r (expressed as a decimal), the balance $B(t)$ after t years will be**
> $$B(t) = P(1 + rt) \quad \text{dollars}$$

Interest that is computed on the principal plus the previous interest is called **compound interest.** If the annual interest rate is r and interest is compounded k times per year, the year is divided into k equal **interest periods** and the interest rate during each is $\frac{r}{k}$. The balance at the end of t years is given by the following formula.

Compound interest

> **If P dollars is invested at an annual interest rate r and interest is compounded k times per year, the balance $B(t)$ after t years will be**
>
> $$B(t) = P \left(1 + \frac{r}{k}\right)^{kt} \quad \text{dollars}$$

If the number of times interest is compounded is allowed to increase without bound, the interest is said to be **compounded continuously.** The balance at the end of t years is given by the following formula.

Continuously compounded interest

> **If P dollars is invested at an annual interest rate r and interest is compounded continuously, the balance $B(t)$ after t years will be**
>
> $$B(t) = Pe^{rt} \quad \text{dollars}$$

The use of these formulas is illustrated in the next example.

EXAMPLE 5.1

Suppose \$1,000 is invested at an annual interest rate of 8 percent. Compute the balance after 10 years in each of the following cases.

(a) The interest is simple interest.
(b) The interest is compounded quarterly.
(c) The interest is compounded continuously.

SOLUTION

(a) Use the formula $B(t) = P(1 + rt)$ with $P = 1,000$, $r = 0.08$, and $t = 10$ to get
$$B(10) = 1,000(1.8) = \$1,800$$

(b) Use the formula $B(t) = P\left(1 + \frac{r}{k}\right)^{kt}$ with $P = 1,000$, $r = 0.08$, $k = 4$, and $t = 10$ to get
$$B(10) = 1,000(1.02)^{40} = \$2,208.04$$

(c) Use the formula $B(t) = Pe^{rt}$ with $P = 1,000$, $r = 0.08$, and $t = 10$ to get
$$B(10) = 1,000e^{0.8} = \$2,225.54$$

Doubling time Investors often want to know how long it will take for a given investment to grow to a particular size. In the next example, you will see
how to calculate the time it takes for money earning compound interest to double.

EXAMPLE 5.2

Derive a formula for the amount of time it takes for money to double
if the annual interest rate is r and interest is compounded k times per
year.

SOLUTION

If P is the principal, the balance after t years is $B(t) = P\left(1 + \dfrac{r}{k}\right)^{kt}$.

Your goal is to find the value of t for which $B(t) = 2P$. That is, you
want to solve the equation

$$2P = P\left(1 + \frac{r}{k}\right)^{kt}$$

for t. To do this, divide both sides by P to get

$$2 = \left(1 + \frac{r}{k}\right)^{kt}$$

and then take the natural logarithm of each side to get

$$\ln 2 = kt \ln\left(1 + \frac{r}{k}\right)$$

or
$$t = \frac{\ln 2}{k \ln (1 + r/k)}$$

Using a similar argument you can derive a formula for the
doubling time when interest is compounded continuously. This derivation is left as a problem for you to do. Here is a summary of the two
results.

Doubling time **If money is invested at an annual interest rate r and interest is
compounded k times per year,**

$$\text{Doubling time} = \frac{\ln 2}{k \ln (1 + r/k)}$$

**If money is invested at an annual interest rate r and interest is
compounded continuously,**

$$\text{Doubling time} = \frac{\ln 2}{r}$$

In the next example, the doubling times for two investments are compared using these formulas.

EXAMPLE 5.3

How quickly will money double if it is invested at an annual interest rate of 6 percent and interest is compounded

(a) semiannually   (b) continuously

SOLUTION

(a) Use the first formula with $k = 2$ and $r = 0.06$ to get

$$\text{Doubling time} = \frac{\ln 2}{2 \ln (1.03)} = 11.72 \quad \text{years}$$

(b) Use the second formula with $r = 0.06$ to get

$$\text{Doubling time} = \frac{\ln 2}{0.06} = 11.55 \quad \text{years}$$

Effective interest rate

Money deposited in a bank offering interest at an annual rate of 6 percent will increase in value by more than 6 percent in a year if the interest is compounded more than once. This is because interest compounded during one period will itself earn interest during subsequent periods. The actual percentage by which an investment grows during a year is called the **effective interest rate,** while the corresponding annual compound interest rate is called the **nominal interest rate.** In other words, the effective interest rate is the *simple* interest rate that is equivalent to the nominal compound interest rate. In the next example, you will see how to derive a formula for the effective interest rate.

EXAMPLE 5.4

Find a formula for the effective interest rate if the nominal rate is r and interest is compounded k times per year.

SOLUTION

If P dollars is invested at a nominal rate r and interest is compounded k times per year, the balance at the end of 1 year will be

$$B = P \left(1 + \frac{r}{k}\right)^k$$

On the other hand, if x is the simple interest rate, the corresponding balance at the end of 1 year will be

$$B = P(1 + x)$$

Your goal is to find the value of x for which these two expressions for the balance are equal. To do this, equate the expressions and solve for x as follows.

$$P \left(1 + \frac{r}{k}\right)^k = P(1 + x)$$

$$\left(1 + \frac{r}{k}\right)^k = 1 + x$$

$$x = \left(1 + \frac{r}{k}\right)^k - 1$$

Using a similar argument you can derive a formula for the effective interest rate when interest is compounded continuously. This derivation is left as a problem for you to do. Here is a summary of the two results.

Effective interest rate

If interest is compounded k times per year at an annual interest rate r, then

$$\text{Effective interest rate} = \left(1 + \frac{r}{k}\right)^k - 1$$

If interest is compounded continuously at an annual interest rate r, then

$$\text{Effective interest rate} = e^r - 1$$

The next example illustrates how you can use the effective interest rates to compare two investments.

EXAMPLE 5.5

One bank offers interest at an annual rate of 6.1 percent compounded quarterly, and a competing bank offers interest at an annual rate of 6 percent compounded continuously. Which bank should you choose?

SOLUTION

Since the effective interest rate at the first bank is

$$\left(1 + \frac{0.061}{4}\right)^4 - 1 = 0.0624$$

or 6.24 percent, and the effective interest rate at the second bank is only

$$e^{0.06} - 1 = 0.0618$$

or 6.18 percent, you should choose the first bank.

Present value

The **present value of B dollars payable t years from now** is the amount P that should be invested today so that it will be worth B dollars at the end of t years. To derive a formula for the present value, simply solve the compound interest formula for the principal P. For example, if interest is compounded k times per year, the balance after t years is

$$B = P \left(1 + \frac{r}{k}\right)^{kt}$$

and the present value of B dollars payable t years from now is

$$P = B \left(1 + \frac{r}{k}\right)^{-kt}$$

If interest is compounded continuously, the balance is

$$B = Pe^{rt}$$

and the present value is $P = Be^{-rt}$

Present value of future money

If interest is compounded k times per year at an annual interest rate r, the present value of B dollars payable t years from now is

$$P = B \left(1 + \frac{r}{k}\right)^{-kt} \qquad \text{dollars}$$

If interest is compounded continuously at an annual interest rate r, the present value of B dollars payable t years from now is

$$P = Be^{-rt} \qquad \text{dollars}$$

The concept of present value is illustrated in the next example.

EXAMPLE 5.6

How much should you invest now at an annual interest rate of 8 percent so that your balance 20 years from now will be $10,000 if the interest is compounded

(a) quarterly   (b) continuously

SOLUTION

(a) The amount you should invest is the present value of $10,000 payable 20 years from now. Using the formula $P = B\left(1 + \dfrac{r}{k}\right)^{-kt}$ with $B = 10,000$, $r = 0.08$, $k = 4$, and $t = 20$, you find that this amount is

$$P = 10,000(1.02)^{-80} = \$2,051.10$$

(b) Using the present-value formula $P = Be^{-rt}$ with $B = 10,000$, $r = 0.08$, and $t = 20$, you find that the amount you should invest is

$$P = 10,000e^{-1.6} = \$2,018.97$$

Optimal holding time   Suppose you own an asset whose value increases with time. The longer you hold the asset, the more it will be worth. However, if there comes a time after which money invested at the prevailing interest rate grows more quickly than the asset, you will be better off selling the asset and investing the proceeds. You determine the most profitable time to sell the asset and invest the proceeds by comparing the prevailing interest rate with the percentage rate of growth of the asset. The technique is illustrated in the next example.

EXAMPLE 5.7

Suppose you own a parcel of land whose market price t years from now will be $V(t) = 20,000e^{\sqrt{t}}$ dollars. If the prevailing interest rate remains constant at 7 percent compounded continuously, when will it be most profitable to sell the land?

SOLUTION

The percentage rate of change of the market price of the land (expressed in decimal form) is

$$\frac{V'(t)}{V(t)} = \frac{10,000t^{-1/2}e^{\sqrt{t}}}{20,000e^{\sqrt{t}}} = \frac{1}{2\sqrt{t}}$$

This will be equal to the prevailing interest rate when

$$\frac{1}{2\sqrt{t}} = 0.07 \quad \text{or} \quad t = \left[\frac{1}{2(0.07)}\right]^2 = 51.02$$

Moreover,

$$\frac{1}{2\sqrt{t}} > 0.07 \text{ when } 0 < t < 51.02$$

and

$$\frac{1}{2\sqrt{t}} < 0.07 \text{ when } t > 51.02$$

which says that the percentage rate of growth of the value of the land is greater than the prevailing interest rate when $0 < t < 51.02$ and less than the prevailing interest rate when $t > 51.02$. It follows that you should sell the land 51.02 years from now.

Economists have a different way of solving problems like the one in Example 5.7. They determine the optimal time to sell an asset by finding the time at which the *present value* of the selling price of the asset is greatest. That is, they maximize today's dollar-equivalent of the selling price. Here is Example 5.7 solved as an economist would do it. Notice that the final answer is exactly the same as the one you obtained before.

EXAMPLE 5.8

Suppose you own a parcel of land whose market price t years from now will be $V(t) = 20{,}000e^{\sqrt{t}}$ dollars. If the prevailing interest rate remains constant at 7 percent compounded continuously, when will the present value of the market price of the land be greatest?

SOLUTION

In t years, the market price of the land will be $V(t) = 20{,}000e^{\sqrt{t}}$. The present value of this amount is

$$P(t) = V(t)e^{-0.07t} = 20{,}000e^{-0.07t+\sqrt{t}}$$

The goal is to maximize $P(t)$ for $t \geq 0$. The derivative of P is

$$P'(t) = 20{,}000e^{-0.07t+\sqrt{t}}\left(-0.07 + \frac{1}{2\sqrt{t}}\right)$$

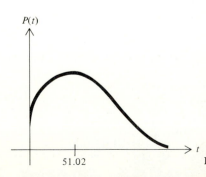

Figure 5.1 Present value: $P(t) = 20{,}000e^{-0.07t+\sqrt{t}}$.

which is zero if and only if

$$-0.07 + \frac{1}{2\sqrt{t}} = 0 \quad \text{or} \quad t = \left[\frac{1}{2(0.07)}\right]^2 = 51.02$$

(and which is undefined when $t = 0$).

Since $P'(t)$ is positive if $0 < t < 51.02$ and negative if $t > 51.02$, it follows that the graph of P is increasing for $0 < t < 51.02$ and decreasing for $t > 51.02$ as shown in Figure 5.1 and that $t = 51.02$ is the time at which the present value will be greatest.

Problems

1. Suppose \$5,000 is invested at an annual interest rate of 5 percent. Compute the balance after 20 years in each of the following cases.
   (a) The interest is simple interest.
   (b) The interest is compounded semiannually.
   (c) The interest is compounded continuously.

2. Suppose \$5,000 is invested at an annual interest rate of 6 percent. Compute the balance after 20 years in each of the following cases.
   (a) The interest is simple interest.
   (b) The interest is compounded semiannually.
   (c) The interest is compounded continuously.

3. How quickly will money double if it is invested at an annual interest rate of 8 percent and interest is compounded
   (a) quarterly   (b) continuously

4. How quickly will money double if it is invested at an annual simple interest rate of 8 percent?

5. Derive a formula for the amount of time it takes for money to double if it earns simple interest at an annual rate r.

6. Derive the formula for the amount of time it takes for money to double if the annual interest rate is r and interest is compounded continuously.

7. How quickly will money triple if it is invested at an annual interest rate of 6 percent compounded semiannually?

8. How quickly will money triple if it is invested at an annual interest rate of 6 percent compounded continuously?

9. Derive a formula for the amount of time it takes for money to triple if the annual interest rate is r and interest is compounded k times per year.

10. Derive a formula for the amount of time it takes for money to

triple if the annual interest rate is r and interest is compounded continuously.

11. Determine how quickly $1,000 will grow to $2,500 if the annual interest rate is 6 percent and interest is compounded
    (a) quarterly   (b) continuously

12. Determine how quickly $600 will grow to $1,000 if the annual interest rate is 10 percent and interest is compounded
    (a) semiannually   (b) continuously

13. Find the effective interest rate if the nominal interest rate is 6 percent per year and interest is compounded
    (a) quarterly   (b) continuously

14. Which is the better investment: 8.2 percent per year compounded quarterly or 8.1 percent per year compounded continuously?

15. Which is the better investment: 10.25 percent per year compounded semiannually or 10.20 percent per year compounded continuously?

16. Derive the formula for the effective interest rate if the nominal rate is r and interest is compounded continuously.

17. A bank compounds interest quarterly. What (nominal) interest rate does the bank offer if $1,000 grows to $2,203.76 in 8 years?

18. A bank compounds interest continuously. What (nominal) interest rate does it offer if $1,000 grows to $2,054.44 in 12 years?

19. A certain bank offers an interest rate of 6 percent a year compounded annually. A competing bank compounds its interest continuously. What (nominal) interest rate should the competing bank offer so that the effective interest rates of the two banks will be equal?

20. How much should you invest now at an annual interest rate of 6 percent so that your balance 10 years from now will be $5,000 if interest is compounded
    (a) quarterly   (b) continuously

21. How much should you invest now at an annual interest rate of 8 percent so that your balance 15 years from now will be $20,000 if interest is compounded
    (a) quarterly   (b) continuously

22. Find the present value of $8,000 payable 10 years from now if the annual interest rate is 6.25 percent and interest is compounded
    (a) semiannually   (b) continuously

Optimal holding time

23. Suppose you own a parcel of land whose value t years from now will be $V(t) = 8{,}000e^{\sqrt{t}}$ dollars. If the prevailing interest rate remains constant at 6 percent per year compounded continuously, when should you sell the land to maximize its present value?

Optimal holding time

24. Suppose your family owns a rare book whose value t years from now will be $V(t) = 200e^{\sqrt{2t}}$ dollars. If the prevailing interest rate remains constant at 6 percent per year compounded continuously, when will it be most profitable for your family to sell the book?

Optimal holding time

25. Suppose you own a stamp collection that is currently worth $1,200 and whose value increases linearly at the rate of $200 per year. If the prevailing interest rate remains constant at 8 percent per year compounded continuously, when will it be most profitable for you to sell the collection?

The amount of an annuity

26. At the end of each interest period you deposit $100 in an account that pays interest at an annual rate of 8 percent compounded quarterly. How much will you have in the account just after you make your 4th deposit? (In accounting, such a sequence of equal payments or deposits made at regular intervals is called an **annuity** and the accumulated sum after a specified period of time is known as the **amount** of the annuity.)

The amount of an annuity

27. At the end of each interest period you deposit $50 in an account that pays interest at an annual rate of 6 percent compounded semiannually. How much will you have in the account just after you make your 4th deposit?

The present value of an annuity

28. A bank offers interest at an annual rate of 5 percent compounded continuously. How much should you deposit today so that you will be able to make withdrawals of $2,000 at the end of each of the next 3 years, after which nothing will be left in the account? (*Hint:* Use the formula for present value.)

The present value of an annuity

29. How much should you invest now (at the beginning of an interest period) at an annual interest rate of 6 percent compounded annually to enable you to make withdrawals of $500 at the end of each of the next 4 years, after which nothing will be left in the account?

CHAPTER SUMMARY AND PROFICIENCY TEST

Important terms, symbols, and formulas

The number e: $\left(1 + \dfrac{1}{n}\right)^n \rightarrow e \approx 2.718$ as n increases without bound

Exponential function: $f(x) = a^x$ $(a > 0)$

Integer powers: $a^n = a \cdot a \cdots a$ (n factors)

Fractional powers: $a^{n/m} = (\sqrt[m]{a})^n$

Negative powers: $a^{-x} = \dfrac{1}{a^x}$

Zero power: $a^0 = 1$

Laws of exponents: $a^r a^s = a^{r+s}$ $\dfrac{a^r}{a^s} = a^{r-s}$ $(a^r)^s = a^{rs}$

Exponential growth: $Q(t) = Q_0 e^{kt}$

Exponential decay: $Q(t) = Q_0 e^{-kt}$

Learning curve: $Q(t) = B - Ae^{-kt}$

Logistic curve: $Q(t) = \dfrac{B}{1 + Ae^{-Bkt}}$

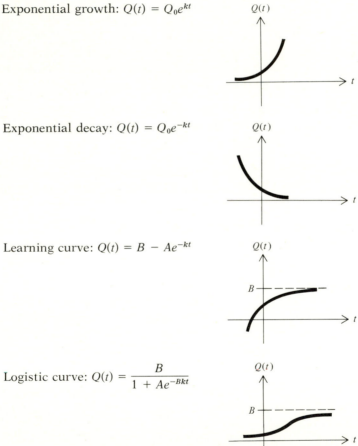

The natural logarithm: $y = \ln x$ if and only if $x = e^y$

Inverse relationship: $e^{\ln x} = x$ and $\ln e^x = x$

Properties of logarithms:

$$\ln uv = \ln u + \ln v \qquad \ln \frac{u}{v} = \ln u - \ln v \qquad \ln u^v = v \ln u$$

$$\frac{d}{dx}(\ln x) = \frac{1}{x} \quad \text{and} \quad \frac{d}{dx}[\ln h(x)] = \frac{h'(x)}{h(x)}$$

$$\frac{d}{dx}(e^x) = e^x \quad \text{and} \quad \frac{d}{dx}[e^{h(x)}] = h'(x)e^{h(x)}$$

Logarithmic differentiation
Simple interest formula: $B(t) = P(1 + rt)$
Interest compounded k times per year:

$$\text{Balance} = B(t) = P\left(1 + \frac{r}{k}\right)^{kt}$$

$$\text{Doubling time} = \frac{\ln 2}{k \ln (1 + r/k)}$$

$$\text{Effective rate} = \left(1 + \frac{r}{k}\right)^k - 1$$

$$\text{Present value} = B\left(1 + \frac{r}{k}\right)^{-kt}$$

Continuously compounded interest:

$$\text{Balance} = B(t) = Pe^{rt}$$

$$\text{Doubling time} = \frac{\ln 2}{r}$$

$$\text{Effective rate} = e^r - 1$$
$$\text{Present value} = Be^{-rt}$$

Proficiency test

1. Sketch the following functions.

   (a) $f(x) = 5e^{-x}$ (b) $f(x) = 5 - 2e^{-x}$

2. Find $f(9)$ if $f(x) = 30 + Ae^{-kx}$, $f(0) = 50$, and $f(3) = 40$.

3. The value of a certain industrial machine is decreasing exponentially. If the machine was originally worth $50,000 and was worth $20,000 when it was 5 years old, how much will it be worth when it is 10 years old?

4. It is estimated that if x thousand dollars is spent on advertising, approximately $Q(x) = 50 - 40e^{-0.1x}$ thousand units of a certain commodity will be sold.

   (a) Sketch the relevant portion of this sales function.
   (b) How many units will be sold if no money is spent on advertising?
   (c) How many units will be sold if $8,000 is spent on advertising?
   (d) How much should be spent on advertising to generate sales of 35,000 units?
   (e) According to this model, what is the most optimistic sales projection?

5. It is estimated that t years from now the population of a certain country will be $P(t) = \dfrac{120}{4 + 8e^{-0.05t}}$ million.

   (a) What is the current population?

   (b) What will the population be 20 years from now?

   (c) What will happen to the population in the long run?

6. Evaluate the following expressions without using tables or a calculator.

   (a) $\ln e^5$ (b) $e^{\ln 2}$ (c) $e^{3 \ln 4 - \ln 2}$

7. Solve for x.

   (a) $8 = 2e^{0.04x}$ 　　　　　　　　(b) $5 = 1 + 4e^{-6x}$

   (c) $4 \ln x = 8$ 　　　　　　　　　(d) $5^x = e^3$

8. The number of bacteria in a certain culture grows exponentially. If 5,000 bacteria were initially present and 8,000 were present 10 minutes later, how long will it take for the number of bacteria to double?

9. Differentiate the following functions.

   (a) $f(x) = 2e^{3x+5}$ 　　　　　　　(b) $f(x) = x^2 e^{-x}$

   (c) $f(x) = \ln \sqrt{x^2 + 4x + 1}$ 　　(d) $f(x) = x \ln x^2$

   (e) $f(x) = \dfrac{x}{\ln 2x}$

10. An environmental study of a certain suburban community suggests that t years from now, the average level of carbon monoxide in the air will be $Q(t) = 4e^{0.03t}$ parts per million.

    (a) At what rate will the carbon monoxide level be changing 2 years from now?

    (b) At what percentage rate will the carbon monoxide level be changing t years from now? Does this percentage rate depend on t or is it constant?

11. A manufacturer can produce cameras at a cost of $40 apiece and estimates that if they are sold for p dollars apiece, consumers will buy approximately $D(p) = 800e^{-0.01p}$ cameras per week. At what price should the manufacturer sell the cameras to maximize profit?

12. Determine where the given function is increasing, decreasing, concave upward, and concave downward. Find the relative extrema and inflection points and draw the graph.

    (a) $f(x) = xe^{-2x}$ 　　　　　　　(b) $f(x) = e^x - e^{-x}$

    (c) $f(x) = \dfrac{4}{1 + e^{-x}}$ 　　　　　(d) $f(x) = \ln (x^2 + 1)$

13. Find $f'(x)$ by logarithmic differentiation.

    (a) $f(x) = \sqrt{(x^2 + 1)(x^2 + 2)}$

    (b) $f(x) = x^{x^2}$, for $x > 0$

14. How quickly will $2,000 grow to $5,000 when invested at an annual interest rate of 8 percent if interest is compounded
    (a) quarterly   (b) continuously

15. Which is the better investment: 8.25 percent per year compounded quarterly or 8.20 percent per year compounded continuously?

16. How much should you invest now at an annual interest rate of 6.25 percent so that your balance 10 years from now will be $2,000 if interest is compounded
    (a) semiannually   (b) continuously

17. Suppose you own a coin collection whose value t years from now will be $V(t) = 2,000e^{\sqrt{t}}$ dollars. If the prevailing interest rate remains constant at 7 percent per year compounded continuously, when will it be most profitable for you to sell the collection?

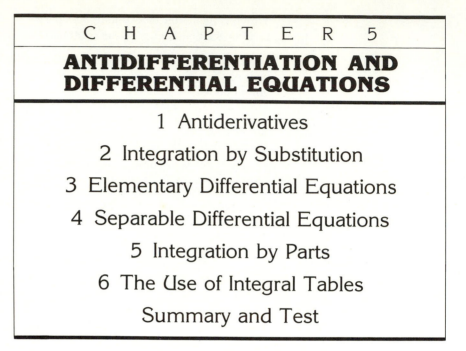

C H A P T E R 5

ANTIDIFFERENTIATION AND DIFFERENTIAL EQUATIONS

1 Antiderivatives

2 Integration by Substitution

3 Elementary Differential Equations

4 Separable Differential Equations

5 Integration by Parts

6 The Use of Integral Tables

Summary and Test

1 ANTIDERIVATIVES

In many problems, the derivative of a function is known and the goal is to find the function itself. For example: a sociologist who knows the rate at which the population is growing may wish to use this information to predict future population levels; a physicist who knows the speed of a moving body may wish to calculate the future position of the body; an economist who knows the rate of inflation may wish to estimate future prices.

The process of obtaining a function from its derivative is called **antidifferentiation** or **integration.**

Antiderivative
> **A function F whose derivative equals f is said to be an antiderivative (or indefinite integral) of f.**

Later in this section you will learn techniques you can use to find antiderivatives. Once you have found what you believe to be an antiderivative of a function, you can always check your answer by differentiating. Here is an example.

214

EXAMPLE 1.1

Verify that $F(x) = \frac{1}{3}x^3 + 5x + 2$ is an antiderivative of $f(x) = x^2 + 5$.

SOLUTION

Differentiate F and you will find that

$$F'(x) = x^2 + 5 = f(x)$$

as required.

A function has more than one antiderivative. For example, x^3 is an antiderivative of $3x^2$. But so is $x^3 + 12$, since the derivative of the constant 12 is zero. In general, if F is one antiderivative of f, any function obtained by adding a constant to F is also an antiderivative of f. In fact, it turns out that all the antiderivatives of f can be obtained by adding constants to a given antiderivative.

The antiderivatives of a function

If F and G are antiderivatives of f, then there is a constant C such that $G(x) = F(x) + C$.

There is a simple geometric explanation for the fact that any two antiderivatives of the same function must differ by a constant. If F is an antiderivative of f, then $F'(x) = f(x)$. This says that for each value of x, $f(x)$ is the slope of the tangent to the graph of $F(x)$. If G is another antiderivative of f, the slope of its tangent is also $f(x)$. Hence the graph of G is "parallel" to the graph of F and can be obtained by translating the graph of F vertically. That is, there is some constant C for which $G(x) = F(x) + C$. The situation is illustrated in Figure 1.1 which shows several antiderivatives of the function $f(x) = 3x^2$.

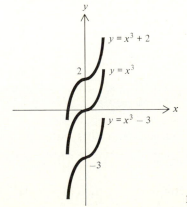

Figure 1.1 Some antiderivatives of $3x^2$.

Notation It is customary to write

$$\int f(x)\, dx = F(x) + C$$

to express the fact that every antiderivative of $f(x)$ is of the form $F(x) + C$. For example, you can express the fact that every antiderivative of $3x^2$ is of the form $x^3 + C$ by writing

$$\int 3x^2\, dx = x^3 + C$$

The symbol $\int$ is called an **integral sign** and indicates that you are to find the antiderivative of the function following it. At first, the symbol dx may seem mysterious. Its role is to indicate that x is the variable with respect to which the integration is to be performed. Analogous notation is used if the function is expressed in terms of a variable other than x. For example, $\int 3t^2\, dt = t^3 + C$. In the expression $\int 3px^2\, dx$, the dx tells you that x rather than p is the variable. Thus, $\int 3px^2\, dx = px^3 + C$. (Do you see how to evaluate $\int 3px^2\, dp$?)

Here are some general rules for integration that can be obtained by stating rules for differentiation in reverse.

Integration of power functions According to the power rule $\dfrac{d}{dx}(x^n) = nx^{n-1}$, you differentiate a power function by reducing its power by 1 and multiplying by the original power. Here is the rule stated in reverse as a rule for integrating power functions.

Integration of power functions

> **For $n \neq -1$,**
>
> $$\int x^n\, dx = \frac{1}{n+1}x^{n+1} + C$$
>
> **That is, to integrate x^n (for $n \neq -1$), increase the power of x by 1 and divide by the new power.**

To convince yourself that this rule is valid, simply differentiate $\dfrac{x^{n+1}}{n+1}$ and observe that you get x^n. The rule holds for all values of n except, of course, for $n = -1$, in which case $\dfrac{1}{n+1}$ is undefined. This special case will be discussed shortly. In the meantime, see if you can think of an antiderivative of x^{-1} on your own.

   The rule for integrating power functions is illustrated in the next example.

EXAMPLE 1.2

Find the following integrals.

(a) $\int x^{3/5}\,dx$ (b) $\int 1\,dx$ (c) $\int \dfrac{1}{\sqrt{x}}\,dx$

SOLUTION

(a) Increasing the power of x by 1 and then dividing by the new power, you get

$$\int x^{3/5}\,dx = \tfrac{5}{8}x^{8/5} + C$$

(b) Since $1 = x^0$, it follows that

$$\int 1\,dx = x + C$$

(c) Begin by rewriting $\dfrac{1}{\sqrt{x}}$ as $x^{-1/2}$. Then,

$$\int \frac{1}{\sqrt{x}}\,dx = \int x^{-1/2}\,dx = 2x^{1/2} + C = 2\sqrt{x} + C$$

The integral of $\dfrac{1}{x}$ Were you able to discover an antiderivative of x^{-1}? That is, were you able to think of a function whose derivative is $\dfrac{1}{x}$? The natural logarithm $\ln x$ is such a function, and so it appears that $\displaystyle\int \frac{1}{x}\,dx = \ln x + C$. Actually, this is true only when x is positive, since $\ln x$ is not defined for negative values of x. When x is negative, it turns out that the function $\ln |x|$ is an antiderivative of $\dfrac{1}{x}$. To see this, observe that if x is negative, $|x| = -x$ and so

$$\frac{d}{dx}(\ln |x|) = \frac{d}{dx}[\ln(-x)] = \left(\frac{1}{-x}\right)(-1) = \frac{1}{x}$$

Since $|x| = x$ when x is positive, the situation can be summarized using a single formula as follows.

The integral of $\frac{1}{x}$

$$\int \frac{1}{x}\,dx = \ln|x| + C$$

The integral of e^x Integration of the exponential function e^x is trivial since e^x is its own derivative.

The integral of e^x

$$\int e^x\,dx = e^x + C$$

Constant multiple rule and sum rule It is easy to rewrite the constant multiple rule and the sum rule for differentiation as rules for integration.

The constant multiple rule for integrals

For any constant c,

$$\int c\,f(x)\,dx = c \int f(x)\,dx$$

That is, the integral of a constant times a function is equal to the constant times the integral of the function.

The sum rule for integrals

$$\int [f(x) + g(x)]\,dx = \int f(x)\,dx + \int g(x)\,dx$$

That is, the integral of a sum is the sum of the individual integrals.

Here is an example.

EXAMPLE 1.3

Find $\int \left(3e^x + \dfrac{2}{x} - \dfrac{1}{2}x^2\right) dx$.

SOLUTION

$$\int \left(3e^x + \frac{2}{x} - \frac{1}{2}x^2\right) dx = 3 \int e^x\,dx + 2 \int \frac{1}{x}\,dx - \frac{1}{2} \int x^2\,dx$$
$$= 3e^x + 2\ln|x| - \tfrac{1}{6}x^3 + C$$

Notice that instead of writing a separate constant for each of the three antiderivatives computed in Example 1.3, you can simply add a single constant C at the end.

Integration of products and quotients You may have noticed that no general rules have been given for the integration of products and quotients. This is because there are no such general rules. Occasionally, you will be able to rewrite a product or a quotient in a form in which you can integrate it using the techniques you have learned in this section. Here is an example.

EXAMPLE 1.4

Find $\int \dfrac{3x^5 + 2x - 5}{x^3}\, dx$

SOLUTION

Perform the indicated division to get

$$\frac{3x^5 + 2x - 5}{x^3} = 3x^2 + \frac{2}{x^2} - \frac{5}{x^3} = 3x^2 + 2x^{-2} - 5x^{-3}$$

and then integrate term by term to get

$$\int \frac{3x^5 + 2x - 5}{x^3}\, dx = \int (3x^2 + 2x^{-2} - 5x^{-3})\, dx$$
$$= x^3 - 2x^{-1} + \tfrac{5}{2}x^{-2} + C$$

Practical applications Here are two problems in which the rate of change of a quantity is known and the goal is to find an expression for the quantity itself. Since the rate of change is the derivative of the quantity, you find the expression for the quantity itself by antidifferentiation.

EXAMPLE 1.5

It is estimated that x months from now the population of a certain town will be changing at a rate of $2 + 6\sqrt{x}$ people per month. The current population is 5,000. What will the population be 9 months from now?

SOLUTION

Let $P(x)$ denote the population of the town x months from now. Then the derivative of P is the rate of change of the population with respect to time. That is,

$$\frac{dP}{dx} = 2 + 6\sqrt{x}$$

It follows that the population function P is an antiderivative of $2 + 6\sqrt{x}$. That is,

$$P(x) = \int (2 + 6\sqrt{x})\, dx = 2x + 4x^{3/2} + C$$

for some constant C. To determine C, use the information that at present (when $x = 0$) the population is 5,000. That is,

$$5,000 = 2(0) + 4(0)^{3/2} + C \quad \text{or} \quad C = 5,000$$

Hence, $\qquad P(x) = 2x + 4x^{3/2} + 5,000$

In 9 months, the population will be

$$P(9) = 2(9) + 4(27) + 5,000 = 5,126$$

EXAMPLE 1.6

A manufacturer has found that marginal cost is $3q^2 - 60q + 400$ dollars per unit when q units have been produced. The total cost of producing the first 2 units is $900. What is the total cost of producing the first 5 units?

SOLUTION

Recall that the marginal cost is the derivative of the total cost function C. Hence, C must be an antiderivative of $3q^2 - 60q + 400$. That is,

$$C(q) = \int (3q^2 - 60q + 400) \, dq = q^3 - 30q^2 + 400q + K$$

for some constant K. (The letter K is used for the constant to avoid confusion with the cost function C.)

The value of K is determined by the fact that $C(2) = 900$. In particular,

$$900 = (2)^3 - 30(2)^2 + 400(2) + K \quad \text{or} \quad K = 212$$

Hence, $\qquad C(q) = q^3 - 30q^2 + 400q + 212$

and the cost of producing the first 5 units is

$$C(5) = (5)^3 - 30(5)^2 + 400(5) + 212 = \$1,587$$

Geometric application In the next example you will see how to use integration to find the equation of a curve whose slope is known.

EXAMPLE 1.7

Find the equation of the function f whose tangent has slope $3x^2 + 1$ for each value of x and whose graph passes through the point $(2, 6)$.

SOLUTION

The slope of the tangent is the derivative of f. That is

$$f'(x) = 3x^2 + 1$$

and so f is the antiderivative

$$f(x) = \int (3x^2 + 1) \, dx = x^3 + x + C$$

To find C, use the fact that the graph of f passes through $(2, 6)$. That is, substitute $x = 2$ and $f(2) = 6$ into the equation for $f(x)$ and solve for C to get

$$6 = (2)^3 + 2 + C \qquad \text{or} \qquad C = -4$$

Hence the desired function is

$$f(x) = x^3 + x - 4$$

Problems In Problems 1 through 16, find the indicated integral.

1. $\displaystyle \int x^5 \, dx$

2. $\displaystyle \int x^{3/4} \, dx$

3. $\displaystyle \int \frac{1}{x^2} \, dx$

4. $\displaystyle \int \sqrt{x} \, dx$

5. $\displaystyle \int 5 \, dx$

6. $\displaystyle \int 3e^x \, dx$

7. $\displaystyle \int (3x^2 - 5x + 2) \, dx$

8. $\displaystyle \int (x^{1/2} - 3x^{2/3} + 6) \, dx$

9. $\displaystyle \int \left(3\sqrt{x} - \frac{2}{x^3} + \frac{1}{x} \right) dx$

10. $\displaystyle \int \left(\frac{1}{2x} - \frac{2}{x^2} + \frac{3}{\sqrt{x}} \right) dx$

11. $\displaystyle \int \left(2e^x + \frac{6}{x} \right) dx$

12. $\displaystyle \int \frac{x + 1}{x^3} \, dx$

13. $\displaystyle \int \frac{x^2 + 2x + 1}{x^2} \, dx$

14. $\displaystyle \int x^3 \left(2x + \frac{1}{x} \right) dx$

15. $\displaystyle \int \sqrt{x}(x^2 - 1) \, dx$

16. $\displaystyle \int x(2x + 1)^2 \, dx$

Population growth 17. It is estimated that t months from now the population of a certain town will be changing at a rate of $4 + 5t^{2/3}$ people per month. If the current population is 10,000, what will the population be 8 months from now?

Retail prices 18. In a certain section of the country, the price of large Grade A eggs is currently $1.60 per dozen. Studies indicate that x weeks from now, the price will be changing at a rate of $0.2 + 0.003x^2$ cents per week. How much will eggs cost 10 weeks from now?

Depreciation 19. The resale value of a certain industrial machine decreases at a rate that changes with time. When the machine is t years old, the rate at which its value is changing is $220(t - 10)$ dollars per year. If the machine was bought new for \$12,000, how much will it be worth 10 years later?

The speed of a moving object

20. An object is moving so that its speed after t minutes is $3 + 2t + 6t^2$ meters per minute. How far does the object travel during the 2nd minute?

Marginal cost

21. A manufacturer has found that marginal cost is $6q + 1$ dollars per unit when q units have been produced. The total cost (including overhead) of producing the 1st unit is \$130. What is the total cost of producing the first 10 units?

Marginal profit

22. A manufacturer estimates marginal revenue to be $100q^{-1/2}$ dollars per unit when the level of production is q units. The corresponding marginal cost has been found to be $0.4q$ dollars per unit. Suppose the manufacturer's profit is \$520 when the level of production is 16 units. What is the manufacturer's profit when the level of production is 25 units?

Marginal profit

23. The marginal profit (marginal revenue minus marginal cost) of a certain company is $100 - 2q$ dollars per unit when q units are produced. If the company's profit is \$700 when 10 units are produced, what is the company's maximum possible profit?

Air pollution

24. An environmental study of a certain community suggests that t years from now the level of carbon monoxide in the air will be changing at a rate of $0.1t + 0.1$ parts per million per year. If the current level of carbon monoxide in the air is 3.4 parts per million, what will the level be 3 years from now?

25. Find the equation of the function whose tangent has slope $4x + 1$ for each value of x and whose graph passes through the point $(1, 2)$.

26. Find the equation of the function whose tangent has slope $3x^2 + 6x - 2$ for each value of x and whose graph passes through the point $(0, 6)$.

27. Find the equation of the function whose tangent has slope $x^3 - \dfrac{2}{x^2} + 2$ for each value of x and whose graph passes through the point $(1, 3)$.

28. Find the equation of a function whose graph has a relative minimum when $x = 1$ and a relative maximum when $x = 4$.

In Problems 29 through 34, find the indicated integral. In each case you may want to start with an educated guess, which you can later modify.

29. $\displaystyle\int e^{3x}\,dx$ (*Hint:* Try $e^{3x} + C$.)

30. $\displaystyle\int 3e^{5x}\,dx$

31. $\displaystyle\int (2x + 3)^5\,dx$ [*Hint:* Try $\frac{1}{6}(2x + 3)^6$.]

32. $\displaystyle\int 5(3x + 1)^4\,dx$

33. $\displaystyle\int \frac{1}{2x + 1}\,dx$ 34. $\displaystyle\int \frac{2}{3x + 2}\,dx$

2 INTEGRATION BY SUBSTITUTION

The integral version of the chain rule is known as **integration by substitution.** To refresh your memory, here is a typical application of the chain rule for differentiation.

According to the chain rule, the derivative of the function $(x^2 + 3x + 5)^9$ is

$$\frac{d}{dx}\,[(x^2 + 3x + 5)^9] = 9(x^2 + 3x + 5)^8(2x + 3)$$

Notice that this derivative is a product and that one of its factors, $2x + 3$, is the derivative of an expression, $x^2 + 3x + 5$, that occurs in the other factor. More precisely, the product is of the form

$$g(u)\,\frac{du}{dx}$$

where, in this case, $g(u) = 9u^8$ and $u = x^2 + 3x + 5$.

You can integrate many products of the form $g(u)\,\dfrac{du}{dx}$ by applying the chain rule in reverse. Specifically, if G is an antiderivative of g, then

$$\int g(u)\,\frac{du}{dx}\,dx = G(u) + C$$

since, by the chain rule,

$$\frac{d}{dx}\,[G(u)] = G'(u)\,\frac{du}{dx} = g(u)\,\frac{du}{dx}$$

Integration by substitution

$$\int g(u)\,\frac{du}{dx}\,dx = G(u) + C$$

where G is an antiderivative of g.

That is, to integrate a product of the form $g(u)\,\dfrac{du}{dx}$, find $\displaystyle\int g(u)\,du$ and then replace u in the answer by the corresponding expression involving the variable x.

Here is an example.

EXAMPLE 2.1

Find $\int 9(x^2 + 3x + 5)^8(2x + 3)\ dx$.

SOLUTION

Notice that $2x + 3$ is the derivative of $x^2 + 3x + 5$. This suggests that you should let $u = x^2 + 3x + 5$ and $g(u) = 9u^8$. Then,

$$9(x^2 + 3x + 5)^8(2x + 3) = g(u)\,\frac{du}{dx}$$

and you can use the method of substitution to integrate this product.

First integrate $g(u) = 9u^8$ to get

$$\int 9u^8\ du = u^9 + C$$

and then replace u by $x^2 + 3x + 5$ in the answer to conclude that

$$\int 9(x^2 + 3x + 5)^8(2x + 3)\ dx = (x^2 + 3x + 5)^9 + C$$

The product to be integrated in the next example is not exactly of the form $g(u)\,\dfrac{du}{dx}$. However, it is a constant multiple of such a function and you integrate it by combining the method of substitution with the constant multiple rule.

EXAMPLE 2.2

Find $\int x^3 e^{x^4+2}\ dx$.

SOLUTION

The derivative of $x^4 + 2$ is not x^3, but rather $4x^3$, and so you cannot apply the method of substitution to the product $x^3 e^{x^4+2}$ directly. You can, however, apply substitution to the related product $4x^3 e^{x^4+2}$ to get

$$\int 4x^3 e^{x^4+2}\ dx = e^{x^4+2} + C$$

This suggests that you use the constant multiple rule to get

$$\int x^3 e^{x^4+2}\ dx = \tfrac{1}{4}\int 4x^3 e^{x^4+2}\ dx = \tfrac{1}{4}e^{x^4+2} + C$$

Change of variables   Integration by substitution may be thought of as a technique for simplifying an integral by changing the variable of integration. In particular, you start with $\int g(u) \dfrac{du}{dx} \, dx$, an integral in which the variable of integration is x, and transform it into $\int g(u) \, du$, a simpler integral in which the variable of integration is u. In this transformation, the expression $\dfrac{du}{dx} \, dx$ in the original integral is replaced in the simplified integral by the symbol du. You can remember this relationship between $\dfrac{du}{dx} \, dx$ and du by pretending that $\dfrac{du}{dx}$ is a quotient and writing

$$\frac{du}{dx} \, dx = du$$

These observations suggest an alternative version of the method of substitution in which the variable u is formally substituted for an appropriate expression in x, and the original integral is transformed into one in which the variable of integration is u. The expression for du is found by computing the derivative $\dfrac{du}{dx}$ and then multiplying by dx. This procedure is illustrated in the next example using the integral from Example 2.1.

EXAMPLE 2.3

Find $\int 9(x^2 + 3x + 5)^8(2x + 3) \, dx$.

SOLUTION

Let $u = x^2 + 3x + 5$. Then $\dfrac{du}{dx} = 2x + 3$ and so $du = (2x + 3) \, dx$. Substituting $u = x^2 + 3x + 5$ and $du = (2x + 3) \, dx$, you get

$$\int 9(x^2 + 3x + 5)^8(2x + 3) \, dx = \int 9u^8 \, du = u^9 + C$$
$$= (x^2 + 3x + 5)^9 + C$$

Many people find this formal method of substitution more appealing than the original method used in Examples 2.1 and 2.2. They like the fact that it involves straightforward manipulation of symbols, and they appreciate the convenience of the notation. This formal method is also somewhat more versatile, as you shall see presently. However, for most of the integrals you will encounter, both methods work well and you should feel free to use the one with which you are most comfortable.

Here is another example in which the constant multiple rule is combined with integration by substitution.

EXAMPLE 2.4

Find $\int x^3 \sqrt{x^4 + 2} \, dx$.

SOLUTION

Let $u = x^4 + 2$. Then $\dfrac{du}{dx} = 4x^3$ and so $du = 4x^3 \, dx$. Hence,

$$\int x^3 \sqrt{x^4 + 2} \, dx = \int \tfrac{1}{4} u^{1/2} \, du$$

$$= \tfrac{1}{4} \int u^{1/2} \, du$$

$$= \tfrac{1}{4}(\tfrac{2}{3}) u^{3/2} + C$$
$$= \tfrac{1}{6}(x^4 + 2)^{3/2} + C$$

The next example is designed to show you the versatility of the formal method of substitution. It deals with an integral that does not seem to be of the form $\int g(u)\dfrac{du}{dx} \, dx$ but that nevertheless can be simplified significantly by a clever change of variables.

EXAMPLE 2.5

Find $\int \dfrac{x}{x + 1} \, dx$.

SOLUTION

There seems to be no easy way to integrate this quotient as it stands. But watch what happens if you make the substitution $u = x + 1$. Then $du = dx$ and $x = u - 1$. Hence,

$$\int \frac{x}{x + 1} \, dx = \int \frac{u - 1}{u} \, du = \int 1 \, du - \int \frac{1}{u} \, du$$

$$= u - \ln |u| + C = x + 1 - \ln |x + 1| + C$$

Problems   In Problems 1 through 22, find the indicated integral.

1. $\int (2x + 6)^5 \, dx$ 2. $\int e^{5x} \, dx$

3. $\int \sqrt{4x - 1} \, dx$

4. $\int \dfrac{1}{3x + 5} \, dx$

5. $\int e^{1-x} \, dx$

6. $\int [(x - 1)^5 + 3(x - 1)^2 + 5] \, dx$

7. $\int xe^{x^2} \, dx$

8. $\int 2xe^{x^2-1} \, dx$

9. $\int x(x^2 + 1)^5 \, dx$

10. $\int 3x \sqrt{x^2 + 8} \, dx$

11. $\int x^2(x^3 + 1)^{3/4} \, dx$

12. $\int x^5 e^{1-x^6} \, dx$

13. $\int \dfrac{2x^4}{x^5 + 1} \, dx$

14. $\int \dfrac{x^2}{(x^3 + 5)^2} \, dx$

15. $\int (x + 1)(x^2 + 2x + 5)^{12} \, dx$

16. $\int (3x^2 - 1)e^{x^3-x} \, dx$

17. $\int \dfrac{3x^4 + 12x^3 + 6}{x^5 + 5x^4 + 10x + 12} \, dx$

18. $\int \dfrac{10x^3 - 5x}{\sqrt{x^4 - x^2 + 6}} \, dx$

19. $\int \dfrac{\ln 5x}{x} \, dx$

20. $\int \dfrac{1}{x \ln x} \, dx$

21. $\int \dfrac{2x \ln (x^2 + 1)}{x^2 + 1} \, dx$

22. $\int \dfrac{e^{\sqrt{x}}}{\sqrt{x}} \, dx$

In Problems 23 through 27, use an appropriate change of variables to find the indicated integral.

23. $\int \dfrac{x}{x - 1} \, dx$

24. $\int x \sqrt{x + 1} \, dx$

25. $\int \dfrac{x}{(x - 5)^6} \, dx$

26. $\int (x + 1)(x - 2)^9 \, dx$

27. $\int \dfrac{x + 3}{(x - 4)^2} \, dx$

28. Find a function whose tangent has slope $x \sqrt{x^2 + 5}$ for each value of x and whose graph passes through the point (2, 10).

29. Find a function whose tangent has slope $\dfrac{2x}{1 - 3x^2}$ for each value of x and whose graph passes through the point (0, 5).

30. A tree has been transplanted and after x years is growing at a rate of $1 + \dfrac{1}{(x + 1)^2}$ meters per year. After 2 years it has reached a height of 5 meters. How tall was it when it was transplanted?

Depreciation 31. The resale value of a certain industrial machine decreases at a rate that changes with time. When the machine is t years old, the rate at which its value is changing is $-960e^{-t/5}$ dollars per year. If the machine was bought new for $5,000, how much will it be worth 10 years later?

Population growth 32. It is projected that t years from now the population of a certain country will be changing at the rate of $e^{0.02t}$ million per year. If the current population is 50 million, what will the population be 10 years from now?

Land value 33. It is estimated that x years from now the value of an acre of farmland will be increasing at the rate of $\dfrac{0.4x^3}{\sqrt{0.2x^4 + 8,000}}$ dollars per year. If the land is currently worth $500 per acre, how much will it be worth in 10 years?

3 ELEMENTARY DIFFERENTIAL EQUATIONS

Any equation that contains a derivative is called a **differential equation.** For example, the equations

$$\frac{dy}{dx} = 3x^2 + 5 \qquad \frac{dP}{dt} = kP \qquad \text{and} \qquad \left(\frac{dy}{dx}\right)^2 + 3\,\frac{dy}{dx} + 2y = e^x$$

are all differential equations.

Many practical situations, especially those involving rates, can be described mathematically by differential equations. For example, the assumption that population grows at a rate proportional to its size can be expressed by the differential equation $\dfrac{dP}{dt} = kP$, where P denotes the population size, t stands for time, and k is the constant of proportionality. In economics, statements about marginal cost and marginal revenue can be formulated as differential equations. Here are two examples.

EXAMPLE 3.1

Write a differential equation describing the fact that the rate at which people hear about a new increase in postal rates is proportional to the number of people in the country who have not heard about it.

SOLUTION

Let t denote time, Q the number of people who have heard about the rate increase, and B the total population of the country. Then,

$$\begin{array}{l}\text{Rate at which people} \\ \text{hear about the increase}\end{array} = \frac{dQ}{dt}$$

and $\qquad \begin{array}{l}\text{Number of people who have} \\ \text{not heard about the increase}\end{array} = B - Q$

Hence the desired differential equation is

$$\frac{dQ}{dt} = k(B - Q)$$

where k is the constant of proportionality.

EXAMPLE 3.2

Write a differential equation describing the fact that when environmental factors impose an upper bound on its size, population grows at a rate that is jointly proportional to its current size and the difference between its upper bound and current size.

SOLUTION

Let t denote time, P the size of the population, and B the upper bound imposed on the population by the environment. Then,

$$\text{Rate of population growth} = \frac{dP}{dt}$$

and

$$\begin{matrix} \text{Difference between upper} \\ \text{bound and population} \end{matrix} = B - P$$

Since "jointly proportional" means "proportional to the product" (as you saw in Chapter 1, Section 5), it follows that the desired differential equation is

$$\frac{dP}{dt} = kP(B - P)$$

where k is the constant of proportionality.

General and particular solutions

Any function that satisfies a differential equation is said to be a **solution** of that equation. Here is an example to illustrate this concept.

EXAMPLE 3.3

Verify that the function $y = e^x - x$ is a solution of the differential equation $\frac{dy}{dx} - y = x - 1$.

SOLUTION

You must subtract the function y from its derivative $\frac{dy}{dx}$ and show that the result is equal to $x - 1$. Since

$$\frac{dy}{dx} = e^x - 1$$

you get $\qquad \dfrac{dy}{dx} - y = (e^x - 1) - (e^x - x) = x - 1$

as required.

You can easily check that the function $y = 3e^x - x$ is also a solution of the differential equation in Example 3.3. In fact, every function of the form $y = Ce^x - x$, where C is a constant, is a solution of this equation. Moreover, it turns out that *every* solution of the equation is of this form. For this reason, the function $y = Ce^x - x$ is said to be the **general solution** of the differential equation in Example 3.3. A solution obtained by replacing C by a specific number is sometimes called a **particular solution.**

EXAMPLE 3.4

Find the particular solution of the differential equation $\dfrac{dy}{dx} - y = x - 1$ that satisfies the condition that $y = 4$ when $x = 0$.

SOLUTION

Use the given condition to determine the numerical value of the constant C in the general solution $y = Ce^x - x$. In particular, substitute $y = 4$ and $x = 0$ into the general solution to get $C = 4$ and conclude that the desired particular solution is $y = 4e^x - x$.

Differential equations of the form $\dfrac{dy}{dx} = f(x)$ Every time you find an integral, you are actually solving a special type of differential equation. The differential equation in this case is of the form $\dfrac{dy}{dx} = f(x)$, and its general solution is $y = F(x) + C$, where F is an antiderivative of f and C is a constant.

Differential equations of the form $\dfrac{dy}{dx} = f(x)$ are particularly easy to solve because the derivative of the quantity in question is given explicitly as a function of the independent variable. (Notice that the differential equation $\dfrac{dP}{dt} = kP$ describing population growth is not of this form because the derivative of P is expressed in terms of P itself rather than t.) Here are two practical problems involving differential equations of the form $\dfrac{dy}{dx} = f(x)$.

EXAMPLE 3.5

The resale value of a certain industrial machine decreases over a 10-year period at a rate that depends on the age of the machine. When the machine is x years old, the rate at which its value is changing is $220(x - 10)$ dollars per year. Express the value of the machine as a function of its age and initial value. If the machine was originally worth $12,000, how much will it be worth when it is 10 years old?

SOLUTION

Let $V(x)$ denote the value of the machine when it is x years old. The derivative $\dfrac{dV}{dx}$ is equal to the rate $220(x - 10)$ at which the value is changing. Hence, you can begin with the differential equation

$$\frac{dV}{dx} = 220(x - 10) = 220x - 2{,}200$$

To find V, solve this differential equation by integration.

$$V(x) = \int (220x - 2{,}200)\, dx = 110x^2 - 2{,}200x + C$$

Notice that C is equal to $V(0)$, the initial value of the machine. A more descriptive symbol for this constant is V_0. Using this notation you can write

$$V(x) = 110x^2 - 2{,}200x + V_0$$

If $V_0 = 12{,}000$, then $V(x) = 110x^2 - 2{,}200x + 12{,}000$ and the value after 10 years is

$$V(10) = 11{,}000 - 22{,}000 + 12{,}000 = \$1{,}000$$

Graphs showing the rate of depreciation and the resale value of the machine are sketched in Figure 3.1. Can you explain why the negative

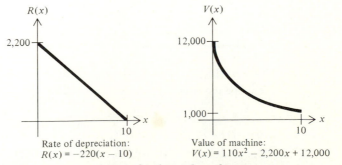

Rate of depreciation:
$R(x) = -220(x - 10)$

Value of machine:
$V(x) = 110x^2 - 2{,}200x + 12{,}000$

Figure 3.1 Depreciation of industrial machinery.

of the rate of change $220(x - 10)$ is used to represent the rate of depreciation?

EXAMPLE 3.6

An oil well that yields 300 barrels of crude oil a month will run dry in 3 years. It is estimated that t months from now the price of crude oil will be $P(t) = 18 + 0.3\sqrt{t}$ dollars per barrel. If the oil is sold as soon as it is extracted from the ground, what will the total future revenue from the well be?

SOLUTION

Let R denote revenue. Then

$$\begin{array}{c}\text{Rate of change} \\ \text{of revenue}\end{array} = \left(\begin{array}{c}\text{number of dollars} \\ \text{received per barrel}\end{array}\right)\left(\begin{array}{c}\text{number of barrels} \\ \text{sold per month}\end{array}\right)$$

where $\qquad\qquad$ Rate of change of revenue $= \dfrac{dR}{dt}$

Number of dollars received per barrel $= P(t) = 18 + 0.3\sqrt{t}$

and $\qquad\qquad$ Number of barrels sold per month $= 300$

It follows that

$$\frac{dR}{dt} = P(t)(300) \qquad \text{or} \qquad \frac{dR}{dt} = 5,400 + 90\sqrt{t}$$

The solution of this differential equation is

$$R(t) = \int (5,400 + 90\sqrt{t})\, dt = 5,400t + 60t^{3/2} + C$$

Since $R(0) = 0$, it follows that $C = 0$, and so

$$R(t) = 5,400t + 60t^{3/2}$$

Since the well will run dry in 36 months, the total future revenue will be

$$R(36) = 5,400(36) + 60(216) = \$207,360$$

Notice that in setting up the differential equation in Example 3.6, you were essentially using the chain rule

$$\frac{dR}{dt} = \frac{dR}{dB}\frac{dB}{dt} \qquad \text{(dollars per month)}$$

where B denotes the number of barrels extracted, R denotes the revenue,

$$\frac{dR}{dB} = 18 + 0.3\sqrt{t} \qquad \text{(dollars per barrel)}$$

and $$\frac{dB}{dt} = 300 \qquad \text{(barrels per month)}$$

You may find it helpful to use the chain rule in this way when setting up similar differential equations in the future.

Problems   In Problems 1 through 12, write a differential equation describing the given situation. (Do not try to solve the differential equation at this time.)

Growth of bacteria

1. The number of bacteria in a culture grows at a rate that is proportional to the number present.

Radioactive decay

2. A sample of radium decays at a rate that is proportional to its size.

Investment growth

3. An investment grows at a rate equal to 7 percent of its size.

Concentration of drugs

4. The rate at which the concentration of a drug in the bloodstream decreases is proportional to the concentration.

Population growth

5. The population of a certain town increases at a constant rate of 500 people per year.

Marginal cost

6. A manufacturer's marginal cost is $60 per unit.

Temperature change

7. The rate at which the temperature of an object changes is proportional to the difference between its own temperature and the temperature of the surrounding medium.

Dissolution of sugar

8. After being placed in a container of water, sugar dissolves at a rate proportional to the amount of undissolved sugar remaining in the container.

Recall from memory

9. When a person is asked to recall a set of facts, the rate at which the facts are recalled is proportional to the number of relevant facts in the person's memory that have not yet been recalled.

The spread of an epidemic

10. The rate at which an epidemic spreads through a community is jointly proportional to the number of people who have caught the disease and the number who have not.

Corruption in government

11. The rate at which people are implicated in a government scandal is jointly proportional to the number of people already implicated and the number of people involved who have not yet been implicated.

12. The rate at which a rumor spreads through a community is jointly proportional to the number of people in the community who have heard the rumor and the number who have not.

13. Verify that the function $y = Ce^{kx}$ is a solution of the differential equation $\dfrac{dy}{dx} = ky$.

14. Verify that the function $Q = B - Ce^{-kt}$ is a solution of the differential equation $\dfrac{dQ}{dt} = k(B - Q)$.

15. Verify that the function $y = C_1e^x + C_2xe^x$ is a solution of the differential equation $\dfrac{d^2y}{dx^2} - 2\dfrac{dy}{dx} + y = 0$.

16. Verify that the function $y = \dfrac{1}{20}x^4 - \dfrac{C_1}{x} + C_2$ is a solution of the differential equation $x\dfrac{d^2y}{dx^2} + 2\dfrac{dy}{dx} = x^3$.

In Problems 17 through 22, find the general solution of the given differential equation.

17. $\dfrac{dy}{dx} = 3x^2 + 5x - 6$

18. $\dfrac{dP}{dt} = \sqrt{t} + e^{-t}$

19. $\dfrac{dV}{dx} = \dfrac{2}{x + 1}$

20. $\dfrac{dA}{dt} = 2te^{t^2+5}$

21. $\dfrac{d^2P}{dt^2} = 50$ (*Hint:* Integrate twice.)

22. $\dfrac{d^2y}{dx^2} = 3x^2 + 5x - 6$

In Problems 23 through 28, find the particular solution of the given differential equation that satisfies the given condition.

23. $\dfrac{dy}{dx} = e^{5x}$; $y = 1$ when $x = 0$

24. $\dfrac{dy}{dx} = 5x^4 - 3x^2 - 2$; $y = 4$ when $x = 1$

25. $\dfrac{dV}{dt} = 16t(t^2 + 1)^3$; $V = 1$ when $t = 0$

26. $\dfrac{d^2y}{dt^2} = 3t^2 + 2t - 1$; $y = 3$ and $\dfrac{dy}{dt} = 0$ when $t = 1$ (*Hint:* Integrate twice.)

27. $\dfrac{d^2A}{dt^2} = e^{-t/2}$; $A = 2$ and $\dfrac{dA}{dt} = 1$ when $t = 0$

28. $\dfrac{d^2H}{dt^2} = -32$; $H = H_0$ and $\dfrac{dH}{dt} = S_0$ when $t = 0$

Depreciation 29. The resale value of a certain industrial machine decreases at a rate that depends on its age. When the machine is t years old, the rate at which its value is changing is $-960e^{-t/5}$ dollars per year.
   (a) Express the value of the machine in terms of its age and initial value.
   (b) If the machine was originally worth $5,200, how much will it be worth when it is 10 years old?

Marginal cost 30. At a certain factory, the marginal cost is $3(q - 4)^2$ dollars per unit when the level of output is q units.
   (a) Express the total production cost in terms of the overhead (the cost of producing no units) and the number of units produced.
   (b) What is the cost of producing 14 units if the overhead is $436?

Population growth 31. Population statistics indicate that x years after 1970 a certain county was growing at a rate of approximately $1{,}500x^{-1/2}$ people per year. In 1979 the population of the county was 39,000.
   (a) What was the population in 1970?
   (b) If this pattern of population growth continues in the future, how many people will be living in the county in 1995?

Retail prices 32. In a certain section of the country, the price of chicken is currently $3 per kilogram. It is estimated that x weeks from now the price will be increasing at a rate of $3\sqrt{x + 1}$ cents per week. How much will chicken cost 8 weeks from now?

Air pollution 33. In a certain Los Angeles suburb, a reading of air pollution levels taken at 7:00 A.M. shows the ozone level to be 0.25 parts per million. A 12-hour forecast of air conditions predicts that t hours later the ozone level will be changing at a rate of $\dfrac{0.24 - 0.03t}{\sqrt{36 + 16t - t^2}}$ parts per million per hour.
   (a) Express the ozone level as a function of t.
   (b) At what time will the peak ozone level occur? What will the ozone level be at this time?

Production of oil

34. A certain oil well that yields 400 barrels of crude oil a month will run dry in 2 years. The price of crude oil is currently $18 per barrel and is expected to rise at a constant rate of 3 cents per barrel per month. If the oil is sold as soon as it is extracted from the ground, what will the total future revenue from the well be?

Farming

35. It is estimated that t days from now a farmer's crop will be increasing at a rate of $0.3t^2 + 0.6t + 1$ bushels per day. By how much will the value of the crop increase during the next 5 days if the market price remains fixed at $3 per bushel?

Water pollution

36. It is estimated that t years from now the population of a certain lakeside community will be changing at a rate of $0.6t^2 + 0.2t + 0.5$ thousand per year. Environmentalists have found that the level of pollution in the lake increases at the rate of approximately 5 units per 1,000 people. If the level of pollution in the lake is currently 60 units, what will the pollution level be 2 years from now?

Stopping distance

37. After its brakes are applied, a certain sports car decelerates at a constant rate of 28 feet per second per second.
   (a) Express the distance the car travels in terms of its speed at the moment of braking and the amount of time that has elapsed since that moment. (*Hint:* Acceleration is the second derivative of distance. Let $D(t)$ denote the distance the car has traveled after t seconds, and solve the differential equation $\dfrac{d^2D}{dt^2} = -28$ by integrating twice.)
   (b) Compute the stopping distance if the car was going 60 miles per hour when the brakes were applied. (*Hint:* 60 miles per hour = 88 feet per second.)

Spy story

38. The hero of a popular spy story (who defused the bomb in 5 minutes and survived Problem 16 in Chapter 3, Section 1) is driving the sports car in Problem 37 at a speed of 60 miles per hour on Highway 1 in the remote republic of San Dimas. Suddenly he sees a camel in the road 199 feet in front of him. After a reaction time of 0.7 second, he steps on the brakes. Will he stop before hitting the camel?

4 SEPARABLE DIFFERENTIAL EQUATIONS

Many useful differential equations can be formally rewritten so that all the terms containing the independent variable appear on one side of the equation and all the terms containing the dependent variable appear on the other. Differential equations with this special property are said to be **separable** and can be solved by the following procedure involving two integrations.

Separable differential equations

A differential equation that can be written in the form

$$g(y)\, dy = f(x)\, dx$$

is said to be separable. Its general solution is obtained by integrating both sides of this equation. That is,

$$\int g(y)\, dy = \int f(x)\, dx$$

A proof that this procedure works will be given later in this section. First, here are some examples to illustrate how the procedure is used. In these examples, you will see how to derive some of the exponential models that were introduced in Chapter 4, Section 2.

Exponential growth and decay

In Chapter 4, Section 4, you saw that if a quantity grows exponentially, its rate of change is proportional to its size. Using separable differential equations you can now establish the converse of this result.

EXAMPLE 4.1

Show that a quantity that grows at a rate proportional to its size grows exponentially.

SOLUTION

Let Q denote the quantity and t denote time, and begin with the differential equation

$$\frac{dQ}{dt} = kQ$$

where k is the constant of proportionality. To separate the variables, pretend that the derivative $\frac{dQ}{dt}$ is actually a quotient and write

$$\frac{1}{Q}\, dQ = k\, dt$$

Now integrate both sides of this equation to get

$$\int \frac{1}{Q}\, dQ = \int k\, dt$$

or

$$\ln |Q| = kt + C$$

Solve this equation for $|Q|$ by applying the exponential function to each side to get

$$|Q| = e^{kt+C} = e^{C} e^{kt}$$

Since $|Q| = \pm Q$, depending on whether Q is positive or negative, you can drop the absolute value sign and write

$$Q = \pm e^C e^{kt}$$

Finally, since the constant $\pm e^C$ is the value of Q when $t = 0$, you can introduce Q_0 to stand for this constant and, using functional notation, write

$$Q(t) = Q_0 e^{kt}$$

which is precisely the equation describing exponential growth.

You can use a similar argument to show that a quantity that decreases at a rate proportional to its size decreases exponentially. For practice, work out the details of this argument.

A technical detail   The solution of Example 4.1 is not quite correct as stated. In particular, the argument that the equation

$$|Q| = e^C e^{kt}$$

can be rewritten as $\qquad Q = \pm e^C e^{kt}$

where $\pm e^C$ is a *constant* is valid only if Q never changes sign. If Q were to change sign, the expression $\pm e^C$ would be positive part of the time and negative part of the time and hence would not be constant. Fortunately, it turns out that the continuity of Q guarantees that this cannot happen and that the expression $\pm e^C$ is really a constant. Feel free to ignore this technical point when solving similar separable differential equations, and simply assume that the function does not change sign.

Learning curves   As you saw in Chapter 4, Section 2, the graphs of functions of the form $Q(t) = B - Ae^{-kt}$ are called learning curves because functions of this form often describe the relationship between the efficiency with which an individual performs a task and the amount of training or experience the individual has had. In general, any quantity that grows at a rate that is proportional to the difference between its size and a fixed upper bound can be represented by a function of this form. This is illustrated in the next example.

EXAMPLE 4.2

The rate at which people hear about a new postal rate increase is proportional to the number of people in the country who have not heard about it. Express the number of people who have heard about the increase as a function of time.

SOLUTION

Let t denote time, Q the number of people who have heard about the increase, and B the total population of the country. Then (as you saw in Example 3.1 of this chapter),

$$\frac{dQ}{dt} = k(B - Q)$$

Separate the variables by writing

$$\frac{1}{B - Q}\, dQ = k\, dt$$

and integrate to get $\qquad \displaystyle\int \frac{1}{B - Q}\, dQ = \int k\, dt$

or $\qquad\qquad -\ln |B - Q| = kt + C$

(Be sure you see where the minus sign came from.) This time you can drop the absolute value sign immediately since $B - Q$ cannot be negative in this context. Hence,

$$-\ln (B - Q) = kt + C$$

$$\ln (B - Q) = -kt - C$$

$$B - Q = e^{-kt - C}$$

or $\qquad\qquad\qquad Q = B - e^{-C}e^{-kt}$

Denoting the constant e^{-C} by A and using functional notation, you can conclude that

$$Q(t) = B - Ae^{-kt}$$

which is precisely the general equation of a learning curve. For reference, the graph of Q is sketched in Figure 4.1.

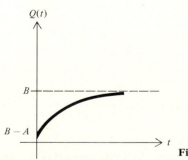

Figure 4.1 A learning curve: $Q(t) = B - Ae^{-kt}$.

Logistic curves   In Chapter 4, Section 2, you learned that the graphs of functions of the form $Q(t) = \dfrac{B}{1 + Ae^{-Bkt}}$ are called logistic curves. They are used as models of population growth when environmental factors impose an upper bound on the possible size of the population, and they also describe such phenomena as the spread of rumors and epidemics. The calculation in the next example can be used to show that any quantity that satisfies a differential equation of the form $\dfrac{dQ}{dt} = kQ(B - Q)$ can be represented by a logistic curve.

EXAMPLE 4.3

The rate at which an epidemic spreads through a community is jointly proportional to the number of residents who have been infected and the number of susceptible residents who have not. Express the number of residents who have been infected as a function of time.

SOLUTION

Let t denote time, Q the number of residents who have been infected, and B the total number of susceptible residents. Then the number of susceptible residents who have not been infected is $B - Q$, and the differential equation describing the spread of the epidemic is

$$\frac{dQ}{dt} = kQ(B - Q)$$

where k is the constant of proportionality. This is a separable differential equation whose solution is

$$\int \frac{1}{Q(B - Q)}\, dQ = \int k\, dt$$

The trick to finding the integral on the left-hand side is to observe that

$$\frac{1}{Q(B - Q)} = \frac{1}{B} \frac{B}{Q(B - Q)} = \frac{1}{B}\left(\frac{1}{Q} + \frac{1}{B - Q}\right)$$

Then,

$$\int \frac{1}{Q(B - Q)}\, dQ = \frac{1}{B}\int \frac{1}{Q}\, dQ + \frac{1}{B}\int \frac{1}{B - Q}\, dQ$$

$$= \frac{1}{B}\ln|Q| - \frac{1}{B}\ln|B - Q| + C$$

$$= \frac{1}{B}\ln\frac{Q}{B - Q} + C$$

Hence the solution of the differential equation is

$$\frac{1}{B} \ln \frac{Q}{B - Q} + C = kt$$

which you can solve for Q as follows.

$$\ln \frac{Q}{B - Q} = Bkt - BC$$

$$\frac{Q}{B - Q} = e^{-BC}e^{Bkt}$$

$$Q = Be^{-BC}e^{Bkt} - Qe^{-BC}e^{Bkt}$$

$$Q + Qe^{-BC}e^{Bkt} = Be^{-BC}e^{Bkt}$$

$$Q(1 + e^{-BC}e^{Bkt}) = Be^{-BC}e^{Bkt}$$

$$Q = \frac{Be^{-BC}e^{Bkt}}{1 + e^{-BC}e^{Bkt}}$$

To make this formula more attractive, divide numerator and denominator by $e^{-BC}e^{Bkt}$ to get

$$Q = \frac{B}{e^{BC}e^{-Bkt} + 1}$$

Finally, denote the constant e^{BC} by A, and use functional notation to get

$$Q(t) = \frac{B}{1 + Ae^{-Bkt}}$$

which is, as promised, the general equation of a logistic curve.

For reference, the graph of Q is sketched in Figure 4.2. It is not hard to show that the inflection point occurs when $Q(t) = \frac{B}{2}$. (See Problem 28 at the end of this section.) This corresponds to the fact that the epidemic is spreading most rapidly when half of the susceptible residents have been infected.

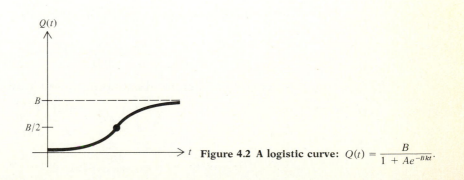

Figure 4.2 A logistic curve: $Q(t) = \dfrac{B}{1 + Ae^{-Bkt}}$.

Dilution problem The next example is typical of an important class of dilution problems that lead to separable differential equations.

EXAMPLE 4.4

The residents of a certain community have voted to discontinue the fluoridation of their water supply. The local reservoir currently holds 200 million gallons of fluoridated water that contains 1,600 pounds of fluoride. The fluoridated water is flowing out of the reservoir at the rate of 4 million gallons per day and is being replaced at the same rate by unfluoridated water. At all times the remaining fluoride is evenly distributed in the reservoir. Express the amount of fluoride in the reservoir as a function of time.

SOLUTION

Begin with the following relationship:

$$\begin{array}{l}\text{Rate of change of} \\ \text{fluoride with respect} \\ \text{to time}\end{array} = \left(\begin{array}{l}\text{concentration of} \\ \text{fluoride in the} \\ \text{water}\end{array}\right)\left(\begin{array}{l}\text{rate of flow} \\ \text{of fluoridated} \\ \text{water}\end{array}\right)$$

Let Q denote the number of pounds of fluoride in the reservoir and t the number of days that have elapsed. Then,

$$\begin{array}{l}\text{Rate of change of fluoride} \\ \text{with respect to time}\end{array} = \frac{dQ}{dt} \quad \text{(pounds per day)}$$

$$\begin{array}{l}\text{Concentration} \\ \text{of fluoride in} \\ \text{the water}\end{array} = \frac{\text{number of pounds of fluoride in reservoir}}{\text{number of million gallons of water in reservoir}}$$

$$= \frac{Q}{200} \quad \text{(pounds per million gallons)}$$

and

$$\begin{array}{l}\text{Rate of flow of} \\ \text{fluoridated water}\end{array} = -4 \quad \text{(million gallons per day)}$$

where the minus sign indicates that the water is leaving the reservoir. Hence,

$$\frac{dQ}{dt} = \frac{Q}{200}(-4) = -\frac{Q}{50}$$

Solving this differential equation by separation of variables, you get

$$\int \frac{1}{Q}\,dQ = -\int \frac{1}{50}\,dt$$

$$\ln Q = -\frac{t}{50} + C$$

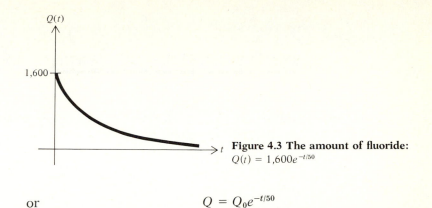

Figure 4.3 The amount of fluoride:
$Q(t) = 1,600e^{-t/50}$

or

$$Q = Q_0 e^{-t/50}$$

Initially, 1,600 pounds of fluoride were in the reservoir. Replacing Q_0 by 1,600 and using functional notation, you can conclude that

$$Q(t) = 1,600e^{-t/50}$$

That is, the amount of fluoride in the reservoir decreases exponentially. The situation is illustrated in Figure 4.3.

Why the method of separation of variables works

It is not hard to see why the method of separation of variables works. Before the variables were separated, the differential equation

$$g(y)\, dy = f(x)\, dx$$

was

$$\frac{dy}{dx} = \frac{f(x)}{g(y)}$$

or, equivalently,

$$g(y)\frac{dy}{dx} - f(x) = 0$$

On the other hand, if G is an antiderivative of g and F an antiderivative of f, it follows from the chain rule that

$$\frac{d}{dx}[G(y) - F(x)] = G'(y)\frac{dy}{dx} - F'(x) = g(y)\frac{dy}{dx} - f(x)$$

Hence,

$$\frac{d}{dx}[G(y) - F(x)] = 0$$

But constants are the only functions whose derivatives are identically zero, and so

$$G(y) - F(x) = C$$

for some constant C. That is,

$$G(y) = F(x) + C$$

or, equivalently,
$$\int g(y)\, dy = \int f(x)\, dx$$

which is precisely what you were trying to show.

Problems
In Problems 1 through 8, find the general solution of the given separable differential equation.

1. $\dfrac{dy}{dx} = 3y$ 2. $\dfrac{dy}{dx} = y^2$

3. $\dfrac{dy}{dx} = e^y$ 4. $\dfrac{dy}{dx} = e^{x+y}$

5. $\dfrac{dy}{dx} = \dfrac{x}{y}$ 6. $\dfrac{dy}{dx} = \dfrac{y}{x}$

7. $\dfrac{dy}{dx} = y + 10$ 8. $\dfrac{dy}{dx} = 80 - y$

In Problems 9 through 12, find the particular solution of the given differential equation that satisfies the given condition.

9. $\dfrac{dy}{dx} = 0.05y$; $y = 500$ when $x = 0$

10. $\dfrac{dy}{dx} = \dfrac{x}{y^2}$; $y = 3$ when $x = 2$ 11. $\dfrac{dy}{dx} = 4x^3y^2$; $y = 2$ when $x = 1$

12. $\dfrac{dy}{dx} = 5(8 - y)$; $y = 6$ when $x = 0$

Do Problems 13 through 26 by solving appropriate separable differential equations.

Investment
13. A \$1,000 investment grows at a rate equal to 7 percent of its size. Express the value of the investment as a function of time.

Drug concentration
14. The rate at which the concentration of a drug in the bloodstream decreases is proportional to the concentration. Express the concentration of the drug in the bloodstream as a function of time.

Exponential decay
15. Show that a quantity that decays at a rate proportional to its size decays exponentially.

16. Show that if a differentiable function is equal to its own derivative, then the function must be of the form $y = Ce^x$.

Recall from memory
17. Psychologists believe that when a person is asked to recall a set of facts, the rate at which the facts are recalled is proportional to

the number of relevant facts in the subject's memory that have not yet been recalled. Express the number of facts that have been recalled as a function of time and draw the graph.

Dissolution of sugar 18. After being placed in a container of water, sugar dissolves at a rate proportional to the amount of undissolved sugar remaining in the container. Express the amount of sugar that has been dissolved as a function of time and draw the graph.

Newton's law of heating 19. The rate at which the temperature of an object changes is proportional to the difference between its own temperature and that of the surrounding medium. A cold drink is removed from a refrigerator on a hot summer day and placed in an 80-degree room. Express the temperature of the drink as a function of time if the temperature of the drink was 40 degrees when it left the refrigerator and 50 degrees 20 minutes later.

Newton's law of cooling 20. The rate at which the temperature of an object changes is proportional to the difference between its own temperature and that of the surrounding medium. Express the temperature of the object as a function of time and draw the graph if the temperature of the object is greater than that of the surrounding medium.

Fick's law 21. When a cell is placed in a liquid containing a solute, the solute passes through the cell wall by diffusion. As a result, the concentration of the solute inside the cell changes, increasing if the concentration of the solute outside the cell is greater than the concentration inside and decreasing if the opposite is true. A biological law known as **Fick's law** asserts that the concentration of the solute inside the cell changes at a rate that is jointly proportional to the area of the cell wall and the difference between the concentration of the solute inside and outside the cell. Assuming that the concentration of the solute outside the cell is constant and greater than the concentration inside, derive a formula for the concentration of the solute inside the cell.

Dilution 22. A tank contains 200 gallons of clear water. Brine (salt water) containing 2 pounds of salt per gallon flows into the tank at a rate of 5 gallons per minute, and the mixture, which is stirred so that the salt is evenly distributed at all times, runs out of the tank at the same rate. Express the amount of salt in the tank as a function of time and draw the graph.

Dilution 23. A tank currently holds 200 gallons of brine that contains 3 pounds of salt per gallon. Brine containing 2 pounds of salt per gallon flows into the tank at a rate of 5 gallons per minute, while the mixture, which is kept uniform, runs out of the tank at the same rate. Express the amount of salt in the tank as a function of time and draw the graph.

Air purification 24. A 2,400-cubic-feet room contains an activated charcoal air filter through which air passes at a rate of 400 cubic feet per minute. The ozone in the air is absorbed by the charcoal as the air flows through the filter, and the purified air is recirculated in the room. Assuming that the remaining ozone is evenly distributed throughout the room at all times, determine how long it takes the filter to remove 50 percent of the ozone from the room.

The spread of a rumor 25. The rate at which a rumor spreads through a community is jointly proportional to the number of residents who have heard the rumor and the number who have not. If $\frac{1}{10}$ of the residents heard the rumor initially and $\frac{1}{4}$ had heard after 2 hours, what fraction had heard after 4 hours?

Corruption in government 26. The number of people implicated in a certain major government scandal increases at a rate jointly proportional to the number of people already implicated and the number involved who have not yet been implicated. Suppose that 7 people were implicated when a Washington newspaper first made the scandal public, that 9 more were implicated over the next 3 months, and that another 12 were implicated during the following 3 months. Approximately how many people are involved in the scandal? (*Warning: This problem will test your algebraic ingenuity!*)

The spread of an epidemic 27. The rate at which an epidemic spreads through a community is jointly proportional to the number of residents who have been infected and the number of susceptible residents who have not. Show that the epidemic is spreading most rapidly when one-half of the susceptible residents have been infected. (*Hint:* You do not have to solve a differential equation to do this. Just start with a formula for the *rate* at which the epidemic is spreading and use calculus to maximize this rate.)

Logistic curves 28. Show that if a quantity Q satisfies the differential equation $\dfrac{dQ}{dt} = kQ(B - Q)$, where k and B are positive constants, then the rate of change $\dfrac{dQ}{dt}$ is greatest when $Q(t) = \dfrac{B}{2}$. What does this result tell you about the inflection point of a logistic curve? Explain.

5 INTEGRATION BY PARTS In this section you will see a technique you can use to integrate certain products, $f(x)g(x)$, in which one of the factors, say $g(x)$, can be easily integrated and the other, $f(x)$, can be simplified by differentiation.

The technique is called **integration by parts,** and, as you will see, it is a restatement of the product rule for differentiation.

Integration by parts

$$\int f(x)g(x)\ dx = f(x)G(x) - \int f'(x)G(x)\ dx$$

where G is an antiderivative of g.

To evaluate $\int f(x)g(x)\ dx$ using this technique, first integrate g and multiply the result by f to get

$$f(x)G(x)$$

where G is an antiderivative of g. Then multiply the antiderivative G by the derivative of f and subtract the integral of this product from the result of the first step to get

$$f(x)G(x) - \int f'(x)G(x)\ dx$$

This expression will be equal to the original integral $\int f(x)g(x)\ dx$, and, if you are lucky, the new integral $\int f'(x)G(x)\ dx$ will be easier to find than the original one. Here are some examples.

EXAMPLE 5.1

Find $\int xe^x\ dx$.

SOLUTION

In this case, both factors, x and e^x, are easy to integrate. Both are also easy to differentiate, but the process of differentiation simplifies x more than e^x. This suggests that you should try integration by parts with

$$g(x) = e^x \qquad \text{and} \qquad f(x) = x$$

Then, $\qquad\qquad G(x) = e^x \qquad \text{and} \qquad f'(x) = 1$

and so

$$\int xe^x\ dx = xe^x - \int 1e^x\ dx = xe^x - e^x + C = (x - 1)e^x + C$$

EXAMPLE 5.2

Find $\int x\sqrt{x + 5}\ dx$.

SOLUTION

Again, both factors in the product are easy to integrate and differentiate. However, the factor x is simplified by differentiation, whereas the derivative of $\sqrt{x + 5}$ is even more complicated than $\sqrt{x + 5}$ itself. This suggests that you should try integration by parts, with

$$g(x) = \sqrt{x + 5} \qquad \text{and} \qquad f(x) = x$$

Then, $\qquad G(x) = \frac{2}{3}(x + 5)^{3/2} \qquad$ and $\qquad f'(x) = 1$

and so

$$\int x\sqrt{x + 5}\ dx = \frac{2}{3}x(x + 5)^{3/2} - \int \frac{2}{3}(x + 5)^{3/2}\ dx$$

$$= \frac{2}{3}x(x + 5)^{3/2} - \frac{4}{15}(x + 5)^{5/2} + C$$

In the next example, you will see how to use integration by parts to integrate the natural logarithm $\ln x$.

EXAMPLE 5.3

Find $\displaystyle\int \ln x\ dx$.

SOLUTION

The trick is to write $\ln x$ as $1(\ln x)$ and to let

$$g(x) = 1 \qquad \text{and} \qquad f(x) = \ln x$$

Then, $\qquad G(x) = x \qquad$ and $\qquad f'(x) = \dfrac{1}{x}$

and so

$$\int \ln x\ dx = x \ln x - \int x \left(\frac{1}{x}\right) dx = x \ln x - \int 1\ dx = x \ln x - x + C$$

Sometimes integration by parts leads to a new integral that also must be integrated by parts. This situation is illustrated in the next example.

EXAMPLE 5.4

Find $\displaystyle\int x^2 e^x\ dx$.

SOLUTION

It is natural to try integration by parts with

$$g(x) = e^x \quad \text{and} \quad f(x) = x^2$$

Then, $$G(x) = e^x \quad \text{and} \quad f'(x) = 2x$$

and so $$\int x^2 e^x \, dx = x^2 e^x - 2 \int x e^x \, dx$$

To find $\int x e^x \, dx$, you have to integrate by parts again, this time with

$$g(x) = e^x \quad \text{and} \quad f(x) = x$$

Then, $$G(x) = e^x \quad \text{and} \quad f'(x) = 1$$

and it follows that

$$\int x^2 e^x \, dx = x^2 e^x - 2(x e^x - e^x) + C = (x^2 - 2x + 2)e^x + C$$

Why integration by parts works

Integration by parts is actually nothing more than a restatement of what happens when the product rule is used to differentiate $f(x)G(x)$, where G is an antiderivative of g. In particular,

$$\frac{d}{dx}[f(x)G(x)] = f'(x)G(x) + f(x)G'(x) = f'(x)G(x) + f(x)g(x)$$

Expressed in terms of integrals, this says

$$f(x)G(x) = \int f'(x)G(x) \, dx + \int f(x)g(x) \, dx$$

or $$\int f(x)g(x) \, dx = f(x)G(x) - \int f'(x)G(x) \, dx$$

which is precisely the formula for integration by parts.

Problems

In Problems 1 through 21, use integration by parts to find the given integral.

1. $\int x e^{-x} \, dx$

2. $\int x e^{2x} \, dx$

3. $\int (1 - x)e^x \, dx$

4. $\int (3 - 2x)e^{-x} \, dx$

5. $\int x \ln x \, dx$

6. $\int x \ln 2x \, dx$

7. $\int x\sqrt{x-6} \, dx$

8. $\int x\sqrt{1-x} \, dx$

9. $\int x(x+1)^8 \, dx$

10. $\int (x+1)(x+2)^6 \, dx$

11. $\int \dfrac{x}{\sqrt{x+2}} \, dx$

12. $\int \dfrac{x}{\sqrt{2x+1}} \, dx$

13. $\int x^2 e^{-x} \, dx$

14. $\int x^2 e^{3x} \, dx$

15. $\int x^3 e^x \, dx$

16. $\int x^3 e^{2x} \, dx$

17. $\int x^2 \ln x \, dx$

18. $\int x(\ln x)^2 \, dx$

19. $\int x^3 e^{x^2} \, dx$ [*Hint:* Let $f(x) = x^2$.]

20. $\int x^3 (x^2-1)^{10} \, dx$ [*Hint:* Let $f(x) = x^2$.]

21. $\int x^7 (x^4+5)^8 \, dx$

Motion 22. After t seconds, an object is moving at a speed of $te^{-t/2}$ meters per second. Express the distance the object travels as a function of time.

Efficiency 23. After t hours on the job, a factory worker can produce $100te^{-0.5t}$ units per hour. How many units does the worker produce during the first 3 hours?

Fundraising 24. After t weeks, contributions in response to a local fundraising campaign were coming in at the rate of $2{,}000te^{-0.2t}$ dollars per week. How much money was raised during the first 5 weeks?

25. (a) Use integration by parts to derive the formula

$$\int x^n e^{ax} \, dx = \frac{1}{a} x^n e^{ax} - \frac{n}{a} \int x^{n-1} e^{ax} \, dx$$

(b) Use the formula in part (a) to find $\int x^3 e^{5x} \, dx$.

6 THE USE OF INTEGRAL TABLES

Most of the integrals you will encounter in the social, managerial, and life sciences can be evaluated using the techniques you have learned so far in this chapter. From time to time, however, an integral will turn up that cannot be handled by these techniques. For such occasions, it is helpful to know how to use a **table of integrals.**

A table of integrals is a list of integration formulas. Extensive tables listing several hundred formulas can be found in most mathematics handbooks, and condensed versions appear in many calculus texts. Here is a tiny sampling of the formulas that appear in a table of integrals.

A small table of integrals

$$1. \quad \int \frac{dx}{p^2 - x^2} = \frac{1}{2p} \ln \left| \frac{p + x}{p - x} \right|$$

$$2. \quad \int \frac{dx}{x(ax + b)} = \frac{1}{b} \ln \left| \frac{x}{ax + b} \right|$$

$$3. \quad \int \frac{dx}{\sqrt{x^2 \pm p^2}} = \ln \left| x + \sqrt{x^2 \pm p^2} \right|$$

$$4. \quad \int x^n e^{ax} \, dx = \frac{1}{a} x^n e^{ax} - \frac{n}{a} \int x^{n-1} e^{ax} \, dx$$

In these formulas, the letters a, b, p, and n denote constants. The term p^2 in the first and third formulas may be any *positive* constant (since any positive number is the square of its square root). The compact fractional notation $\frac{dx}{p^2 - x^2}$ in the first formula is an abbreviation for $\frac{1}{p^2 - x^2} \, dx$. Similar notation occurs in some of the other formulas. Also, to keep the formulas simple, the constant C is omitted from each of the integrals in the table.

For convenience, most tables of integrals are divided into sections. Integrals containing similar expressions are grouped together in the same section. For example, the first formula would be found in a section entitled "Expressions Containing $p^2 - x^2$," the second in a section called "Expressions Containing $ax + b$," and the fourth in the section "Expressions Containing Exponential and Logarithmic Functions."

The use of these integration formulas is illustrated in the following examples.

EXAMPLE 6.1

Find $\int \frac{1}{x(3x - 6)} \, dx$.

SOLUTION

Apply the second formula with $a = 3$ and $b = -6$ to get

$$\int \frac{1}{x(3x - 6)} \, dx = -\frac{1}{6} \ln \left| \frac{x}{3x - 6} \right| + C$$

EXAMPLE 6.2

Find $\int \dfrac{1}{6 - 3x^2}\, dx$.

SOLUTION

If the coefficient of x^2 were 1 instead of 3, you could use the first formula. This suggests that you should divide numerator and denominator by 3 to get

$$\int \frac{1}{6 - 3x^2}\, dx = \frac{1}{3} \int \frac{1}{2 - x^2}\, dx$$

and then apply the first formula with $p = \sqrt{2}$ to conclude that

$$\int \frac{1}{6 - 3x^2}\, dx = \frac{1}{3}\left(\frac{1}{2\sqrt{2}}\right) \ln \left|\frac{\sqrt{2} + x}{\sqrt{2} - x}\right| + C$$

$$= \frac{\sqrt{2}}{12} \ln \left|\frac{\sqrt{2} + x}{\sqrt{2} - x}\right| + C$$

EXAMPLE 6.3

Find $\int \dfrac{1}{3x^2 - 6}\, dx$.

SOLUTION

Since

$$\frac{1}{3x^2 - 6} = -\frac{1}{6 - 3x^2}$$

you can apply the first formula as in Example 6.2 to get

$$\int \frac{1}{3x^2 - 6}\, dx = -\frac{\sqrt{2}}{12} \ln \left|\frac{\sqrt{2} + x}{\sqrt{2} - x}\right| + C$$

EXAMPLE 6.4

Find $\int \dfrac{1}{x^2 + 2x}\, dx$.

SOLUTION

Factor $\dfrac{1}{x^2 + 2x}$ as

$$\frac{1}{x^2 + 2x} = \frac{1}{x(x + 2)}$$

and apply the second formula with $a = 1$ and $b = 2$ to get

$$\int \frac{1}{x^2 + 2x}\, dx = \frac{1}{2} \ln \left|\frac{x}{x + 2}\right| + C$$

EXAMPLE 6.5

Find $\int \dfrac{1}{3x^2 + 6}\, dx$.

SOLUTION

It is natural to try to match this integral to the one in the first formula by writing

$$\int \frac{1}{3x^2 + 6}\, dx = -\frac{1}{3} \int \frac{1}{-2 - x^2}\, dx$$

However, since -2 is negative, it cannot be written as the square, p^2, of any real number p, and so the formula does not apply.

There is a formula for integrals of the form $\int \dfrac{1}{x^2 + p^2}\, dx$ that can be used in this case. You can find it in a table of integrals under a heading like "Expressions Containing $x^2 \pm p^2$." However, the antiderivative will be written in terms of what are called inverse trigonometric functions and cannot be expressed in more elementary terms.

EXAMPLE 6.6

Find $\int \dfrac{1}{\sqrt{4x^2 - 9}}\, dx$.

SOLUTION

To put this integral in the form of the third formula, divide numerator and denominator by 2 to get

$$\int \frac{1}{\sqrt{4x^2 - 9}}\, dx = \frac{1}{2} \int \frac{1}{\sqrt{x^2 - \frac{9}{4}}}\, dx$$

Then apply the formula with $p^2 = \frac{9}{4}$, using minus signs in place of the symbol $\pm$, to get

$$\int \frac{1}{\sqrt{4x^2 - 9}}\, dx = \frac{1}{2} \ln \left| x + \sqrt{x^2 - \frac{9}{4}} \right| + C$$

The fourth formula expresses an integral in terms of a simpler integral of the same type. If the formula is subsequently applied to the new integral, further simplification may occur. Successive applications of the formula usually lead to an integral that can be found by elementary methods. A formula of this type is called a **recursion formula.** The use of a recursion formula is illustrated in the next example.

EXAMPLE 6.7

Find $\int x^2 e^{5x}\, dx$.

SOLUTION

Apply the fourth formula with $n = 2$ and $a = 5$ to get

$$\int x^2 e^{5x}\, dx = \tfrac{1}{5}x^2 e^{5x} - \tfrac{2}{5}\int x e^{5x}\, dx$$

Now apply the fourth formula again, this time with $n = 1$ and $a = 5$, to get

$$\int x e^{5x}\, dx = \tfrac{1}{5}x e^{5x} - \tfrac{1}{5}\int e^{5x}\, dx = \tfrac{1}{5}x e^{5x} - \tfrac{1}{25}e^{5x} + C$$

Combine these results to conclude that

$$\int x^2 e^{5x}\, dx = \tfrac{1}{5}x^2 e^{5x} - \tfrac{2}{25}x e^{5x} + \tfrac{2}{125}e^{5x} + C$$

No special formula was really needed to find the integral in Example 6.7. You could have found the integral quite easily using integration by parts. Indeed, if the formula had not been so conveniently displayed on page 251, you would have been much better off integrating by parts directly than hunting through a table of integrals for the appropriate formula. Try not to succumb to the temptation to rely excessively on tables when computing integrals. Many of the integrals you will encounter can be found quite easily without the aid of formulas. Moreover, before you can use a formula, you must find it, and this can be time-consuming. In general, it is good strategy to use a table of integrals only as a last resort.

Problems In Problems 1 through 10, use one of the integration formulas listed in this section to find the given integral.

1. $\int \dfrac{1}{x(2x - 3)}\, dx$

2. $\int \dfrac{3}{4x(x - 5)}\, dx$

3. $\int \dfrac{1}{\sqrt{x^2 + 25}}\, dx$

4. $\int \dfrac{1}{\sqrt{9x^2 - 4}}\, dx$

5. $\int \dfrac{1}{4 - x^2}\, dx$

6. $\int \dfrac{1}{3x^2 - 9}\, dx$

7. $\int \dfrac{1}{3x^2 + 2x}\, dx$

8. $\int \dfrac{4}{x^2 - x}\, dx$

9. $\displaystyle\int x^2 e^{3x}\, dx$ 10. $\displaystyle\int x^3 e^{-x}\, dx$

Locate a table of integrals and use it to find the integrals in Problems 11 through 18.

11. $\displaystyle\int \frac{x}{2 - x^2}\, dx$ 12. $\displaystyle\int \frac{x + 3}{\sqrt{2x + 4}}\, dx$

13. $\displaystyle\int (\ln 2x)^2\, dx$ 14. $\displaystyle\int (x^2 + 1)^{3/2}\, dx$

15. $\displaystyle\int \frac{1}{3x\sqrt{2x + 5}}\, dx$ 16. $\displaystyle\int \frac{x}{\sqrt{4 - x^2}}\, dx$

17. $\displaystyle\int \frac{1}{2 - 3e^{-x}}\, dx$ 18. $\displaystyle\int \frac{1}{\sqrt{3x^2 - 6x + 2}}\, dx$

19. One table of integrals lists the formula

$$\int \frac{dx}{\sqrt{x^2 \pm p^2}} = \ln\left|\frac{x + \sqrt{x^2 \pm p^2}}{p}\right|$$

while another table lists

$$\int \frac{dx}{\sqrt{x^2 \pm p^2}} = \ln\left|x + \sqrt{x^2 \pm p^2}\right|$$

Can you reconcile this apparent contradiction?

20. The following two formulas appear in a table of integrals:

$$\int \frac{dx}{p^2 - x^2} = \frac{1}{2p} \ln\left|\frac{p + x}{p - x}\right|$$

and

$$\int \frac{dx}{a + bx^2} = \frac{1}{2\sqrt{-ab}} \ln\left|\frac{a + x\sqrt{-ab}}{a - x\sqrt{-ab}}\right| \qquad (\text{for } -ab \geq 0)$$

(a) Use the second formula to derive the first.

(b) Apply both formulas to the integral $\displaystyle\int \frac{1}{9 - 4x^2}\, dx$. Which do you find easier to use in this problem?

CHAPTER SUMMARY AND PROFICIENCY TEST

Important terms, symbols, and formulas

Antiderivative; indefinite integral

$$\int f(x)\, dx = F(x) + C$$

Power rule: $\int x^n \, dx = \dfrac{1}{n+1} x^{n+1} + C \quad$ (for $n \neq -1$)

Constant multiple rule: $\int cf(x) \, dx = c \int f(x) \, dx$

Sum rule: $\int [f(x) + g(x)] \, dx = \int f(x) \, dx + \int g(x) \, dx$

$\int e^x \, dx = e^x + C \qquad \int \dfrac{1}{x} \, dx = \ln |x| + C$

Integration by substitution: $\int g(u) \dfrac{du}{dx} \, dx = G(u) + C$ where G is an antiderivative of g

Integration by parts: $\int f(x)g(x) \, dx = f(x)G(x) - \int f'(x)G(x) \, dx$, where G is an antiderivative of g

Differential equation

General solution; particular solution

Separable differential equation: If $g(y) \, dy = f(x) \, dx$, then $\int g(y) \, dy = \int f(x) \, dx$

Exponential models:

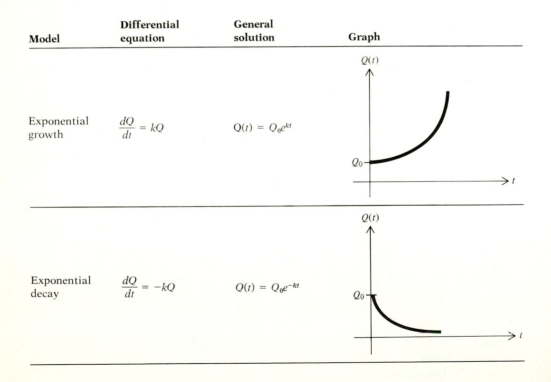

Model	Differential equation	General solution	Graph
Exponential growth	$\dfrac{dQ}{dt} = kQ$	$Q(t) = Q_0 e^{kt}$	
Exponential decay	$\dfrac{dQ}{dt} = -kQ$	$Q(t) = Q_0 e^{-kt}$	

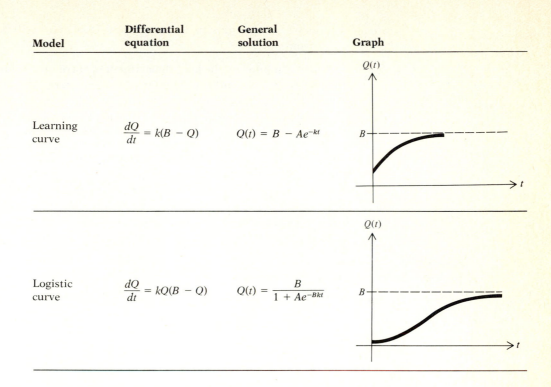

Model	Differential equation	General solution	Graph
Learning curve	$\dfrac{dQ}{dt} = k(B - Q)$	$Q(t) = B - Ae^{-kt}$	
Logistic curve	$\dfrac{dQ}{dt} = kQ(B - Q)$	$Q(t) = \dfrac{B}{1 + Ae^{-Bkt}}$	

Proficiency test   In Problems 1 through 14, find the indicated integral.

1. $\displaystyle \int \left(x^5 - 3x^2 + \frac{1}{x^2} \right) dx$

2. $\displaystyle \int \left(x^{2/3} - \frac{1}{x} + 5 + \sqrt{x} \right) dx$

3. $\displaystyle \int \sqrt{3x + 1}\ dx$

4. $\displaystyle \int (3x + 1)\sqrt{3x^2 + 2x + 5}\ dx$

5. $\displaystyle \int (x + 2)(x^2 + 4x + 2)^5\ dx$

6. $\displaystyle \int \frac{x + 2}{x^2 + 4x + 2}\ dx$

7. $\displaystyle \int (x - 5)^{12}\ dx$

8. $\displaystyle \int x(x - 5)^{12}\ dx$

9. $\displaystyle \int 5e^{3x}\ dx$

10. $\displaystyle \int 5xe^{3x}\ dx$

11. $\displaystyle \int x \ln 3x\ dx$

12. $\displaystyle \int \ln 3x\ dx$

13. $\displaystyle \int \frac{\ln 3x}{x}\ dx$

14. $\displaystyle \int x^3(x^2 + 1)^8\ dx$

15. Find the equation of the curve whose tangent has slope $x(x^2 + 1)^3$ for each value of x and that passes through the point $(1, 5)$.

16. It is estimated that x weeks from now, the number of commuters using a new subway line will be increasing at the rate of $18x^2 + 500$ per week.

Currently 8,000 commuters use the subway. How many will be using it 5 weeks from now?

17. Statistics compiled by the local department of corrections indicate that x years from now the number of inmates in county prisons will be increasing at the rate of $280e^{0.2x}$ per year. Currently 2,000 inmates are housed in county prisons. How many inmates should the county expect 10 years from now?

18. A manufacturer estimates marginal revenue to be $200q^{-1/2}$ dollars per unit when the level of production is q units. The corresponding marginal cost has been found to be $0.4q$ dollars per unit. If the manufacturer's profit is $2,000 when the level of production is 25 units, what is the profit when the level of production is 36 units?

In Problems 19 through 22, find the general solution of the given differential equation.

19. $\dfrac{dy}{dx} = x^3 - 3x^2 + 5$

20. $\dfrac{dy}{dx} = 0.02y$

21. $\dfrac{dy}{dx} = k(80 - y)$

22. $\dfrac{dy}{dx} = y(1 - y)$

In Problems 23 through 26, find the particular solution of the given differential equation that satisfies the given condition.

23. $\dfrac{dy}{dx} = 5x^4 - 3x^2 - 2$; $y = 4$ when $x = 1$

24. $\dfrac{dy}{dx} = 0.06y$; $y = 100$ when $x = 0$

25. $\dfrac{dy}{dx} = 3 - y$; $y = 2$ when $x = 0$

26. $\dfrac{d^2y}{dx^2} = 2$; $y = 5$ and $\dfrac{dy}{dx} = 3$ when $x = 0$

27. The resale value of a certain industrial machine decreases at a rate proportional to the difference between its current value and its scrap value of $5,000. The machine was bought new for $40,000 and was worth $30,000 after 4 years. How much will it be worth when it is 8 years old?

28. A tank currently holds 200 gallons of brine that contains 3 pounds of salt per gallon. Clear water flows into the tank at the rate of 4 gallons per minute, while the mixture, which is kept uniform, runs out of the tank at the rate of 5 gallons per minute. How much salt is in the tank at the end of 100 minutes?

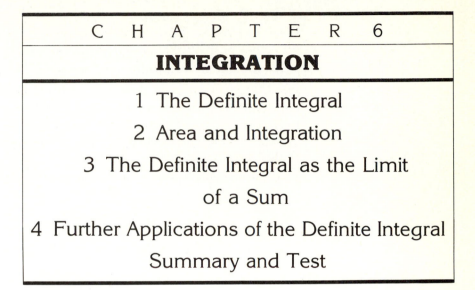

1 THE DEFINITE INTEGRAL

Suppose you know the rate, $f(x) = \dfrac{dF}{dx}$, at which a certain quantity F is changing. To find the actual amount by which the quantity will change between $x = a$ and $x = b$, you first find F by antidifferentiation and then compute the difference

$$\text{Change in } F \text{ between } x = a \text{ and } x = b \;=\; F(b) - F(a)$$

The numerical result of such a computation is called a **definite integral** of the original function f and is denoted by the symbol $\int_a^b f(x)\, dx$.

The definite integral

The definite integral of f from a to b is the difference

$$\int_a^b f(x)\, dx = F(b) - F(a)$$

where F is an antiderivative of f.

The symbol $\int_a^b f(x)\, dx$ is read "the (definite) integral of $f(x)$ from a to b." The numbers a and b are called **limits of integration.** In compu-

259

tations involving definite integrals, it is often convenient to use the symbol $F(x) \Big|_a^b$ to stand for $F(b) - F(a)$.

Many mathematicians give a different definition of the definite integral. They define it to be a "limiting value" of a certain sum and then use this definition to prove that the computation of definite integrals is related to antidifferentiation by the formula $\int_a^b f(x)\, dx = F(b) - F(a)$. This relationship between antiderivatives and sums is known as the **fundamental theorem of calculus** and will be discussed in Section 3.

Here are three practical problems whose solutions are definite integrals.

EXAMPLE 1.1

A study indicates that x months from now the population of a certain town will be increasing at the rate of $2 + 6\sqrt{x}$ people per month. By how much will the population of the town increase during the next 4 months?

SOLUTION

Let $P(x)$ denote the population of the town x months from now. Then $\dfrac{dP}{dx} = 2 + 6\sqrt{x}$, and the amount by which the population will increase during the next 4 months is

$$P(4) - P(0) = \int_0^4 (2 + 6\sqrt{x})\, dx$$

$$= (2x + 4x^{3/2} + C) \Big|_0^4$$

$$= (40 + C) - (0 + C) = 40 \text{ people}$$

Notice what happens to the constant C in the evaluation of a definite integral. It appears in the expressions for both $F(b)$ and $F(a)$ and is eventually eliminated by the subtraction. You may, therefore, omit the constant C altogether when evaluating definite integrals.

EXAMPLE 1.2

At a certain factory, the marginal cost is $3(q - 4)^2$ dollars per unit when the level of production is q units. By how much will the total manufacturing cost increase if the level of production is raised from 6 units to 10 units?

SOLUTION

Let $C(q)$ denote the total cost of producing q units. Then

$$\frac{dC}{dq} = 3(q - 4)^2$$

and

$$\text{Increase in cost} = C(10) - C(6) = \int_6^{10} 3(q - 4)^2 \, dq$$

$$= (q - 4)^3 \, \Big|_6^{10} = 216 - 8 = \$208$$

EXAMPLE 1.3

In a certain community, the demand for gasoline is increasing exponentially at the rate of 5 percent per year. If the current demand is 4 million gallons per year, how much gasoline will be consumed in the community during the next 3 years?

SOLUTION

Let $Q(t)$ denote the total consumption of gasoline in the community over the next t years. Then,

$$\frac{dQ}{dt} = \text{rate of consumption} = 4e^{0.05t} \text{ million gallons per year}$$

and

$$\text{Consumption during the next 3 years} = Q(3) - Q(0)$$

$$= \int_0^3 4e^{0.05t} \, dt$$

$$= 80e^{0.05t} \, \Big|_0^3$$

$$= 80(e^{0.15} - 1)$$

$$= 12.95 \text{ million gallons}$$

The evaluation of definite integrals by substitution

In the next example, you will see how the method of substitution can be used to evaluate a definite integral.

EXAMPLE 1.4

Evaluate $\int_0^1 8x(x^2 + 1)^3 \, dx$.

SOLUTION

Let $u = x^2 + 1$. Then $du = 2x\,dx$, and so

$$\int 8x(x^2 + 1)^3\,dx = \int 4u^3\,du = u^4$$

The limits of integration, 0 and 1, refer to the variable x and not to u. You may, therefore, proceed in one of two ways. Either you can rewrite the antiderivative in terms of x, or you can find the values of u that correspond to $x = 0$ and $x = 1$.

If you choose the first alternative, you find that

$$\int 8x(x^2 + 1)^3\,dx = u^4 = (x^2 + 1)^4$$

and so $\quad \displaystyle\int_0^1 8x(x^2 + 1)^3\,dx = (x^2 + 1)^4 \,\Big|_0^1 = 16 - 1 = 15$

If you choose the second alternative, you use the fact that $u = x^2 + 1$ to conclude that $u = 1$ when $x = 0$, and $u = 2$ when $x = 1$. Hence,

$$\int_0^1 8x(x^2 + 1)^3\,dx = u^4 \,\Big|_1^2 = 16 - 1 = 15$$

Probably the most efficient approach is to adopt the second alternative and write the solution compactly as follows.

$$\int_0^1 8x(x^2 + 1)^3\,dx = \int_1^2 4u^3\,du = u^4 \,\Big|_1^2 = 16 - 1 = 15$$

Here is a summary of the method of substitution for definite integrals.

Integration by substitution

$$\int_a^b g[u(x)]\frac{du}{dx}\,dx = \int_{u(a)}^{u(b)} g(u)\,du$$

Here is one more example.

EXAMPLE 1.5

Evaluate $\displaystyle\int_1^e \frac{\ln x}{x}\,dx$.

SOLUTION

Let $u = \ln x$. Then $du = \dfrac{1}{x}\,dx$, $u(1) = 0$, and $u(e) = 1$. Hence,

$$\int_1^e \frac{\ln x}{x}\, dx = \int_0^1 u\, du = \frac{1}{2} u^2 \Big|_0^1 = \frac{1}{2}$$

The evaluation of definite integrals by parts

The formula for integration by parts can be rephrased for definite integrals as follows.

Integration by parts

$$\int_a^b f(x)g(x)\, dx = f(x)G(x) \Big|_a^b - \int_a^b f'(x)G(x)\, dx$$

where G is an antiderivative of g.

Here is an example.

EXAMPLE 1.6

Evaluate $\displaystyle\int_0^{\ln 2} xe^x\, dx$.

SOLUTION

Use integration by parts with

$$g(x) = e^x \quad \text{and} \quad f(x) = x$$

Then, $$G(x) = e^x \quad \text{and} \quad f'(x) = 1$$

$$\int_0^{\ln 2} xe^x\, dx = xe^x \Big|_0^{\ln 2} - \int_0^{\ln 2} e^x\, dx = (xe^x - e^x) \Big|_0^{\ln 2}$$
$$= (2\ln 2 - 2) - (0 - 1) = 2\ln 2 - 1$$

Problems   In Problems 1 through 20 evaluate the given definite integral.

1. $\displaystyle\int_0^1 (x^4 - 3x^3 + 1)\, dx$

2. $\displaystyle\int_{-1}^0 (3x^5 - 3x^2 + 2x - 1)\, dx$

3. $\displaystyle\int_2^5 (2 + 2u + 3u^2)\, du$

4. $\displaystyle\int_1^9 \left(\sqrt{t} - \frac{1}{\sqrt{t}}\right) dt$

5. $\displaystyle\int_1^3 \left(1 + \frac{1}{x} + \frac{1}{x^2}\right) dx$

6. $\displaystyle\int_{\ln 1/2}^{\ln 2} (e^u - e^{-u})\, du$

7. $\displaystyle\int_{-3}^{-1} \frac{t + 1}{t^3}\, dt$

8. $\displaystyle\int_0^6 x^2(x - 1)\, dx$

9. $\displaystyle\int_1^2 (2x - 4)^5\, dx$

10. $\displaystyle\int_{-3}^0 (2x + 6)^4\, dx$

11. $\displaystyle\int_0^4 \frac{1}{\sqrt{6u + 1}}\, du$

12. $\displaystyle\int_1^2 \frac{x^2}{(x^3 + 1)^2}\, dx$

13. $\displaystyle\int_{-1}^1 (u^3 + u)\sqrt{u^4 + 2u^2 + 6}\, du$

14. $\displaystyle\int_0^1 \frac{6x}{x^2 + 1}\, dx$

15. $\displaystyle\int_2^{e+1} \frac{x}{x - 1}\, dx$

16. $\displaystyle\int_1^2 (u + 1)(u - 2)^9\, du$

17. $\displaystyle\int_1^{e^2} \ln t\, dt$

18. $\displaystyle\int_{1/2}^{e/2} \ln 2t\, dt$

19. $\displaystyle\int_{-2}^2 xe^{-x}\, dx$

20. $\displaystyle\int_0^1 u^2 e^{2u}\, du$

Population growth 21. A study indicates that x months from now the population of a certain town will be increasing at a rate of $5 + 3x^{2/3}$ people per month. By how much will the population of the town increase over the next 8 months?

Distance and speed 22. An object is moving so that its speed after t minutes is $5 + 2t + 3t^2$ meters per minute. How far does the object travel during the 2nd minute?

Depreciation 23. The resale value of a certain industrial machine decreases over a 10-year period at a rate that changes with time. When the machine is x years old, the rate at which its value is changing is $220(x - 10)$ dollars per year. By how much does the machine depreciate during the 2nd year?

Admission to events 24. The promoters of a county fair estimate that t hours after the gates open at 9:00 A.M. visitors will be entering the fair at the rate of $-4(t + 2)^3 + 54(t + 2)^2$ people per hour. How many people will enter the fair between 10:00 A.M. and noon?

Marginal cost 25. At a certain factory, the marginal cost is $6(q - 5)^2$ dollars per unit when the level of production is q units. By how much will the total manufacturing cost increase if the level of production is raised from 10 units to 13 units?

Oil production 26. A certain oil well that yields 400 barrels of crude oil a month will run dry in 2 years. The price of crude oil is currently $18 per barrel and is expected to rise at a constant rate of 3 cents per barrel per month. If the oil is sold as soon as it is extracted from the ground, what will the total future revenue from the well be?

Farming 27. It is estimated that t days from now a farmer's crop will be increasing at a rate of $0.3t^2 + 0.6t + 1$ bushels per day. By how much will the value of the crop increase during the next 5 days if the market price remains fixed at $3 per bushel?

Energy consumption   28. It is estimated that the demand for oil is increasing exponentially at the rate of 10 percent per year. If the demand for oil is currently 30 billion barrels per year, how much oil will be consumed during the next 10 years?

Sales revenue   29. It is estimated that the demand for a manufacturer's product is increasing exponentially at a rate of 2 percent per year. If the current demand is 5,000 units per year and if the price remains fixed at $400 per unit, how much revenue will the manufacturer receive from the sale of the product over the next 2 years?

Efficiency   30. After t hours on the job, a factory worker can produce $100te^{-0.5t}$ units per hour. How many units does a worker who arrives on the job at 8:00 A.M. produce between 10:00 A.M. and noon?

31. (a) Show that $\displaystyle\int_a^b f(x)\,dx + \int_b^c f(x)\,dx = \int_a^c f(x)\,dx$.

(b) Use the formula in part (a) to evaluate $\displaystyle\int_{-1}^1 |x|\,dx$. (*Hint:* Evaluate $\displaystyle\int_{-1}^0 |x|\,dx$ and $\displaystyle\int_0^1 |x|\,dx$ and combine the results.)

(c) Evaluate $\displaystyle\int_0^4 (1 + |x - 3|)^2\,dx$.

Even and odd functions   32. (a) Show that if F is an antiderivative of f, then
$$\int_a^b f(-x)\,dx = -F(-b) + F(-a)$$

(*Hint:* Use the method of substitution with $u = -x$.)

(b) A function f is said to be **even** if $f(-x) = f(x)$. (For example, $f(x) = x^2$ is even.) Use Problem 31 and part (a) to show that if f is even, then
$$\int_{-a}^a f(x)\,dx = 2\int_0^a f(x)\,dx$$

(c) Use part (b) to evaluate $\displaystyle\int_{-1}^1 |x|\,dx$ and $\displaystyle\int_{-2}^2 x^2\,dx$.

(d) A function f is said to be **odd** if $f(-x) = -f(x)$. Use Problem 31 and part (a) to show that if f is odd, then
$$\int_{-a}^a f(x)\,dx = 0$$

(e) Evaluate $\displaystyle\int_{-12}^{12} x^3\,dx$.

2 AREA AND INTEGRATION   There is a surprising connection between definite integrals and the geometric concept of area. If $f(x)$ is continuous and nonnegative on

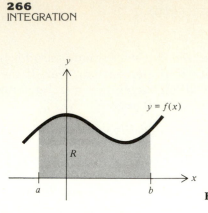

Figure 2.1 The area under the curve $y = f(x)$.

the interval $a \leq x \leq b$ and R is the region under the graph of f between $x = a$ and $x = b$ shown in Figure 2.1, then the area of R is simply the definite integral $\int_a^b f(x)\, dx$.

The area under a curve

> **If $f(x)$ is continuous and nonnegative on the interval $a \leq x \leq b$ and R is the region bounded by the graph of f, the vertical lines $x = a$ and $x = b$, and the x axis, then**
>
> $$\text{Area of } R = \int_a^b f(x)\, dx$$

The use of this formula is illustrated in the next example for a region whose area you already know.

EXAMPLE 2.1

Find the area of the region bounded by the lines $y = 2x$ and $x = 2$, and the x axis.

SOLUTION

The region in question is the triangle in Figure 2.2, and its area is clearly 4.

To compute this area using calculus, apply the integral formula with $f(x) = 2x$. Take $b = 2$ since the region is bounded on the right by the line $x = 2$, and take $a = 0$ since, on the left, the boundary consists of the single point $(0, 0)$, which is part of the vertical line $x = 0$. You will find, as expected, that

$$\text{Area} = \int_0^2 2x\, dx = x^2 \Big|_0^2 = 4$$

Here is another example.

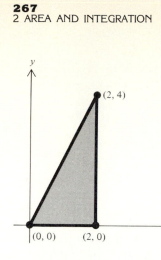

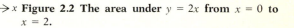

Figure 2.2 The area under $y = 2x$ from $x = 0$ to $x = 2$.

EXAMPLE 2.2

Find the area of the region bounded by the curve $y = -x^2 + 4x - 3$ and the x axis.

SOLUTION

From the factored form of the polynomial

$$y = -x^2 + 4x - 3 = -(x - 3)(x - 1)$$

you see that the x intercepts of the curve are $(1, 0)$ and $(3, 0)$. From the corresponding graph (shown in Figure 2.3) you see that the region in question is below the curve $y = -x^2 + 4x - 3$ and extends from $x = 1$ to $x = 3$. Hence,

$$\text{Area} = \int_1^3 (-x^2 + 4x - 3) \, dx = \left(-\tfrac{1}{3}x^3 + 2x^2 - 3x \right) \Big|_1^3$$

$$= (-9 + 18 - 9) - (-\tfrac{1}{3} + 2 - 3) = \tfrac{4}{3}$$

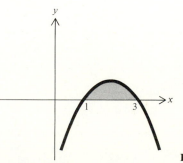

Figure 2.3 The area under $y = -x^2 + 4x - 3$.

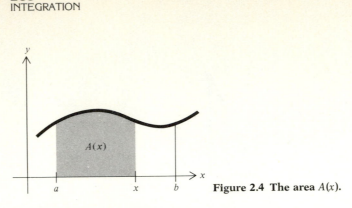

Figure 2.4 The area $A(x)$.

Why the integral formula for area works

To see why the integral formula for area works, suppose $f(x)$ is continuous and nonnegative on the interval $a \leq x \leq b$. For any value of x in this interval, let $A(x)$ denote the area of the region under the graph of f between a and x as shown in Figure 2.4.

Your goal is to show that $A(b) = \displaystyle\int_a^b f(x)\, dx$. The key step is to establish that the derivative $A'(x)$ of the area function is equal to $f(x)$. To do this, consider the difference quotient

$$\frac{A(x + \Delta x) - A(x)}{\Delta x}$$

The expression $A(x + \Delta x) - A(x)$ in the numerator is just the area under the curve between x and $x + \Delta x$. If Δx is small, this area is approximately the same as the area of the rectangle whose height is $f(x)$ and whose width is Δx, as indicated in Figure 2.5. That is,

$$A(x + \Delta x) - A(x) \approx f(x)\, \Delta x$$

or, equivalently, $\qquad \dfrac{A(x + \Delta x) - A(x)}{\Delta x} \approx f(x)$

As Δx approaches zero, the error resulting from this approximation approaches zero and it follows that

$$\frac{A(x + \Delta x) - A(x)}{\Delta x} \to f(x) \qquad \text{as} \qquad \Delta x \to 0$$

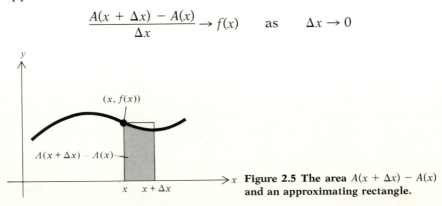

Figure 2.5 The area $A(x + \Delta x) - A(x)$ and an approximating rectangle.

But, by the definition of the derivative, this quotient also approaches $A'(x)$ as Δx approaches zero. Hence,

$$A'(x) = f(x)$$

and so

$$\int_a^b f(x)\,dx = A(b) - A(a)$$

But $A(a)$ is the area under the curve between $x = a$ and $x = a$, which is clearly zero. Hence,

$$\int_a^b f(x)\,dx = A(b)$$

and the formula is verified.

Probability density functions

One of the most important applications in the social, managerial, and life sciences of the integral formula for area is the computation of probabilities. Here is a simplified outline of a typical situation.

Suppose that you are the quality-control manager for a company that manufactures electronic components and that you are interested in predicting what fraction of all the components manufactured by the company can be expected to have life spans within a certain range. That is, you are interested in determining the **probability** that a component selected at random will have a life span x in some range $a \le x \le b$. To begin, you select a large number of components at random, test them in the laboratory, and record the life span of each. Using the resulting data, you can construct a positive continuous function $f(x)$ with the following property: The probability, $P(a \le x \le b)$, that the life span of a randomly selected component will be between a and b is the area under the graph of f from $x = a$ to $x = b$. The situation is illustrated in Figure 2.6.

A function with this property is said to be a **probability density function** for the variable x. The construction of probability density functions from experimental data involves techniques beyond the scope of this book and is explained in most probability and statistics

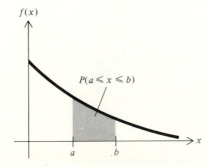

Figure 2.6 A probability density function.

texts. The purpose of this discussion is to show you how to use integral calculus to compute probabilities once the appropriate probability density function is known. Here is an example.

EXAMPLE 2.3

The probability density function for the life span of electronic components produced by a certain company is $f(x) = 0.02e^{-0.02x}$, where x denotes the life span (in months) of a randomly selected component.

(a) What is the probability that the life span of a component selected at random will be between 20 and 30 months?
(b) What is the probability that the life span of a component selected at random will be less than or equal to 20 months?
(c) What is the probability that the life span of a component selected at random will be greater than 20 months?

SOLUTION

(a) The desired probability $P(20 \leq x \leq 30)$ is the area (Figure 2.7a) under the graph of the density function between $x = 20$ and $x = 30$. Using the integral formula to compute this area, you get

$$P(20 \leq x \leq 30) = \int_{20}^{30} 0.02e^{-0.02x} \, dx = -e^{-0.02x} \Big|_{20}^{30}$$
$$= -0.5488 + 0.6703 = 0.1215$$

That is, about 12.15 percent of the components manufactured by the company will have life spans of between 20 and 30 months.

(b) The desired probability is $P(0 \leq x \leq 20)$ which is the area (Figure 2.7b) under the density function between $x = 0$ and $x = 20$. That is,

$$P(0 \leq x \leq 20) = \int_{0}^{20} 0.02e^{-0.02x} \, dx = -e^{-0.02x} \Big|_{0}^{20}$$
$$= -0.6703 + 1 = 0.3297$$

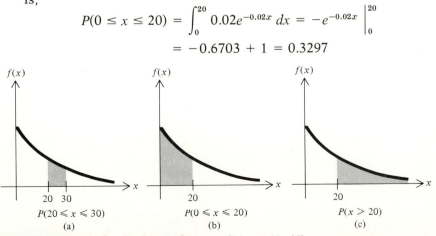

Figure 2.7 Areas under the density function $f(x) = 0.02e^{-0.02x}$.

That is, roughly $\frac{1}{3}$ of the components will fail during the first 20 months.

(c) The fraction of components whose life span is greater than 20 months is 1 minus the fraction whose life span is less than or equal to 20 months. Hence,

$$P(x > 20) = 1 - P(0 \leq x \leq 20) = 1 - 0.3297 = 0.6703$$

In geometric terms, this is the area (Figure 2.7c) under the density function to the right of $x = 20$.

A variable x (like the one in Example 2.3) that has a probability density function of the form $f(x) = ke^{-kx}$ is said to have an **exponential distribution.** It can be shown that the constant k in the density function for such a variable is the reciprocal of the average value of x. For instance, in Example 2.3 the average life span of the electronic components is $\dfrac{1}{k} = \dfrac{1}{0.02} = 50$ months. Quantities that may have exponential distributions include the life span of electrical appliances, the duration of telephone calls, and the interval between the arrivals of successive planes at an airport.

The area between two curves In some practical problems, you may have to compute the area *between* two curves. Suppose $f(x)$ and $g(x)$ are nonnegative functions and that $f(x) \geq g(x)$ on the interval $a \leq x \leq b$ as shown in Figure 2.8.

To find the area of the region R between the curves from $x = a$ to $x = b$, you simply subtract the area under the lower curve $y = g(x)$ from the area under the upper curve $y = f(x)$. That is,

$$\text{Area of } R = \int_a^b f(x)\,dx - \int_a^b g(x)\,dx = \int_a^b [f(x) - g(x)]\,dx$$

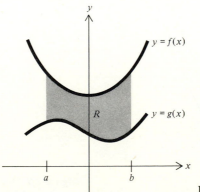

Figure 2.8 The area between two curves.

It can be shown that this formula remains valid, even if the functions f and g are not assumed to be nonnegative.

The area between
two curves

> If $f(x)$ and $g(x)$ are continuous on the interval $a \le x \le b$ with $f(x) \ge g(x)$, and if R is the region bounded by the graphs of f and g and the vertical lines $x = a$ and $x = b$, then
>
> $$\text{Area of } R = \int_a^b [f(x) - g(x)]\, dx$$

Here are two examples.

EXAMPLE 2.4

Find the area of the region bounded by the curves $y = x^2 + 1$ and $y = 2x - 2$ between $x = -1$ and $x = 2$.

SOLUTION

So that you can visualize the situation, begin by sketching the region as shown in Figure 2.9. Then apply the integral formula with $f(x) = x^2 + 1$, $g(x) = 2x - 2$, $a = -1$, and $b = 2$ to get

$$\text{Area} = \int_{-1}^{2} [(x^2 + 1) - (2x - 2)]\, dx = \int_{-1}^{2} (x^2 - 2x + 3)\, dx$$

$$= (\tfrac{1}{3}x^3 - x^2 + 3x) \Big|_{-1}^{2} = \tfrac{14}{3} - (-\tfrac{13}{3}) = 9$$

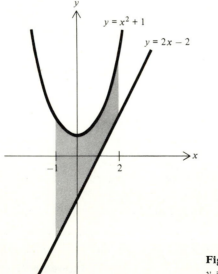

Figure 2.9 The area between $y = x^2 + 1$ and $y = 2x - 2$ from -1 to 2.

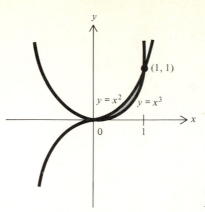

Figure 2.10 The area between $y = x^3$ **and** $y = x^2$.

EXAMPLE 2.5

Find the area of the region bounded by the curves $y = x^3$ and $y = x^2$.

SOLUTION

Graph the curves (Figure 2.10), and solve their equations simultaneously to find the points of intersection, $(0, 0)$ and $(1, 1)$. The region in question is bounded above by the curve $y = x^2$ and below by the curve $y = x^3$ and extends from $x = 0$ to $x = 1$. Hence,

$$\text{Area} = \int_0^1 (x^2 - x^3)\, dx = \left(\tfrac{1}{3}x^3 - \tfrac{1}{4}x^4 \right) \Big|_0^1 = \tfrac{1}{12}$$

A business application The total net earnings generated by an industrial machine over a period of years is the difference between the total revenue generated by the machine and the total cost of operating and servicing the machine. In the next example, you will see how the total net earnings of a machine can be interpreted as the area between two curves.

EXAMPLE 2.6

When it is x years old, a certain industrial machine generates revenue at the rate of $R(x) = 5,000 - 20x^2$ dollars per year and results in costs that accumulate at the rate of $C(x) = 2,000 + 10x^2$ dollars per year.

(a) For how many years is the use of the machine profitable?
(b) What are the total net earnings generated by the machine during the period of time in part (a)?

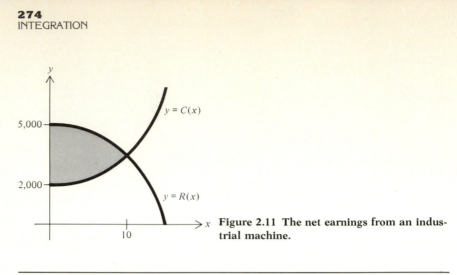

Figure 2.11 The net earnings from an industrial machine.

SOLUTION

Begin by sketching the functions R and C as shown in Figure 2.11.

(a) Use of the machine will be profitable as long as the rate at which revenue is generated is greater than the rate at which costs accumulate, that is, until $R(x) = C(x)$. Setting $R(x)$ equal to $C(x)$ and solving for x, you get

$$5{,}000 - 20x^2 = 2{,}000 + 10x^2$$

$$30x^2 = 3{,}000$$

$$x^2 = 100$$

$$x = \pm 10$$

and you can conclude that the use of the machine will be profitable until the machine is 10 years old.

(b) The functions $R(x)$ and $C(x)$ represent the rates of change of total revenue and total cost, respectively, and hence their difference, $R(x) - C(x)$, represents the rate of change of the total net earnings generated by the machine. It follows that the total net earnings for the period between $x = 0$ and $x = 10$ is the definite integral

$$\int_0^{10} [R(x) - C(x)]\, dx = \int_0^{10} [(5{,}000 - 20x^2) - (2{,}000 + 10x^2)]\, dx$$

$$= \int_0^{10} (3{,}000 - 30x^2)\, dx$$

$$= (3{,}000x - 10x^3)\Big|_0^{10} = \$20{,}000$$

Notice that this definite integral representing the total net earnings can be interpreted geometrically as the area between the curves $y = R(x)$ and $y = C(x)$ from $x = 0$ to $x = 10$.

Problems

1. Use calculus to find the area of the triangle bounded by the line $y = 4 - 3x$ and the coordinate axes.

2. Use calculus to find the area of the triangle with vertices $(-4, 0)$, $(2, 0)$, and $(2, 6)$.

3. Use calculus to find the area of the rectangle with vertices $(1, 0)$, $(-2, 0)$, $(-2, 5)$, and $(1, 5)$.

4. Use calculus to find the area of the trapezoid bounded by the lines $y = x + 6$ and $x = 2$ and the coordinate axes.

5. Find the area of the region bounded by the curve $y = \sqrt{x}$, the lines $x = 4$ and $x = 9$, and the x axis.

6. Find the area of the region bounded by the curve $y = 4x^3$, the line $x = 2$, and the x axis.

7. Find the area of the region bounded by the curve $y = 1 - x^2$ and the x axis.

8. Find the area of the region bounded by the curve $y = -x^2 - 6x - 5$ and the x axis.

9. Find the area of the region bounded by the curve $y = e^x$, the lines $x = 0$ and $x = \ln \frac{1}{2}$, and the x axis.

10. Find the area of the region bounded by the curve $y = x^2 - 2x$ and the x axis. (*Hint:* Reflect the region across the x axis and integrate the corresponding function.)

Product reliability

11. The probability density function for the life span of the light bulbs manufactured by a certain company is $f(x) = 0.01e^{-0.01x}$, where x denotes the life span (in hours) of a randomly selected bulb.
    (a) What is the probability that the life span of a randomly selected bulb will be between 50 and 60 hours?
    (b) What is the probability that the life span of a randomly selected bulb will be less than or equal to 60 hours?
    (c) What is the probability that the life span of a randomly selected bulb will be greater than 60 hours?

Duration of telephone calls

12. The probability density function for the duration of telephone calls in a certain city is $f(x) = 0.5e^{-0.5x}$, where x denotes the duration (in minutes) of a randomly selected call.
    (a) What percentage of the calls last between 2 and 3 minutes?
    (b) What percentage of the calls last 2 minutes or less?
    (c) What percentage of the calls last more than 2 minutes?

Airplane arrivals

13. The probability density function for the time interval between the arrivals of successive planes at a certain airport is $f(x) = $

$0.2e^{-0.2x}$, where x is the time (in minutes) between the arrivals of a randomly selected pair of successive planes.
   (a) What is the probability that two successive planes selected at random will arrive within 5 minutes of one another?
   (b) What is the probability that two successive planes selected at random will arrive more than 6 minutes apart?

14. Find the area of the region bounded by the curves $y = x^2 + 3$ and $y = 1 - x^2$ between $x = -2$ and $x = 1$.

15. Find the area of the region bounded by the curves $y = x^2 + 5$ and $y = -x^2$, the line $x = 3$, and the y axis.

16. Find the area of the region bounded by the curve $y = e^x$ and the lines $y = 1$ and $x = 1$.

17. Find the area of the region bounded by the curve $y = x^2$ and the line $y = x$.

18. Find the area of the region bounded by the curve $y = x^2$ and the line $y = 4$.

19. Find the area of the region bounded by the curves $y = \sqrt{x}$ and $y = x^2$.

20. (a) Find the area of the region to the right of the y axis that is bounded by the curves $y = x$ and $y = x^3$.
    (b) Find the total area of the region bounded by the curves $y = x$ and $y = x^3$.

21. Find the area of the region bounded by the curves $y = x^3 - 6x^2$ and $y = -x^2$.

22. (a) Find the area of the region to the right of the y axis that is bounded above by the curve $y = 4 - x^2$ and below by the line $y = 3$.
    (b) Find the area of the region to the right of the y axis that lies below the line $y = 3$ and is bounded by the curve $y = 4 - x^2$, the line $y = 3$, and the coordinate axes. (*Hint:* Subtract one area from another.)

23. Find the area of the region that lies below the curve $y = x^2 + 4$ and is bounded by this curve, the line $y = -x + 10$, and the coordinate axes.

Industrial machinery   24. When it is x years old, a certain industrial machine generates revenue at the rate of $R(x) = 6,025 - 10x^2$ dollars per year and results in costs that accumulate at the rate of $C(x) = 4,000 + 15x^2$ dollars per year.
    (a) For how many years is the use of the machine profitable?

(b) What are the total net earnings generated by the machine during the period of time in part (a)?

Investment 25. You have a certain amount of money to invest in one of two competing investment plans. After x years, the first plan will generate income at the rate of $R_1(x) = 50 + 3x^2$ dollars per year while the second will generate income at the constant rate of $R_2(x) = 200$ dollars per year.

(a) If you invest your money in the second plan, how much more income will you have earned by the end of 5 years than you would have with the first plan?

(b) Interpret your answer to part (a) as the area of a region between two curves.

Efficiency 26. After x hours on the job, one factory worker is producing $Q_1(x) = 60 - 2(x - 1)^2$ units per hour while a second worker is producing $Q_2(x) = 50 - 5x$ units per hour.

(a) If both arrive on the job at 8:00 A.M., how many more units will the first worker have produced by noon than the second worker?

(b) Interpret your answer in part (a) as the area of a region between two curves.

3 THE DEFINITE INTEGRAL AS THE LIMIT OF A SUM

In this section you will see an important relationship between antiderivatives and sums known as the **fundamental theorem of calculus.** It can be established using the following geometric argument based on the interpretation of definite integrals as areas.

Suppose that $f(x)$ is nonnegative and continuous on the interval $a \le x \le b$. You can approximate the area under the graph of f between $x = a$ and $x = b$ as follows: First divide the interval $a \le x \le b$ into n equal subintervals of width Δx and let x_j denote the beginning of the jth subinterval. Then draw n rectangles such that the base of the jth rectangle is the jth subinterval and the height of the jth rectangle is $f(x_j)$. The situation is illustrated in Figure 3.1.

The area of the jth rectangle is $f(x_j) \Delta x$ and is an approximation to the area under the curve from $x = x_j$ to $x = x_{j+1}$. The sum of the areas of all n rectangles is

$$f(x_1) \Delta x + f(x_2) \Delta x + \cdots + f(x_n) \Delta x$$

This sum is an approximation to the total area under the curve from $x = a$ to $x = b$ and hence an approximation to the corresponding definite integral

$$\int_a^b f(x)\, dx$$

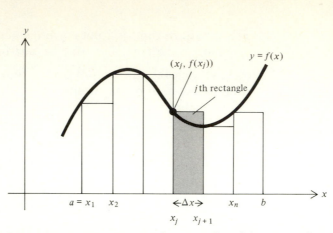

Figure 3.1 An approximation by rectangles of the area under a curve.

That is, $\quad f(x_1)\,\Delta x + f(x_2)\,\Delta x + \cdots + f(x_n)\,\Delta x \approx \int_a^b f(x)\,dx$

As Figure 3.2 suggests, the sum of the areas of the rectangles approaches the actual area under the curve as the number of rectangles increases without bound. That is, as n increases without bound,

$$f(x_1)\,\Delta x + f(x_2)\,\Delta x + \cdots + f(x_n)\,\Delta x \to \int_a^b f(x)\,dx = F(b) - F(a)$$

where F is an antiderivative of f.

This is the relationship between sums and integrals that is known as the fundamental theorem of calculus. Although this argument establishes it only for nonnegative functions, it actually holds for any function that is continuous on the interval $a \le x \le b$.

You can write the relationship between definite integrals and sums more compactly if you use the following **summation notation.** (The use of summation notation is discussed in more detail in the algebra review in Section A of the appendix.) To describe the sum

$$f(x_1)\,\Delta x + f(x_2)\,\Delta x + \cdots + f(x_n)\,\Delta x$$

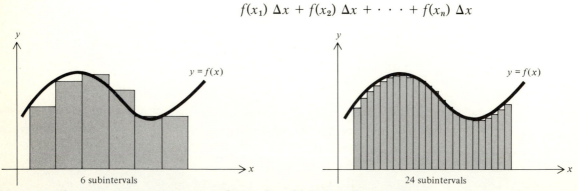

Figure 3.2 The approximation improves as the number of subintervals increases.

specify the general term $f(x_j) \, \Delta x$ and use the symbol

$$\sum_{j=1}^{n} f(x_j) \, \Delta x$$

to indicate that n terms of this form are to be added together, starting with the term in which $j = 1$ and ending with the term in which $j = n$. Thus,

$$\sum_{j=1}^{n} f(x_j) \, \Delta x = f(x_1) \, \Delta x + f(x_2) \, \Delta x + \cdots + f(x_n) \, \Delta x$$

and the fundamental theorem of calculus can be stated compactly as follows.

The fundamental theorem of calculus

Suppose f is continuous on the interval $a \leq x \leq b$ which is divided into n equal subintervals of length Δx by $x_1, x_2, \ldots, x_n$, where x_j is the beginning of the jth subinterval. Then, as n increases without bound,

$$\sum_{j=1}^{n} f(x_j) \, \Delta x \rightarrow \int_{a}^{b} f(x) \, dx = F(b) - F(a)$$

where F is any antiderivative of f.

Actually this is a somewhat restricted version of a more general characterization of definite integrals. The relationship between definite integrals and sums still holds if $f(x_j)$ in the jth term of the sum is replaced by $f(x_j')$, where x_j' is *any* point whatsoever in the jth subinterval. Moreover, the n subintervals need not have equal width, as long as the width of the largest eventually approaches zero as n increases. For most applications, however, the restricted characterization is quite sufficient, and you will have no need to use the more general result in this text.

Here are two economic examples illustrating the use of the relationship between integrals and sums. Actually, these two problems could be formulated as differential equations and solved without this relationship. (In fact, you did one of them this way in Chapter 5, Example 3.6.) There are, however, many important problems that you cannot solve easily without using the characterization of the definite integral as the limit of a sum, and some of these will be presented in the next section.

Total revenue In the following example, the relationship between integrals and sums is used to compute total revenue.

EXAMPLE 3.1

A certain oil well that yields 300 barrels of crude oil a month will run dry in 3 years. It is estimated that t months from now the price of crude oil will be $P(t) = 18 + 0.3\sqrt{t}$ dollars per barrel. If the oil is sold as soon as it is extracted from the ground, what will the total future revenue from the well be?

SOLUTION

To approximate the revenue during the 36-month period, divide the interval $0 \le t \le 36$ into n equal subintervals of length Δt and let t_j denote the beginning of the jth subinterval as shown in Figure 3.3.

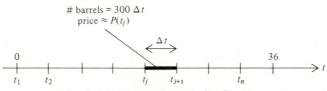

Figure 3.3 The division into subintervals of a 36-month period.

During each subinterval, $300\,\Delta t$ barrels of crude oil are produced. Moreover, if Δt is small, the price of crude oil throughout the jth subinterval will be approximately $P(t_j)$ dollars per barrel, the price that was in effect at the beginning of the subinterval. Hence,

$$\text{Revenue from } j\text{th subinterval} \approx 300P(t_j)\,\Delta t$$

and

$$\text{Total revenue} \approx \sum_{j=1}^{n} 300P(t_j)\,\Delta t$$

As n increases, the length Δt of the subintervals decreases and the approximation improves. In fact,

$$\sum_{j=1}^{n} 300P(t_j)\,\Delta t \to \text{total revenue}$$

as n increases without bound. But according to the characterization of the definite integral as the limit of a sum,

$$\sum_{j=1}^{n} 300P(t_j)\,\Delta t \to \int_0^{36} 300P(t)\,dt$$

as n increases without bound. Hence,

$$\text{Total revenue} = \int_0^{36} 300P(t)\,dt = 300\int_0^{36}(18 + 0.3\sqrt{t})\,dt$$

$$= 300(18t + 0.2t^{3/2})\,\Big|_0^{36}$$

$$= \$207{,}360$$

Inventory storage costs   In the next example, you will see how to calculate the total cost resulting from the storage of unused inventory.

EXAMPLE 3.2

A retailer receives a shipment of 10,000 kilograms of rice that will be used up over a 5-month period at a constant rate of 2,000 kilograms per month. If storage costs are 1 cent per kilogram per month, how much will the retailer pay in storage costs over the next 5 months?

SOLUTION

Let $Q(t)$ denote the number of kilograms of rice in storage after t months. Then $Q(t) = 10,000 - 2,000t$.

Divide the interval $0 \leq t \leq 5$ into n equal subintervals of length Δt and let t_j denote the beginning of the jth subinterval as shown in Figure 3.4.

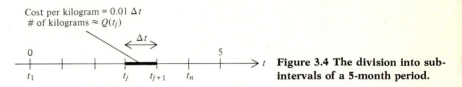

Figure 3.4 The division into sub-intervals of a 5-month period.

During the jth subinterval,

$$\text{Cost per kilogram} = (\text{cost per kilogram per month})(\text{number of months})$$
$$= 0.01 \, \Delta t$$

and $$\text{Number of kilograms in storage} \approx Q(t_j)$$

Hence,

$$\text{Storage cost during } j\text{th subinterval} \approx 0.01 Q(t_j) \, \Delta t$$

and so $$\text{Total storage cost} \approx \sum_{j=1}^{n} 0.01 Q(t_j) \, \Delta t$$

The approximation improves as n increases without bound, and it follows from the characterization of the definite integral as the limit of a sum that

$$\text{Total storage cost} = \int_0^5 0.01 Q(t) \, dt$$

$$= \int_0^5 0.01(10,000 - 2,000t) \, dt$$

$$= \int_0^5 (100 - 20t) \, dt$$

$$= \$250$$

Verify that this is the same as the cost of storing $\frac{10,000}{2} = 5,000$ kilograms of rice for the entire 5 months.

Problems Use the characterization of the definite integral as the limit of a sum to solve the following problems.

Distance and speed

1. An object is moving so that its speed after t minutes is $S(t) = 1 + 4t + 3t^2$ meters per minute. How far does the object travel during the 3rd minute?

Growth

2. A tree has been transplanted and after x years is growing at a rate of $f(x) = 0.5 + \dfrac{1}{(x + 1)^2}$ meters per year. By how much does the tree grow during the 2nd year?

Land values

3. It is estimated that t years from now the value of a certain parcel of land will be increasing at a rate of $r(t)$ dollars per year. Find an expression for the amount by which the value of the land will increase during the next 5 years.

Admission to events

4. The promoters of a county fair estimate that t hours after the gates open at 9:00 A.M., visitors will be entering the fair at a rate of $r(t)$ people per hour. Find an expression for the number of people who will enter the fair between 11:00 A.M. and 1:00 P.M.

Sales revenue

5. A bicycle manufacturer expects that x months from now consumers will be buying 5,000 bicycles a month at a price of $P(x) = 80 + 3\sqrt{x}$ dollars per bicycle. What is the total revenue the manufacturer can expect from the sale of the bicycles over the next 16 months?

Sales revenue

6. A bicycle manufacturer expects that x months from now consumers will be buying $f(x) = 5,000 + 60\sqrt{x}$ bicycles per month at a price of $P(x) = 80 + 3\sqrt{x}$ dollars per bicycle. What is the total revenue the manufacturer can expect from the sale of the bicycles over the next 16 months?

Sales revenue

7. A manufacturer expects that x months from now consumers will be buying $n(x)$ lamps per month at a price of $p(x)$ dollars per lamp. Find an expression for the total revenue the manufacturer can expect from the sale of the lamps over the next 12 months.

Oil production

8. Suppose that t months from now an oil well will be producing crude oil at a rate of $r(t)$ barrels per month and that the price of crude oil will be $p(t)$ dollars per barrel. Assuming that the oil is sold as soon as it is extracted from the ground, find an expression for the total revenue from the oil well over the next 2 years.

Farming

9. It is estimated that t days from now a farmer's crop will be increasing at the rate of $0.3t^2 + 0.6t + 1$ bushels per day. By how much will the value of the crop increase during the next 5 days if the market price remains fixed at $3 per bushel?

Water pollution 10. It is estimated that t years from now the population of a certain lakeside community will be changing at the rate of $0.6t^2 + 0.2t + 0.5$ thousand people per year. Environmentalists have found that the level of pollution in the lake increases at the rate of approximately 5 units per 1,000 people. By how much will the pollution in the lake increase during the next 2 years?

Storage cost 11. The owner of a chain of health food restaurants receives a shipment of 12,000 pounds of soybeans that will be used at a constant rate of 300 pounds per week. If the cost of storing the soybeans is 0.2 cent per pound per week, how much will the restaurant owner have to pay in storage costs over the next 40 weeks?

Storage cost 12. A manufacturer receives N units of a certain raw material that are initially placed in storage and then withdrawn and used at a constant rate until the supply is exhausted 1 year later. If storage costs remain fixed at p dollars per unit per year, find an expression for the total storage cost the manufacturer will pay during the year. Show that this is the same as the cost of storing $\dfrac{N}{2}$ units for the entire year.

4 FURTHER APPLICATIONS OF THE DEFINITE INTEGRAL

In this section you will see some further applications of the characterization of the definite integral as the limit of a sum.

The average value of a function

In many practical situations one is interested in the **average value** of a continuous function over an interval, such as the average level of air pollution over a 24-hour period, the average speed of a truck during a 3-hour trip, the average productivity of a worker during a production run, and the average blood pressure of a patient during an operation. Here is a simple formula involving a definite integral that you can use to compute averages of this type.

The average value of a function

The average value of a continuous function $f(x)$ over the interval $a \leq x \leq b$ is given by the formula

$$\text{Average value} = \frac{1}{b-a} \int_a^b f(x)\,dx$$

To see why this formula is valid, imagine that the interval $a \leq x \leq b$ is divided into n equal subintervals with x_j denoting the beginning of the jth subinterval. The numerical average of the corresponding function values $f(x_1), f(x_2), \ldots, f(x_n)$ is

$$\frac{f(x_1) + f(x_2) + \cdots + f(x_n)}{n}$$

As n increases, this numerical average becomes increasingly sensitive to fluctuations in f and therefore approximates with increasing accuracy the average value of f over the entire interval $a \le x \le b$. That is,

$$\frac{f(x_1) + f(x_2) + \cdots + f(x_n)}{n} \to \begin{array}{l} \text{average value of } f \\ \text{over } a \le x \le b \end{array}$$

as n increases without bound.

To write this as an integral, observe that if the interval $a \le x \le b$ is divided into n equal subintervals, then the length Δx of each is $\dfrac{b - a}{n}$. Hence,

$$\frac{f(x_1) + f(x_2) + \cdots + f(x_n)}{n} = \frac{1}{b - a}\left[f(x_1)\frac{b - a}{n} + f(x_2)\frac{b - a}{n} \right.$$
$$\left. + \cdots + f(x_n)\frac{b - a}{n} \right]$$
$$= \frac{1}{b - a}\sum_{j=1}^{n} f(x_j)\, \Delta x$$

where $\Delta x = \dfrac{b - a}{n}$. As n increases without bound, the sum $\displaystyle\sum_{j=1}^{n} f(x_j)\, \Delta x$ approaches the definite integral $\displaystyle\int_a^b f(x)\, dx$, and it follows that

$$\begin{array}{l} \text{Average value of } f(x) \\ \text{over } a \le x \le b \end{array} = \frac{1}{b - a}\int_a^b f(x)\, dx$$

The use of the integral formula for average value is illustrated in the next example.

EXAMPLE 4.1

For several weeks, the highway department has been recording the speed of freeway traffic flowing past a certain downtown exit. The data suggest that between the hours of 1:00 P.M. and 6:00 P.M. on a normal weekday, the speed of the traffic at the exit is approximately $S(t) = 2t^3 - 21t^2 + 60t + 40$ kilometers per hour, where t is the number of hours past noon. Compute the average speed of the traffic between the hours of 1:00 P.M. and 6:00 P.M.

SOLUTION

Your goal is to find the average value of $S(t)$ on the interval $1 \leq t \leq 6$. Using the integral formula, you get

$$\text{Average speed} = \frac{1}{6-1} \int_1^6 (2t^3 - 21t^2 + 60t + 40)\, dt$$

$$= \tfrac{1}{5}(\tfrac{1}{2}t^4 - 7t^3 + 30t^2 + 40t)\Big|_1^6$$

$$= \tfrac{1}{5}(456 - 63.5) = 78.5 \qquad \text{kilometers per hour}$$

Geometric interpretation of average value   The integral formula for average value has an interesting geometric interpretation. To see this, multiply both sides of the equation by $b - a$ to get

$$(b - a)(\text{average value}) = \int_a^b f(x)\, dx$$

If $f(x)$ is nonnegative, the integral on the right-hand side of this equation is equal to the area under the graph of f from $x = a$ to $x = b$ (Figure 4.1a). The product on the left-hand side is the area of a rectangle whose width is $b - a$ and whose height is the average value of f over the interval $a \leq x \leq b$ (Figure 4.1b). It follows that the average value of $f(x)$ over the interval $a \leq x \leq b$ is equal to the height of the rectangle whose base is this interval and whose area is the same as the area under the graph of $f(x)$ from $x = a$ to $x = b$.

The amount of an annuity   An **annuity** is a sequence of equal payments made at regular intervals over a specified period of time. Once a payment is made it earns interest at a fixed rate until the expiration or **term** of the annuity. The total amount of money (deposits plus interest) accumulated in this way is

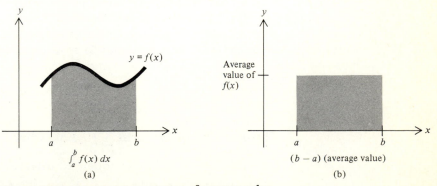

Figure 4.1 Geometric interpretation of average value.

known as the **amount** of the annuity. The amount of an annuity can be expressed as a sum which can often be closely approximated by a definite integral. Here is an example.

EXAMPLE 4.2

On the 1st of each month you deposit $100 in an account yielding interest at the annual rate of 8 percent compounded continuously. Use a definite integral to estimate how much you will have in your account at the end of 2 years (immediately before you make your 25th deposit).

SOLUTION

Recall that P dollars invested at 8 percent compounded continuously will be worth $Pe^{0.08t}$ dollars t years later.

Let t_j denote the time (in years) of your jth deposit. This deposit will remain in the account $2 - t_j$ years and hence will grow to $100e^{0.08(2-t_j)}$ dollars. The situation is illustrated in Figure 4.2.

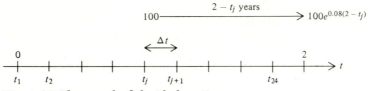

Figure 4.2 The growth of the jth deposit.

The amount of the annuity (your total balance at the end of the 2-year term) is the sum of 24 terms of the form $100e^{0.08(2-t_j)}$, one for each of the 24 deposits. That is,

$$\text{Amount of annuity} = \sum_{j=1}^{24} 100e^{0.08(2-t_j)}$$

Since the times, t_j, of the deposits are known, you could compute this sum directly, but the resulting calculations would be cumbersome. A better approach is to approximate the sum by a definite integral. To do this, observe that $\Delta t = \frac{1}{12}$ (since the time between deposits is 1 month or $\frac{1}{12}$ of a year) and rewrite the sum as

$$\text{Amount of annuity} = \sum_{j=1}^{24} 100e^{0.08(2-t_j)} = \sum_{j=1}^{24} 1{,}200e^{0.08(2-t_j)}\,\Delta t$$

This is a sum of the form $\sum_{j=1}^{n} f(t_j)\,\Delta t$, where $t_1, t_2, \ldots, t_n$ divide the interval $0 \leq t \leq 2$ into n equal subintervals, which can be approximated by the definite integral $\int_0^2 f(t)\,dt$. Hence,

$$\text{Amount of annuity} \approx \int_0^2 1{,}200e^{0.08(2-t)}\,dt$$

$$= 1{,}200e^{0.16}\int_0^2 e^{-0.08t}\,dt$$

$$= -\frac{1{,}200}{0.08}\,e^{0.16}\left(e^{-0.08t}\right)\Big|_0^2$$

$$= -15{,}000e^{0.16}(e^{-0.16} - 1)$$

$$= -15{,}000 + 15{,}000e^{0.16}$$

$$= \$2{,}602.66$$

**The present value
of an annuity**

In Chapter 4, Section 5, you saw that the present value of B dollars payable t years from now is the amount P that should be invested today so that it will be worth B dollars at the end of t years. If interest is compounded continuously, this amount is given by the formula

$$P = Be^{-rt}$$

where r is the interest rate expressed as a decimal.

The present value of an annuity that consists of n payments of R dollars is the amount of money that must be invested today to generate the same sequence of payments. This means that withdrawals of R dollars can be made from the investment for each of the next n periods, after which the investment will be used up. In the next example, you will learn how to use a definite integral to estimate the present value of an annuity.

EXAMPLE 4.3

Use a definite integral to estimate the present value of an annuity that pays \$100 per month for each of the next 2 years if the prevailing interest rate remains fixed at 8 percent per year compounded continuously.

SOLUTION

Let t_j denote the time (in years) of the jth payment of \$100. Then, as is illustrated in Figure 4.3,

$$\text{Present value of } j\text{th payment} = 100e^{-0.08t_j}$$

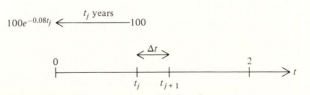

Figure 4.3 The present value of the jth payment.

and $\qquad$ Present value of annuity $= \displaystyle\sum_{j=1}^{24} 100e^{-0.08t_j}$

Since $\Delta t = \frac{1}{12}$, you can rewrite this sum as

$$\text{Present value of annuity} = \sum_{j=1}^{24} 1{,}200e^{-0.08t_j}\, \Delta t$$

which can be approximated by the definite integral $\displaystyle\int_0^2 1{,}200e^{-0.08t}\,dt.$ Hence,

$$\text{Present value of annuity} \approx \int_0^2 1{,}200e^{-0.08t}\,dt = -\frac{1{,}200}{0.08} e^{-0.08t} \Big|_0^2$$

$$= -15{,}000(e^{-0.16} - 1)$$
$$= \$2{,}217.84$$

The present value of an income stream

The present value of an investment scheme or business venture that generates income continuously at a certain rate over a specified period of time is the amount of money that must be deposited today at the prevailing interest rate to generate the same income stream over the same specified period of time. The calculation of the present value of a continuous income stream is illustrated in the next example.

EXAMPLE 4.4

The management of a national chain of ice-cream parlors is selling a 5-year franchise to operate its newest outlet in Madison, Wisconsin. Past experience in similar localities suggests that t years from now the franchise will be generating profit at the rate of $f(t) = 14{,}000 + 490t$ dollars per year. If the prevailing annual interest rate remains fixed during the next 5 years at 7 percent compounded continuously, what is the present value of the franchise?

SOLUTION

To approximate the present value of the franchise, divide the 5-year time interval $0 \le t \le 5$ into n equal subintervals of length Δt and let t_j denote the beginning of the jth interval as shown in Figure 4.4. Then,

$$\text{Profit from the } j\text{th subinterval} \approx f(t_j)\, \Delta t$$

and $\qquad$ Present value of the profit from the jth subinterval $\approx e^{-0.07t_j}f(t_j)\, \Delta t$

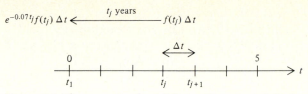

Figure 4.4 The present value of the profit generated during the jth subinterval.

hence $\qquad$ Present value of franchise $\approx \sum_{j=1}^{n} e^{-0.07t_j} f(t_j) \, \Delta t$

The approximation improves as n increases, and it follows that

$$\text{Present value of franchise} = \int_0^5 e^{-0.07t} f(t) \, dt$$

$$= \int_0^5 (14{,}000e^{-0.07t} + 490te^{-0.07t}) \, dt$$

If you evaluate this integral (using integration by parts to integrate the term $490te^{-0.07t}$), you should find that

$$\text{Present value of franchise} = (-200{,}000e^{-0.07t} - 7{,}000te^{-0.07t}$$

$$\left. -100{,}000e^{-0.07t}) \right|_0^5$$

$$= \$63{,}929.49$$

Take another look at Example 4.3 in which the present value of an annuity was approximated by a definite integral. Verify that the integral that was used gives the present value of a continuous income stream generating income at the constant rate of $1,200 per year. This shows that the approximation in Example 4.3 can be interpreted as the approximation of the present value of a (discrete) annuity by that of a continuous income stream.

Survival and renewal functions $\qquad$ In the next example, a **survival function** gives the fraction of individuals in a group or population that can be expected to remain in the group for any specified period of time. A **renewal function** giving the rate at which new members arrive is also known, and the goal is to predict the size of the group at some future time. Problems of this type arise in many fields, including sociology, demography, and ecology.

EXAMPLE 4.5

A new county mental health clinic has just opened. Statistics compiled at similar facilities suggest that the fraction of patients who will still be receiving treatment at the clinic t months after their initial visit is given by the function $f(t) = e^{-t/20}$. The clinic initially accepts 300 people for treatment and plans to accept new patients at a rate of 10 per month. Approximately how many people will be receiving treatment at the clinic 15 months from now?

SOLUTION

Since $f(15)$ is the fraction of patients whose treatment continues at least 15 months, it follows that of the current 300 patients, only $300f(15)$ will still be receiving treatment 15 months from now.

To approximate the number of *new* patients who will be receiving treatment 15 months from now, divide the 15-month time interval $0 \le t \le 15$ into n equal subintervals of length Δt and let t_j denote the beginning of the jth subinterval. Since new patients are accepted at the rate of 10 per month, the number of new patients accepted during the jth subinterval is $10\Delta t$. Fifteen months from now, approximately $15 - t_j$ months will have elapsed since these $10\Delta t$ new patients had their initial visits, and so approximately $10\Delta t f(15 - t_j)$ of them will still be receiving treatment at that time. (See Figure 4.5.) It follows that the total number of new patients still receiving treatment 15 months from now can be approximated by the sum

$$\sum_{j=1}^{n} 10f(15 - t_j)\, \Delta t$$

Adding this to the number of current patients who will still be receiving treatment in 15 months, you get

$$P \approx 300f(15) + \sum_{j=1}^{n} 10f(15 - t_j)\, \Delta t$$

where P is the total number of patients who will be receiving treatment 15 months from now.

As n increases, the approximation improves, and it follows that

$$P = 300f(15) + \int_{0}^{15} 10f(15 - t)\, dt$$

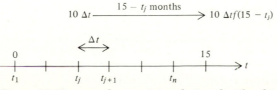

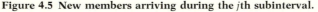

Figure 4.5 New members arriving during the jth subinterval.

Since $f(t) = e^{-t/20}$, it follows that $f(15) = e^{-3/4}$ and $f(15 - t) = e^{-(15-t)/20} = e^{-3/4}e^{t/20}$. Hence,

$$P = 300e^{-3/4} + 10e^{-3/4} \int_0^{15} e^{t/20}\, dt$$

$$= 300e^{-3/4} + 200e^{-3/4}e^{t/20} \Big|_0^{15}$$

$$= 300e^{-3/4} + 200(1 - e^{-3/4})$$

$$= 100e^{-3/4} + 200$$

$$= 247.24$$

That is, 15 months from now, the clinic will be treating approximately 247 patients.

The flow of blood through an artery

Biologists have found that the speed of blood in an artery is a function of the distance of the blood from the artery's central axis. According to Poiseuille's law, the speed (in centimeters per second) of blood that is r centimeters from the central axis of the artery is $S(r) = k(R^2 - r^2)$, where R is the radius of the artery and k is a constant. In the next example, you will see how to use this information to compute the rate (in cubic centimeters per second) at which blood passes through the artery.

EXAMPLE 4.6

Find an expression for the rate (in cubic centimeters per second) at which blood flows through an artery of radius R if the speed of blood r centimeters from the central axis is $S(r) = k(R^2 - r^2)$, where k is a constant.

SOLUTION

To approximate the volume of blood that flows through a cross section of the artery per second, divide the interval $0 \le r \le R$ into n equal subintervals of width Δr and let r_j denote the beginning of the jth subinterval. These subintervals determine n concentric rings as illustrated in Figure 4.6.

If Δr is small,   Area of jth ring $\approx 2\pi r_j\, \Delta r$

where $2\pi r_j$ is the circumference of the circle of radius r_j that forms the inner boundary of the ring, and Δr is the width of the ring.

If you multiply the area of the jth ring (square centimeters) by the speed (centimeters per second) of the blood flowing through this ring, you will get the rate (cubic centimeters per second) at which blood

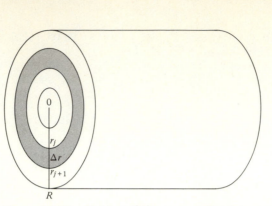

Figure 4.6 The division of the cross section into concentric rings.

flows through the jth ring. Since the speed of blood flowing through the jth ring is approximately $S(r_j)$, it follows that

$$\begin{pmatrix}\text{Rate of flow through} \\ \text{jth ring}\end{pmatrix} \approx \begin{pmatrix}\text{area of} \\ \text{jth ring}\end{pmatrix}\begin{pmatrix}\text{speed of blood} \\ \text{through jth ring}\end{pmatrix}$$

$$\approx 2\pi r_j \,\Delta r\, S(r_j)$$
$$= 2\pi r_j \,\Delta r\, k(R^2 - r_j^2) = 2\pi k(R^2 r_j - r_j^3) \,\Delta r$$

The rate of flow of blood through the entire cross section is the sum of n such terms, one for each of the n concentric rings. That is,

$$\text{Rate of flow} \approx \sum_{j=1}^{n} 2\pi k(R^2 r_j - r_j^3) \,\Delta r$$

As n increases without bound, this sum approaches the corresponding definite integral, and so

$$\text{Rate of flow} = \int_0^R 2\pi k(R^2 r - r^3) \, dr$$

$$= 2\pi k \left(\frac{R^2}{2} r^2 - \frac{1}{4} r^4\right)\Bigg|_0^R$$

$$= \frac{\pi k R^4}{2} \qquad \text{cubic centimeters per second}$$

Problems In Problems 1 through 4, find the average value of the given function over the specified interval.

1. $f(x) = x$; $0 \le x \le 4$

2. $f(x) = 2x - x^2$; $0 \le x \le 2$

3. $f(x) = (x + 2)^2$; $-4 \le x \le 0$

4. $f(x) = \dfrac{1}{x}$; $1 \le x \le 2$

Temperature
5. Records indicate that t hours past midnight, the temperature at the local airport was $f(t) = -0.3t^2 + 4t + 10$ degrees Celsius. What was the average temperature at the airport between 9:00 A.M. and noon?

Food prices
6. Records indicate that t months after the beginning of the year, the price of ground beef in local supermarkets was $P(t) = 0.09t^2 - 0.2t + 1.6$ dollars per pound. What was the average price of ground beef during the first 3 months of the year?

Efficiency
7. After t months on the job, a postal clerk can sort mail at the rate of $Q(t) = 700 - 400e^{-0.5t}$ letters per hour. What is the average rate at which the clerk sorts mail during the first 3 months on the job?

Bacterial growth
8. The number of bacteria present in a certain culture after t minutes of an experiment was $Q(t) = 2{,}000e^{0.05t}$. What was the average number of bacteria present during the first 5 minutes of the experiment?

Speed and distance
9. A car is driven so that after t hours its speed is $S(t)$ miles per hour.
   (a) Write down a definite integral that gives the average speed of the car during the first N hours.
   (b) Write down a definite integral that gives the total distance the car travels during the first N hours.
   (c) Discuss the relationship between the integrals in parts (a) and (b).

Inventory
10. An inventory of 60,000 kilograms of a certain commodity is used at a constant rate and is exhausted after 1 year. What is the average inventory for the year?

The amount of an annuity
11. On the 1st of each month you deposit $200 in an account yielding interest at the annual rate of 6 percent compounded continuously. Use a definite integral to estimate how much you will have in your account at the end of 5 years (immediately before you make your 61st deposit).

The amount of an annuity
12. Use a definite integral to estimate the amount of an annuity that consists of semiannual deposits of $500 for a term of 10 years if the annual interest rate is 10 percent compounded continuously.

The amount of an annuity
13. The authors of a popular new cookbook have just received their first royalty check for $2,000. They plan to deposit it, and all subsequent royalty checks, in a bank offering an annual interest rate of 8 percent compounded continuously. If they receive the royalty checks quarterly and if sales of the book continue at the same level, approximately how much will the authors have in the bank

3 years from now (immediately before they deposit the 13th check)?

The amount of an income stream

14. Money is transferred continuously into an account at the constant rate of $6,000 per year. The account earns interest at the annual rate of 6 percent compounded continuously. How much will be in the account at the end of 10 years?

The present value of an annuity

15. Use a definite integral to estimate the present value of an investment plan that guarantees monthly payments of $200 for each of the next 5 years if the prevailing interest rate remains fixed at 6 percent per year compounded continuously.

The present value of an annuity

16. Use a definite integral to estimate how much you should deposit today in an account yielding interest at an annual rate of 8 percent compounded continuously so that you can make monthly withdrawals of $500 for each of the next 6 years, after which nothing will be left in the account.

The present value of an investment

17. An investment scheme will generate income at a constant rate of $1,200 per year for 5 years. If the $1,200 is dispensed continuously throughout the year and if the prevailing annual interest rate remains fixed at 6 percent compounded continuously, what is the present value of the investment scheme?

The present value of a franchise

18. The management of a national chain of fast-food outlets is selling a 10-year franchise in Cleveland, Ohio. Past experience in similar localities suggests that t years from now the franchise will be generating profit at the rate of $f(t) = 10,000 + 500t$ dollars per year. If the prevailing annual interest rate remains fixed at 10 percent compounded continuously, what is the present value of the franchise?

Spy story

19. Having been left partially disabled after a head-on collision with a camel (see Chapter 5, Section 3, Problem 38), the hero of a popular spy story has been retired from the Secret Service. As compensation for many long years of dedicated public service, the government has offered the spy a choice between a 10-year pension of 5,000 pounds sterling per year or a flat sum of 35,000 pounds sterling to be paid immediately. Assuming that an interest rate of 10 percent compounded continuously will be available at banks throughout this period, decide which offer the spy should accept. (*Hint:* Compare the flat sum of 35,000 pounds with the present value of the pension. Assume that the pension is paid continuously.)

Present value

20. A certain investment scheme generates income over a period of N years. After t years the scheme will be generating income at the

rate of $f(t)$ dollars per year. Derive an expression for the present value of this investment scheme if the prevailing annual interest rate remains fixed at $100r$ percent compounded continuously.

Computer dating
21. The operators of a new computer dating service estimate that the fraction of people who will retain their memberships in the service for at least t months is given by the function $f(t) = e^{-t/10}$. There are 8,000 charter members and the operators expect to attract 200 new members per month. How many members will the service have 10 months from now?

Association membership
22. A national consumers' association has compiled statistics suggesting that the fraction of its members who are still active t months after joining is given by the function $f(t) = e^{-0.2t}$. A new local chapter has 200 charter members and expects to attract new members at the rate of 10 per month. How many members can the chapter expect to have at the end of 8 months?

Group membership
23. Let $f(t)$ denote the fraction of the membership of a certain group that will remain in the group for at least t years. Suppose that the group has just been formed with an initial membership of P_0 and that t years from now new members will be added to the group at the rate of $r(t)$ per year. Find an expression for the size of the group N years from now.

Poiseuille's law
24. Calculate the rate (in cubic centimeters per second) at which blood flows through an artery of radius 0.1 centimeter if the speed of the blood r centimeters from the central axis is $8 - 800r^2$ centimeters per second.

Fluid flow
25. Find an expression for the rate (in cubic centimeters per second) at which a fluid flows through a cylindrical pipe of radius R if the speed of the fluid r centimeters from the central axis of the pipe is $S(r)$ centimeters per second.

The area of a disk
26. Use integral calculus to find a formula for the area of a circular disk of radius R. (*Hint:* Divide the disk into n concentric rings as in Example 4.6.)

Population density
27. The population density r miles from the center of a certain city is $D(r) = 5,000e^{-0.1r}$ people per square mile. How many people live within 3 miles of the center of the city? (*Hint:* Divide a circular disk of radius 3 into concentric rings.)

Population density
28. The population density r miles from the center of a certain city is $D(r) = 25,000e^{-0.05r}$ people per square mile. How many people live between 1 and 2 miles from the center of the city?

Radioactive waste
29. A certain atomic plant produces radioactive waste in the form of

strontium 90 at a constant rate of 500 pounds per year. The half-life of strontium 90 is 28 years. How much of this radioactive waste from the atomic power plant will be present after 140 years?

CHAPTER SUMMARY AND PROFICIENCY TEST

Important terms, symbols, and formulas

Definite integral: $\int_a^b f(x)\,dx$

Integration by substitution: $\int_a^b g[u(x)]\dfrac{du}{dx}\,dx = \int_{u(a)}^{u(b)} g(u)\,du$

Integration by parts: $\int_a^b f(x)g(x)\,dx = f(x)G(x)\Big|_a^b - \int_a^b f'(x)G(x)\,dx$, where G is an antiderivative of g

Area under a curve:

Area of $R = \displaystyle\int_a^b f(x)\,dx$

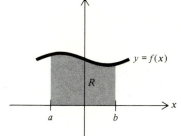

Probability density function: $P(a \le x \le b) = \displaystyle\int_a^b f(x)\,dx$

Area between two curves:

Area of $R = \displaystyle\int_a^b [f(x) - g(x)]\,dx$

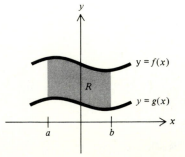

Fundamental theorem of calculus: $\displaystyle\sum_{j=1}^{n} f(x_j)\,\Delta x \rightarrow \int_a^b f(x)\,dx = F(b) - F(a)$ as n increases without bound, where F is an antiderivative of f, and $x_1, x_2, \cdots, x_n$ are as shown below

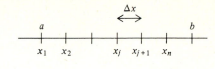

Average value of f over $a \leq x \leq b = \dfrac{1}{b-a}\displaystyle\int_a^b f(x)\,dx$

Proficiency test   In Problems 1 through 6, evaluate the given definite integral.

1. $\displaystyle\int_0^1 (5x^4 - 8x^3 + 1)\,dx$

2. $\displaystyle\int_1^4 (\sqrt{x} + x^{-3/2})\,dx$

3. $\displaystyle\int_{-1}^2 30(5x - 2)^2\,dx$

4. $\displaystyle\int_0^1 2xe^{x^2-1}\,dx$

5. $\displaystyle\int_0^1 (x - 3)(x^2 - 6x + 2)^3\,dx$

6. $\displaystyle\int_{-1}^1 xe^x\,dx$

7. A study indicates that x months from now the population of a certain town will be increasing at the rate of $10 + 2\sqrt{x}$ people per month. By how much will the population of the town increase over the next 9 months?

8. It is estimated that t days from now a farmer's crop will be increasing at a rate of $0.3t^2 + 0.6t + 1$ bushels per day. By how much will the value of the crop increase during the next 6 days if the market price remains fixed at \$2 per bushel?

9. Find the area of the region bounded by the curve $y = 3x^2 + 2$, the lines $x = -1$ and $x = 3$, and the x axis.

10. Find the area of the region bounded by the curve $y = \dfrac{1}{x^2}$, the x axis, and the vertical lines $x = 1$ and $x = 4$.

11. Find the area of the region bounded by the curve $y = 2 + x - x^2$ and the x axis.

12. Find the area of the region bounded by the curve $y = x^4$ and the line $y = x$.

13. The probability density function for the duration of telephone calls in a certain city is $f(x) = 0.4e^{-0.4x}$, where x denotes the duration (in minutes) of a randomly selected call.

(a) What percentage of the calls last between 1 and 2 minutes?

(b) What percentage of the calls last 2 minutes or less?

(c) What percentage of the calls last more than 2 minutes?

14. When it is x years old, a certain industrial machine generates revenue at the rate of $R(x) = 4,575 - 5x^2$ dollars per year and results in costs that accumulate at the rate of $C(x) = 1,200 + 10x^2$ dollars per year.

(a) For how many years is the use of the machine profitable?

(b) What are the total net earnings generated by the machine during the period of time in part (a)?

15. A retailer expects that x months from now consumers will be buying 50 cameras a month at a price of $P(x) = 40 + 3\sqrt{x}$ dollars per camera. Use the characterization of the definite integral as the limit of a sum to find the total revenue the retailer can expect from the sale of the cameras over the next 9 months.

16. Economists predict that x months from now the demand for beef will be $D(x)$ pounds per month and the price will be $P(x)$ dollars per pound. Use the characterization of the definite integral as the limit of a sum to find an expression for the total amount that consumers will spend on beef this year.

17. Records indicate that t months after the beginning of the year, the price of chicken in local supermarkets was $P(t) = 0.06t^2 - 0.2t + 1.2$ dollars per pound. What was the average price of chicken during the first 6 months of the year?

18. On the 1st of each month you deposit $100 in an account yielding interest at the annual rate of 8 percent compounded continuously. Use a definite integral to estimate how much you will have in the account at the end of 5 years (immediately before you make your 61st deposit).

19. What is the present value of an investment scheme that will generate income continuously at a constant rate of $1,000 per year for 10 years if the prevailing annual interest rate remains fixed at 7 percent compounded continuously?

20. In a certain community the fraction of the homes placed on the market that remain unsold for at least t weeks is approximately $f(t) = e^{-0.2t}$. If 200 homes are currently on the market and if additional homes are placed on the market at the rate of 8 per week, approximately how many homes will be on the market 10 weeks from now?

21. The population density r miles from the center of a certain city is $D(r) = 6,000e^{-0.1r}$ people per square mile. How many people live between 2 and 3 miles from the center of the city?

1 LIMITS AT INFINITY AND L'HÔPITAL'S RULE

The limit of a function as its variable increases without bound (sometimes called the **limit at infinity**) may give useful information in practical situations. For example, if the variable represents time, such a limit describes what will happen to the function "in the long run." This interpretation is illustrated in Figure 1.1. The curve in Figure 1.1a represents the value of an industrial machine that decreases

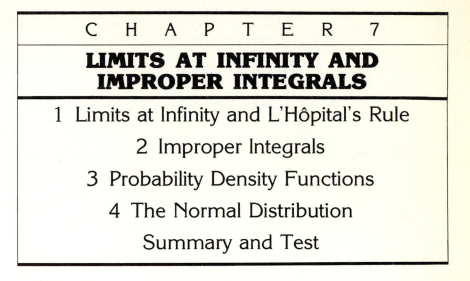

Figure 1.1 (a) Depreciation of machinery. (b) Population growth.

exponentially and approaches $1,000 (scrap value) in the long run. The curve in Figure 1.1b is a logistic curve for bounded population growth. It shows a population that will eventually approach a fixed upper bound B.

Later in this chapter, you will see some important consequences of the application of this limit concept to definite integrals. Limits at infinity will also play a central role in the discussion of infinite series in Chapter 10. The purpose of this introductory section is to show you some of the computational techniques you can use to find the limit of a function as its variable increases without bound.

Limit notation The symbol $\lim_{x \to \infty} f(x)$ is read "the limit of $f(x)$ as x approaches infinity" and denotes the behavior of $f(x)$ as x increases without bound. If $f(x)$ approaches a finite number L, you write

$$\lim_{x \to \infty} f(x) = L$$

If $f(x)$ increases or decreases without bound, you write

$$\lim_{x \to \infty} f(x) = \infty \qquad \text{or} \qquad \lim_{x \to \infty} f(x) = -\infty$$

The situation is illustrated in Figure 1.2.

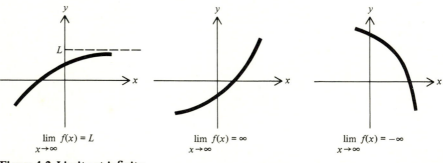

$$\lim_{x \to \infty} f(x) = L \qquad\qquad \lim_{x \to \infty} f(x) = \infty \qquad\qquad \lim_{x \to \infty} f(x) = -\infty$$

Figure 1.2 Limits at infinity.

Review of some basic limits For reference, here is a summary of some important limits that you should already know.

Limits of basic functions

Powers of x: If $n > 0$,

$$\lim_{x \to \infty} x^n = \infty \qquad \textbf{and} \qquad \lim_{x \to \infty} \frac{1}{x^n} = 0$$

Exponential functions: If $k > 0$,

$$\lim_{x \to \infty} e^{kx} = \infty \qquad \textbf{and} \qquad \lim_{x \to \infty} e^{-kx} = 0$$

Natural logarithm:

$$\lim_{x \to \infty} \ln x = \infty$$

The graphs of these basic functions are sketched in Figure 1.3.

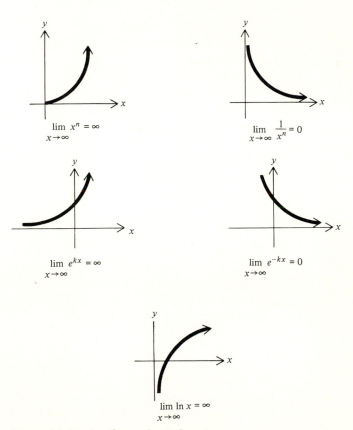

$$\lim_{x \to \infty} x^n = \infty \qquad\qquad \lim_{x \to \infty} \frac{1}{x^n} = 0$$

$$\lim_{x \to \infty} e^{kx} = \infty \qquad\qquad \lim_{x \to \infty} e^{-kx} = 0$$

$$\lim_{x \to \infty} \ln x = \infty$$

Figure 1.3 Limits of some basic functions.

Elementary calculations   The following examples illustrate elementary techniques you can use to find the limits of functions that are built from one or more of the five basic functions. The first example involves a polynomial. In general, the limit of a polynomial as x increases without bound is determined by its term of highest degree, which increases (or decreases) more rapidly than the other terms of lower degree.

The limit at infinity of a polynomial

If $a_n \neq 0$,

$$\lim_{x \to \infty} (a_0 + a_1 x + \cdots + a_n x^n) = \lim_{x \to \infty} a_n x^n$$

That is, to find the limit at infinity of a polynomial, take the limit of the term of highest degree.

Here is an example.

EXAMPLE 1.1

Find $\lim_{x \to \infty} (1 - x^2 + x^3 - 3x^4)$.

SOLUTION

$$\lim_{x \to \infty} (1 - x^2 + x^3 - 3x^4) = \lim_{x \to \infty} (-3x^4) = -\infty$$

One way to find the limit at infinity of a rational function is to first compare the degrees of the numerator and the denominator and divide numerator and denominator by x raised to the smaller of these degrees. This will reduce the problem to one in which most of the terms are of the form a/x^n, which approach zero as x approaches infinity. Later in this section you will see a more powerful technique involving differentiation that can also be used to find the limits of rational functions.

The limit at infinity of a rational function

To find the limit at infinity of a rational function, first compare the degrees of the numerator and the denominator and divide numerator and denominator by x raised to the smaller of these degrees. Then take the limits of the new numerator and denominator.

Here are two examples illustrating the technique.

EXAMPLE 1.2

Find $\lim_{x \to \infty} \dfrac{2x^2 + 3x + 1}{3x^2 - 5x + 2}$.

SOLUTION

Divide numerator and denominator by x^2 to get

$$\lim_{x \to \infty} \frac{2x^2 + 3x + 1}{3x^2 - 5x + 2} = \lim_{x \to \infty} \frac{2 + 3/x + 1/x^2}{3 - 5/x + 2/x^2} = \frac{2 + 0 + 0}{3 - 0 + 0} = \frac{2}{3}$$

EXAMPLE 1.3

Find $\displaystyle\lim_{x \to \infty} \frac{-x^3 + 2x + 1}{x - 3}$.

SOLUTION

Divide numerator and denominator by x to get

$$\lim_{x \to \infty} \frac{-x^3 + 2x + 1}{x - 3} = \lim_{x \to \infty} \frac{-x^2 + 2 + 1/x}{1 - 3/x}$$

Since $\displaystyle\lim_{x \to \infty} \left(-x^2 + 2 + \frac{1}{x} \right) = -\infty$ and $\displaystyle\lim_{x \to \infty} \left(1 - \frac{3}{x} \right) = 1$

it follows that $\displaystyle\lim_{x \to \infty} \frac{-x^3 + 2x + 1}{x - 3} = -\infty$

In the next two examples, the elementary limit properties of the exponential function are used to find the limits of more complicated functions.

EXAMPLE 1.4

Find $\displaystyle\lim_{x \to \infty} \frac{e^{-x}}{1 + e^{2x}}$.

SOLUTION

The limits of the numerator and denominator are, respectively,

$$\lim_{x \to \infty} e^{-x} = 0 \quad \text{and} \quad \lim_{x \to \infty} (1 + e^{2x}) = \infty$$

The behavior of the numerator and that of the denominator reinforce one another. Both tend to drive the quotient to zero as x increases without bound. It follows that

$$\lim_{x \to \infty} \frac{e^{-x}}{1 + e^{2x}} = 0$$

EXAMPLE 1.5

Find $\lim\limits_{x \to \infty} \dfrac{x^2 + 1}{-e^{-x}}$.

SOLUTION

The limits of the numerator and the denominator are, respectively,

$$\lim_{x \to \infty} (x^2 + 1) = \infty \quad \text{and} \quad \lim_{x \to \infty} (-e^{-x}) = 0$$

Again, the behavior of the numerator and that of the denominator reinforce one another. This time, both tend to drive the absolute value of the quotient to infinity as x increases without bound. Moreover, the sign of the quotient is negative, since the numerator is positive and the denominator negative. It follows that

$$\lim_{x \to \infty} \frac{x^2 + 1}{-e^{-x}} = -\infty$$

The indeterminate forms 0/0 and ∞/∞   In each of the four preceding examples, you were able to determine the limit of a quotient easily from the limits of its numerator and denominator. This determination is not always so straightforward. Consider, for example, the quotient x/e^x. As x increases without bound, the numerator approaches infinity, which tends to drive the quotient to infinity, while the denominator also approaches infinity, which tends to drive the quotient to zero. Without further analysis, it is not clear whether the numerator or the denominator will dominate. The quotient might approach zero or infinity or any number in between. In this case one says that the limit of the quotient is of the **indeterminate form** ∞/∞.

Similar uncertainty arises if the limits of the numerator and the denominator are both zero. In this case one says that the limit of the quotient is of the indeterminate form 0/0.

There is a powerful technique, known as L'Hôpital's rule, which you can use to analyze indeterminate forms. The rule says, in effect, that if your attempt to find the limit of a quotient leads to one of the indeterminate forms 0/0 or ∞/∞, take the derivatives of the numerator and the denominator and try again. Here is a summary of the procedure.

L'Hôpital's rule   If $\lim\limits_{x \to \infty} f(x) = 0$ **and** $\lim\limits_{x \to \infty} g(x) = 0$, **then**

$$\lim_{x \to \infty} \frac{f(x)}{g(x)} = \lim_{x \to \infty} \frac{f'(x)}{g'(x)}$$

> **If** $\lim\limits_{x\to\infty} f(x) = \infty$ **and** $\lim\limits_{x\to\infty} g(x) = \infty$, **then**
>
> $$\lim_{x\to\infty} \frac{f(x)}{g(x)} = \lim_{x\to\infty} \frac{f'(x)}{g'(x)}$$

The proof of L'Hôpital's rule is technical and beyond the scope of this text. However, the fact that the rule involves the ratio of the derivatives of f and g should not come as a complete surprise. The value of an indeterminate form depends on the relative rates with which the functions in the numerator and the denominator approach their limits, and these rates are given by the derivatives of the functions.

Advice on the use of L'Hôpital's rule   The use of L'Hôpital's rule is illustrated in the following examples. As you read these examples, pay particular attention to the following two points:

(a) L'Hôpital's rule involves differentiation of the numerator and the denominator *separately*. A common mistake is to differentiate the entire quotient using the quotient rule.
(b) L'Hôpital's rule applies only to quotients whose limits are indeterminate forms $0/0$ or ∞/∞. Limits of the form $0/\infty$ or $\infty/0$, for example, are *not* indeterminate, as you saw in Examples 1.4 and 1.5, and L'Hôpital's rule does not apply to such limits.

EXAMPLE 1.6

Find $\lim\limits_{x\to\infty} \dfrac{x}{e^x}$.

SOLUTION

This limit is of the indeterminate form ∞/∞. Applying L'Hôpital's rule, you get

$$\lim_{x\to\infty} \frac{x}{e^x} = \lim_{x\to\infty} \frac{1}{e^x} = 0$$

EXAMPLE 1.7

Find $\lim\limits_{x\to\infty} \dfrac{e^{1/x} - 1}{1/x}$.

SOLUTION

This limit is of the indeterminate form 0/0. Applying L'Hôpital's rule and then simplifying the resulting quotient, you get

$$\lim_{x\to\infty}\frac{e^{1/x}-1}{1/x} = \lim_{x\to\infty}\frac{(-1/x^2)e^{1/x}}{-1/x^2} \quad \text{(L'Hôpital's rule)}$$

$$= \lim_{x\to\infty}e^{1/x} \quad \text{(algebraic simplification)}$$

$$= e^0 = 1$$

EXAMPLE 1.8

Find $\displaystyle\lim_{x\to\infty}\frac{e^{-x}}{1+e^{-x}}$.

SOLUTION

This limit is of the form 0/1, which is *not* indeterminate and can be calculated directly as follows:

$$\lim_{x\to\infty}\frac{e^{-x}}{1+e^{-x}} = \frac{0}{1} = 0$$

To convince yourself that L'Hôpital's rule does not apply in this case, differentiate the numerator and the denominator. The resulting quotient is equal to 1, and its limit is certainly not zero.

Sometimes one application of L'Hôpital's rule is not enough. In the next example, the use of L'Hôpital's rule leads to a new quotient whose limit is also of an indeterminate form. A second application of L'Hôpital's rule is needed to determine the limit.

EXAMPLE 1.9

Find $\displaystyle\lim_{x\to\infty}\frac{x\ln x}{e^{2x}}$.

SOLUTION

This limit is of the indeterminate form ∞/∞. Applying L'Hôpital's rule (and using the product rule to differentiate the numerator), you get

$$\lim_{x\to\infty}\frac{x\ln x}{e^{2x}} = \lim_{x\to\infty}\frac{1+\ln x}{2e^{2x}}$$

Since the new limit is also of the form ∞/∞, you apply L'Hôpital's rule again, to get

$$\lim_{x \to \infty} \frac{1 + \ln x}{2e^{2x}} = \lim_{x \to \infty} \frac{1/x}{4e^{2x}} = 0$$

and you conclude that

$$\lim_{x \to \infty} \frac{x \ln x}{e^{2x}} = 0$$

In problems involving repeated application of L'Hôpital's rule, don't forget to verify that each new limit is of an indeterminate form before you apply L'Hôpital's rule again. In Example 1.9, for instance, the limit of the quotient $(1/x)/4e^{2x}$ was of the form $0/\infty$, and further application of L'Hôpital's rule would have been invalid.

L'Hôpital's rule gives you an alternative way to find limits of rational functions. Here is the calculation for the rational function from Example 1.2.

EXAMPLE 1.10

Find $\lim\limits_{x \to \infty} \dfrac{2x^2 + 3x + 1}{3x^2 - 5x + 2}$.

SOLUTION

This limit is of the indeterminate form ∞/∞. Repeated application of L'Hôpital's rule gives

$$\lim_{x \to \infty} \frac{2x^2 + 3x + 1}{3x^2 - 5x + 2} = \lim_{x \to \infty} \frac{4x + 3}{6x - 5} = \lim_{x \to \infty} \frac{4}{6} = \frac{2}{3}$$

The indeterminate form $0 \cdot \infty$ The limit of a product in which one of the factors approaches zero and the other approaches infinity is said to be of the indeterminate form $0 \cdot \infty$, since it is not clear which factor will dominate. To evaluate such a limit, try writing the product as a quotient whose limit is of the form $0/0$ or ∞/∞, and apply L'Hôpital's rule. Here is an example.

EXAMPLE 1.11

Find $\lim\limits_{x \to \infty} e^{-x} \ln x$.

SOLUTION

This limit is of the indeterminate form $0 \cdot \infty$ and can be rewritten as

$$\lim_{x \to \infty} \frac{e^{-x}}{1/\ln x} \qquad \text{(of the form } 0/0)$$

or as $\qquad\qquad \lim\limits_{x \to \infty} \dfrac{\ln x}{e^x} \qquad$ (of the form ∞/∞)

Applying L'Hôpital's rule to the simpler second quotient, you get

$$\lim_{x \to \infty} e^{-x} \ln x = \lim_{x \to \infty} \frac{\ln x}{e^x} = \lim_{x \to \infty} \frac{1/x}{e^x} = 0$$

Other indeterminate forms

The limit of an expression of the form $f(x)^{g(x)}$ may be of one of the indeterminate forms 1^∞, ∞^0, or 0^0. To evaluate such a limit, try using logarithms to simplify the situation. The technique is illustrated in the next example.

EXAMPLE 1.12

Find $\lim\limits_{x \to \infty} x^{1/x}$.

SOLUTION

This limit is of the indeterminate form ∞^0. To simplify the problem, let

$$y = x^{1/x}$$

and take the logarithm of both sides of this equation to get

$$\ln y = \frac{1}{x} \ln x$$

As a preliminary calculation find the limit of $\ln y$ as follows:

$$\lim_{x \to \infty} \ln y = \lim_{x \to \infty} \frac{1}{x} \ln x \qquad (0 \cdot \infty)$$

$$= \lim_{x \to \infty} \frac{\ln x}{x} \qquad (\infty/\infty)$$

$$= \lim_{x \to \infty} \frac{1/x}{1} \qquad \text{(L'Hôpital's rule)}$$

$$= 0$$

Having completed this preliminary calculation, you know the limit of the natural logarithm of y but still have to find the limit of the original function y itself. To proceed, observe that since

$$\ln y \to 0$$

it follows from the identity $y = e^{\ln y}$ that

$$y \rightarrow e^0 = 1$$

That is,

$$\lim_{x \to \infty} x^{1/x} = 1$$

As a final illustration of this technique, here is the limit that was used in Chapter 4, Section 1, to define the number e.

EXAMPLE 1.13

Find $\displaystyle\lim_{x \to \infty} \left(1 + \frac{1}{x}\right)^x$.

SOLUTION

This limit is of the indeterminate form 1^∞. To simplify the problem let

$$y = \left(1 + \frac{1}{x}\right)^x$$

Then,

$$\ln y = x \ln \left(1 + \frac{1}{x}\right)$$

and

$$\lim_{x \to \infty} \ln y = \lim_{x \to \infty} x \ln \left(1 + \frac{1}{x}\right) \qquad (\infty \cdot 0)$$

$$= \lim_{x \to \infty} \frac{\ln (1 + 1/x)}{1/x} \qquad (0/0)$$

$$= \lim_{x \to \infty} \frac{\dfrac{(-1/x^2)}{(1 + 1/x)}}{-1/x^2} \qquad \text{(L'Hôpital's rule)}$$

$$= \lim_{x \to \infty} \frac{1}{1 + 1/x} \qquad \text{(algebraic simplification)}$$

$$= 1$$

Since

$$\ln y \rightarrow 1$$

it follows that

$$y \rightarrow e^1 = e$$

That is,

$$\lim_{x \to \infty} \left(1 + \frac{1}{x}\right)^x = e$$

Problems   For each of the following functions, find $\lim\limits_{x\to\infty} f(x)$.

1. $f(x) = x^3 - 4x^2 - 4$

2. $f(x) = 1 - x + 2x^2 - 3x^3$

3. $f(x) = (1 - 2x)(x + 5)$

4. $f(x) = (1 - x^2)^3$

5. $f(x) = \dfrac{x^2 - 2x + 3}{2x^2 + 5x + 1}$

6. $f(x) = \dfrac{1 - 3x^3}{2x^3 - 6x + 2}$

7. $f(x) = \dfrac{2x + 1}{3x^2 + 2x - 7}$

8. $f(x) = \dfrac{x^2 + x - 5}{1 - 2x - x^3}$

9. $f(x) = \dfrac{3x^2 - 6x + 2}{2x - 9}$

10. $f(x) = \dfrac{1 - 2x^3}{x + 1}$

11. $f(x) = 2 - 3e^x$

12. $f(x) = 5 + e^{-3x}$

13. $f(x) = \dfrac{3}{2 + 5e^{-8x}}$

14. $f(x) = \dfrac{e^x}{2 + e^{-x}}$

15. $f(x) = \dfrac{e^{2x}}{3x}$

16. $f(x) = \dfrac{e^{-2x}}{3x}$

17. $f(x) = \dfrac{2x}{e^{-3x}}$

18. $f(x) = \dfrac{2x}{e^{3x}}$

19. $f(x) = \dfrac{x^2}{e^x}$

20. $f(x) = \dfrac{e^{2x}}{x^2}$

21. $f(x) = \dfrac{\sqrt{x}}{e^x}$

22. $f(x) = \dfrac{e^x}{x^{3/2}}$

23. $f(x) = \dfrac{\ln x}{x}$

24. $f(x) = \dfrac{\sqrt{x}}{\ln x}$

25. $f(x) = \dfrac{\ln (x^2 + 1)}{x}$

26. $f(x) = \dfrac{(\ln x)^2}{x}$

27. $f(x) = \dfrac{\ln (2x + 1)}{\ln (3x - 1)}$

28. $f(x) = \dfrac{\ln (x^2 + 4)}{\ln (x - 1)}$

29. $f(x) = \dfrac{e^x}{x \ln x}$

30. $f(x) = \dfrac{x \ln x}{x + \ln x}$

31. $f(x) = \dfrac{x}{e^{\sqrt{x}}}$

32. $f(x) = xe^{-x^2}$

33. $f(x) = e^{-2x} \ln x$

34. $f(x) = x(e^{1/x} - 1)$

35. $f(x) = x^{2/x}$

36. $f(x) = (\ln x)^{1/x}$

37. $f(x) = \left(1 + \dfrac{2}{x}\right)^x$

38. $f(x) = \left(1 + \dfrac{1}{x}\right)^{x^2}$

39. In the next section you will encounter several limits of the form $\lim_{x \to \infty} x^P e^{-kx}$. Show that for any constant P and any positive constant k, this limit is equal to zero. (*Hint:* Write $x^P e^{-kx}$ as a quotient and show that after a sufficient number of applications of L'Hôpital's rule, the limit can be rewritten as

$$\lim_{x \to \infty} \frac{C}{e^{kx}} \quad \text{or} \quad \lim_{x \to \infty} \frac{C}{x^n e^{kx}}$$

where C and n are constants and $n > 0$.)

2 IMPROPER INTEGRALS This section extends the concept of the definite integral to integrals of the form

$$\int_a^\infty f(x)\, dx$$

in which the upper limit of integration is not a finite number. Such integrals are known as **improper integrals** and arise in a variety of practical situations.

Geometric interpretation If f is nonnegative, the improper integral $\int_a^\infty f(x)\, dx$ may be interpreted as the area of the region under the graph of f to the right of $x = a$ (Figure 2.1a). Although this region has infinite extent, its area may be either finite or infinite, depending on how quickly $f(x)$ approaches zero as x increases.

A reasonable strategy for finding the area of such a region is to first use a definite integral to compute the area from $x = a$ to some finite number $x = N$, and then to let N approach infinity in the resulting expression. That is,

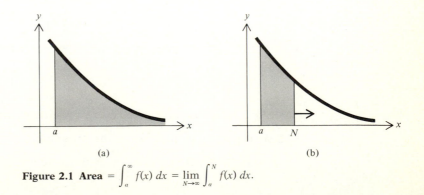

(a)   (b)

Figure 2.1 Area $= \int_a^\infty f(x)\, dx = \lim_{N \to \infty} \int_a^N f(x)\, dx.$

$$\text{Total area} = \lim_{N \to \infty} (\text{area from } a \text{ to } N) = \lim_{N \to \infty} \int_a^N f(x)\, dx$$

This strategy is illustrated in Figure 2.1b and motivates the following definition of the improper integral.

The improper integral

$$\int_a^\infty f(x)\, dx = \lim_{N \to \infty} \int_a^N f(x)\, dx$$

If the limit defining the improper integral is a finite number, the integral is said to **converge**. Otherwise the integral is said to **diverge**. Here are some examples.

EXAMPLE 2.1

Evaluate $\displaystyle\int_1^\infty \frac{1}{x^2}\, dx$.

SOLUTION

First compute the integral from 1 to N and then let N approach infinity. Arrange your work compactly as follows:

$$\int_1^\infty \frac{1}{x^2}\, dx = \lim_{N \to \infty} \int_1^N \frac{1}{x^2}\, dx = \lim_{N \to \infty} \left(-\frac{1}{x} \Big|_1^N \right) = \lim_{N \to \infty} \left(-\frac{1}{N} + 1 \right) = 1$$

EXAMPLE 2.2

Evaluate $\displaystyle\int_1^\infty \frac{1}{x}\, dx$.

SOLUTION

$$\int_1^\infty \frac{1}{x}\, dx = \lim_{N \to \infty} \int_1^N \frac{1}{x}\, dx = \lim_{N \to \infty} \left(\ln |x| \, \Big|_1^N \right) = \lim_{N \to \infty} \ln N = \infty$$

Notice that the improper integral of the function $f(x) = \dfrac{1}{x^2}$ in Example 2.1 converged while that of the function $f(x) = \dfrac{1}{x}$ in Example 2.2 diverged. In geometric terms, this says that the area to the right of $x = 1$ under the curve $y = \dfrac{1}{x^2}$ is finite while the corresponding area under the curve $y = \dfrac{1}{x}$ is infinite. The reason for the difference is that, as x increases, $\dfrac{1}{x^2}$ approaches zero more quickly than does $\dfrac{1}{x}$. (See Figure 2.2.)

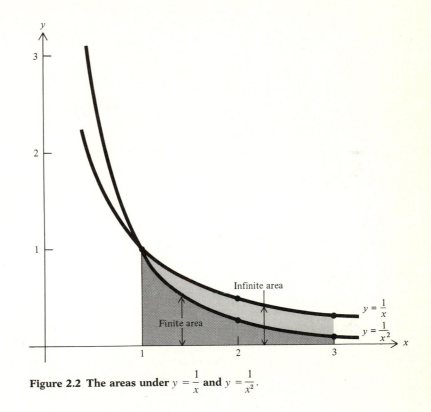

Figure 2.2 The areas under $y = \dfrac{1}{x}$ **and** $y = \dfrac{1}{x^2}$.

The evaluation of improper integrals arising from practical problems often involves limits of the form

$$\lim_{N \to \infty} N^p e^{-kN} \qquad \text{(for } k > 0)$$

These limits are of the indeterminate form $\infty \cdot 0$ and can be found using L'Hôpital's rule. (See Problem 39 of the preceding section.)

A useful limit for improper integrals	**For any power p and positive number k,** $$\lim_{N \to \infty} N^p e^{-kN} = 0$$

The next example involves a limit of this type and also requires the use of integration by parts.

EXAMPLE 2.3

Evaluate $\displaystyle\int_0^\infty xe^{-2x}\,dx$.

SOLUTION

$$\int_0^\infty xe^{-2x}\,dx = \lim_{N\to\infty} \int_0^N xe^{-2x}\,dx$$

$$= \lim_{N\to\infty}\left(-\frac{1}{2}xe^{-2x}\Big|_0^N + \frac{1}{2}\int_0^N e^{-2x}\,dx\right) \quad \text{(integration by parts)}$$

$$= \lim_{N\to\infty}\left(-\frac{1}{2}xe^{-2x} - \frac{1}{4}e^{-2x}\right)\Big|_0^N$$

$$= \lim_{N\to\infty}\left(-\frac{1}{2}Ne^{-2N} - \frac{1}{4}e^{-2N} + 0 + \frac{1}{4}\right)$$

$$= \frac{1}{4} \quad \text{(since } Ne^{-2N} \to 0\text{)}$$

Applications of the improper integral   The following applications of the improper integral generalize applications of the definite integral that you saw in Chapter 6, Section 4. In each, the strategy is to use the characterization of the definite integral as the limit of a sum to construct an appropriate definite integral and then to let the upper limit of integration increase without bound. As you read these examples, you may want to refer back to the corresponding examples in Chapter 6.

Present value   In Example 4.4 of Chapter 6 you saw that the present value of an investment that generates income over a finite period of time is given by a definite integral. The present value of an investment that generates income in perpetuity (i.e., forever) is given by an improper integral. Here is an example.

EXAMPLE 2.4

A donor wishes to make a gift to a private college from which the college will draw $7,000 per year in perpetuity to support the operation of its computer center. Assuming that the prevailing annual interest rate will remain fixed at 14 percent compounded continuously, how much should the donor give the college? That is, what is the present value of the endowment?

SOLUTION

To find the present value of a gift that generates \$7,000 per year for N years, divide the N-year time interval $0 \leq t \leq N$ into n subintervals of length Δt years and let t_j denote the beginning of the jth subinterval (Figure 2.3). Then,

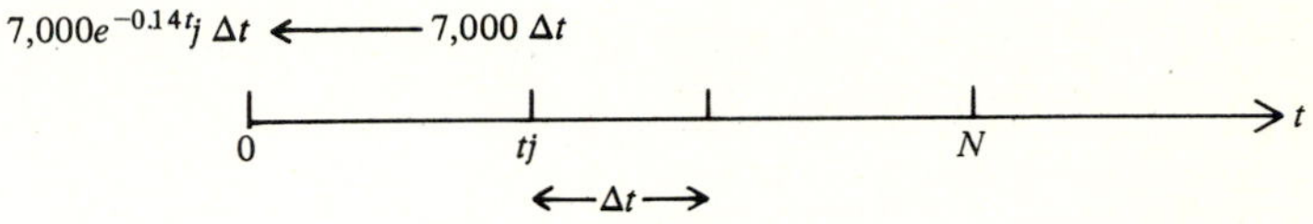

$$7{,}000e^{-0.14t_j}\,\Delta t \longleftarrow 7{,}000\,\Delta t$$

Figure 2.3 The present value of the money generated during the jth subinterval.

$$\begin{array}{l}\text{Amount generated during}\\ j\text{th subinterval}\end{array} \approx 7{,}000\,\Delta t$$

and

$$\begin{array}{l}\text{Present value of}\\ \text{amount generated during}\\ j\text{th subinterval}\end{array} \approx 7{,}000e^{-0.14t_j}\,\Delta t$$

Hence,

$$\begin{array}{l}\text{Present value of}\\ N\text{-year gift}\end{array} = \lim_{n \to \infty} \sum_{j=1}^{n} 7{,}000e^{-0.14t_j}\,\Delta t$$

$$= \int_{0}^{N} 7{,}000e^{-0.14t}\,dt$$

To find the present value of the total gift, take the limit of this integral as N approaches infinity. That is,

$$\begin{array}{l}\text{Present value of}\\ \text{total gift}\end{array} = \lim_{N \to \infty} \int_{0}^{N} 7{,}000e^{-0.14t}\,dt$$

$$= \lim_{N \to \infty} \left(-50{,}000e^{-0.14t}\,\Big|_{0}^{N} \right)$$

$$= \lim_{N \to \infty} -50{,}000(e^{-0.14N} - 1)$$

$$= \$50{,}000$$

Nuclear waste   The next example is similar in structure to the problem involving survival and renewal functions in Example 4.5 of Chapter 6.

EXAMPLE 2.5

It is estimated that t years from now, a certain nuclear power plant will be producing radioactive waste at the rate of $f(t) = 400t$ pounds per year. The waste decays exponentially at the rate of 2 percent per year. What will happen to the accumulation of radioactive waste from the plant in the long run?

SOLUTION

To find the amount of radioactive waste present after N years, divide the N-year interval $0 \leq t \leq N$ into n equal subintervals of length Δt years and let t_j denote the beginning of the jth subinterval (Figure 2.4). Then,

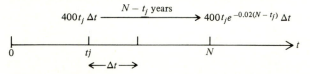

Figure 2.4 Radioactive waste generated during the jth subinterval.

$$\text{Amount of waste produced during } j\text{th subinterval} \approx 400t_j \, \Delta t$$

Since the waste decays exponentially at the rate of 2 percent per year, and since there are $N - t_j$ years between times $t = t_j$ and $t = N$, it follows that

$$\begin{array}{l}\text{Amount of waste produced} \\ \text{during } j\text{th subinterval} \\ \text{still present at } t = N\end{array} \approx 400t_j e^{-0.02(N-t_j)} \, \Delta t$$

Thus,

$$\begin{array}{l}\text{Amount of waste present} \\ \text{after } N \text{ years}\end{array} = \lim_{n \to \infty} \sum_{j=1}^{n} 400t_j e^{-0.02(N-t_j)} \, \Delta t$$

$$= \int_0^N 400t e^{-0.02(N-t)} \, dt$$

$$= 400 e^{-0.02N} \int_0^N t e^{0.02t} \, dt$$

The amount of radioactive waste present in the long run is the limit of this expression as N approaches infinity. That is,

Amount of waste
present in the
long run

$$= \lim_{N \to \infty} 400e^{-0.02N} \int_0^N te^{0.02t} \, dt$$

$$= \lim_{N \to \infty} 400e^{-0.02N}(50te^{0.02t} - 2{,}500e^{0.02t}) \Big|_0^N$$

$$= \lim_{N \to \infty} 400e^{-0.02N}(50Ne^{0.02N} - 2{,}500e^{0.02N} + 2{,}500)$$

$$= \lim_{N \to \infty} 400(50N - 2{,}500 + 2{,}500e^{-0.02N})$$

$$= \infty$$

That is, in the long run the accumulation of radioactive waste from the plant will increase without bound.

Other types of improper integrals

In the next two sections you will see applications of improper integrals to probability. In the course of the discussion, you will encounter improper integrals of the form $\int_{-\infty}^{\infty} f(x) \, dx$. Here, for future reference, is the definition of such an integral.

Improper integrals from $-\infty$ to ∞

$$\int_{-\infty}^{\infty} f(x) \, dx = \lim_{N \to \infty} \int_{-N}^{0} f(x) \, dx + \lim_{N \to \infty} \int_{0}^{N} f(x) \, dx$$

If both limits are finite, the improper integral is said to converge to their sum. Otherwise, the improper integral is said to diverge.

If f is nonnegative, the improper integral $\int_{-\infty}^{\infty} f(x) \, dx$ may be interpreted as the total area under the graph of f (Figure 2.5).

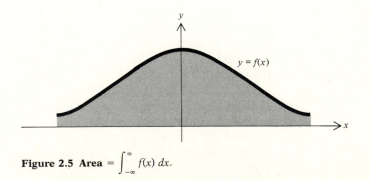

Figure 2.5 Area $= \int_{-\infty}^{\infty} f(x) \, dx.$

Problems In Problems 1 through 24, evaluate the given improper integral.

1. $\int_1^\infty \frac{1}{x^3} dx$
2. $\int_1^\infty x^{-3/2} dx$

3. $\int_1^\infty \frac{1}{\sqrt{x}} dx$
4. $\int_1^\infty x^{-2/3} dx$

5. $\int_3^\infty \frac{1}{2x - 1} dx$
6. $\int_3^\infty \frac{1}{\sqrt[3]{2x - 1}} dx$

7. $\int_3^\infty \frac{1}{(2x - 1)^2} dx$
8. $\int_0^\infty e^{-x} dx$

9. $\int_0^\infty 5e^{-2x} dx$
10. $\int_1^\infty e^{1-x} dx$

11. $\int_1^\infty \frac{x^2}{(x^3 + 2)^2} dx$
12. $\int_1^\infty \frac{x^2}{x^3 + 2} dx$

13. $\int_1^\infty \frac{x^2}{\sqrt{x^3 + 2}} dx$
14. $\int_0^\infty xe^{-x^2} dx$

15. $\int_1^\infty \frac{e^{-\sqrt{x}}}{\sqrt{x}} dx$
16. $\int_0^\infty xe^{-x} dx$

17. $\int_0^\infty 2xe^{-3x} dx$
18. $\int_0^\infty xe^{1-x} dx$

19. $\int_0^\infty 5xe^{10-x} dx$
20. $\int_1^\infty \frac{\ln x}{x} dx$

21. $\int_2^\infty \frac{1}{x \ln x} dx$
22. $\int_2^\infty \frac{1}{x\sqrt{\ln x}} dx$

23. $\int_0^\infty x^2e^{-x} dx$
24. $\int_0^\infty x^3e^{-x^2} dx$

Present value of an investment 25. An investment will generate \$2,400 per year in perpetuity. If the money is dispensed continuously throughout the year and if the prevailing annual interest rate remains fixed at 12 percent compounded continuously, what is the present value of the investment?

Present value 26. An investment will generate income continuously at the constant rate of Q dollars per year in perpetuity. Assuming a fixed annual interest rate of r compounded continuously, use an improper integral to show that the present value of the investment is Q/r dollars.

Present value of
rental property

27. It is estimated that t years from now an apartment complex will be generating profit for its owner at the rate of $f(t) = 10{,}000 + 500t$ dollars per year. If the profit is generated in perpetuity and the prevailing annual interest rate remains fixed at 10 percent compounded continuously, what is the present value of the apartment complex?

Present value of a
franchise

28. The management of a national chain of fast-food outlets is selling a permanent franchise in Seattle, Washington. Past experience in similar localities suggests that t years from now, the franchise will be generating profit at the rate of $f(t) = 12{,}000 + 900t$ dollars per year. If the prevailing interest rate remains fixed at 10 percent compounded continuously, what is the present value of the franchise?

Present value

29. In t years, an investment will be generating $f(t) = A + Bt$ dollars per year, where A and B are constants. If the income is generated in perpetuity and the prevailing annual interest rate of r compounded continuously does not change, show that the present value of this investment is $A/r + B/r^2$.

Nuclear waste

30. A certain nuclear power plant produces radioactive waste at the rate of 600 pounds per year. The waste decays exponentially at the rate of 2 percent per year. How much radioactive waste from the plant will be present in the long run?

Health care

31. The fraction of patients who will still be receiving treatment at a certain mental health clinic t months after their initial visit is $f(t) = e^{-t/20}$. If the clinic accepts new patients at the rate of 10 per month, approximately how many patients will be receiving treatment at the clinic in the long run?

Population growth

32. Demographic studies conducted in a certain city indicate that the fraction of the residents that will remain in the city for at least t years is $f(t) = e^{-t/20}$. The current population of the city is 200,000, and it is estimated that new residents will be arriving at the rate of 100 people per year. If this estimate is correct, what will happen to the population of the city in the long run?

Medicine

33. A hospital patient receives intravenously 5 units of a certain drug per hour. The drug is eliminated exponentially, so that the fraction that remains in the patient's body for t hours is $f(t) = e^{-t/10}$. If the treatment is continued indefinitely, approximately how many units of the drug will be in the patient's body in the long run?

3 PROBABILITY DENSITY FUNCTIONS

Some of the most important applications of integration to the social, managerial, and life sciences are in the areas of probability and statistics. In Chapter 6, Section 2, you were introduced to the technique of integrating probability density functions to compute probabilities. The purpose of this section is to explore in more detail the relationship between integration and probability. Improper integrals will play an important role in the discussion.

Random variables

The life span of a light bulb selected at random from a manufacturer's stock is a quantity that cannot be predicted with certainty. In the terminology of probability and statistics, the process of selecting the bulb at random is called a **random experiment,** and the life span of the bulb is said to be a **random variable.** In general, a random variable is a number associated with the outcome of a random experiment.

A random variable that can take on only integer values is said to be **discrete.** The face value of a randomly selected playing card and the number of times heads comes up in three tosses of a coin are discrete random variables. So is the IQ of a randomly selected university student, since IQs are measured in whole numbers.

A random variable that can take on any value in some interval is said to be **continuous.** Some continuous random variables are the time a randomly selected motorist spends waiting at a traffic light, the time interval between the arrivals of randomly selected successive planes at an airport, and the time it takes a randomly selected subject to learn a particular task. Integral calculus is used in the study of continuous random variables.

Probability

The **probability** of an event that can result from a random experiment is a number between 0 and 1 that specifies the likelihood of the event. In particular, the probability is the fraction of the time the event can be expected to occur if the experiment is repeated a large number of times. For example, the probability that an evenly balanced tossed coin will come up heads is $\frac{1}{2}$ since this event can be expected approximately $\frac{1}{2}$ of the time if the coin is tossed repeatedly. In a group containing 13 men and 10 women, the probability is $\frac{10}{23}$ that a person selected at random is a woman. The probability of an event that is certain to occur is 1, while the probability of an event that cannot possibly occur is zero. For example, if you roll an ordinary die, the probability is 1 that you will get a number between 1 and 6, inclusive, while the probability is zero that you will get a 7.

Events described in terms of random variables

Consider again the random experiment in which a light bulb is selected from a manufacturer's stock. A possible event resulting from this experiment is that the life span of the selected bulb is between 20 and 35 hours. If x is the random variable denoting the life span of a randomly selected bulb, this event can be described by the inequality $20 \leq x \leq 35$ and its probability denoted by the symbol $P(20 \leq x \leq 35)$. Similarly, the probability that the bulb will burn for at least 50 hours is denoted by $P(x \geq 50)$ or $P(50 \leq x < \infty)$.

Probability density functions

A **probability density function** for a continuous random variable x is a nonnegative function f with the property that $P(a \leq x \leq b)$ is the area under the graph of f from $x = a$ to $x = b$. A possible probability density function for the life span of a light bulb is sketched in Figure 3.1. Its shape reflects the fact that most bulbs burn out relatively

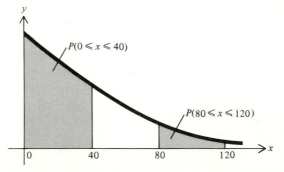

Figure 3.1 Probability density function for the life span of a light bulb.

quickly. For example, the probability that a bulb will fail within the first 40 hours is represented by the area under the curve between $x = 0$ and $x = 40$. This is much greater than the area under the curve between $x = 80$ and $x = 120$, which represents the probability that the bulb will fail between its 80th and 120th hour of use.

The basic property of probability density functions can be restated in terms of the integrals you would use to compute the appropriate areas.

Probability density functions

A probability density function for a continuous random variable x is a nonnegative function f such that

$$P(a \leq x \leq b) = \int_{a}^{b} f(x)\, dx$$

The values of a and b in this formula need not be finite. If either is infinite, the corresponding probability is given by an improper integral. For example, the probability that x is greater than or equal to a is

$$P(x \geq a) = P(a \leq x < \infty) = \int_a^\infty f(x) \, dx$$

The total area under the graph of a probability density function must be equal to 1. This is because the total area represents the probability that x is between $-\infty$ and ∞, which is an event that is certain to occur. This observation can be restated in terms of improper integrals as follows:

A property of probability density functions

If f is a probability density function for a continuous random variable x,

$$\int_{-\infty}^{\infty} f(x) \, dx = 1$$

How to determine the appropriate probability density function for a particular random variable is a central problem in probability. It involves techniques beyond the scope of this book, which can be found in most probability and statistics texts. The purpose of the discussion in the remainder of this chapter is to show you some of the probability density functions that have proved to be most useful, to illustrate their use, and to introduce some important properties of probability density functions in general.

Uniform density functions

A **uniform density function** (Figure 3.2) is constant over a bounded interval $A \leq x \leq B$ and zero outside the interval. A random variable that has a uniform density function is said to be **uniformly distributed.** Roughly speaking, a uniformly distributed random variable is

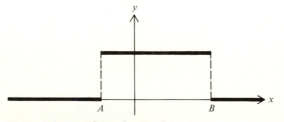

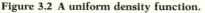

Figure 3.2 A uniform density function.

one for which all the values in some bounded interval are "equally likely." More precisely, a continuous random variable is uniformly distributed if the probability that its value will be in a particular subinterval of the bounded interval is equal to the probability that it will be in any other subinterval that has the same length. An example of a uniformly distributed random variable is the waiting time of a motorist at a traffic light that remains red for, say, 40 seconds at a time. This random variable has a uniform distribution because all waiting times between 0 and 40 seconds are equally likely.

If k is the constant value of a uniform density function $f(x)$ on the interval $A \leq x \leq B$, the value of k is determined by the requirement that the total area under the graph of f must be equal to 1. In particular,

$$1 = \int_{-\infty}^{\infty} f(x)\, dx = \int_{A}^{B} f(x)\, dx \qquad \text{[since } f(x) = 0 \text{ off the interval } A \leq x \leq B\text{]}$$

$$= \int_{A}^{B} k\, dx = kx \Big|_{A}^{B} = k(B - A)$$

and so
$$k = \frac{1}{B - A}$$

This observation leads to the following formula for a uniform density function.

Uniform density function

$$f(x) = \begin{cases} \dfrac{1}{B - A} & \textbf{if } A \leq x \leq B \\ 0 & \textbf{otherwise} \end{cases}$$

Here is a typical application involving a uniform density function.

EXAMPLE 3.1

A certain traffic light remains red for 40 seconds at a time. You arrive (at random) at the light and find it red. Use an appropriate uniform density function to find the probability that you will have to wait at least 15 seconds for the light to turn green.

SOLUTION

Let x denote the time (in seconds) that you must wait. Since all waiting times between 0 and 40 are "equally likely," x is uniformly distributed over the interval $0 \leq x \leq 40$. The corresponding uniform density function is

$$f(x) = \begin{cases} \dfrac{1}{40} & \text{if } 0 \le x \le 40 \\ 0 & \text{otherwise} \end{cases}$$

and the desired probability is

$$P(15 \le x \le 40) = \int_{15}^{40} \frac{1}{40} \, dx = \frac{x}{40} \Big|_{15}^{40} = \frac{40 - 15}{40} = \frac{5}{8}$$

Exponential density functions

An **exponential density function** is a function $f(x)$ that is zero for $x < 0$ and that decreases exponentially for $x \ge 0$. That is, for $x \ge 0$,

$$f(x) = Ae^{-kx}$$

where A is a constant and k a positive constant.

The value of A is determined by the requirement that the total area under the graph of f be equal to 1. Thus,

$$1 = \int_{-\infty}^{\infty} f(x) \, dx = \int_{0}^{\infty} Ae^{-kx} \, dx = \lim_{N \to \infty} \int_{0}^{N} Ae^{-kx} \, dx$$

$$= \lim_{N \to \infty} \left(-\frac{A}{k} e^{-kx} \, \Big|_{0}^{N} \right) = \lim_{N \to \infty} \left(-\frac{A}{k} e^{-kN} + \frac{A}{k} \right) = \frac{A}{k}$$

and so

$$A = k$$

This calculation leads to the following general formula for an exponential density function. The corresponding graph is shown in Figure 3.3.

Exponential density function

$$f(x) = \begin{cases} ke^{-kx} & \textbf{if } x \ge 0 \\ 0 & \textbf{if } x < 0 \end{cases}$$

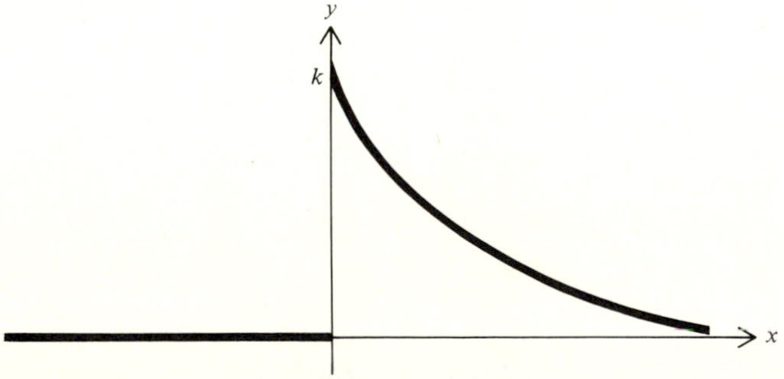

Figure 3.3 An exponential density function.

A random variable that has an exponential density function is said to be **exponentially distributed.** As you can see from the graph in Figure 3.3, the value of an exponentially distributed random variable is much more likely to be small than large. Such random variables include the life span of electronic components, the duration of telephone calls, and the interval between the arrivals of successive planes at an airport. The use of exponential density functions was illustrated in Example 2.3 of Chapter 6. Here is another example.

EXAMPLE 3.2

The probability density function for the duration of telephone calls in a certain city is

$$f(x) = \begin{cases} 0.5e^{-0.5x} & \text{if } x \geq 0 \\ 0 & \text{if } x < 0 \end{cases}$$

where x denotes the duration (in minutes) of a randomly selected call.

(a) Find the probability that a randomly selected call will last between 2 and 3 minutes.
(b) Find the probability that a randomly selected call will last at least 2 minutes.

SOLUTION

(a)
$$P(2 \leq x \leq 3) = \int_2^3 0.5e^{-0.5x}\, dx = -e^{-0.5x} \Big|_2^3$$

$$= -e^{-1.5} + e^{-1} = 0.1447$$

(b) There are two ways to compute this probability. The first method is to evaluate an improper integral.

$$P(x \geq 2) = P(2 \leq x < \infty) = \int_2^\infty 0.5e^{-0.5x}\, dx$$

$$= \lim_{N \to \infty} \int_2^N 0.5e^{-0.5x}\, dx = \lim_{N \to \infty} \left(-e^{-0.5x} \Big|_2^N \right)$$

$$= \lim_{N \to \infty} (-e^{-0.5N} + e^{-1}) = e^{-1} = 0.3679$$

The second method is to compute 1 minus the probability that x is less than 2. That is,

$$P(x \geq 2) = 1 - \int_0^2 0.5e^{-0.5x}\, dx = 1 - \left(-e^{-0.5x} \Big|_0^2 \right)$$

$$= 1 - (-e^{-1} + 1) = e^{-1} = 0.3679$$

Normal density functions

The most widely used probability density functions are the **normal density functions.** You are probably already familiar with their famous "bell-shaped" graphs like the one in Figure 3.4. The discussion of the normal density functions will be postponed until Section 4, which will be devoted entirely to this important topic.

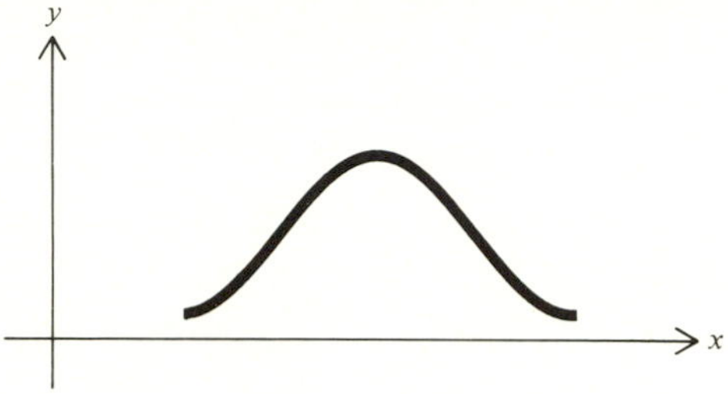

Figure 3.4 A normal density function.

The expected value of a random variable

An important characteristic of a random variable is its "average" value. Familiar averages of random variables include the average highway mileage for a particular car model, the average waiting time at the check-in counter of a certain airline, and the average life span of workers in a hazardous profession. The average value of a random variable x is called its **expected value** or **mean** and is denoted by the symbol $E(x)$. For continuous random variables, you can calculate the expected value from the probability density function using the following formula.

Expected value

> **If x is a continuous random variable with probability density function f, the expected value (or mean) of x is**
>
> $$E(x) = \int_{-\infty}^{\infty} xf(x)\, dx$$

Why the expected value is the average

To see why this integral gives the average value of a continuous random variable, first consider the simpler case in which x is a discrete random variable that takes on the values $x_1, x_2, \ldots, x_n$. If each of these values occurs with equal frequency, the probability of each is $\dfrac{1}{n}$ and the average value of x is

$$\frac{x_1 + x_2 + \cdots + x_n}{n} = x_1\left(\frac{1}{n}\right) + x_2\left(\frac{1}{n}\right) + \cdots + x_n\left(\frac{1}{n}\right)$$

More generally, if the values $x_1, x_2, \ldots, x_n$ occur with probabilities $p_1, p_2, \ldots, p_n$, respectively, the average value of x is the weighted sum

$$x_1 p_1 + x_2 p_2 + \cdots + x_n p_n = \sum_{j=1}^{n} x_j p_j$$

Now consider the continuous case. For simplicity, restrict your attention to a bounded interval $A \le x \le B$. Imagine that this interval is divided into n subintervals of width Δx, and let x_j denote the beginning of the jth subinterval. Then,

$$p_j = \begin{array}{c}\text{probability that } x \text{ is} \\ \text{in } j\text{th subinterval}\end{array} = \begin{array}{c}\text{area under graph} \\ \text{of } f \text{ from } x_j \text{ to } x_{j+1}\end{array} \approx f(x_j)\,\Delta x$$

where $f(x_j)\,\Delta x$ is the area of an approximating rectangle (Figure 3.5).

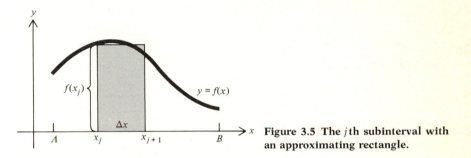

Figure 3.5 The jth subinterval with an approximating rectangle.

To approximate the average value of x over the interval $A \le x \le B$, treat x as if it were a discrete random variable that takes on the values $x_1, x_2, \ldots, x_n$ with probabilities $p_1, p_2, \ldots, p_n$, respectively. Then,

$$\begin{array}{c}\text{Average value of} \\ x \text{ on } A \le x \le B\end{array} \approx \sum_{j=1}^{n} x_j p_j \approx \sum_{j=1}^{n} x_j f(x_j)\,\Delta x$$

The actual average value of x on the interval $A \le x \le B$ is the limit of this approximating sum as n increases without bound. That is,

$$\begin{array}{c}\text{Average value of} \\ x \text{ on } A \le x \le B\end{array} = \lim_{n \to \infty} \sum_{j=1}^{n} x_j f(x_j)\,\Delta x = \int_{A}^{B} x f(x)\,dx$$

An extension of this argument shows that

$$\begin{array}{c}\text{Average value of } x \\ \text{on } -\infty < x < \infty\end{array} = \int_{-\infty}^{\infty} x f(x)\,dx$$

which is the formula for the expected value of a continuous random variable.

Geometric interpretation of the expected value

To get a better feel for the expected value of a continuous random variable x, think of the probability density function for x as describing the distribution of mass on a beam lying on the x axis. Then the expected value of x is the point at which the beam will balance. If the graph of the density function is symmetric, the expected value is the point of symmetry, as illustrated in Figure 3.6.

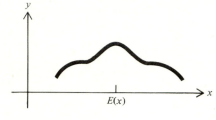

Figure 3.6 The expected value of a symmetric random variable.

Calculation of the expected value

The use of the integral formula to compute the expected value of a continuous random variable is illustrated in the next two examples.

EXAMPLE 3.3

Find the expected value of the uniformly distributed random variable from Example 3.1 with density function

$$f(x) = \begin{cases} \dfrac{1}{40} & \text{if } 0 \le x \le 40 \\ 0 & \text{otherwise} \end{cases}$$

SOLUTION

$$E(x) = \int_{-\infty}^{\infty} xf(x)\, dx = \int_0^{40} \frac{x}{40}\, dx = \frac{x^2}{80}\bigg|_0^{40} = \frac{1{,}600}{80} = 20$$

In the context of Example 3.1, this says that the average waiting time at the red light is 20 seconds, a conclusion that should come as no surprise since the random variable is uniformly distributed between 0 and 40.

EXAMPLE 3.4

Find the expected value of the exponentially distributed random variable from Example 3.2 with density function

$$f(x) = \begin{cases} 0.5e^{-0.5x} & \text{if } x \ge 0 \\ 0 & \text{if } x < 0 \end{cases}$$

SOLUTION

$$E(x) = \int_{-\infty}^{\infty} xf(x)\, dx = \int_{0}^{\infty} 0.5xe^{-0.5x}\, dx$$

$$= \lim_{N \to \infty} \int_{0}^{N} 0.5xe^{-0.5x}\, dx$$

$$= \lim_{N \to \infty} \left(-xe^{-0.5x} \Big|_{0}^{N} + \int_{0}^{N} e^{-0.5x}\, dx \right)$$

$$= \lim_{N \to \infty} \left(-xe^{-0.5x} - 2e^{-0.5x} \Big|_{0}^{N} \right)$$

$$= \lim_{N \to \infty} (-Ne^{-0.5N} - 2e^{-0.5N} + 2)$$

$$= 2$$

That is, the average duration of telephone calls in the city in Example 3.2 is 2 minutes.

The variance of a random variable

The expected value or mean of a random variable tells you the center of its distribution. Another concept that is useful in describing the distribution of a random variable is the **variance,** which tells you how spread out the distribution is. That is, the variance measures the tendency of the values of a random variable to cluster about their mean. Here is the definition.

Variance

> **If x is a continuous random variable with probability density function f, the variance of x is**
>
> $$\text{Var}(x) = \int_{-\infty}^{\infty} [x - E(x)]^2 f(x)\, dx$$

The definition of the variance is not as mysterious as it may seem at first glance. Notice that the formula for the variance is the same as that for the expected value, except that x has been replaced by the expression $[x - E(x)]^2$, which represents the square of the deviation of x from its mean $E(x)$. The variance, therefore, is simply the expected value or average of the squared deviations of the values of x from the mean. If the values of x tend to cluster about the mean as in Figure 3.7a, most of the deviations from the mean will be small and the variance, which is the average of the squares of these deviations, will

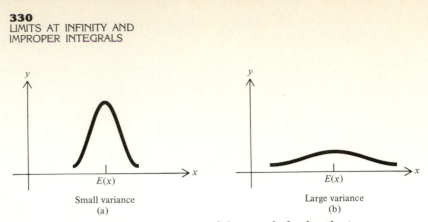

Figure 3.7 The variance as a measure of the spread of a distribution.

also be small. On the other hand, if the values of the random variable are widely scattered as in Figure 3.7b, there will be many large deviations from the mean and the variance will be large.

Calculation of the variance

As you just saw, the formula

$$\text{Var}(x) = \int_{-\infty}^{\infty} [x - E(x)]^2 f(x)\, dx$$

defining the variance is fairly easily interpreted as a measure of the spread of the distribution of x. However, for all but the simplest density functions, this formula is cumbersome to use for the actual calculation of the variance. This is because the term $x - E(x)$ must be squared and multiplied by $f(x)$ before the integration can be performed. Here is an equivalent formula for the variance, which is easier to use for computational purposes.

Variance formula

$$\text{Var}(x) = \int_{-\infty}^{\infty} x^2 f(x)\, dx - [E(x)]^2$$

The derivation of this formula is straightforward. It involves expanding the integrand in the definition of the variance and rearranging the resulting terms until the new formula emerges. As you read the steps, keep in mind that the expected value $E(x)$ is a constant and can be brought outside integrals by the constant multiple rule.

$$\text{Var}(x) = \int_{-\infty}^{\infty} [x - E(x)]^2 f(x)\, dx \qquad \text{(definition of variance)}$$

$$= \int_{-\infty}^{\infty} \{x^2 - 2xE(x) + [E(x)]^2\} f(x)\, dx \qquad \text{(integrand expanded)}$$

$$= \int_{-\infty}^{\infty} x^2 f(x)\, dx - 2E(x) \int_{-\infty}^{\infty} x f(x)\, dx + [E(x)]^2 \int_{-\infty}^{\infty} f(x)\, dx$$

(sum rule and constant multiple rule)

$$= \int_{-\infty}^{\infty} x^2 f(x)\, dx - 2E(x) \cdot E(x) + [E(x)]^2 (1)$$

(definition of $E(x)$ and fact that $\int_{-\infty}^{\infty} f(x)\, dx = 1$)

$$= \int_{-\infty}^{\infty} x^2 f(x)\, dx - [E(x)]^2$$

The use of this formula to compute the variance is illustrated in the next two examples.

EXAMPLE 3.5

Find the variance of the uniformly distributed random variable from Example 3.1 with density function

$$f(x) = \begin{cases} \dfrac{1}{40} & \text{if } 0 \le x \le 40 \\ 0 & \text{otherwise} \end{cases}$$

SOLUTION

The first step is to compute $E(x)$. This was done in Example 3.3, where you found that $E(x) = 20$. Using this value in the variance formula, you get

$$\text{Var}(x) = \int_{-\infty}^{\infty} x^2 f(x)\, dx - [E(x)]^2$$

$$= \int_{0}^{40} \frac{x^2}{40}\, dx - 400 = \frac{x^3}{120}\bigg|_{0}^{40} - 400$$

$$= \frac{64{,}000}{120} - 400 = \frac{16{,}000}{120} = \frac{400}{3}$$

EXAMPLE 3.6

Find the variance of the exponentially distributed random variable from Example 3.2 with density function

$$f(x) = \begin{cases} 0.5e^{-0.5x} & \text{if } x \ge 0 \\ 0 & \text{if } x < 0 \end{cases}$$

SOLUTION

Using the value $E(x) = 2$ obtained in Example 3.4 and integrating by parts twice, you get

$$\text{Var}(x) = \int_{-\infty}^{\infty} x^2 f(x)\, dx - [E(x)]^2$$

$$= \int_0^{\infty} 0.5x^2 e^{-0.5x}\, dx - 4$$

$$= \lim_{N \to \infty} \int_0^N 0.5x^2 e^{-0.5x}\, dx - 4$$

$$= \lim_{N \to \infty} \left(-x^2 e^{-0.5x} \Big|_0^N + 2 \int_0^N x e^{-0.5x}\, dx \right) - 4$$

$$= \lim_{N \to \infty} \left[(-x^2 e^{-0.5x} - 4x e^{-0.5x}) \Big|_0^N + 4 \int_0^N e^{-0.5x}\, dx \right] - 4$$

$$= \lim_{N \to \infty} (-x^2 - 4x - 8)e^{-0.5x} \Big|_0^N - 4$$

$$= \lim_{N \to \infty} (-N^2 - 4N - 8)e^{-0.5N} + 8 - 4$$

$$= 4$$

Problems In Problems 1 through 8, integrate the given probability density function to find the indicated probabilities.

1. $f(x) = \begin{cases} \frac{1}{3} & \text{if } 2 \le x \le 5 \\ 0 & \text{otherwise} \end{cases}$

    (a) $P(2 \le x \le 5)$ (b) $P(3 \le x \le 4)$ (c) $P(x \ge 4)$

2. $f(x) = \begin{cases} \dfrac{x}{2} & \text{if } 0 \le x \le 2 \\ 0 & \text{otherwise} \end{cases}$

    (a) $P(0 \le x \le 2)$ (b) $P(1 \le x \le 2)$ (c) $P(x \le 1)$

3. $f(x) = \begin{cases} \frac{1}{8}(4 - x) & \text{if } 0 \le x \le 4 \\ 0 & \text{otherwise} \end{cases}$

    (a) $P(0 \le x \le 4)$ (b) $P(2 \le x \le 3)$ (c) $P(x \ge 1)$

4. $f(x) = \begin{cases} \frac{3}{32}(4x - x^2) & \text{if } 0 \le x \le 4 \\ 0 & \text{otherwise} \end{cases}$

    (a) $P(0 \le x \le 4)$ (b) $P(1 \le x \le 2)$ (c) $P(x \le 1)$

5. $f(x) = \begin{cases} \dfrac{3}{x^4} & \text{if } 1 \le x < \infty \\ 0 & \text{if } x < 1 \end{cases}$

    (a) $P(1 \le x < \infty)$ (b) $P(1 \le x \le 2)$ (c) $P(x \ge 2)$

6. $f(x) = \begin{cases} \frac{1}{10} e^{-x/10} & \text{if } x \ge 0 \\ 0 & \text{if } x < 0 \end{cases}$

    (a) $P(0 \le x < \infty)$ (b) $P(x \le 2)$ (c) $P(x \ge 5)$

7. $f(x) = \begin{cases} 2xe^{-x^2} & \text{if } x \ge 0 \\ 0 & \text{if } x < 0 \end{cases}$

    (a) $P(x \ge 0)$ (b) $P(1 \le x \le 2)$ (c) $P(x \le 2)$

8. $f(x) = \begin{cases} \frac{1}{4} xe^{-x/2} & \text{if } x \ge 0 \\ 0 & \text{if } x < 0 \end{cases}$

    (a) $P(0 \le x < \infty)$ (b) $P(2 \le x \le 4)$ (c) $P(x \ge 6)$

Traffic flow 9. A certain traffic light remains red for 45 seconds at a time. You arrive (at random) at the light and find it red. Use an appropriate uniform density function to find the probability that the light will turn green within 15 seconds.

Commuting 10. During the morning rush hour, commuter trains from Long Island to Manhattan run every 20 minutes. You arrive (at random) at the station during the rush hour and find no train at the platform. Assuming that the trains are running on schedule, use an appropriate uniform density function to find the probability that you will have to wait at least 8 minutes for your train.

Movie theaters 11. A 2-hour movie runs continuously at a local theater. You leave for the theater without first checking the show times. Use an appropriate uniform density function to find the probability that you will arrive at the theater within 10 minutes (before or after) of the start of the film.

Experimental psychology 12. Suppose x is the length of time (in minutes) that it takes a laboratory rat to traverse a certain maze. If x is exponentially distributed with density function

$$f(x) = \begin{cases} \frac{1}{3}e^{-x/3} & \text{if } x \ge 0 \\ 0 & \text{if } x < 0 \end{cases}$$

find the probability that a randomly selected rat will require more than 3 minutes to traverse the maze.

Customer service 13. The time x (in minutes) that a customer must spend waiting in line at a certain bank is exponentially distributed with density function

$$f(x) = \begin{cases} \frac{1}{4}e^{-x/4} & \text{if } x \geq 0 \\ 0 & \text{if } x < 0 \end{cases}$$

Find the probability that a randomly selected customer at the bank will have to stand in line at least 8 minutes.

Warranty protection 14. The life span x (in months) of a certain electrical appliance is exponentially distributed with density function

$$f(x) = \begin{cases} 0.08e^{-0.08x} & \text{if } x \geq 0 \\ 0 & \text{if } x < 0 \end{cases}$$

The appliance carries a 1-year warranty from the manufacturer. Suppose you purchase one of these appliances, selected at random from the manufacturer's stock. Find the probability that the warranty will expire before your appliance becomes unusable.

15. Find the expected value and variance for the random variable in Problem 1.

16. Find the expected value and variance for the random variable in Problem 2.

17. Find the expected value and variance for the random variable in Problem 3.

18. Find the expected value and variance for the random variable in Problem 5.

19. Find the expected value and variance for the random variable in Problem 6.

Traffic flow 20. Find the average waiting time for cars arriving on red at the traffic light in Problem 9.

Commuting 21. Find the average wait for rush-hour commuters arriving at the station in Problem 10 when no train is at the platform.

Experimental psychology 22. Find the average time required for laboratory rats to traverse the maze in Problem 12.

Customer service 23. Find the average waiting time for customers at the bank in Problem 13.

24. Show that the expected value of a uniformly distributed random variable with density function

$$f(x) = \begin{cases} \dfrac{1}{B - A} & \text{if } A \leq x \leq B \\ 0 & \text{otherwise} \end{cases}$$

is $(A + B)/2$.

25. Show that the variance of the uniformly distributed random variable in Problem 24 is $(B - A)^2/12$.

26. Show that the expected value of an exponentially distributed random variable with density function

$$f(x) = \begin{cases} ke^{-kx} & \text{if } x \geq 0 \\ 0 & \text{if } x < 0 \end{cases}$$

is $1/k$.

27. Show that the variance of the exponentially distributed random variable in Problem 26 is $1/k^2$.

4 THE NORMAL DISTRIBUTION

In this section you will learn how to work with the most important and widely used probability density functions. They are called **normal density functions,** and their graphs are bell-shaped curves. They describe or approximate the distributions of many random variables arising in the social and natural sciences, including heights, weights, test scores, and measurement errors.

Normal density functions

> **A normal density function is a function of the form**
>
> $$f(x) = \frac{1}{\sigma\sqrt{2\pi}} e^{-(x-\mu)^2/2\sigma^2}$$
>
> **where μ (mu) and σ (sigma) are constants and $\sigma > 0$.**

The graph of a normal density function is a symmetric bell-shaped curve (Figure 4.1). A routine calculation shows that the peak of the curve occurs at $x = \mu$ and the inflection points occur σ units to the left and right of the peak. (A special case of this calculation was performed in Example 4.8 of Chapter 4.) As required, the total area

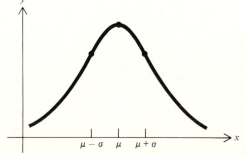

Figure 4.1 A normal density function.

under the graph is 1, but the verification of this fact involves advanced techniques beyond the scope of this text.

The interpretation of μ and σ

The constant μ that appears in the definition of the normal density function turns out to be the mean (or expected value) of the corresponding random variable. This fact can be derived fairly easily from the integral formula for expected value, which was given in Section 3. (Problem 27 at the end of this section asks for a special case of this derivation.) Notice that the fact that μ is the mean is also consistent with the previous observation that the peak (which is the point of symmetry) of the graph of the normal density function occurs at $x = \mu$.

The constant σ in the formula for the normal density function turns out to be the square root of the variance of the corresponding random variable. (This can be derived from the integral formula for the variance given in Section 3, but the evaluation of the integral requires advanced techniques beyond the scope of this text.) The quantity σ is called the **standard deviation** of the random variable and, like the variance, is a measure of the dispersion of the distribution.

The mean and standard deviation

If x **is a normally distributed random variable with density function**

$$f(x) = \frac{1}{\sigma\sqrt{2\pi}} e^{-(x-\mu)^2/2\sigma^2}$$

then $\mu =$ **mean of** $x = E(x)$

and $\sigma =$ **standard deviation of** $x = \sqrt{\mathrm{Var}(x)}$

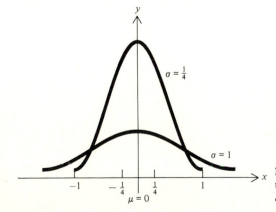

Figure 4.2 Normal density functions with $\mu = 0$ and $\sigma = 1$ and $\sigma = \frac{1}{4}$.

Figure 4.2 shows the graphs of two normal density functions that have the same mean $\mu = 0$ and standard deviations $\sigma = 1$ and $\sigma = 1/4$.

With knowledge of the mean and the standard deviation of a normal random variable x, you can use the following guidelines to get a rough idea of the distribution of x.

Probability guidelines for normal distributions

> 1. **Approximately 68 percent of all the values of x lie between $\mu - \sigma$ and $\mu + \sigma$ (i.e., within 1 standard deviation of the mean).**
> 2. **Approximately 95 percent of all the values of x lie between $\mu - 2\sigma$ and $\mu + 2\sigma$ (i.e., within 2 standard deviations of the mean).**
> 3. **Approximately 99.7 percent of all the values of x lie between $\mu - 3\sigma$ and $\mu + 3\sigma$ (i.e., within 3 standard deviations of the mean).**

Later in this section you will see how to verify these guidelines. What is important to remember when applying the guidelines is that they hold for *every* normally distributed random variable, no matter what the values of μ and σ happen to be. The guidelines are illustrated in Figure 4.3.

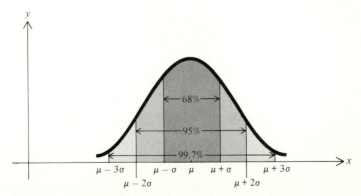

Figure 4.3 Probability guidelines for normal distributions.

The use of these guidelines is illustrated in the next example.

EXAMPLE 4.1

The weights of newborn babies in a certain community are normally distributed with mean $\mu = 7.1$ pounds and standard deviation $\sigma = 0.4$ pound. Use the probability guidelines for normal distributions to find intervals containing the weights of 68 percent, 95 percent, and 99.7 percent of the newborn babies in the community.

SOLUTION

Since $\quad \mu - \sigma = 7.1 - 0.4 = 6.7 \quad$ and $\quad \mu + \sigma = 7.1 + 0.4 = 7.5$

it follows that approximately 68 percent of the babies will weigh between 6.7 and 7.5 pounds. Since

$$\mu - 2\sigma = 7.1 - 0.8 = 6.3 \quad \text{and} \quad \mu + 2\sigma = 7.1 + 0.8 = 7.9$$

it follows that approximately 95 percent of the babies will weigh between 6.3 and 7.9 pounds. Since

$$\mu - 3\sigma = 7.1 - 1.2 = 5.9 \quad \text{and} \quad \mu + 3\sigma = 7.1 + 1.2 = 8.3$$

it follows that approximately 99.7 percent of the babies will weigh between 5.9 and 8.3 pounds.

The standard normal density function

The simplest of the normal density functions is the one with mean $\mu = 0$ and standard deviation $\sigma = 1$. It is known as the **standard normal density function,** and the letter z is frequently used to denote random variables with this special distribution.

The standard normal density function

The standard normal density function is the function

$$f(z) = \frac{1}{\sqrt{2\pi}} e^{-z^2/2}$$

The importance of the standard normal density function is due to its relative simplicity and to the fact that any other normal density function can be transformed to the standard normal by a routine change of variables. Thus, once you know how to compute probabilities using this particular density function, you will be able to compute probabilities associated with any normal density function.

How to use the
standard normal
table

If z is a standard normal random variable, the probability that z lies between a and b is the area under the standard normal curve between $z = a$ and $z = b$, and is given by the integral

$$P(a \le z \le b) = \frac{1}{\sqrt{2\pi}} \int_a^b e^{-z^2/2} \, dz$$

Unfortunately, the integrand $e^{-z^2/2}$ does not have an elementary antiderivative, and numerical methods must be used to approximate the integral.

Using numerical methods and electronic computers, statisticians have constructed highly accurate tables of areas under the standard normal curve. One such table is reproduced in Table III at the back of the book. It gives areas under the standard normal curve to the left of specified positive values of z or, equivalently, probabilities of the form $P(z \le a)$. To get other probabilities from the table, you use the symmetry of the normal curve and the fact that the total area under the curve is 1. Here is an example illustrating the use of the table.

EXAMPLE 4.2

Suppose z is a standard normal random variable. Use the normal table to find the following probabilities.

(a) $P(z \le 1.25)$ (b) $P(z \ge 0.03)$
(c) $P(z \le -1.87)$ (d) $P(-1.87 \le z \le 1.25)$

SOLUTION

(a) $P(z \le 1.25)$ is the area under the standard normal curve to the left of 1.25 (Figure 4.4). To look up this area in the table, go down

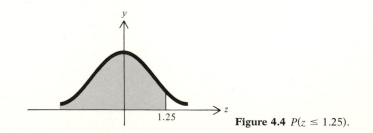

Figure 4.4 $P(z \le 1.25)$.

the first column until you get to 1.2, and then move to the right until you reach the column labeled .05. The number .8944 in this location is the desired probability. That is,

$$P(z \le 1.25) = 0.8944$$

(b) $P(z \geq 0.03)$ is the area under the curve to the right of 0.03 (Figure 4.5). This is equal to 1 (the total area under the curve) minus the

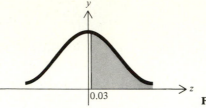

0.03

Figure 4.5 $P(z \geq 0.03) = 1 - P(z \leq 0.03)$.

area to the left of 0.03. That is,

$$P(z \geq 0.03) = 1 - P(z \leq 0.03) = 1 - 0.5120 = 0.4880$$

(c) $P(z \leq -1.87)$ is the area under the curve to the left of -1.87 (Figure 4.6). Since the curve is symmetric about $z = 0$, this is the

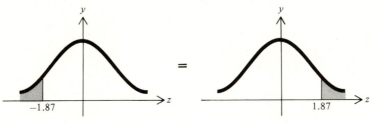

Figure 4.6 $P(z \leq -1.87) = P(z \geq 1.87) = 1 - P(z \leq 1.87)$.

same as the area to the right of 1.87, which you can find as in part (b). That is,

$$P(z \leq -1.87) = P(z \geq 1.87) = 1 - P(z \leq 1.87)$$

$$= 1 - 0.9693 = 0.0307$$

(d) $P(-1.87 \leq z \leq 1.25)$ is the area under the curve between -1.87 and 1.25. This is equal to the area to the left of 1.25 minus the area to the left of -1.87 (Figure 4.7). Hence,

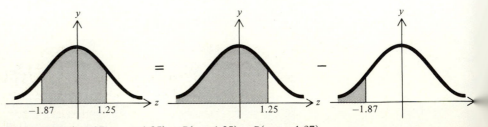

Figure 4.7 $P(-1.87 \leq z \leq 1.25) = P(z \leq 1.25) - P(z \leq -1.87)$.

$$P(-1.87 \le z \le 1.25) = P(z \le 1.25) - P(z \le -1.87)$$
$$= 0.8944 - 0.0307 = 0.8637$$

Transformation of normal random variables

To compute probabilities for an arbitrary normal random variable x, you make a change of variables that transforms x to the standard normal variable z. The transformation shifts the original normal curve so that the new mean is zero and "compresses" or "expands" it so that the new standard deviation is 1. In particular, if x has a normal distribution with mean μ and standard deviation σ, then the variable

$$z = \frac{x - \mu}{\sigma}$$

has the standard normal distribution, and you can use the following formula to compute probabilities for the original variable x.

How to compute normal probabilities

If x is a random variable that has a normal distribution with mean μ and standard deviation σ,

$$P(a \le x \le b) = P\left(\frac{a - \mu}{\sigma} \le z \le \frac{b - \mu}{\sigma}\right)$$

where z has the standard normal distribution.

The verification of this formula involves the method of substitution for definite integrals from Chapter 6, Section 1. In particular, you begin with the integral formula

$$P(a \le x \le b) = \frac{1}{\sigma\sqrt{2\pi}} \int_a^b e^{-(x-\mu)^2/2\sigma^2} \, dx$$

and let $z = \dfrac{x - \mu}{\sigma}$. Then $dz = \dfrac{1}{\sigma} dx$ or $dx = \sigma \, dz$. Moreover, $z(a) = \dfrac{a - \mu}{\sigma}$ and $z(b) = \dfrac{b - \mu}{\sigma}$. Substituting in the original integral you get

$$P(a \le x \le b) = \frac{1}{\sqrt{2\pi}} \int_{(a-\mu)/\sigma}^{(b-\mu)/\sigma} e^{-z^2/2} \, dz$$

which is precisely the integral of the standard normal density function from $\dfrac{a - \mu}{\sigma}$ to $\dfrac{b - \mu}{\sigma}$. Thus,

$$P(a \leq x \leq b) = P\left(\frac{a - \mu}{\sigma} \leq z \leq \frac{b - \mu}{\sigma}\right)$$

where z has the standard normal distribution.

The use of this formula is illustrated in the following examples.

EXAMPLE 4.3

Suppose x is a normally distributed random variable with mean $\mu = 20$ and standard deviation $\sigma = 4$. Find the following probabilities.

(a) $P(15 \leq x \leq 21)$ (b) $P(x \leq 26)$

SOLUTION

(a) Convert to standard normal values and use the normal table (Table III) at the back of the book. In particular,

$$P(15 \leq x \leq 21) = P\left(\frac{15 - \mu}{\sigma} \leq z \leq \frac{21 - \mu}{\sigma}\right)$$

$$= P\left(\frac{15 - 20}{4} \leq z \leq \frac{21 - 20}{4}\right)$$

$$= P(-1.25 \leq z \leq 0.25)$$

$$= P(z \leq 0.25) - P(z \leq -1.25)$$

$$= 0.5987 - 0.1056 = 0.4931$$

(b) $P(x \leq 26) = P\left(z \leq \dfrac{26 - 20}{4}\right) = P(z \leq 1.5) = 0.9332$

EXAMPLE 4.4

The weights of newborn babies in a certain community are normally distributed with mean $\mu = 7.1$ pounds and standard deviation $\sigma = 0.4$ pound. What percentage of the babies weigh between 6.7 and 7.2 pounds?

SOLUTION

Let x denote the weight (in pounds) of a randomly selected newborn baby. The percentage of babies that weigh between 6.7 and 7.2 pounds is obtained from the corresponding probability:

$$P(6.7 \leq x \leq 7.2) = P\left(\frac{6.7 - 7.1}{0.4} \leq z \leq \frac{7.2 - 7.1}{0.4}\right)$$

$$= P(-1 \leq z \leq 0.25)$$

$$= P(z \leq 0.25) - P(z \leq -1)$$

$$= 0.5987 - 0.1587 = 0.44$$

That is, 44 percent of the newborn babies weigh between 6.7 and 7.2 pounds.

EXAMPLE 4.5

Suppose x is a normally distributed random variable with mean μ and standard deviation σ. Find $P(\mu - 3\sigma \leq x \leq \mu + 3\sigma)$.

SOLUTION

$$P(\mu - 3\sigma \leq x \leq \mu + 3\sigma) = P\left[\frac{(\mu - 3\sigma) - \mu}{\sigma} \leq z \leq \frac{(\mu + 3\sigma) - \mu}{\sigma}\right]$$

$$= P\left(\frac{-3\sigma}{\sigma} \leq z \leq \frac{3\sigma}{\sigma}\right)$$

$$= P(-3 \leq z \leq 3)$$

$$= P(z \leq 3) - P(z \leq -3)$$

$$= 0.9987 - 0.0013 = 0.9974$$

This says that approximately 99.7 percent of the values of x lie within 3 standard deviations of the mean, which is one of the three probability guidelines for normal distributions introduced earlier in this section. The other two are established by similar computations.

Problems

Measurement

1. The measurement errors resulting from the use of a certain scale are normally distributed with mean $\mu = 0$ ounces and standard deviation $\sigma = 0.1$ ounce. Use the probability guidelines for normal distributions to find intervals containing the errors resulting from 68 percent, 95 percent, and 99.7 percent of all measurements made on this scale.

Packaging

2. The amounts (in ounces) of diet cola in bottles filled by a certain machine are normally distributed with mean $\mu = 33.8$ ounces and standard deviation $\sigma = 0.4$ ounce. Use the probability guidelines for normal distributions to find intervals containing the number of ounces in 68 percent, 95 percent, and 99.7 percent of the bottles filled by this machine.

Educational testing   3. In the early 1980s, the scores on the Graduate Record Examination of verbal ability were normally distributed with mean $\mu = 479$ and standard deviation $\sigma = 129$. Use the probability guidelines for normal distributions to find intervals containing 68 percent, 95 percent, and 99.7 percent of the scores on this exam during this period of time.

In Problems 4 through 15, use the normal table at the back of the book to find the indicated probabilities for a standard normal random variable z.

4. $P(z \leq 1.68)$ 5. $P(z \leq 0.06)$

6. $P(z \geq 2.03)$ 7. $P(z \geq 0.40)$

8. $P(z \leq -0.98)$ 9. $P(z \leq -2.13)$

10. $P(z \geq -1.96)$ 11. $P(z \geq -3.02)$

12. $P(-0.64 \leq z \leq 1.85)$ 13. $P(-1.34 \leq z \leq 2.06)$

14. $P(-2 \leq z \leq 2)$ 15. $P(-1.2 \leq z \leq 1.2)$

16. Suppose x is a normally distributed random variable with mean $\mu = 75$ and standard deviation $\sigma = 5$. Find the following probabilities.
    (a) $P(70 \leq x \leq 85)$ (b) $P(x \leq 65)$

17. Suppose x is a normally distributed random variable with mean $\mu = 100$ and standard deviation $\sigma = 10$. Find the following probabilities.
    (a) $P(94 \leq x \leq 104)$ (b) $P(x \geq 88)$

18. Suppose x is a normally distributed random variable with mean $\mu = 4$ and standard deviation $\sigma = 0.2$. Find the following probabilities.
    (a) $P(3.9 \leq x \leq 4.3)$ (b) $P(4.2 \leq x \leq 4.4)$

19. Suppose x is a normally distributed random variable with mean $\mu = 1$ and standard deviation $\sigma = 0.1$. Find the following probabilities.
    (a) $P(0.94 \leq x \leq 1.02)$ (b) $P(0.90 \leq x \leq 0.93)$

Quality control   20. The capacities (in liters) of the jars produced by a certain company are normally distributed with mean $\mu = 2$ liters and standard deviation $\sigma = 0.05$ liter. Find the probability that a randomly selected jar will hold between 1.96 and 2.04 liters.

Vital statistics   21. The heights of adults in a certain population are normally distributed with mean $\mu = 69$ inches and standard deviation $\sigma = 2.4$ inches. What percentage of the adults in the population are over 6 feet tall?

Factory output   **22.** The daily output at a certain factory is normally distributed with mean $\mu = 500$ tons and standard deviation $\sigma = 20$ tons. Find the probability that the output for a randomly selected day is less than 497 tons.

Measurement   **23.** The measurement errors resulting from the use of a certain scale are normally distributed with mean $\mu = 0$ ounces and standard deviation $\sigma = 0.1$ ounce. Find the probability that the measured weight of a randomly selected object differs from the true weight of the object by more than 0.12 ounce.

Merchandising   **24.** A supermarket stocks 30 pounds of peanuts each week. If the weekly demand for peanuts is normally distributed with mean $\mu = 24$ pounds and standard deviation $\sigma = 5$ pounds, find the probability that the supermarket will run out of peanuts during a randomly selected week.

25. Use the normal table to verify that approximately 68 percent of all the values of a normal random variable lie within 1 standard deviation of the mean.

26. Use the normal table to verify that approximately 95 percent of all the values of a normal random variable lie within 2 standard deviations of the mean.

27. Use the integral formula for expected value from Section 3 to show that the mean of a standard normal random variable is 0.

28. Assume without proof that

$$\int_{-\infty}^{\infty} e^{-z^2/2} \, dz = \sqrt{2\pi}$$

and use this fact together with the integral formula for the variance from Section 3 to show that the variance of a standard normal random variable is 1.

CHAPTER SUMMARY AND PROFICIENCY TEST

Important terms, symbols, and formulas

Limits at infinity: $\lim\limits_{x \to \infty} f(x)$

Limits of powers of x: If $n > 0$,

$$\lim_{x \to \infty} x^n = \infty \qquad \text{and} \qquad \lim_{x \to \infty} \frac{1}{x^n} = 0$$

Limits of exponential functions: If $k > 0$,

$$\lim_{x \to \infty} e^{kx} = \infty \qquad \text{and} \qquad \lim_{x \to \infty} e^{-kx} = 0$$

Limit of the natural logarithm: $\lim\limits_{x\to\infty} \ln x = \infty$

Limits of powers of x times exponential functions: If $k > 0$,

$$\lim_{x\to\infty} x^n e^{-kx} = 0$$

Limits of polynomials: If $a_n \neq 0$,

$$\lim_{x\to\infty} (a_0 + a_1 x + \cdots + a_n x^n) = \lim_{x\to\infty} a_n x^n$$

Limits of rational functions: Divide numerator and denominator by x^k, where k is the smaller of the degrees of the numerator and denominator (or use L'Hôpital's rule).

Indeterminate forms: $0/0$, ∞/∞, $0 \cdot \infty$, 1^∞, ∞^0, 0^0

L'Hôpital's rule: If $\lim\limits_{x\to\infty} f(x) = 0$ and $\lim\limits_{x\to\infty} g(x) = 0$, or if $\lim\limits_{x\to\infty} f(x) = \infty$ and $\lim\limits_{x\to\infty} g(x) = \infty$, then

$$\lim_{x\to\infty} \frac{f(x)}{g(x)} = \lim_{x\to\infty} \frac{f'(x)}{g'(x)}$$

Improper integrals:

$$\int_a^\infty f(x)\,dx = \lim_{N\to\infty} \int_a^N f(x)\,dx$$

$$\int_{-\infty}^\infty f(x)\,dx = \lim_{N\to\infty} \int_{-N}^0 f(x)\,dx + \lim_{N\to\infty} \int_0^N f(x)\,dx$$

Discrete random variable; continuous random variable

Probability density function: $P(a \le x \le b) = \int_a^b f(x)\,dx$

Uniform density function: $f(x) = \begin{cases} \dfrac{1}{B - A} & \text{if } A \le x \le B \\ 0 & \text{otherwise} \end{cases}$

Exponential density function: $f(x) = \begin{cases} ke^{-kx} & \text{if } x \ge 0 \\ 0 & \text{if } x < 0 \end{cases}$

Expected value (mean): $E(x) = \int_{-\infty}^\infty x f(x)\,dx$

Variance: $\text{Var}(x) = \int_{-\infty}^\infty [x - E(x)]^2 f(x)\,dx = \int_{-\infty}^\infty x^2 f(x)\,dx - [E(x)]^2$

Normal density function:

$$f(x) = \frac{1}{\sigma\sqrt{2\pi}}\, e^{-(x-\mu)^2/2\sigma^2}$$

where μ = mean = $E(x)$ and σ = standard deviation = $\sqrt{\text{Var}(x)}$.

Probability guidelines for normal distributions:

1. $P(\mu - \sigma \le x \le \mu + \sigma) \approx 0.68$
2. $P(\mu - 2\sigma \le x \le \mu + 2\sigma) \approx 0.95$
3. $P(\mu - 3\sigma \le x \le \mu + 3\sigma) \approx 0.997$

Standard normal density function: $\mu = 0$, $\sigma = 1$

Normal table

Computation of normal probabilities:

$$P(a \le x \le b) = P\left(\frac{a - \mu}{\sigma} \le z \le \frac{b - \mu}{\sigma}\right)$$

where z has the standard normal distribution.

Proficiency test   In Problems 1 through 12, find $\lim\limits_{x \to \infty} f(x)$.

1. $f(x) = (x^2 + 1)(2 - x^4)$

2. $f(x) = \dfrac{1 + x^2 - 3x^3}{2x^3 + 5}$

3. $f(x) = \dfrac{(x + 1)(2 - x)}{x + 3}$

4. $f(x) = \dfrac{2}{3 + 4e^{-5x}}$

5. $f(x) = \dfrac{x^2 + 1}{e^{2x}}$

6. $f(x) = \dfrac{x^2 - 1}{e^{-2x}}$

7. $f(x) = \dfrac{1 + e^{-x}}{e^{-x}}$

8. $f(x) = \dfrac{\ln(3x + 1)}{\ln(x^2 - 4)}$

9. $f(x) = \dfrac{e^{\sqrt{x}}}{2x}$

10. $f(x) = x^2(e^{1/x} - 1)$

11. $f(x) = x^{1/x^2}$

12. $f(x) = \left(1 + \dfrac{2}{x}\right)^{3x}$

In Problems 13 through 21, evaluate the given improper integral.

13. $\displaystyle\int_0^\infty \dfrac{1}{\sqrt[3]{1 + 2x}}\, dx$

14. $\displaystyle\int_0^\infty (1 + 2x)^{-3/2}\, dx$

15. $\displaystyle\int_0^\infty \dfrac{3x}{x^2 + 1}\, dx$

16. $\displaystyle\int_0^\infty 3e^{-5x}\, dx$

17. $\displaystyle\int_0^\infty xe^{-2x}\, dx$

18. $\displaystyle\int_0^\infty 2x^2 e^{-x^3}\, dx$

19. $\displaystyle\int_0^\infty x^2 e^{-2x}\, dx$

20. $\displaystyle\int_2^\infty \dfrac{1}{x(\ln x)^2}\, dx$

21. $\displaystyle\int_0^\infty x^5 e^{-x^3}\, dx$

22. The publishers of a national magazine have found that the fraction of subscribers who remain subscribers for at least t years is $f(t) = e^{-t/10}$. Currently the magazine has 20,000 subscribers and estimates that new subscriptions will be sold at the rate of 1,000 per year. Approximately how many subscribers will the magazine have in the long run?

23. It is estimated that t years from now, a certain investment will be generating income at the rate of $f(t) = 8,000 + 400t$ dollars per year. If the income is generated in perpetuity and the prevailing annual interest rate remains fixed at 10 percent compounded continuously, find the present value of the investment.

24. Demographic studies conducted in a certain city indicate that the fraction of the residents that will remain in the city for at least t years is $f(t) = e^{-t/20}$. The current population of the city is 100,000, and it is estimated that t years from now, new people will be arriving at the rate of $100t$ people per year. If this estimate is correct, what will happen to the population of the city in the long run?

In Problems 25 through 27, integrate the given probability density functions to find the indicated probabilities.

25. $f(x) = \begin{cases} \frac{1}{3} & \text{if } 1 \leq x \leq 4 \\ 0 & \text{otherwise} \end{cases}$
   (a) $P(1 \leq x \leq 4)$ (b) $P(2 \leq x \leq 3)$ (c) $P(x \leq 2)$

26. $f(x) = \begin{cases} \frac{2}{9}(3 - x) & \text{if } 0 \leq x \leq 3 \\ 0 & \text{otherwise} \end{cases}$
   (a) $P(0 \leq x \leq 3)$ (b) $P(1 \leq x \leq 2)$

27. $f(x) = \begin{cases} 0.2e^{-0.2x} & \text{if } x \geq 0 \\ 0 & \text{if } x < 0 \end{cases}$
   (a) $P(x \geq 0)$ (b) $P(1 \leq x \leq 4)$ (c) $P(x \geq 5)$

28. Find the expected value and variance for the random variable in Problem 25.

29. Find the expected value and variance for the random variable in Problem 26.

30. Find the expected value and variance for the random variable in Problem 27.

31. A bakery turns out a fresh batch of chocolate chip cookies every 45 minutes. You arrive (at random) at the bakery, hoping to buy a fresh cookie. Use an appropriate uniform density function to find the probability that you arrive within 5 minutes (before or after) of the time that the cookies come out of the oven.

32. The time x (in minutes) between the arrivals of successive cars at a toll booth is exponentially distributed with density function

$$f(x) = \begin{cases} 0.5e^{-0.5x} & \text{if } x \geq 0 \\ 0 & \text{if } x < 0 \end{cases}$$

    (a) Find the probability that a randomly selected pair of successive cars will arrive at the toll booth at least 6 minutes apart.

    (b) Find the average time interval between the arrivals of successive cars at the toll booth.

33. The highway mileage of a certain compact car is a normally distributed random variable with mean $\mu = 32$ miles per gallon and standard deviation $\sigma = 2$ miles per gallon. Use the probability guidelines for normal distributions to find intervals containing the highway mileages of 68 percent, 95 percent, and 99.7 percent of these cars.

34. Suppose z is a standard normal random variable. Use the normal table to find the following probabilities.

    (a) $P(z \leq 0.65)$ (b) $P(z \geq 2.01)$

    (c) $P(z \geq 0)$ (d) $P(z \geq -1.44)$

    (e) $P(-0.52 \leq z \leq 1.75)$ (f) $P(-1.5 \leq z \leq 1.5)$

35. Suppose x is a normally distributed random variable with mean $\mu = 5$ and standard deviation $\sigma = 0.1$. Find the following probabilities.

    (a) $P(4.95 \leq x \leq 5.12)$ (b) $P(x \geq 5)$

    (c) $P(4.90 \leq x \leq 4.98)$

36. In the early 1980s, scores on the Graduate Record Examination of quantitative ability were normally distributed with mean $\mu = 518$ and standard deviation $\sigma = 135$. What percentage of the scores were between 491 and 626?

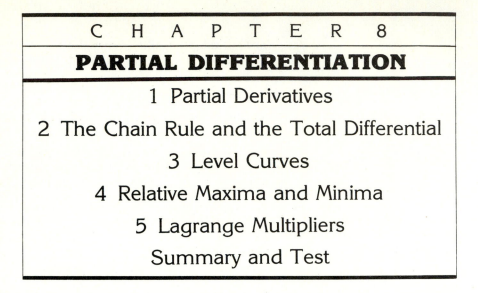

C H A P T E R 8

PARTIAL DIFFERENTIATION

1 Partial Derivatives

2 The Chain Rule and the Total Differential

3 Level Curves

4 Relative Maxima and Minima

5 Lagrange Multipliers

Summary and Test

1 PARTIAL DERIVATIVES

In many practical situations, the value of one quantity may depend on the values of two or more others. For example, the amount of water in a reservoir may depend on the amount of rainfall and on the amount of water consumed by local residents. The demand for butter may depend on the price of butter and on the price of margarine. The output at a certain factory may depend on the amount of capital invested in the plant and on the size of the labor force. Relationships of this sort often can be represented mathematically by functions having more than one independent variable. Here is an example.

EXAMPLE 1.1

A liquor store in Minneapolis carries two brands of inexpensive white table wine, one from California and the other from New York. The consumer demand for each brand depends not only on its own price but also on the price of the competing brand. Sales figures indicate that if the California wine sells for x dollars per bottle and the New York wine for y dollars per bottle, the demand for the California wine will be

$$D_1 = 300 - 20x + 30y \quad \text{bottles per month}$$

and the demand for the New York wine will be

$$D_2 = 200 + 40x - 10y \qquad \text{bottles per month}$$

Express the liquor store's total monthly revenue from the sale of these wines as a function of the prices x and y.

SOLUTION

Let R denote the total monthly revenue. Then,

R = (number of bottles of California wine sold)(price per bottle)
    + (number of bottles of New York wine sold)(price per bottle)
  = $(300 - 20x + 30y)(x) + (200 + 40x - 10y)(y)$
  = $300x + 200y + 70xy - 20x^2 - 10y^2$

Functional notation
The functional notation you have been using for functions of a single variable can be extended to functions of several variables. For instance, you can use functional notation to write the revenue function in Example 1.1 as

$$R(x, y) = 300x + 200y + 70xy - 20x^2 - 10y^2$$

The use of functional notation is illustrated further in the next example.

EXAMPLE 1.2

Compute $f(2, 3, -1)$ if $f(r, s, t) = \dfrac{3r^2 + 5s}{r - t}$.

SOLUTION

Substitute $r = 2$, $s = 3$, and $t = -1$ into the formula for f to get

$$f(2, 3, -1) = \frac{3(2)^2 + 5(3)}{2 - (-1)} = 9$$

Partial derivatives
In many problems involving functions of several variables, the goal is to find the rate of change of the function with respect to one of its variables when all the others are held constant. That is, the goal is to differentiate the function with respect to the particular variable in question while keeping all the other variables fixed. This process is known as **partial differentiation,** and the resulting derivative is said to be a **partial derivative** of the function.

Partial derivatives

> **Suppose $f(x_1, x_2, \ldots, x_n)$ is a function of n variables. The partial derivative of f with respect to its jth variable x_j is denoted by f_{x_j} and is defined to be the function obtained by differentiating f with respect to x_j, treating all the other variables as constants.**

No new rules are needed for the computation of partial derivatives. To compute f_{x_j}, simply differentiate f with respect to the single variable x_j, pretending that all the other variables are constants. Here is an example.

EXAMPLE 1.3

Find the partial derivatives f_x, f_y, and f_z if $f(x, y, z) = x^2 + 2xy^2 + yz^3$.

SOLUTION

To compute f_x, think of f as a function of x and differentiate the sum term by term, treating y and z as constants to get

$$f_x(x, y, z) = 2x + 2y^2 + 0 = 2x + 2y^2$$

To compute f_y, pretend that x and z are constants and differentiate with respect to y to get

$$f_y(x, y, z) = 0 + 4xy + z^3 = 4xy + z^3$$

To compute f_z, treat x and y as the constants and differentiate with respect to z to get

$$f_z(x, y, z) = 0 + 0 + 3yz^2 = 3yz^2$$

When a dependent variable such as z is used to denote a function $f(x_1, x_2, \ldots, x_n)$, the symbol $\dfrac{\partial z}{\partial x_j}$ is usually used instead of f_{x_j} to denote the partial derivative of the function with respect to x_j. This notation is used in the next example.

EXAMPLE 1.4

Find the partial derivatives $\dfrac{\partial z}{\partial x}$ and $\dfrac{\partial z}{\partial y}$ if $z = (x^2 + xy + y)^5$.

SOLUTION

Holding y fixed and using the chain rule to differentiate z with respect to x, you get

$$\frac{\partial z}{\partial x} = 5(x^2 + xy + y)^4(2x + y)$$

Holding x fixed and using the chain rule to differentiate z with respect to y, you get

$$\frac{\partial z}{\partial y} = 5(x^2 + xy + y)^4(x + 1)$$

Marginal analysis

In economics, the term **marginal analysis** refers to the practice of using a derivative to estimate the change in the value of a function resulting from a 1-unit increase in one of its variables. In Chapter 2, Section 3, you saw some examples of marginal analysis involving ordinary derivatives of functions of one variable. Here are two examples illustrating marginal analysis for functions of several variables.

EXAMPLE 1.5

Suppose the daily output Q of a factory depends on the amount K of capital (measured in units of $1,000) invested in the plant and equipment, and also on the size L of the labor force (measured in worker-hours). In economics, the partial derivatives $\frac{\partial Q}{\partial K}$ and $\frac{\partial Q}{\partial L}$ are known as the **marginal products** of capital and labor, respectively. Give economic interpretations of these two marginal products.

SOLUTION

The marginal product of labor $\frac{\partial Q}{\partial L}$ is the rate at which output Q changes with respect to labor L for a fixed level K of capital investment. Hence, $\frac{\partial Q}{\partial L}$ is approximately the change in output that will result if capital investment is held fixed and labor is increased by 1 worker-hour.

Similarly, the marginal product of capital $\frac{\partial Q}{\partial K}$ is approximately the change in output that will result if the size of the labor force is held fixed and capital investment is increased by $1,000.

EXAMPLE 1.6

It is estimated that the weekly output at a certain plant is given by the function $f(x, y) = 1{,}200x + 500y + x^2y - x^3 - y^3$ units, where x is the number of skilled workers and y the number of unskilled workers employed at the plant. Currently the work force consists of 30 skilled workers and 60 unskilled workers. Use marginal analysis to estimate the change in the weekly output that will result from the addition of 1 more skilled worker if the number of unskilled workers is not changed.

SOLUTION

The partial derivative

$$f_x(x, y) = 1,200 + 2xy - 3x^2$$

is the rate of change of output with respect to the number of skilled workers. For any values of x and y, this is an approximation to the number of additional units that will be produced each week if the number of skilled workers is increased from x to $x + 1$ while the number of unskilled workers is kept fixed at y. In particular, if the work force is increased from 30 skilled and 60 unskilled workers to 31 skilled and 60 unskilled workers, the resulting change in output is approximately

$$f_x(30, 60) = 1,200 + 2(30)(60) - 3(30)^2 = 2,100 \qquad \text{units}$$

For practice, compute the change in output exactly by subtracting appropriate values of f.

Second-order partial derivatives Partial derivatives can themselves be differentiated. The resulting functions are called **second-order partial derivatives.** If $z = f(x, y)$, the partial derivative of f_x with respect to x is denoted by

$$f_{xx} \qquad \text{or} \qquad \frac{\partial^2 z}{\partial x^2}$$

and the partial derivative of f_x with respect to y is denoted by

$$f_{xy} \qquad \text{or} \qquad \frac{\partial^2 z}{\partial y \, \partial x}$$

Similar notation is used to denote the partial derivatives of f_y. Here is an example.

EXAMPLE 1.7

Compute the four second-order partial derivatives of the function $f(x, y) = xy^3 + 5xy^2 + 2x + 1$.

SOLUTION

Since
$$f_x = y^3 + 5y^2 + 2$$

it follows that

$$f_{xx} = 0 \qquad \text{and} \qquad f_{xy} = 3y^2 + 10y$$

Since $$f_y = 3xy^2 + 10xy$$

it follows that

$$f_{yx} = 3y^2 + 10y \quad \text{and} \quad f_{yy} = 6xy + 10x$$

The two derivatives f_{xy} and f_{yx} are sometimes called the **mixed second-order partial derivatives** of f. Notice that the mixed partial derivatives in Example 1.7 were equal. This is not an accident. It turns out that for virtually all the functions you will encounter in practical work, the mixed partial derivatives will be equal. That is, you will get the same answer if you first differentiate f with respect to x and then differentiate the resulting function with respect to y as you would if you perform the differentiation in the opposite order.

In the next example, you will see how a second-order partial derivative can give useful information in a practical situation.

EXAMPLE 1.8

Suppose the output Q of a factory depends on the amount K of capital invested in the plant and equipment, and also on the size L of the labor force, measured in worker-hours. Give an economic interpretation of the sign of the second-order partial derivative $\frac{\partial^2 Q}{\partial L^2}$.

SOLUTION

If $\frac{\partial^2 Q}{\partial L^2}$ is negative, the marginal product of labor $\frac{\partial Q}{\partial L}$ decreases as L increases. This implies that for a fixed level of capital investment, the effect on output of the addition of 1 worker-hour of labor is greater when the work force is small than when the work force is large.

Similarly, if $\frac{\partial^2 Q}{\partial L^2}$ is positive, it follows that for a fixed level of capital investment, the effect on output of the addition of 1 worker-hour of labor is greater when the work force is large than when it is small.

For most factories operating with adequate work forces, the derivative $\frac{\partial^2 Q}{\partial L^2}$ is generally negative. Can you give an economic explanation for this fact?

Geometric interpretation   Functions of two variables can be represented graphically as surfaces drawn on three-dimensional coordinate systems. In particular, if $z = f(x, y)$, the ordered pair (x, y) can be identified with a point in the xy plane and the corresponding function value $z = f(x, y)$ can be thought of as associating a height to this point. The graph of f is the surface

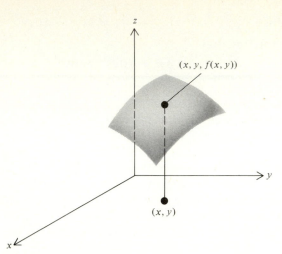

Figure 1.1 The graph of $z = f(x, y)$.

consisting of all points (x, y, z) in three-dimensional space whose height z is equal to $f(x, y)$. The situation is illustrated in Figure 1.1.

The partial derivatives of a function of two variables can be interpreted geometrically as follows. For each fixed number y_0, the points (x, y_0, z) form a vertical plane whose equation is $y = y_0$. If $z = f(x, y)$ and if y is kept fixed at $y = y_0$, then the corresponding points $(x, y_0, f(x, y_0))$ form a curve in three-dimensional space that is the intersection of the surface $z = f(x, y)$ with the plane $y = y_0$. At each point on this curve, the partial derivative $\frac{\partial z}{\partial x}$ is simply the slope of the line in the plane $y = y_0$ that is tangent to the curve at the point in question. The situation is illustrated in Figure 1.2a.

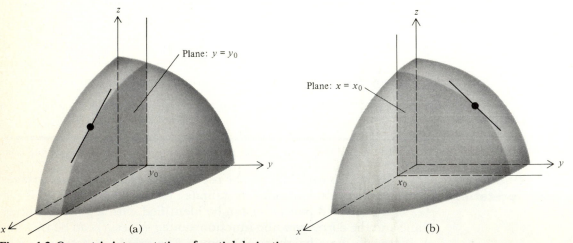

Figure 1.2 Geometric interpretation of partial derivatives.

Similarly, if x is kept fixed at $x = x_0$, the corresponding points $(x_0, y, f(x_0, y))$ form a curve that is the intersection of the surface $z = f(x, y)$ with the vertical plane $x = x_0$. At each point on this curve, the partial derivative $\frac{\partial z}{\partial y}$ is the slope of the tangent in the plane $x = x_0$. The situation is illustrated in Figure 1.2b.

Problems

Retail sales
1. A paint store carries two brands of latex paint. Sales figures indicate that if the first brand is sold for x_1 dollars per gallon and the second for x_2 dollars per gallon, the demand for the first brand will be $D_1(x_1, x_2) = 200 - 10x_1 + 20x_2$ gallons per month, and the demand for the second brand will be $D_2(x_1, x_2) = 100 + 5x_1 - 10x_2$ gallons per month.
   (a) Express the paint store's total monthly revenue from the sale of the paint as a function of the prices x_1 and x_2.
   (b) Compute the revenue in part (a) if the first brand is sold for $6 per gallon and the second for $5 per gallon.

Production cost
2. A manufacturer can produce electric typewriters at a cost of $80 apiece and manual typewriters at a cost of $20 apiece.
   (a) Express the manufacturer's total monthly production cost as a function of the number of electric typewriters and the number of manual typewriters produced.
   (b) Compute the total monthly cost if 500 electric and 800 manual typewriters are produced.
   (c) The manufacturer wants to increase the output of electric typewriters by 50 a month from the level in part (b). What corresponding change should be made in the monthly output of manual typewriters so that the total monthly cost will not change?

Productivity
3. Using x skilled workers and y unskilled workers, a manufacturer can produce $Q(x, y) = 10x^2y$ units per day. Currently there are 20 skilled workers and 40 unskilled workers on the job.
   (a) How many units are currently being produced each day?
   (b) By how much will the daily production level change if 1 more skilled worker is added to the current work force?
   (c) By how much will the daily production level change if 1 more unskilled worker is added to the current work force?
   (d) By how much will the daily production level change if 1 more skilled worker *and* 1 more unskilled worker are added to the current work force?

In Problems 4 through 17, compute all the first-order partial derivatives of the given function.

4. $f(x, y) = 2xy^5 + 3x^2y + x^2$

5. $f(x, y) = (3x + 2y)^5$

6. $z = 5x^2y + 2xy + 3y^2$

7. $z = (x + xy + y)^3$

8. $f(x, y, z) = x^2yz + 3xy^2 + xz^5$

9. $w = (3x + 2y + xz^2)^9$

10. $z = \dfrac{x}{y}$

11. $f(r, s) = \dfrac{r^2}{s^3}$

12. $f(x, y) = xye^x$

13. $z = xe^{xy}$

14. $z = \dfrac{u + v}{u - v}$

15. $f(x, y) = \dfrac{xy + 1}{xy^2}$

16. $f(r, s) = r \ln s$

17. $z = x \ln xy$

Consumer demand
18. The monthly demand for a certain brand of toasters is given by a function $f(x, y)$, where x is the amount of money (measured in units of $1,000) spent on advertising and y is the selling price (in dollars) of the toasters. Give economic interpretations of the partial derivatives f_x and f_y. Under normal economic conditions, what will the sign of each of these partial derivatives be?

Marginal analysis
19. At a certain factory, the daily output is $Q = 60K^{1/2}L^{1/3}$ units, where K denotes the capital investment measured in units of $1,000 and L the size of the labor force measured in worker-hours. Suppose the current capital investment is $900,000 and that 1,000 worker-hours of labor are used each day. Use marginal analysis to estimate the effect of an additional capital investment of $1,000 on the daily output if the size of the labor force is not changed.

Marginal analysis
20. A bicycle dealer has found that if 10-speed bicycles are sold for x dollars apiece and the price of gasoline is $10y$ cents per gallon, approximately $f(x, y) = 200 - 10\sqrt{x} + 4(y + 7)^{3/2}$ bicycles will be sold each month. Currently the bicycles sell for $121 apiece and the price of gasoline is 90 cents per gallon. Use marginal analysis to estimate the effect on the monthly sale of bicycles if the price of gasoline is increased to $1 per gallon while the price of bicycles is held fixed.

Marginal analysis
21. A publishing house has found that in a certain city *each* of its salespeople will sell approximately $\dfrac{r^2}{2,000p} + \dfrac{s^2}{100} - s$ sets of encyclopedias per month, where s denotes the total number of salespeople employed, p the price of a set of the encyclopedias, and r the amount of money spent each month on local advertising. Currently the publisher employs 10 salespeople, spends $6,000 per month on local advertising, and sells the encyclopedias for $800 per set. The cost of producing the encyclopedias is $80 per set,

and each salesperson earns $600 per month. Use marginal analysis to estimate the change in the publisher's total monthly profit that will result if 1 more salesperson is hired.

Consumer demand 22. Two competing brands of power lawnmowers are sold in the same town. The price of the first brand is x dollars per mower, the price of the second brand is y dollars per mower, and the average per capita income of the community is z dollars per year. The local demand for the first brand of mower is given by a function $D(x, y, z)$.
(a) How would you expect the demand for the first brand of mower to be affected by an increase in x? By an increase in y? By an increase in z?
(b) Translate your answers to part (a) into conditions on the signs of the partial derivatives of D.
(c) If $D(x, y, z) = a + bx + cy + dz$, what can you say about the signs of the coefficients $b, c,$ and d if your conclusions in part (a) are to hold?

Substitute commodities 23. In economics, two commodities are said to be **substitute commodities** if the demand Q_1 for the first increases as the price p_2 of the second increases, and if the demand Q_2 for the second increases as the price p_1 of the first increases.
(a) Give an example of a pair of substitute commodities.
(b) If two commodities are substitutes, what must be true of the partial derivatives $\dfrac{\partial Q_1}{\partial p_2}$ and $\dfrac{\partial Q_2}{\partial p_1}$?
(c) Suppose the demand functions for two commodities are
$$Q_1 = 3,000 + \frac{400}{p_1 + 3} + 50p_2 \quad \text{and} \quad Q_2 = 2,000 - 100p_1 +$$
$\dfrac{500}{p_2 + 4}$. Are the commodities substitutes?

Complementary commodities 24. Two commodities are said to be **complementary commodities** if the demand Q_1 for the first decreases as the price p_2 of the second increases, and if the demand Q_2 for the second decreases as the price p_1 of the first increases.
(a) Give an example of a pair of complementary commodities.
(b) If two commodities are complementary, what must be true of the partial derivatives $\dfrac{\partial Q_1}{\partial p_2}$ and $\dfrac{\partial Q_2}{\partial p_1}$?
(c) Suppose the demand functions for two commodities are
$$Q_1 = 2,000 + \frac{400}{p_1 + 3} - 50p_2 \quad \text{and} \quad Q_2 = 2,000 - 100p_2 +$$
$\dfrac{500}{p_1 + 4}$. Are the commodities complementary?

In Problems 25 through 30, compute all the second-order partial derivatives of the given function.

25. $f(x, y) = 5x^4y^3 + 2xy$

26. $f(x, y) = \dfrac{x + 1}{y - 1}$

27. $f(x, y) = e^{x^2y}$

28. $f(x, y) = \ln(x^2 + y^2)$

29. $f(x, y) = \sqrt{x^2 + y^2}$

30. $f(x, y) = x^2ye^x$

Marginal productivity
31. Suppose the output Q of a factory depends on the amount K of capital investment measured in units of $1,000 and on the size L of the labor force measured in worker-hours. Give an economic interpretation of the second-order partial derivative $\dfrac{\partial^2 Q}{\partial K^2}$.

Marginal productivity
32. At a certain factory, the output is $Q = 120K^{1/2}L^{1/3}$, where K denotes the capital investment measured in units of $1,000 and L the size of the labor force measured in worker-hours.
 (a) Determine the sign of the second-order partial derivative $\dfrac{\partial^2 Q}{\partial L^2}$ and give an economic interpretation.
 (b) Determine the sign of the second-order partial derivative $\dfrac{\partial^2 Q}{\partial K^2}$ and give an economic interpretation.

Law of diminishing returns
33. Suppose the daily output Q of a factory depends on the amount K of capital investment and on the size L of the labor force. A **law of diminishing returns** states that in certain circumstances, there is a value L_0 such that the marginal product of labor will be increasing for $L < L_0$ and decreasing for $L > L_0$.
 (a) Translate this law of diminishing returns into statements about the sign of a certain second-order partial derivative.
 (b) Discuss the economic factors that might account for this phenomenon.

2 THE CHAIN RULE AND THE TOTAL DIFFERENTIAL

In many practical situations, a particular quantity is given as a function of two or more variables, each of which can be thought of as a function of yet another variable, and the goal is to find the rate of change of the quantity with respect to this other variable. For example, the demand for a certain commodity may depend on the price of the commodity itself and on the price of a competing commodity, both of which are increasing with time, and the goal may be to find the rate of change of the demand with respect to time. You can solve problems of this type by using a generalization of the chain rule that was introduced in Chapter 2, Section 4.

The chain rule   Recall that if z is a function of x, and x is a function of t, then z can be regarded as a function of t, and the rate of change of z with respect to t is given by the chain rule

$$\frac{dz}{dt} = \frac{dz}{dx}\frac{dx}{dt}$$

Here is the corresponding rule for functions of two variables.

Chain rule for   **Suppose z is a function of x and y, each of which is a function of t.**
partial derivatives   **Then z can be regarded as a function of t and**

$$\frac{dz}{dt} = \frac{\partial z}{\partial x}\frac{dx}{dt} + \frac{\partial z}{\partial y}\frac{dy}{dt}$$

Observe that the expression for $\dfrac{dz}{dt}$ is the sum of two terms, each of which can be interpreted using the chain rule for a function of one variable. In particular,

$$\frac{\partial z}{\partial x}\frac{dx}{dt} = \text{rate of change of } z \text{ with respect to } t \text{ for fixed } y$$

and $$\frac{\partial z}{\partial y}\frac{dy}{dt} = \text{rate of change of } z \text{ with respect to } t \text{ for fixed } x$$

The chain rule for partial derivatives says that the total rate of change of z with respect to t is the sum of these two "partial" rates of change.

Here are two examples illustrating the use of the chain rule for partial derivatives.

EXAMPLE 2.1

Find $\dfrac{dz}{dt}$ if $z = x^2 + 3xy + 1$, $x = 2t + 1$, and $y = t^2$.

SOLUTION

By the chain rule

$$\frac{dz}{dt} = \frac{\partial z}{\partial x}\frac{dx}{dt} + \frac{\partial z}{\partial y}\frac{dy}{dt} = (2x + 3y)(2) + 3x(2t)$$

which you can rewrite in terms of t by substituting $x = 2t + 1$ and $y = t^2$ to get

$$\frac{dz}{dt} = 4(2t + 1) + 6t^2 + 3(2t + 1)(2t) = 18t^2 + 14t + 4$$

For practice, check this answer by first substituting $x = 2t + 1$ and

$y = t^2$ into the formula for z and then differentiating directly with respect to t.

EXAMPLE 2.2

A liquor store carries two brands of inexpensive white wine, one from California and the other from New York. Sales figures indicate that if the California wine is sold for x dollars per bottle and the New York wine for y dollars per bottle, the demand for the California wine will be

$$Q(x, y) = 300 - 20x^2 + 30y \qquad \text{bottles per month}$$

It is estimated that t months from now the price of the California wine will be

$$x = 2 + 0.05t \qquad \text{dollars per bottle}$$

and the price of the New York wine will be

$$y = 2 + 0.1\sqrt{t} \qquad \text{dollars per bottle}$$

At what rate will the demand for the California wine be changing 4 months from now?

SOLUTION

Your goal is to find $\dfrac{dQ}{dt}$ when $t = 4$. Using the chain rule you get

$$\frac{dQ}{dt} = \frac{\partial Q}{\partial x}\frac{dx}{dt} + \frac{\partial Q}{\partial y}\frac{dy}{dt} = -40x(0.05) + 30(0.05t^{-1/2})$$

When $t = 4$, $\qquad\qquad x = 2 + 0.05(4) = 2.2$

and hence, $\dfrac{dQ}{dt} = -40(2.2)(0.05) + 30(0.05)(0.5) = -3.65$

That is, 4 months from now the monthly demand for the California wine will be decreasing at the rate of 3.65 bottles per month.

The total differential
In Chapter 3, Section 4, you learned how to use the differential of a function to approximate the change in the function resulting from a small change in its independent variable. In particular, you saw that

$$\Delta y \approx \frac{dy}{dx}\,\Delta x$$

where Δx is a small change in the variable x and Δy is the corresponding change in the function y. The expression $dy = \dfrac{dy}{dx}\,\Delta x$ that was used to approximate Δy was called the differential of y. Here is the analogous approximation formula for functions of two variables.

Approximation formula

Suppose z is a function of x and y. If Δx denotes a small change in x and Δy a small change in y, the corresponding change in z is

$$\Delta z \approx \frac{\partial z}{\partial x} \Delta x + \frac{\partial z}{\partial y} \Delta y$$

Observe that the expression used to approximate Δz is the sum of two terms, each of which is essentially a one-variable differential. In particular,

$$\frac{\partial z}{\partial x} \Delta x \approx \text{change in } z \text{ due to the change in } x \text{ for fixed } y$$

and $\quad \dfrac{\partial z}{\partial y} \Delta y \approx \text{change in } z \text{ due to the change in } y \text{ for fixed } x$

The approximation formula says that the total change in z is approximately equal to the sum of these two partial changes.

The sum of the two one-variable differentials that appears in the approximation formula is called the **total differential** of z and is denoted by the symbol dz. Notice the similarity between the formula for the total differential dz and the chain rule.

The total differential

If z is a function of x and y, the total differential of z is

$$dz = \frac{\partial z}{\partial x} \Delta x + \frac{\partial z}{\partial y} \Delta y$$

The use of the total differential to approximate the change in a function is illustrated in the next example.

EXAMPLE 2.3

At a certain factory, the daily output is $Q = 60K^{1/2}L^{1/3}$ units, where K denotes the capital investment measured in units of $1,000 and L the size of the labor force measured in worker-hours. The current capital investment is $900,000 and 1,000 worker-hours of labor are used each day. Estimate the change in output that will result if capital investment is increased by $1,000 and labor is increased by 2 worker-hours.

SOLUTION

Apply the approximation formula with $K = 900$, $L = 1,000$, $\Delta K = 1$, and $\Delta L = 2$ to get

$$\Delta Q \approx \frac{\partial Q}{\partial K} \Delta K + \frac{\partial Q}{\partial L} \Delta L$$
$$= 30K^{-1/2}L^{1/3} \Delta K + 20K^{1/2}L^{-2/3} \Delta L$$
$$= 30(\tfrac{1}{30})(10)(1) + 20(30)(\tfrac{1}{100})(2)$$
$$= 22$$

That is, output will increase by approximately 22 units.

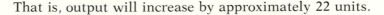

Problems In Problems 1 through 8, use the chain rule to find $\frac{dz}{dt}$. Check your answer by writing z explicitly as a function of t and differentiating directly with respect to t.

1. $z = x + 2y$; $x = 3t$, $y = 2t + 1$

2. $z = 3x^2 + xy$; $x = t + 1$, $y = 1 - 2t$

3. $z = \dfrac{x}{y}$; $x = t^2$, $y = 3t$

4. $z = \dfrac{y}{x}$; $x = 2t$, $y = t^3$

5. $z = \dfrac{x + y}{x - y}$; $x = t^3 + 1$, $y = 1 - t^3$

6. $z = (2x + 3y)^2$; $x = 2t$, $y = 3t$

7. $z = (x - y^2)^3$; $x = t^2$, $y = 2t$

8. $z = xy$; $x = e^t$, $y = e^{-t}$

In Problems 9 through 13, use the chain rule to find $\frac{dz}{dt}$ for the specified value of t.

9. $z = 2x + 3y$; $x = t^2$, $y = 5t$; $t = 2$

10. $z = x^2y$; $x = 3t + 1$, $y = t^2 - 1$; $t = 1$

11. $z = \dfrac{3x}{y}$; $x = t$, $y = t^2$; $t = 3$

12. $z = x^{1/2}y^{1/3}$; $x = 2t$, $y = 2t^2$; $t = 2$

13. $z = xy$; $x = e^{2t}$, $y = e^{3t}$; $t = 0$

Consumer demand 14. A paint store carries two brands of latex paint. Sales figures indicate that if the first brand is sold for x dollars per gallon and the second for y dollars per gallon, the demand for the first brand will be $Q(x, y) = 200 - 10x^2 + 20y$ gallons per month. It is estimated that t months from now the price of the first brand will be $x = 5 + 0.02t$ dollars per gallon and the price of the second brand will be $y = 6 + 0.4\sqrt{t}$ dollars per gallon. At what rate will the demand for the first brand of paint be changing 9 months from now?

Consumer demand 15. A bicycle dealer has found that if 10-speed bicycles are sold for x dollars apiece and the price of gasoline is y cents per gallon, approximately $f(x, y) = 200 - 24\sqrt{x} + 4(0.1y + 5)^{3/2}$ bicycles will be sold each month. It is estimated that t months from now the bicycles will be selling for $129 + 5t$ dollars apiece and the price of gasoline will be $80 + 10\sqrt{3t}$ cents per gallon. At what rate will the monthly demand for the bicycles be changing 3 months from now?

Manufacturing 16. At a certain factory, the output is $Q = 120K^{1/2}L^{1/3}$ units where K denotes the capital investment measured in units of $1,000 and L the size of the labor force measured in worker-hours. The current capital investment is $400,000 and 1,000 worker-hours of labor are currently used. Use the total differential of Q to estimate the change in output that will result if capital investment is increased by $500 and labor is increased by 4 worker-hours.

Manufacturing 17. The output at a certain plant is $Q(x, y) = 0.08x^2 + 0.12xy + 0.03y^2$ units per day, where x is the number of hours of skilled labor used and y is the number of hours of unskilled labor used. Currently 80 hours of skilled labor and 200 hours of unskilled labor are used each day. Use the total differential of Q to estimate the change in output that will result if an additional $\frac{1}{2}$ hour of skilled labor is used along with an additional 2 hours of unskilled labor.

Publishing 18. An editor estimates that if x thousand dollars is spent on development and y thousand on promotion, approximately $Q(x, y) = 20x^{3/2}y$ copies of a new book will be sold. Current plans call for the expenditure of $36,000 on development and $25,000 on promotion. Use the total differential of Q to estimate the change in sales that will result if the amount spent on development is increased by $500 and the amount spent on promotion is decreased by $500.

Retail sales 19. A grocer's daily profit from the sale of two brands of orange juice is $P(x,y) = (x - 30)(70 - 5x + 4y) + (y - 40)(80 + 6x - 7y)$ cents, where x is the price per can of the first brand and y is the price per can of the second. Currently the first brand sells for 50 cents per can and the second for 52 cents per can. Use the total differential of P to estimate the change in the daily profit that will result if the grocer raises the price of the first brand by 1 cent per can and raises the price of the second brand by 2 cents per can.

Measurement 20. A rectangle is 20 centimeters long and 30 centimeters wide. Use a total differential to estimate the amount by which the area will increase if the length is increased by 0.8 centimeter and the width by 0.6 centimeter.

Landscaping 21. A rectangular garden that is 30 yards long and 40 yards wide is surrounded by a concrete path that is 0.8 yard wide. Use a total differential to estimate the area of the concrete path.

Packaging 22. A soft-drink can is 12 centimeters tall and has a radius of 3 centimeters. The manufacturer is planning to reduce the height of the can by 0.2 centimeter and the radius by 0.3 centimeter. Use a total differential to estimate how much less drink consumers will find in each can after the new cans are introduced. (*Hint:* The volume of a cylinder of radius r and height h is $\pi r^2 h$.)

23. Suppose $z = f(x, y)$, $x = at$, and $y = bt$, where a and b are constants. Think of z as a function of t and find an expression for the second derivative $\dfrac{d^2 z}{dt^2}$ in terms of the constants a and b and the second-order partial derivatives f_{xx}, f_{yy}, and f_{xy}.

3 LEVEL CURVES

There are many situations in which one is interested in the possible combinations of variables x and y for which a function $f(x, y)$ will be equal to a certain constant. For example, a manufacturer whose output depends on the numbers of skilled and unskilled workers in the labor force may wish to determine the possible combinations of skilled and unskilled workers that will result in a certain desired level of output.

The combinations of x and y for which $f(x, y)$ is equal to a fixed number often can be represented geometrically as the points on a curve in the xy plane. Such a curve is said to be a **level curve** of f. If the function f represents the output of a factory and the variables x and y represent inputs (such as skilled and unskilled labor), the level curves of f are sometimes called **constant-production curves** or **isoquants**.

Level curve

For any constant C, the points (x, y) for which $f(x, y) = C$ form a curve in the xy plane that is said to be a level curve of f.

Here are two examples.

EXAMPLE 3.1

Sketch the level curve $f(x, y) = 4$ if $f(x, y) = x^2 - y$.

SOLUTION

Rewrite the equation $f(x, y) = 4$ as

$$x^2 - y = 4$$

and solve for y to get

$$y = x^2 - 4 = (x - 2)(x + 2)$$

The graph of this polynomial in the xy plane (Figure 3.1) is the desired level curve $f(x, y) = 4$.

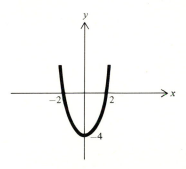

Figure 3.1 The level curve $f = 4$ for $f(x, y) = x^2 - y$.

EXAMPLE 3.2

Find the level curve of the function $f(x, y) = xy$ that passes through the point (2, 3).

SOLUTION

The equations of the level curves of f are all of the form

$$xy = C \qquad \text{or} \qquad y = \frac{C}{x}$$

where C is a constant. To find the equation of the particular level curve that passes through (2, 3), substitute $x = 2$ and $y = 3$ into the general equation of the level curve and solve for C. You find that $C = 6$ and can conclude that the desired level curve is the graph of the function $y = \frac{6}{x}$ as shown in Figure 3.2.

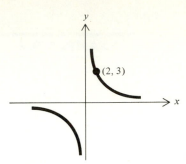

Figure 3.2 **The level curve** $f = 6$ **for** $f(x, y) = xy$.

Geometric interpretation of level curves

Here is another way you can visualize level curves, which is sometimes useful. Think of $z = f(x, y)$ as the equation of a surface in three-dimensional space. The level curve $f(x, y) = C$ is the projection onto the xy plane of the curve formed by the intersection of the surface $z = f(x, y)$ with the horizontal plane $z = C$. The situation is illustrated in Figure 3.3.

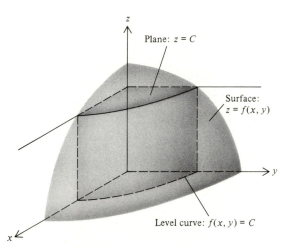

Figure 3.3 **Geometric interpretation of a level curve.**

The slope of a level curve

The slope of the line that is tangent to the level curve $f(x, y) = C$ at a particular point is given by the derivative $\dfrac{dy}{dx}$. This derivative is the rate of change of y with respect to x on the level curve. That is, $\dfrac{dy}{dx}$ is approximately the amount by which the y coordinate of a point on the level curve changes when the x coordinate is increased by 1. (See Figure 3.4.) For example, if f represents output and x and y represent

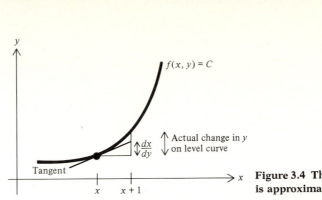

Figure 3.4 The slope of the tangent is approximately the change in *y*.

the levels of skilled and unskilled labor, respectively, the slope $\dfrac{dy}{dx}$ of the tangent to the level curve $f(x, y) = C$ is an estimate of the amount by which the manufacturer should change the level of unskilled labor y to compensate for a 1-unit increase in the level of skilled labor x so that output will remain unchanged.

How to compute $\dfrac{dy}{dx}$ One way to compute the slope $\dfrac{dy}{dx}$ of (the tangent to) a level curve is to solve the equation $f(x, y) = C$ for y in terms of x, and then differentiate the resulting expression with respect to x. Unfortunately, it is often difficult or even impossible to solve the equation $f(x, y) = C$ explicitly for y. In such cases, you can either differentiate the equation $f(x, y) = C$ implicitly (as in Chapter 3, Section 3) or use the following formula involving the partial derivatives of f.

Formula for the slope of a level curve

If the level curve $f(x, y) = C$ is the graph of a differentiable function of x, the slope of its tangent is given by the formula

$$\frac{dy}{dx} = -\frac{f_x}{f_y}$$

All three methods are illustrated in the next example.

EXAMPLE 3.3

Find the slope $\dfrac{dy}{dx}$ of the level curve $f(x, y) = C$ if $f(x, y) = xy$ by:

(a) Solving for y in terms of x and differentiating the resulting expression.
(b) Differentiating the equation $f(x, y) = C$ implicitly.
(c) Applying the formula.

SOLUTION

(a) Solve the equation $xy = C$ for y to get

$$y = \frac{C}{x}$$

and differentiate to get $\quad \dfrac{dy}{dx} = -\dfrac{C}{x^2}$

(b) Use the product rule to differentiate both sides of the equation

$$xy = C$$

with respect to x, keeping in mind that y is really a function of x. You will get

$$x \frac{dy}{dx} + y = 0 \quad \text{or} \quad \frac{dy}{dx} = -\frac{y}{x}$$

which is equivalent to the answer in part (a) since $y = \dfrac{C}{x}$.

(c) Since $f_x = y$ and $f_y = x$, the formula gives

$$\frac{dy}{dx} = -\frac{f_x}{f_y} = -\frac{y}{x}$$

which is the same answer you obtained in part (b).

Here is another example.

EXAMPLE 3.4

Find the slope of the level curve $f(x, y) = 13$ at the point $(1, 2)$ if $f(x, y) = x^2 + 2xy + y^3$.

SOLUTION

In this case, there is no obvious way to solve for y explicitly in terms of x. Hence you must either differentiate implicitly or use the formula.

Differentiating the equation

$$x^2 + 2xy + y^3 = 13$$

implicitly with respect to x, you get

$$2x + 2x \frac{dy}{dx} + 2y + 3y^2 \frac{dy}{dx} = 0$$

or

$$\frac{dy}{dx} = -\frac{2x + 2y}{2x + 3y^2}$$

(For practice, check that you get the same answer using the formula.)
Evaluating the derivative when $x = 1$ and $y = 2$, you conclude that at the point $(1, 2)$,

$$\text{Slope of level curve} = \frac{dy}{dx} = -\frac{2(1) + 2(2)}{2(1) + 3(2)^2} = -\frac{3}{7}$$

Why the formula works

The proof that $\frac{dy}{dx} = -\frac{f_x}{f_y}$ is short, but subtle. Here is the argument.

Suppose the equation $f(x, y) = C$ implicitly defines y as a differentiable function of x. Think of f as a function of the single variable x and differentiate both sides of the equation $f = C$ with respect to x to get

$$\frac{df}{dx} = \frac{dC}{dx} \qquad \text{or} \qquad \frac{df}{dx} = 0$$

But, according to the chain rule,

$$\frac{df}{dx} = \frac{\partial f}{\partial x}\frac{dx}{dx} + \frac{\partial f}{\partial y}\frac{dy}{dx} = \frac{\partial f}{\partial x} + \frac{\partial f}{\partial y}\frac{dy}{dx}$$

Substitute this expression for $\frac{df}{dx}$ into the equation $\frac{df}{dx} = 0$ and conclude, as expected, that

$$\frac{\partial f}{\partial x} + \frac{\partial f}{\partial y}\frac{dy}{dx} = 0 \qquad \text{or} \qquad \frac{dy}{dx} = -\frac{\partial f/\partial x}{\partial f/\partial y} = -\frac{f_x}{f_y}$$

Constant-production curves

The next example illustrates how the slope of a constant-production curve can be used to make decisions concerning the allocation of labor.

EXAMPLE 3.5

Using x skilled workers and y unskilled workers, a manufacturer can produce $f(x, y) = x^2 y$ units per day. Currently the work force consists of 16 skilled workers and 32 unskilled workers, and the manufacturer is planning to hire 1 additional skilled worker. Use calculus to estimate the corresponding change that the manufacturer should make in the level of unskilled labor so that the total output will remain the same.

SOLUTION

The present output is $f(16, 32) = 8{,}192$ units. The combinations of x and y for which output will remain at this level are the coordinates of

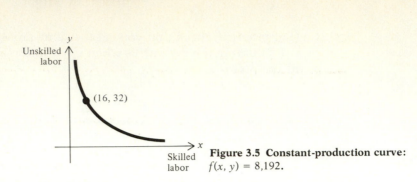

Figure 3.5 **Constant-production curve:** $f(x, y) = 8,192$.

the points that lie on the constant-production curve $f(x, y) = 8,192$ (Figure 3.5).

For any value of x, the slope of this curve is an estimate of the change in y (unskilled labor) that should be made to offset a 1-unit increase in x (skilled labor) so that the level of output will remain the same. That is,

$$\text{Change in unskilled labor} \approx \frac{dy}{dx} = -\frac{f_x}{f_y} = -\frac{2xy}{x^2} = -\frac{2y}{x}$$

Evaluating $\frac{dy}{dx}$ for the current labor force, $x = 16$ and $y = 32$, you find that

$$\frac{dy}{dx} = -\frac{2(32)}{16} = -4$$

and you can conclude that to compensate for the proposed increase in the level of skilled labor, the manufacturer should reduce the level of unskilled labor by approximately 4 workers.

Problems   In Problems 1 through 7, sketch the indicated level curves.

1. $f(x, y) = x + 2y$; $f = 1, f = 2, f = 3$

2. $f(x, y) = x^2 + y$; $f = 0, f = 4, f = 9$

3. $f(x, y) = x^2 - 4x - y$; $f = -4, f = 5$

4. $f(x, y) = \dfrac{x}{y}$; $f = -2, f = 2$

5. $f(x, y) = xy$; $f = 1, f = -1, f = 2, f = -2$

6. $f(x, y) = ye^x$; $f = 0, f = 1$ 7. $f(x, y) = xe^y$; $f = 1, f = e$

In Problems 8 through 11, find an equation of the level curve $f = C$ that passes through the given point.

8. $f(x, y) = x^2 + 2x + y$; $(1, 3)$ 9. $f(x, y) = x^2 + y^2 + 1$; $(2, 0)$

10. $f(x, y) = \dfrac{e^y}{x}$; $(e, 0)$ 11. $f(x, y) = x \ln y$; $(\sqrt{2}, 1)$

In Problems 12 through 17, use the formula $\dfrac{dy}{dx} = -\dfrac{f_x}{f_y}$ to find the slope of the level curve $f = C$, where C is a constant. Check your answer by differentiating implicitly.

12. $f(x, y) = x^2 - 4x - y$ 13. $f(x, y) = \dfrac{x}{y}$

14. $f(x, y) = xe^y$ 15. $f(x, y) = x^2y + 2y^3 - 3x - 2y$

16. $f(x, y) = x + 1 - 2xy^2$ 17. $f(x, y) = x \ln y$

In Problems 18 through 22, find the slope of the indicated level curve at the specified point.

18. $f(x, y) = x^2 - y^3$; $f = 1$; $(3, 2)$

19. $f(x, y) = x^2 + xy + y^3$; $f = 8$; $(0, 2)$

20. $f(x, y) = x^2y - 3xy + 5$; $f = 9$; $(1, -2)$

21. $f(x, y) = (x^2 + y)^3$; $f = 8$; $(-1, 1)$

22. $f(x, y) = \dfrac{e^y}{x}$; $f = 2$; $(\tfrac{1}{2}, 0)$

Allocation of labor   23. Using x hours of skilled labor and y hours of unskilled labor, a manufacturer can produce $f(x, y) = 10xy^{1/2}$ units. Currently the manufacturer uses 30 hours of skilled labor and 36 hours of unskilled labor and is planning to use 1 additional hour of skilled labor. Use calculus to estimate the corresponding change that the manufacturer should make in the level of unskilled labor so that the total output will remain the same.

Allocation of labor   24. Suppose the manufacturer in Problem 23 currently uses 30 hours of skilled labor and 36 hours of unskilled labor and is planning to use 1 additional hour of *unskilled* labor. Use calculus to estimate the corresponding change that should be made in the level of skilled labor so that the total output will remain the same. (*Hint:* Use the derivative $\dfrac{dx}{dy}$.)

Allocation of resources   25. At a certain factory, the daily output is $Q = 200K^{1/2}L^{1/3}$ units, where K denotes the capital investment measured in units of $1,000 and L the size of the labor force measured in worker-hours. The current level of capital investment is $60,000 and the current size of the labor force is 10,000 worker-hours. The manufacturer is planning to increase the capital investment by $1,000.

Use calculus to estimate the corresponding decrease that the manufacturer can make in the size of the labor force without affecting the daily output.

Allocation of resources

26. At a certain factory, output Q is related to inputs x and y by the function $Q = 2x^3 + 3x^2y + y^3$. If the current levels of input are $x = 20$ and $y = 10$, use calculus to estimate the change in input x that should be made to offset a 1-unit increase in input y so that output will be maintained at its current level.

4 RELATIVE MAXIMA AND MINIMA

In this section you will learn how to use partial derivatives to find the relative maxima and minima of functions of two variables.

In geometric terms, a **relative maximum** of a function $f(x, y)$ is a peak; a point on the surface $z = f(x, y)$ that is higher than any nearby point on the surface. A **relative minimum** is the bottom of a valley; a point that is lower than any nearby point on the surface. For example, when $(x, y) = (a, b)$, the function sketched in Figure 4.1a has a relative minimum while the function sketched in Figure 4.1b has a relative maximum.

Critical points   The points (a, b) for which both $f_x(a, b) = 0$ and $f_y(a, b) = 0$ are said to be **critical points** of f. Like the critical points for functions of one variable, these critical points play an important role in the study of relative maxima and minima.

To see the connection between critical points and relative extrema, suppose $f(x, y)$ has a relative maximum at (a, b). Then the curve

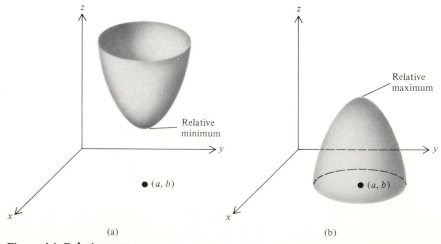

(a)   (b)

Figure 4.1 Relative extrema.

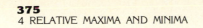

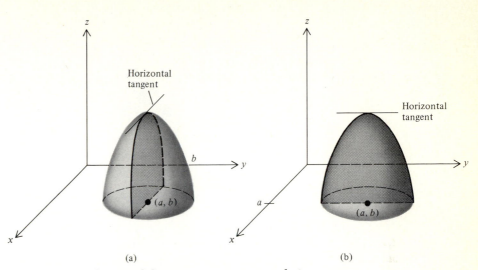

(a)   (b)

Figure 4.2 The partial derivatives are zero at a relative extremum.

formed by intersecting the surface $z = f(x, y)$ with the vertical plane $y = b$ has a relative maximum and hence a horizontal tangent when $x = a$. (See Figure 4.2a.) Since the partial derivative $f_x(a, b)$ is the slope of this tangent, it follows that $f_x(a, b) = 0$. Similarly, the curve formed by intersecting the surface $z = f(x, y)$ with the plane $x = a$ has a relative maximum when $y = b$ (see Figure 4.2b), and so $f_y(a, b) = 0$. This shows that a point at which a function of two variables has a relative maximum must be a critical point. A similar argument shows that a point at which a function of two variables has a relative minimum must also be a critical point.

Here is a more precise statement of the situation.

Critical points and relative extrema

> **A point** (a, b) **for which both**
>
> $$f_x(a, b) = 0 \qquad \text{and} \qquad f_y(a, b) = 0$$
>
> **is said to be a critical point of** f.
>
> **If the first-order partial derivatives of** f **are defined at all points in some region in the** xy **plane, then all the relative extrema of** f **in the region can occur only at critical points.**

Although all the relative extrema of a function must occur at critical points, not every critical point of a function is necessarily a relative extremum. Consider, for example, the function $f(x, y) = y^2 - x^2$ whose graph, which resembles a saddle, is sketched in Figure 4.3. In this case, $f_x(0, 0) = 0$ because the surface has a relative *maximum* (and hence a horizontal tangent) "in the x direction," and $f_y(0, 0) = 0$

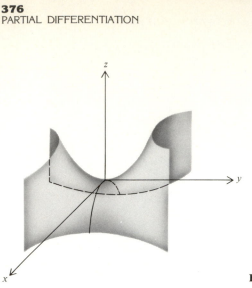

Figure 4.3 The surface $z = y^2 - x^2$.

because the surface has a relative *minimum* (and hence a horizontal tangent) "in the y direction." Hence $(0, 0)$ is a critical point of f, but it is not a relative extremum. For a critical point to be a relative extremum, the nature of the extremum must be the same *in all directions*. A critical point that is neither a relative maximum nor a relative minimum is called a **saddle point.**

The second derivative test

Here is a procedure involving second-order partial derivatives that you can use to decide whether a given critical point is a relative maximum, a relative minimum, or a saddle point. This procedure is the two-variable analog of the second derivative test for functions of a single variable that you saw in Chapter 3, Section 2.

The second derivative test

Suppose f **is a function of two variables** x **and** y, **and that all the second-order partial derivatives of** f **are continuous. Let**

$$D = f_{xx}f_{yy} - (f_{xy})^2$$

and suppose (a, b) **is a critical point of** f.

If $D(a, b) < 0$, **then** f **has a saddle point at** (a, b).

If $D(a, b) > 0$ **and** $f_{xx}(a, b) < 0$, **then** f **has a relative maximum at** (a, b).

If $D(a, b) > 0$ **and** $f_{xx}(a, b) > 0$, **then** f **has a relative minimum at** (a, b).

If $D(a, b) = 0$, **the test is inconclusive and** f **may have either a relative extremum or a saddle point at** (a, b).

The sign of the quantity D that appears in the second derivative test tells you whether the function has a relative extremum at (a, b): if $D(a, b)$ is positive, (a, b) is a relative extremum and if $D(a, b)$ is negative, (a, b) is a saddle point.

Moreover, if $D(a, b)$ is positive (guaranteeing that (a, b) is a relative extremum), the function has either a relative maximum *in all directions* or a relative minimum *in all directions*. To decide which, you restrict your attention to *one* direction (the x direction) and use the second derivative test for functions of a single variable to conclude that f has a relative maximum at (a, b) if $f_{xx}(a, b)$ is negative and a relative minimum at (a, b) if $f_{xx}(a, b)$ is positive.

The proof of the second derivative test involves ideas that are beyond the scope of this text and will be omitted. Here are some examples illustrating the use of this test.

EXAMPLE 4.1

Classify the critical points of the function $f(x, y) = x^2 + y^2$.

SOLUTION

Since
$$f_x = 2x \quad \text{and} \quad f_y = 2y$$
the only critical point of f is $(0, 0)$. To test this point, use the second-order partial derivatives
$$f_{xx} = 2 \quad f_{yy} = 2 \quad \text{and} \quad f_{xy} = 0$$
to get
$$D(x, y) = f_{xx}f_{yy} - (f_{xy})^2 = 2(2) - 0 = 4$$
That is, $D(x, y) = 4$ for all points (x, y) and, in particular,
$$D(0, 0) = 4 > 0$$
Hence, f has a relative extremum at $(0, 0)$. Moreover, since
$$f_{xx}(0, 0) = 2 > 0$$
it follows that the relative extremum at $(0, 0)$ is a relative minimum.
For reference, the graph of f is sketched in Figure 4.4.

EXAMPLE 4.2

Classify the critical points of the function $f(x, y) = y^2 - x^2$.

SOLUTION

Since
$$f_x = -2x \quad \text{and} \quad f_y = 2y$$
the only critical point of f is $(0, 0)$. To test this point, use the second-order partial derivatives

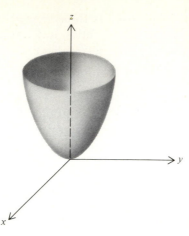

Figure 4.4 The surface $z = x^2 + y^2$.

$$f_{xx} = -2 \qquad f_{yy} = 2 \qquad \text{and} \qquad f_{xy} = 0$$

to get $\qquad D(x, y) = f_{xx}f_{yy} - (f_{xy})^2 = -2(2) - 0 = -4$

That is, $D(x, y) = -4$ for all points (x, y) and, in particular,

$$D(0, 0) = -4 < 0$$

It follows that f must have a saddle point at $(0, 0)$.
The graph of f is shown in Figure 4.5.

Here is a slightly more complicated example.

EXAMPLE 4.3

Classify the critical points of the function $f(x, y) = x^3 - y^3 + 6xy$.

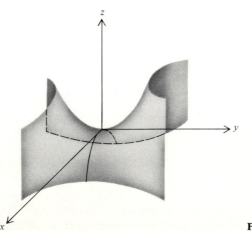

Figure 4.5 The surface $z = y^2 - x^2$.

SOLUTION

Since $\qquad f_x = 3x^2 + 6y \qquad$ and $\qquad f_y = -3y^2 + 6x$

you find the critical points of f by solving simultaneously the two equations

$$3x^2 + 6y = 0 \qquad \text{and} \qquad -3y^2 + 6x = 0$$

From the first equation, you get

$$y = -\frac{x^2}{2}$$

which you can substitute into the second equation to get

$$-\frac{3x^4}{4} + 6x = 0 \qquad \text{or} \qquad -x(x^3 - 8) = 0$$

The solutions of this equation are $x = 0$ and $x = 2$. These are the x coordinates of the critical points of f. To get the corresponding y coordinates, substitute these values of x into the equation $y = -\frac{x^2}{2}$ (or into either one of the two original equations). You will find that $y = 0$ when $x = 0$ and $y = -2$ when $x = 2$. It follows that the critical points of f are $(0, 0)$ and $(2, -2)$.

The second-order partial derivatives of f are

$$f_{xx} = 6x \qquad f_{yy} = -6y \qquad \text{and} \qquad f_{xy} = 6$$

Hence,

$$D(x, y) = f_{xx}f_{yy} - (f_{xy})^2 = -36xy - 36 = -36(xy + 1)$$

Since $\qquad\qquad\qquad D(0, 0) = -36 < 0$

it follows that f has a saddle point at $(0, 0)$. Since

$$D(2, -2) = 108 > 0 \qquad \text{and} \qquad f_{xx}(2, -2) = 12 > 0$$

it follows that f has a relative minimum at $(2, -2)$.

Practical optimization problems In the next example, you will see how to apply the theory of relative extrema to solve an optimization problem from economics. Actually, you will be trying to find the *absolute* maximum of a certain function. It turns out, however, that the absolute and relative maxima of this function coincide. In fact, in the majority of two-variable optimization problems in the social sciences, the relative extrema and absolute extrema coincide. For this reason, the theory of absolute extrema for functions of two variables will not be developed in this text, and you may assume that the relative extremum you find as the solution of a practical optimization problem is actually the absolute extremum.

EXAMPLE 4.4

The only grocery store in a small rural community carries two brands of frozen orange juice, a local brand that it obtains at a cost of 30 cents per can and a well-known national brand that it obtains at a cost of 40 cents per can. The grocer estimates that if the local brand is sold for x cents per can and the national brand for y cents per can, approximately $70 - 5x + 4y$ cans of the local brand and $80 + 6x - 7y$ cans of the national brand will be sold each day. How should the grocer price each brand to maximize the profit from the sale of the juice? (Assume that the absolute maximum and the relative maximum of the profit function are the same.)

SOLUTION

Since

$$\text{Profit} = \frac{\text{profit from the sale}}{\text{of the local brand}} + \frac{\text{profit from the sale}}{\text{of the national brand}}$$

it follows that the total daily profit from the sale of the juice is given by the function

$$f(x, y) = (x - 30)(70 - 5x + 4y) + (y - 40)(80 + 6x - 7y)$$

Using the product rule to compute the partial derivatives of f, you get

$$f_x = -10x + 10y - 20 \quad \text{and} \quad f_y = 10x - 14y + 240$$

Set these partial derivatives equal to zero to get

$$-10x + 10y - 20 = 0 \quad \text{and} \quad 10x - 14y + 240 = 0$$

and solve the resulting equations simultaneously to get

$$x = 53 \quad \text{and} \quad y = 55$$

It follows that (53, 55) is the only critical point of f.
   Use the second-order partial derivatives

$$f_{xx} = -10 \quad f_{yy} = -14 \quad \text{and} \quad f_{xy} = 10$$

to get $D(x, y) = f_{xx}f_{yy} - (f_{xy})^2 = -10(-14) - 100 = 40$

Since $D(53, 55) = 40 > 0 \quad \text{and} \quad f_{xx}(53, 55) = -10 < 0$

it follows that f has a (relative) maximum when $x = 53$ and $y = 55$. That is, the grocer can maximize profit by selling the local brand of juice for 53 cents per can and the national brand for 55 cents per can.

Problems In Problems 1 through 7, find the critical points of the given function and classify them as relative maxima, relative minima, or saddle points.

1. $f(x, y) = 5 - x^2 - y^2$ 2. $f(x, y) = 2x^2 - 3y^2$

3. $f(x, y) = xy$ 4. $f(x, y) = xy + \dfrac{8}{x} + \dfrac{8}{y}$

5. $f(x, y) = 2x^3 + y^3 + 3x^2 - 3y - 12x - 4$

6. $f(x, y) = (x - 1)^2 + y^3 - 3y^2 - 9y + 5$

7. $f(x, y) = x^3 + y^2 - 6xy + 9x + 5y + 2$

In Problems 8 through 12, assume that the absolute maximum and the relative maximum of the profit function are the same.

Retail price   8. A liquor store carries two competing brands of inexpensive wine, one from California and the other from New York. The owner of the store can obtain both wines at a cost of $2 per bottle and estimates that if the California wine is sold for x dollars per bottle and the New York wine for y dollars per bottle, consumers will buy approximately $40 - 50x + 40y$ bottles of the California wine and $20 + 60x - 70y$ bottles of the New York wine each day. How should the owner price the wines to generate the largest possible profit?

Pricing   9. The telephone company is planning to introduce two new types of executive communications systems that it hopes to sell to its largest commercial customers. It is estimated that if the first type of system is priced at x hundred dollars per system and the second type at y hundred dollars per system, approximately $40 - 8x + 5y$ consumers will buy the first type and $50 + 9x - 7y$ will buy the second type. If the cost of manufacturing the first type is $1,000 per system and the cost of the second type is $3,000 per system, how should the telephone company price the systems to generate the largest possible profit?

Allocation of funds   10. A manufacturer is planning to sell a new product at the price of $150 per unit and estimates that if x thousand dollars is spent on development and y thousand dollars is spent on promotion, consumers will buy approximately $\dfrac{320y}{y + 2} + \dfrac{160x}{x + 4}$ units of the product. If manufacturing costs for this product are $50 per unit, how much should the manufacturer spend on development and how much on promotion to generate the largest possible profit from the sale of this product? [*Hint:* Profit = (number of units)(price per unit − cost per unit) − total amount spent on development and promotion.]

Profit under monopoly 11. A manufacturer with exclusive rights to a sophisticated new industrial machine is planning to sell a limited number of the machines to both foreign and domestic firms. The price the manufacturer can expect to receive for the machines will depend on the number of machines made available. (For example, if only a few of the machines are placed on the market, competitive bidding among prospective purchasers will tend to drive the price up.) It is estimated that if the manufacturer supplies x machines to the domestic market and y machines to the foreign market, the machines will sell for $60 - \dfrac{x}{5} + \dfrac{y}{20}$ thousand dollars apiece at home and for $50 - \dfrac{y}{10} + \dfrac{x}{20}$ thousand dollars apiece abroad. If the manufacturer can produce the machines at a cost of $10,000 apiece, how many should be supplied to each market to generate the largest possible profit?

Profit under monopoly 12. A manufacturer with exclusive rights to a new industrial machine is planning to sell a limited number of them and estimates that if x machines are supplied to the domestic market and y to the foreign market, the machines will sell for $150 - \dfrac{x}{6}$ thousand dollars apiece at home and for $100 - \dfrac{y}{10}$ thousand dollars apiece abroad.

(a) How many machines should the manufacturer supply to the domestic market to generate the largest possible profit at home?
(b) How many machines should the manufacturer supply to the foreign market to generate the largest possible profit abroad?
(c) How many machines should the manufacturer supply to each market to generate the largest possible *total* profit?
(d) Is the relationship between the answers in parts (a), (b), and (c) accidental? Explain. Does a similar relationship hold in Problem 11? What accounts for the difference between these two problems in this respect?

Level curves 13. Sometimes one can classify the critical points of a function by inspecting its level curves. In each of the following cases, determine the nature of the critical point of f at $(0, 0)$.

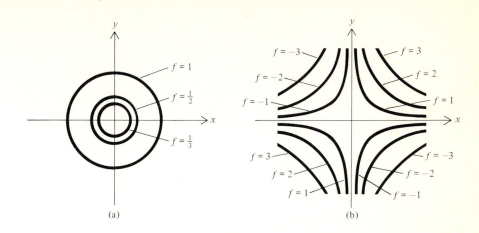

(a)   (b)

14. Let $f(x, y) = x^2 + y^2 - 4xy$. Show that f does *not* have a relative minimum at its critical point $(0, 0)$, even though it does have a relative minimum at $(0, 0)$ in both the x and y directions. (*Hint:* Consider the direction defined by the line $y = x$. That is, substitute x for y in the formula for f and analyze the resulting function of x.)

5 LAGRANGE MULTIPLIERS

In many applied problems, a function of two variables is to be optimized subject to a restriction or **constraint** on the variables. For example, an editor, constrained to stay within a fixed budget of $60,000, may wish to decide how to divide this money between development and promotion in order to maximize the future sales of a new book. If x denotes the amount of money allocated to development, y the amount allocated to promotion, and $f(x, y)$ the corresponding number of books that will be sold, the editor would like to maximize the sales function $f(x, y)$ subject to the budgetary constraint that $x + y = 60,000$.

For a geometric interpretation of the process of optimizing a function of two variables subject to a constraint, think of the function itself as a surface in three-dimensional space and of the constraint (which is an equation involving x and y) as a curve in the xy plane. When you find the maximum or minimum of the function subject to the given constraint, you are restricting your attention to the portion of the surface that lies directly above the constraint curve. The highest point on this portion of the surface is the constrained maximum, and the lowest point is the constrained minimum. The situation is illustrated in Figure 5.1.

You have already seen some constrained optimization problems in Chapter 3 of this text. (See Chapter 3, Section 1, Example 1.2, for instance.) The technique you used in Chapter 3 to solve such a problem

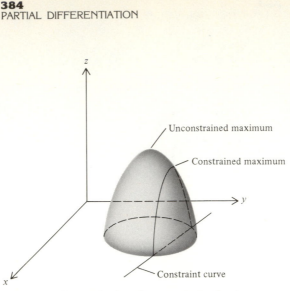

Figure 5.1 Constrained and unconstrained extrema.

was to reduce it to a problem involving only one variable by solving the constraint equation for one of the variables and then substituting the resulting expression into the function that was to be optimized. The success of this technique depended on your ability to solve the constraint equation for one of the variables, which, in many problems, is difficult or even impossible. In this section you will see a more versatile technique called the **method of Lagrange multipliers,** in which the introduction of a *third* variable allows you to solve constrained optimization problems without first solving the constraint equation for one of the variables. Here is a statement of the method.

The method of Lagrange multipliers

Suppose $f(x, y)$ **and** $g(x, y)$ **are functions whose first-order partial derivatives exist. To find the relative maximum and relative minimum of** $f(x, y)$ **subject to the constraint that** $g(x, y) = k$ **for some constant** k**, introduce a new variable** λ **(called a Lagrange multiplier) and solve the following three equations simultaneously:**

$$f_x(x, y) = \lambda g_x(x, y) \qquad f_y(x, y) = \lambda g_y(x, y) \qquad g(x, y) = k$$

The desired relative extrema will be found among the resulting points (x, y)**.**

In the next example, the method of Lagrange multipliers will be used to solve the problem from Chapter 3, Section 1, Example 1.2.

EXAMPLE 5.1

The highway department is planning to build a picnic area for motorists along a major highway. It is to be rectangular with an area of

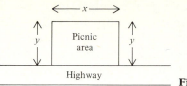

Figure 5.2 Rectangular picnic area.

5,000 square meters and is to be fenced off on the three sides not adjacent to the highway. What is the least amount of fencing that will be needed to complete the job?

SOLUTION

Label the sides of the picnic area as indicated in Figure 5.2 and let f denote the amount of fencing required. Then,

$$f(x, y) = x + 2y$$

The goal is to minimize f subject to the constraint that the area must be 5,000; that is, subject to the constraint

$$xy = 5,000$$

Let $g(x, y) = xy$ and use the partial derivatives

$$f_x = 1 \qquad f_y = 2 \qquad g_x = y \qquad \text{and} \qquad g_y = x$$

to write the three Lagrange equations

$$1 = \lambda y \qquad 2 = \lambda x \qquad \text{and} \qquad xy = 5,000$$

From the first and second equations you get

$$\lambda = \frac{1}{y} \qquad \text{and} \qquad \lambda = \frac{2}{x}$$

(since $y \neq 0$ and $x \neq 0$), which implies that

$$\frac{1}{y} = \frac{2}{x} \qquad \text{or} \qquad x = 2y$$

Now substitute $x = 2y$ into the third equation to get

$$2y^2 = 5,000 \qquad \text{or} \qquad y = \pm 50$$

and finally use $y = 50$ in the equation $x = 2y$ to get $x = 100$. It follows that $x = 100$ and $y = 50$ are the values that minimize the function $f(x, y) = x + 2y$ subject to the constraint that $xy = 5,000$. That is, the optimal picnic area is 100 meters wide (along the highway) and extends 50 meters back from the road.

Strictly speaking, the solution of the preceding example was incomplete since no attempt was made to verify that the optimal values $x = 100$ and $y = 50$ actually minimized the function f. There is a version of the second derivative test that could have been used to show that the point $(100, 50)$ was indeed a relative minimum of the function f. And, with a little more work, it could have been shown that this relative minimum was really the absolute minimum. The techniques needed to carry out this analysis are discussed in more advanced texts. In this text, you may assume that the maximum or minimum you get using the method of Lagrange multipliers is the extremum you are seeking.

Here is one more example.

EXAMPLE 5.2

Find the maximum and minimum values of the function $f(x, y) = xy$ subject to the constraint $x^2 + y^2 = 8$.

SOLUTION

Let $g(x, y) = x^2 + y^2$ and use the partial derivatives

$$f_x = y \qquad f_y = x \qquad g_x = 2x \qquad \text{and} \qquad g_y = 2y$$

to get the three Lagrange equations

$$y = 2\lambda x \qquad x = 2\lambda y \qquad \text{and} \qquad x^2 + y^2 = 8$$

Neither x nor y can be zero if all three of these equations are to hold (do you see why?), and so you can rewrite the first two equations as

$$2\lambda = \frac{y}{x} \qquad \text{and} \qquad 2\lambda = \frac{x}{y}$$

which implies that $\qquad \dfrac{y}{x} = \dfrac{x}{y} \qquad \text{or} \qquad x^2 = y^2$

Now substitute $x^2 = y^2$ into the third equation to get

$$2x^2 = 8 \qquad \text{or} \qquad x = \pm 2$$

If $x = 2$, it follows from the equation $x^2 = y^2$ that $y = 2$ or $y = -2$. Similarly, if $x = -2$, it follows that $y = 2$ or $y = -2$. Hence, the four points at which the constrained extrema can occur are $(2, 2), (2, -2), (-2, 2)$, and $(-2, -2)$. Since

$$f(2, 2) = 4 \qquad f(2, -2) = -4 \qquad f(-2, 2) = -4 \qquad \text{and} \qquad f(-2, -2) = 4$$

it follows that when $x^2 + y^2 = 8$, the maximum value of $f(x, y)$ is 4, which occurs at the points $(2, 2)$ and $(-2, -2)$, and the minimum value is -4, which occurs at $(2, -2)$ and $(-2, 2)$.

For practice, check these answers by solving this optimization problem using the methods of Chapter 2.

Notice that in each of the preceding examples, the first two Lagrange equations were used to eliminate the new variable λ, and then the resulting expression relating x and y was substituted into the third equation. For most constrained optimization problems you will encounter, this particular sequence of steps will lead you quickly to the desired solution.

Why the method of Lagrange multipliers works

Although a rigorous explanation of why the method of Lagrange multipliers works involves advanced ideas beyond the scope of this text, there is a rather simple geometric argument that you should find convincing. Suppose the constraint curve $g(x, y) = k$ and the level curves $f(x, y) = C$ are drawn in the xy plane as shown in Figure 5.3.

To maximize $f(x, y)$ subject to the constraint $g(x, y) = k$, you must find the highest level curve of f that intersects the constraint curve. As the sketch in Figure 5.3 suggests, this critical intersection will occur at a point at which the constraint curve is tangent to a level curve; that is, at which the slope of the constraint curve $g(x, y) = k$ is equal to the slope of a level curve $f(x, y) = C$. According to the formula you learned in Section 3 of this chapter,

$$\text{Slope of constraint curve} = -\frac{g_x}{g_y}$$

and

$$\text{Slope of level curve} = -\frac{f_x}{f_y}$$

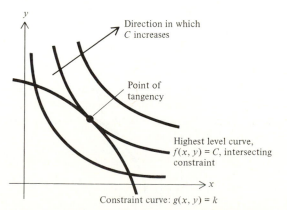

Figure 5.3 labels:
- Direction in which C increases
- Point of tangency
- Highest level curve, $f(x, y) = C$, intersecting constraint
- Constraint curve: $g(x, y) = k$

Figure 5.3 Increasing level curves and the constraint curve.

Hence the condition that the slopes be equal can be expressed by the equation

$$-\frac{f_x}{f_y} = -\frac{g_x}{g_y} \qquad \text{or, equivalently} \qquad \frac{f_x}{g_x} = \frac{f_y}{g_y}$$

If you let λ denote this common ratio, you have

$$\lambda = \frac{f_x}{g_x} \qquad \text{and} \qquad \lambda = \frac{f_y}{g_y}$$

from which you get the first two Lagrange equations

$$f_x = \lambda g_x \qquad \text{and} \qquad f_y = \lambda g_y$$

The third Lagrange equation

$$g(x, y) = k$$

is simply a statement of the fact that the point in question actually lies on the constraint curve.

Suppose all three of the Lagrange equations are satisfied at a certain point (a, b). Then f will reach its constrained *maximum* at (a, b) if the *highest* level curve that intersects the constraint curve does so at this point. On the other hand, if the *lowest* level curve that intersects the constraint curve does so at (a, b), then f will achieve its constrained *minimum* at this point. The situation is illustrated further in Figure 5.4, which shows the constraint curve $x^2 + y^2 = 8$ and the optimal level curves $xy = 4$ and $xy = -4$ from Example 5.2. Notice that in this case there are four points at which the constraint curve is

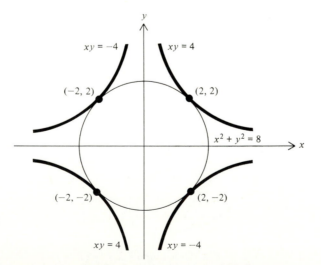

Figure 5.4 Constraint curve and optimal level curves.

tangent to a level curve. Two of these points, $(2, 2)$ and $(-2, -2)$, maximize f subject to the given constraint while the other two, $(2, -2)$ and $(-2, 2)$, minimize f.

Practical optimization problems

In the next example, the method of Lagrange multipliers is used to solve a constrained optimization problem from economics.

EXAMPLE 5.3

An editor has been allotted $60,000 to spend on the development and promotion of a new book. It is estimated that if x thousand dollars is spent on development and y thousand on promotion, approximately $20x^{3/2}y$ copies of the book will be sold. How much money should the editor allocate to development and how much to promotion in order to maximize sales?

SOLUTION

The goal is to maximize the function $f(x, y) = 20x^{3/2}y$ subject to the constraint $g(x, y) = 60$, where $g(x, y) = x + y$. The corresponding Lagrange equations are

$$30x^{1/2}y = \lambda \qquad 20x^{3/2} = \lambda \qquad \text{and} \qquad x + y = 60$$

From the first two equations you get

$$30x^{1/2}y = 20x^{3/2}$$

Since the maximum value of f clearly does not occur when $x = 0$, you may assume that $x \neq 0$ and divide both sides of this equation by $30x^{1/2}$ to get

$$y = \tfrac{2}{3}x \qquad \text{or} \qquad x = \tfrac{3}{2}y$$

Substituting this expression into the third equation, you get

$$\tfrac{5}{2}y = 60$$

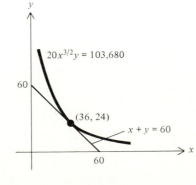

Figure 5.5 Budgetary constraint and optimal sales level.

from which it follows that

$$y = 24 \quad \text{and} \quad x = 36$$

That is, to maximize sales, the editor should spend \$36,000 on development and \$24,000 on promotion. If this is done, approximately $f(36, 24) = 103,680$ copies of the book will be sold.

A graph showing the relationship between the budgetary constraint and the level curve for optimal sales is sketched in Figure 5.5.

The significance of the Lagrange multiplier λ

You can solve most constrained optimization problems by the method of Lagrange multipliers without actually obtaining a numerical value for the multiplier λ. In some problems, however, you may want to compute λ. This is because λ has the following useful interpretation.

The Lagrange multiplier

> **Suppose M is the maximum (or minimum) value of $f(x, y)$ subject to the constraint $g(x, y) = k$. The Lagrange multiplier λ is the rate of change of M with respect to k. That is,**
>
> $$\lambda = \frac{dM}{dk}$$
>
> **Hence,** $\quad \lambda \approx$ change in M due to a 1-unit increase in k

The proof of this fact is discussed in more advanced texts. Here is an example illustrating its use.

EXAMPLE 5.4

Suppose the editor in Example 5.3 is allotted \$61,000 instead of \$60,000 to spend on the development and promotion of the new book. Estimate how the additional \$1,000 will affect the maximum sales level.

SOLUTION

In Example 5.3, you solved the three Lagrange equations

$$30x^{1/2}y = \lambda \quad 20x^{3/2} = \lambda \quad \text{and} \quad x + y = 60$$

to conclude that the maximum value M of $f(x, y)$ subject to the constraint $x + y = 60$ occurred when $x = 36$ and $y = 24$. If you substitute these values of x and y into the second (or first) Lagrange equation, you get

$$\lambda = 20(36)^{3/2} = 4,320$$

Since $\lambda = \dfrac{dM}{dk}$, it follows that the 1-unit increase in k from $k = 60$ to $k = 61$ will increase the maximal sales M by approximately 4,320 copies.

Problems Use the method of Lagrange multipliers to solve the following problems.

1. Find the maximum value of the function $f(x, y) = xy$ subject to the constraint $x + y = 1$.

2. Find the maximum and minimum values of the function $f(x, y) = xy$ subject to the constraint $x^2 + y^2 = 1$.

3. Find the minimum value of the function $f(x, y) = x^2 + y^2$ subject to the constraint $xy = 1$.

4. Find the minimum value of the function $f(x, y) = x^2 + 2y^2 - xy$ subject to the constraint $2x + y = 22$.

5. Find the minimum value of the function $f(x, y) = x^2 - y^2$ subject to the constraint $x^2 + y^2 = 4$.

6. Find the maximum and minimum values of the function $f(x, y) = 8x^2 - 24xy + y^2$ subject to the constraint $x^2 + y^2 = 1$.

Construction 7. A farmer wishes to fence off a rectangular pasture along the bank of a river. The area of the pasture is to be 3,200 square meters, and no fencing is needed along the river bank. Find the dimensions of the pasture that will require the least amount of fencing.

Construction 8. There are 320 meters of fencing available to enclose a rectangular field. How should the fencing be used so that the enclosed area is as large as possible?

Postal service 9. According to postal regulations, the girth plus length of parcels sent by fourth-class mail may not exceed 72 inches. What is the largest possible volume of a rectangular parcel with two square sides that can be sent by fourth-class mail?

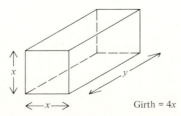

Girth = $4x$

Packaging 10. Use the fact that 12 fluid ounces is (approximately) 6.89π cubic inches to find the dimensions of the 12-ounce beer can that can be constructed using the least amount of metal. (Recall that the volume of a cylinder of radius r and height h is $\pi r^2 h$, that the circumference of a circle of radius r is $2\pi r$, and that the area of a circle of radius r is πr^2.)

Packaging 11. A cylindrical can is to hold 4π cubic inches of frozen orange juice. The cost per square inch of constructing the metal top and bottom is twice the cost per square inch of constructing the cardboard side. What are the dimensions of the least expensive can?

Allocation of funds 12. A manufacturer has $8,000 to spend on the development and promotion of a new product. It is estimated that if x thousand dollars is spent on development and y thousand is spent on promotion, sales will be approximately $50x^{1/2}y^{3/2}$ units. How much money should the manufacturer allocate to development and how much to promotion to maximize sales?

Allocation of funds 13. If x thousand dollars is spent on labor and y thousand dollars is spent on equipment, the output at a certain factory will be $Q(x, y) = 60x^{1/3}y^{2/3}$ units. If $120,000 is available, how should this be allocated between labor and equipment to generate the largest possible output?

Allocation of funds 14. A manufacturer is planning to sell a new product at the price of $150 per unit and estimates that if x thousand dollars is spent on development and y thousand dollars is spent on promotion, approximately $\dfrac{320y}{y + 2} + \dfrac{160x}{x + 4}$ units of the product will be sold. The cost of manufacturing the product is $50 per unit. If the manufacturer has a total of $8,000 to spend for development and promotion, how should this money be allocated to generate the largest possible profit?

Marginal analysis 15. Use the Lagrange multiplier λ to estimate the change in the maximum output of the factory in Problem 13 that would result if the money available for labor and equipment was increased by $1,000.

Marginal analysis 16. Suppose the manufacturer in Problem 14 decides to spend $9,000 instead of $8,000 on the development and promotion of the new product. Use the Lagrange multiplier λ to estimate how this change will affect the maximum possible profit.

Allocation of
unrestricted funds

17. (a) If unlimited funds are available, how much should the manufacturer in Problem 14 spend on development and how much on promotion in order to generate the largest possible profit? (*Hint:* Use the methods of Section 4.)

  (b) What is the value of the Lagrange multiplier that corresponds to the optimal budget in part (a)? Explain your answer in light of the interpretation of λ as $\dfrac{dM}{dk}$.

  (c) Your answer to part (b) should suggest another method for solving the problem in part (a). Solve the problem using this new method.

CHAPTER SUMMARY AND PROFICIENCY TEST

Important terms, symbols, and formulas

Partial derivative: $f_x,\, f_y;\ \dfrac{\partial z}{\partial x},\ \dfrac{\partial z}{\partial y}$

Second-order partial derivative: $f_{xx},\, f_{xy},\, f_{yx},\, f_{yy};\ \dfrac{\partial^2 z}{\partial x^2},\ \dfrac{\partial^2 z}{\partial y\,\partial x},\ \dfrac{\partial^2 z}{\partial x\,\partial y},\ \dfrac{\partial^2 z}{\partial y^2}$

Mixed second-order partial derivatives: $f_{xy} = f_{yx}$

Chain rule: $\dfrac{dz}{dt} = \dfrac{\partial z}{\partial x}\dfrac{dx}{dt} + \dfrac{\partial z}{\partial y}\dfrac{dy}{dt}$

Approximation formula; total differential: $\Delta z \approx \dfrac{\partial z}{\partial x}\,\Delta x + \dfrac{\partial z}{\partial y}\,\Delta y = dz$

Level curve; isoquant: $f(x, y) = C$

Slope of a level curve: $\dfrac{dy}{dx} = -\dfrac{f_x}{f_y}$

Relative maximum; relative minimum; saddle point

Critical point: $f_x = f_y = 0$

Second derivative test at a critical point: Let $D = f_{xx}f_{yy} - (f_{xy})^2$
  If $D < 0$, f has a saddle point.
  If $D > 0$ and $f_{xx} < 0$, f has a relative maximum.
  If $D > 0$ and $f_{xx} > 0$, f has a relative minimum.

Method of Lagrange multipliers: To find the relative extrema of $f(x, y)$ subject to $g(x, y) = k$, solve the equations

$$f_x = \lambda g_x \qquad f_y = \lambda g_y \qquad \text{and} \qquad g = k$$

The Lagrange multiplier: $\lambda = \dfrac{dM}{dk} \approx$ change in M due to a 1-unit increase in k, where M is the optimal value of $f(x, y)$ subject to $g(x, y) = k$

Proficiency test

1. For each of the following functions, compute the first-order partial derivatives f_x and f_y.

    (a) $f(x, y) = 2x^3y + 3xy^2 + \dfrac{y}{x}$

    (b) $f(x, y) = (xy^2 + 1)^5$

    (c) $f(x, y) = xye^{xy}$

2. For each of the following functions, compute the second-order partial derivatives $f_{xx}, f_{yy}, f_{xy},$ and f_{yx}.

    (a) $f(x, y) = x^2 + y^3 - 2xy^2$

    (b) $f(x, y) = e^{x^2+y^2}$

    (c) $f(x, y) = x \ln y$

3. At a certain factory, the daily output is approximately $40K^{1/3}L^{1/2}$ units, where K denotes the capital investment measured in units of \$1,000 and L denotes the size of the labor force measured in worker-hours. Suppose the current capital investment is \$125,000 and that 900 worker-hours of labor are used each day. Use marginal analysis to estimate the effect that an additional capital investment of \$1,000 will have on the daily output if the size of the labor force is not changed.

4. Use the chain rule to find $\dfrac{dz}{dt}$.

    (a) $z = x^3 - 3xy^2; x = 2t, y = t^2$

    (b) $z = x \ln y; x = 2t, y = e^t$

5. A grocery store carries two brands of diet cola. Sales figures indicate that if the first brand is sold for x cents per can and the second brand for y cents per can, consumers will buy $Q(x, y) = 240 + 0.1y^2 - 0.2x^2$ cans of the first brand per week. Currently the first brand sells for 45 cents per can, and the second brand sells for 48 cents per can. Use the total differential to estimate how the demand for the first brand of diet cola will change if the price of the first brand is increased by 2 cents per can while the price of the second brand is decreased by 1 cent per can.

6. For each of the following functions, sketch the indicated level curves.

    (a) $f(x, y) = x^2 - y; f = 2, f = -2$

    (b) $f(x, y) = 6x + 2y; f = 0, f = 1, f = 2$

7. For each of the following functions, find the slope of the indicated level curve at the specified point.

    (a) $f(x, y) = x^2 - y^3; f = 2, (1, -1)$

    (b) $f(x, y) = xe^y; f = 2, (2, 0)$

8. Using x skilled workers and y unskilled workers, a manufacturer can produce $Q(x, y) = 60x^{1/3}y^{2/3}$ units per day. Currently the manufacturer employs 10 skilled workers and 40 unskilled workers and is planning to hire 1 additional skilled worker. Use calculus to estimate the corresponding

change that the manufacturer should make in the level of unskilled labor so that the total output will remain the same.

9. Find the critical points of each of the following functions and classify them as relative maxima, relative minima, or saddle points.
   (a) $f(x, y) = x^3 + y^3 + 3x^2 - 18y^2 + 81y + 5$
   (b) $f(x, y) = x^2 + y^3 + 6xy - 7x - 6y$

10. Use the method of Lagrange multipliers to find the maximum and minimum values of the function $f(x, y) = x^2 + 2y^2 + 2x + 3$ subject to the constraint $x^2 + y^2 = 4$.

11. Use the method of Lagrange multipliers to prove that of all rectangles with a given perimeter the square has the largest area.

12. A manufacturer is planning to sell a new product at the price of $350 per unit and estimates that if x thousand dollars is spent on development and y thousand dollars is spent on promotion, consumers will buy approximately $\dfrac{250y}{y + 2} + \dfrac{100x}{x + 5}$ units of the product. If manufacturing costs for this product are $150 per unit, how much should the manufacturer spend on development and how much on promotion to generate the largest possible profit if unlimited funds are available?

13. Suppose the manufacturer in Problem 12 has only $11,000 to spend on the development and promotion of the new product. How should this money be allocated to generate the largest possible profit?

14. Suppose the manufacturer in Problem 13 decides to spend $12,000 instead of $11,000 on the development and promotion of the new product. Use the Lagrange multiplier λ to estimate how this change will affect the maximum possible profit.

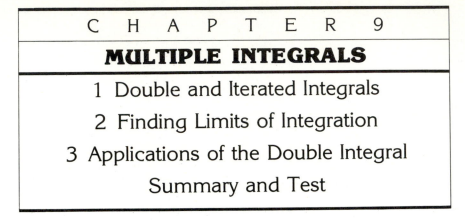

CHAPTER 9

MULTIPLE INTEGRALS

1 Double and Iterated Integrals

2 Finding Limits of Integration

3 Applications of the Double Integral

Summary and Test

---

1 DOUBLE AND ITERATED INTEGRALS

This chapter is about integrals of functions of two variables. As in the one-variable case, integrals will be viewed in two ways: as limits of sums and as antiderivatives. The characterization of integrals as limits of sums is used primarily to set up integrals corresponding to practical problems, while antidifferentiation is used to perform the actual calculations.

In many ways, integrals of functions of two variables are very much like integrals of functions of a single variable, and the new ideas you will encounter in this chapter will seem natural if you keep the one-variable situation in mind. Here is a brief review of what you should remember from Chapter 6 about definite integrals of functions of one variable.

Review of the one-variable situation

Suppose $f(x)$ is nonnegative and continuous on the interval $a \leq x \leq b$, which is divided into n equal subintervals of width Δx, with x_j denoting the beginning of the jth subinterval. Then, as shown in Figure 1.1, the area under the graph of f between $x = a$ and $x = b$ is approximately the sum of the areas of n rectangles, where the base of the jth rectangle is the jth subinterval and its height is $f(x_j)$. That is,

396

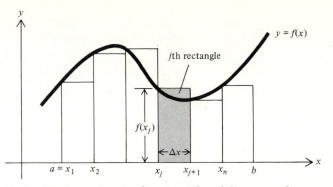

Figure 1.1 Approximation by rectangles of the area under a curve.

$$\text{Area of } j\text{th rectangle} = f(x_j)\,\Delta x$$

and
$$\text{Area under curve} \approx \sum_{j=1}^{n} f(x_j)\,\Delta x$$

If more subintervals are used, the width of each will be smaller and the approximation will usually be more accurate. If the number n of subintervals is allowed to increase without bound, the sum of the areas of the rectangles will approach the actual area under the curve. That is,

$$\text{Area under curve} = \lim_{n \to \infty} \sum_{j=1}^{n} f(x_j)\,\Delta x$$

On the other hand, as you saw in Chapter 6, Section 2, the area under the curve can also be calculated from the formula

$$\text{Area under curve} = F(b) - F(a)$$

where F is any antiderivative of f. These two expressions for the area under the curve must, of course, be equal, and so

$$\lim_{n \to \infty} \sum_{j=1}^{n} f(x_j)\,\Delta x = F(b) - F(a)$$

This relationship between limits of sums and antiderivatives is known as the fundamental theorem of calculus and can be shown to hold for *any* function $f(x)$ that is continuous on the interval $a \leq x \leq b$. For reference, here is a statement of the theorem using the appropriate integral notation.

The fundamental theorem of calculus

Suppose f is continuous on the interval $a \leq x \leq b$, which is divided into n equal subintervals of length Δx by $x_1, x_2, \ldots, x_n$. Then

$$\lim_{n \to \infty} \sum_{j=1}^{n} f(x_j)\, \Delta x = \int_a^b f(x)\, dx = F(b) - F(a)$$

where F is any antiderivative of f.

The double integral

The **double integral** of a function of two variables is defined as the limit of a sum that closely resembles the limit of a sum for the definite integral of a function of one variable.

The double integral

Suppose $f(x, y)$ is continuous on a bounded region R in the xy plane. Use lines drawn parallel to the x and y axes to approximate R by n rectangles of area ΔA, as shown in Figure 1.2, and let (x_j, y_j) be a point in the jth rectangle. Then the double integral of f over R is

$$\iint\limits_R f(x, y)\, dA = \lim_{n \to \infty} \sum_{j=1}^{n} f(x_j, y_j)\, \Delta A$$

where it is understood that the areas ΔA approach zero as n increases without bound.

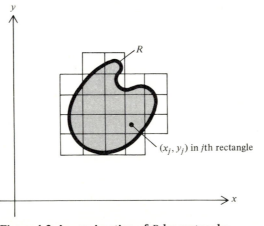

Figure 1.2 Approximation of R by rectangles.

Geometric interpretation

As you know, the definite integral $\int_a^b f(x)\, dx$ of a nonnegative function of one variable gives the area under the curve $y = f(x)$ between $x = a$ and $x = b$ (Figure 1.3a). The double integral of a nonnegative function

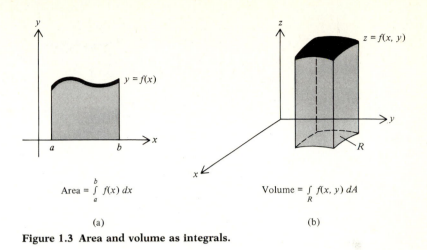

Area = $\int_a^b f(x)\,dx$

Volume = $\int_R f(x, y)\,dA$

(a)

(b)

Figure 1.3 Area and volume as integrals.

of two variables has an analogous interpretation as the volume under a surface (Figure 1.3b).

The volume under a surface

If $f(x, y) \geq 0$ **for all points** (x, y) **in** R, **and if** S **is the solid region bounded above by the surface** $z = f(x, y)$ **and below by** R, **then**

$$\text{Volume of } S = \iint_R f(x, y)\,dA$$

To see why the double integral is equal to the volume, look at Figure 1.4, which shows a surface $z = f(x, y)$ above a region R in the xy plane. Each of the n rectangles used to approximate R is the base of a vertical column that is cut off at the top by the surface $z = f(x, y)$. The jth column is shown in Figure 1.4. Its volume is approximately that of a rectangular solid whose volume is the area ΔA of its base times its (constant) height $f(x_j, y_j)$. Thus,

$$\text{Volume of } j\text{th column} \approx f(x_j, y_j)\,\Delta A$$

(Note that in general this is only an approximation because the top of the column need not actually be horizontal.) The volume of the entire solid is approximately the sum of the volumes of the n columns, and the approximation improves as the number of approximating rectangles increases. Thus,

$$\text{Volume of } S = \lim_{n \to \infty} \sum_{j=1}^{n} f(x_j, y_j)\,\Delta A = \iint_R f(x, y)\,dA$$

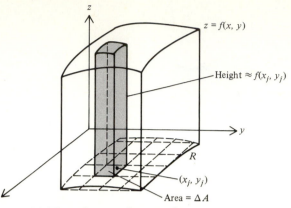

Figure 1.4 **The volume under a surface.**

**A social-science
application**

The following example shows how a double integral can be used to compute the population of a region if the population density is known. This application is presented here to give you a better feel for the definition of the double integral and to indicate how double integrals can arise in nongeometric practical situations. Applications will be discussed in more detail in Section 3 after you have learned how to evaluate double integrals by antidifferentiation.

EXAMPLE 1.1

Suppose R is the region within the boundary of a certain city and $f(x, y)$ the population density (number of people per square mile) at the point (x, y). Use a double integral to express the total population of the city. (Assume x and y are measured in miles.)

SOLUTION

Approximate R by n rectangles of area ΔA (Figure 1.5), and let (x_j, y_j) be a point in the jth rectangle. Then,

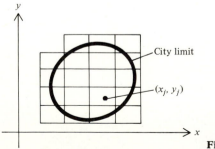

Figure 1.5 **Approximation of R by rectangles.**

$$\text{Population of } j\text{th rectangle} \approx \frac{\text{people}}{\text{square mile}} \times \text{square miles}$$

$$\approx f(x_j, y_j) \, \Delta A$$

$$\text{and} \quad \text{Population of city} = \lim_{n \to \infty} \sum_{j=1}^{n} f(x_j, y_j) \, \Delta A = \iint\limits_{R} f(x, y) \, dA$$

The calculation of double integrals

The characterization of double integrals as limits of sums is useful for interpreting practical situations as integrals, but it is not convenient to use for the actual calculation of double integrals. Double integrals are usually calculated by a process involving repeated partial antidifferentiation. The partial antiderivatives are evaluated at limits of integration that correspond to the region R. In this section you will see how to evaluate double integrals once the limits of integration are known. In Section 2 you will learn how to determine the limits of integration that correspond to a given region.

Iterated integrals

The symbol

$$\int_{a}^{b} \int_{c}^{d} f(x, y) \, dy \, dx$$

is called an **iterated integral** and is an abbreviation for

$$\int_{a}^{b} \left(\int_{c}^{d} f(x, y) \, dy \right) dx$$

To evaluate an iterated integral you first compute the inner definite integral

$$\int_{c}^{d} f(x, y) \, dy$$

taking the antiderivative of f with respect to y while keeping x fixed. The result will be a function of the single variable x, which you then integrate with respect to x between $x = a$ and $x = b$. Here is an example.

EXAMPLE 1.2

Evaluate $\int_{0}^{1} \int_{-1}^{2} xy^2 \, dy \, dx$.

SOLUTION

First perform the inner integration with respect to y, treating x as a constant to get

$$\int_{-1}^{2} xy^2 \, dy = \frac{1}{3} xy^3 \Big|_{y=-1}^{y=2} = \frac{8}{3} x + \frac{1}{3} x = 3x$$

(Note that the limits of integration -1 and 2 refer to the variable y.) Now integrate the result of this calculation with respect to x from $x = 0$ to $x = 1$ to conclude that

$$\int_{0}^{1} \int_{-1}^{2} xy^2 \, dy \, dx = \int_{0}^{1} 3x \, dx = \frac{3}{2} x^2 \Big|_{0}^{1} = \frac{3}{2}$$

To give your solution a more professional appearance, arrange your work compactly as follows:

$$\int_{0}^{1} \int_{-1}^{2} xy^2 \, dy \, dx = \int_{0}^{1} \left(\frac{1}{3} xy^3 \Big|_{y=-1}^{y=2} \right) dx = \int_{0}^{1} 3x \, dx = \frac{3}{2} x^2 \Big|_{0}^{1} = \frac{3}{2}$$

In the preceding example, all four of the limits of integration were constants. In the next example, the limits of integration on the inner integral are not constants but functions of x. The technique for evaluating the integral remains the same.

EXAMPLE 1.3

Evaluate $\int_{0}^{1} \int_{x^2}^{\sqrt{x}} 160xy^3 \, dy \, dx$.

SOLUTION

$$\int_{0}^{1} \int_{x^2}^{\sqrt{x}} 160xy^3 \, dy \, dx = \int_{0}^{1} \left(40xy^4 \Big|_{y=x^2}^{y=\sqrt{x}} \right) dx \quad \text{(since } x \text{ is treated as a constant)}$$

$$= \int_{0}^{1} [40x(\sqrt{x})^4 - 40x(x^2)^4] \, dx$$

$$= \int_{0}^{1} (40x^3 - 40x^9) \, dx$$

$$= (10x^4 - 4x^{10}) \Big|_{0}^{1}$$

$$= 6$$

In Example 1.3, the variable limits of integration on the inner integral were functions of x. These functions, x^2 and $\sqrt{x}$, were substituted for y during the first integration, resulting in an expression containing only the variable x, which was then integrated between the constant limits $x = 0$ and $x = 1$. In general, the outer limits of integration must be constants (so that the final answer will be a constant), while the inner limits may be functions of the variable with respect to which the second integration is to be performed.

In the next example, x is the first variable of integration and y is the second. Notice that in this case the variable limit of integration on the inner integral is a function of y.

EXAMPLE 1.4

Evaluate $\displaystyle\int_0^1 \int_0^y y^2 e^{xy}\, dx\, dy$.

SOLUTION

$$\int_0^1 \int_0^y y^2 e^{xy}\, dx\, dy = \int_0^1 \left(y e^{xy} \Big|_{x=0}^{x=y} \right) dy \quad \left(\text{since } \int e^{xy}\, dx = \frac{1}{y}\, e^{xy} \right)$$

$$= \int_0^1 (y e^{y^2} - y)\, dy$$

$$= \left(\frac{1}{2}\, e^{y^2} - \frac{1}{2}\, y^2 \right) \Big|_0^1$$

$$= \left(\frac{1}{2}\, e - \frac{1}{2} \right) - \left(\frac{1}{2} - 0 \right)$$

$$= \frac{1}{2}\, e - 1$$

In performing the repeated antidifferentiation required to evaluate an iterated integral, you may have to use one or more of the special techniques of integration you learned in Chapter 5. The next example involves integration by parts.

EXAMPLE 1.5

Evaluate $\displaystyle\int_0^1 \int_{1-y}^1 e^y\, dx\, dy$.

SOLUTION

$$\int_0^1 \int_{1-y}^1 e^y\, dx\, dy = \int_0^1 \left(x e^y \Big|_{x=1-y}^{x=1} \right) dy$$

$$= \int_0^1 [e^y - (1-y)e^y]\, dy$$

$$= \int_0^1 y e^y\, dy$$

$$= ye^y \Big|_0^1 - \int_0^1 e^y \, dy \qquad \text{(integration by parts)}$$

$$= (ye^y - e^y) \Big|_0^1$$

$$= 1$$

Problems

Natural resources

1. Suppose R is the region within the boundary of a certain national forest that contains 600,000 trees per square mile. Express the total number of trees in the forest as a double integral. (Assume x and y are measured in miles.)

Mass

2. Suppose mass is distributed on a region R in the xy plane so that the density (mass per unit area) at the point (x, y) is $f(x, y)$. Express the total mass of R as a double integral.

Mass

3. Suppose mass is distributed on a region R in the xy plane so that the density at any point is equal to the distance from that point to the origin $(0, 0)$. Express the total mass of R as a double integral.

Population

4. Suppose R is the region within the boundary of a certain city. Let the origin $(0, 0)$ denote the city center, and suppose the population density r miles from the city center is $12e^{-0.07r}$ thousand people per square mile. Express the total population of the city as a double integral. (*Hint:* First express the population density as a function of x and y, where x and y are measured in miles.)

Area

5. Suppose R is a region in the xy plane. Express the area of R as a double integral.

Water consumption

6. Suppose R is the region within the boundary of a certain suburban community, and let $f(x, y)$ denote the housing density (homes per square mile) at the point (x, y). If each household consumes 1,200 cubic feet of water per month, express the total water consumption in the community as a double integral. (Assume x and y are measured in miles.)

Property tax

7. Suppose R is the region within the boundary of a certain county, and let $f(x, y)$ denote the housing density (homes per square mile) at the point (x, y). If the property tax on each home in the county is $1,400 per year, express the total property tax collected in the county as a double integral. (Assume x and y are measured in miles.)

In Problems 8 through 25, evaluate the given iterated integral.

8. $\displaystyle\int_1^2 \int_0^1 x^2 y \, dy \, dx$

9. $\displaystyle\int_0^1 \int_1^2 x^2 y \, dx \, dy$

10. $\displaystyle\int_2^3 \int_{-1}^1 (x + 2y) \, dy \, dx$

11. $\displaystyle\int_0^{\ln 2} \int_{-1}^0 2x e^y \, dx \, dy$

12. $\displaystyle\int_0^1 \int_0^1 x^2 e^{xy} \, dy \, dx$

13. $\displaystyle\int_1^3 \int_0^1 \frac{2xy}{x^2 + 1} \, dx \, dy$

14. $\displaystyle\int_0^1 \int_1^5 y\sqrt{1 - y^2} \, dx \, dy$

15. $\displaystyle\int_0^4 \int_0^{\sqrt{x}} x^2 y \, dy \, dx$

16. $\displaystyle\int_0^1 \int_{x^2}^x 2xy \, dy \, dx$

17. $\displaystyle\int_0^1 \int_{y-1}^{1-y} (2x + y) \, dx \, dy$

18. $\displaystyle\int_0^1 \int_{e^y}^e \frac{1}{x} \, dx \, dy$

19. $\displaystyle\int_0^2 \int_{x^2}^4 x e^y \, dy \, dx$

20. $\displaystyle\int_0^2 \int_0^{\sqrt{4-y^2}} y \, dx \, dy$

21. $\displaystyle\int_0^1 \int_0^x \frac{2y}{x^3 + 1} \, dy \, dx$

22. $\displaystyle\int_1^2 \int_x^{x^2} e^{y/x} \, dy \, dx$

23. $\displaystyle\int_0^1 \int_0^x x e^y \, dy \, dx$

24. $\displaystyle\int_1^e \int_0^{\ln x} 2 \, dy \, dx$

25. $\displaystyle\int_0^1 \int_y^1 y e^{x+y} \, dx \, dy$

2 FINDING LIMITS OF INTEGRATION

Two types of multiple integrals were introduced in Section 1. The double integral was defined as the limit of a sum and turned out to be well-suited for the analysis of practical problems. The iterated integral was defined in terms of partial antidifferentiation and was relatively easy to evaluate. This section shows the connection between these two types of integrals. In particular, you will learn how to find the limits of integration that correspond to a region R so that the double integral over R can be evaluated as an iterated integral.

The limits of integration for a region come from inequalities that define the region. The procedure will be described for regions of two fundamental types. More complicated regions of integration can usually be divided into two or more of these elementary regions.

Regions described in terms of vertical cross sections

The region R in Figure 2.1 is bounded below by the graph of the function $y = g_1(x)$ and above by the graph of the function $y = g_2(x)$, and it extends from $x = a$ on the left to $x = b$ on the right. The inequalities

$$a \le x \le b \qquad \text{and} \qquad g_1(x) \le y \le g_2(x)$$

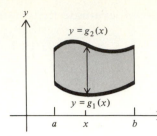

Figure 2.1 Region described by $a \le x \le b$ and $g_1(x) \le y \le g_2(x)$.

can be used to describe such a region. The first inequality specifies the interval in which x must lie, and the second indicates the lower and upper bounds of the vertical cross section of R for each x in this interval. Roughly speaking, the inequalities state that "y goes from $g_1(x)$ to $g_2(x)$ for each x between a and b."

EXAMPLE 2.1

Let R be the region bounded by the curve $y = x^2$ and the line $y = 2x$. Use inequalities to describe R in terms of its vertical cross sections.

SOLUTION

Begin with a sketch of the curve and line as shown in Figure 2.2, iden-

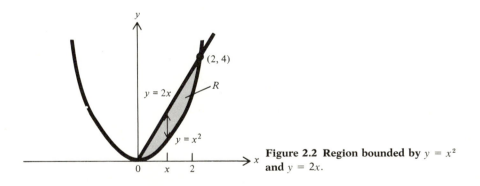

Figure 2.2 Region bounded by $y = x^2$ and $y = 2x$.

tify the region R, and, for reference, draw a vertical cross section. Solve the equations $y = x^2$ and $y = 2x$ simultaneously to find the points of intersection, $(0, 0)$ and $(2, 4)$. Observe that in the region R, the variable x takes on all values from $x = 0$ to $x = 2$ and that for each such value of x, the vertical cross section is bounded below by $y = x^2$ and above by $y = 2x$. Hence R can be described by the inequalities

$$0 \le x \le 2 \quad \text{and} \quad x^2 \le y \le 2x$$

Regions described in terms of horizontal cross sections

The region R in Figure 2.3 is bounded on the left by the graph of $x = h_1(y)$ and on the right by the graph of $x = h_2(y)$, and it extends from $y = c$ on the bottom to $y = d$ on top. It can be described by the inequalities

$$c \le y \le d \quad \text{and} \quad h_1(y) \le x \le h_2(y)$$

where the first inequality specifies the interval in which y must lie, and the second indicates the left-hand and right-hand bounds of a horizontal cross section. Roughly speaking, the inequalities state that "x goes from $h_1(y)$ to $h_2(y)$ for each y between c and d."

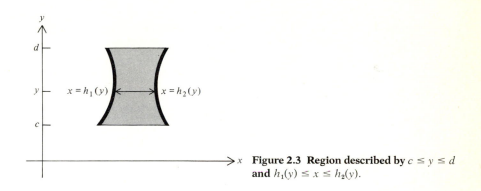

Figure 2.3 Region described by $c \le y \le d$ and $h_1(y) \le x \le h_2(y)$.

EXAMPLE 2.2

Let R be the region from Example 2.1 bounded by the curve $y = x^2$ and the line $y = 2x$. Use inequalities to describe R in terms of its horizontal cross sections.

SOLUTION

As in Example 2.1, sketch the region and find the points of intersection of the line and curve, but this time draw a horizontal cross section (Figure 2.4).

In R the variable y takes on all values from $y = 0$ to $y = 4$, and for each such value of y, the horizontal cross section extends from the line $y = 2x$ on the left to the curve $y = x^2$ on the right. Since the equation of the line can be rewritten as $x = \frac{1}{2}y$ and the equation of the curve as $x = \sqrt{y}$, the inequalities describing R in terms of its horizontal cross sections are

$$0 \le y \le 4 \quad \text{and} \quad \frac{1}{2}y \le x \le \sqrt{y}$$

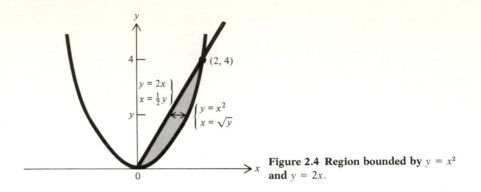

Figure 2.4 Region bounded by $y = x^2$ and $y = 2x$.

Evaluation of double integrals

To evaluate a double integral over a region of one of these types, you use an iterated integral whose limits of integration come from the inequalities describing the region. Here is a more precise statement of the procedure.

Evaluation of double integrals by iteration

> **If R can be described by the inequalities**
>
> $$a \leq x \leq b \qquad \textbf{and} \qquad g_1(x) \leq y \leq g_2(x)$$
>
> **then** $$\iint\limits_R f(x, y)\, dA = \int_a^b \int_{g_1(x)}^{g_2(x)} f(x, y)\, dy\, dx$$
>
> **If R can be described by the inequalities**
>
> $$c \leq y \leq d \qquad \textbf{and} \qquad h_1(y) \leq x \leq h_2(y)$$
>
> **then** $$\iint\limits_R f(x, y)\, dA = \int_c^d \int_{h_1(y)}^{h_2(y)} f(x, y)\, dx\, dy$$

Notice that in each case, the limits of integration on the inner integral are functions of the *second* variable of integration. (In some cases, one or both of these functions may be constant.) The limits on the outer integral are always constants.

A geometric argument justifying this procedure for determining the limits of integration will be given later in this section. Here are a few examples illustrating its use.

EXAMPLE 2.3

Evaluate $\iint\limits_R 40x^2y\, dA$, where R is the region bounded by the curve $y = \sqrt{x}$ and the line $y = x$.

SOLUTION

From the sketch in Figure 2.5, observe that R can be described in terms of vertical cross sections by the inequalities

$$0 \le x \le 1 \quad \text{and} \quad x \le y \le \sqrt{x}$$

Hence

$$\iint_R 40x^2y \, dA = \int_0^1 \int_x^{\sqrt{x}} 40x^2y \, dy \, dx$$

$$= \int_0^1 \left(20x^2y^2 \Big|_{y=x}^{y=\sqrt{x}} \right) dx$$

$$= \int_0^1 (20x^3 - 20x^4) \, dx$$

$$= (5x^4 - 4x^5) \Big|_0^1$$

$$= 1$$

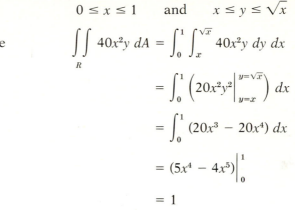

Figure 2.5 Region bounded by $y = \sqrt{x}$ and $y = x$.

Observe that R can also be described in terms of horizontal cross sections by the inequalities

$$0 \le y \le 1 \quad \text{and} \quad y^2 \le x \le y$$

The corresponding iterated integral is

$$\int_0^1 \int_{y^2}^y 40x^2y \, dx \, dy$$

For practice (and to check your answer), evaluate this integral. The answer, of course, should also be 1.

EXAMPLE 2.4

Evaluate $\iint_R (x + y) \, dA$, where R is the triangle with vertices $(0, 0)$, $(0, 1)$, and $(1, 1)$.

SOLUTION

The triangle is sketched in Figure 2.6. It is bounded above by the horizontal line $y = 1$ and below by the line $y = x$ and can be described in terms of vertical cross sections by the inequalities

$$0 \le x \le 1 \quad \text{and} \quad x \le y \le 1$$

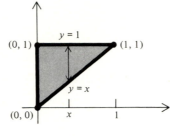

Figure 2.6 Triangle with vertices (0, 0), (0, 1), and (1, 1).

Hence,
$$\iint_R (x + y)\, dA = \int_0^1 \int_x^1 (x + y)\, dy\, dx$$

$$= \int_0^1 \left[\left(xy + \frac{1}{2} y^2 \right) \Big|_{y=x}^{y=1} \right] dx$$

$$= \int_0^1 \left[\left(x + \frac{1}{2} \right) - \left(x^2 + \frac{1}{2} x^2 \right) \right] dx$$

$$= \int_0^1 \left(\frac{1}{2} + x - \frac{3}{2} x^2 \right) dx$$

$$= \left(\frac{1}{2} x + \frac{1}{2} x^2 - \frac{1}{2} x^3 \right) \Big|_0^1$$

$$= \frac{1}{2}$$

For practice, check this answer by evaluating the corresponding iterated integral with the order of integration reversed.

Integration over more complex regions Many regions that are not of one of the fundamental types can be divided into two or more subregions that are. The next example illustrates how to integrate over such a region.

EXAMPLE 2.5

Evaluate $\iint\limits_{R} 1 \, dA$, where R is the region in the first quadrant that lies

under the curve $y = \dfrac{1}{x}$ and is bounded by this curve and the lines

$y = x$, $y = 0$, and $x = 2$.

SOLUTION

The region is sketched in Figure 2.7. Observe that to the left of $x = 1$, vertical cross sections of R are bounded above by the line $y = x$, while to the right of $x = 1$, they are bounded above by the curve $y = \dfrac{1}{x}$. This prevents a simple description of R in terms of vertical cross sections and suggests that R be broken into two subregions, R_1 and R_2, as shown in Figure 2.7.

R_1 can be described by the inequalities

$$0 \le x \le 1 \qquad \text{and} \qquad 0 \le y \le x$$

while R_2 can be described by the inequalities

$$1 \le x \le 2 \qquad \text{and} \qquad 0 \le y \le \frac{1}{x}$$

Hence, $\qquad \iint\limits_{R} 1 \, dA = \iint\limits_{R_1} 1 \, dA + \iint\limits_{R_2} 1 \, dA$

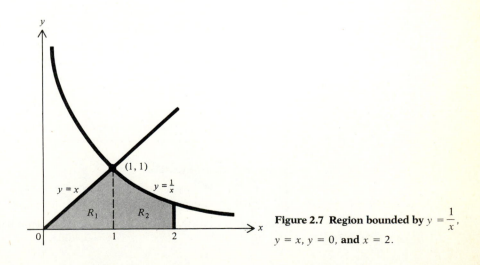

Figure 2.7 Region bounded by $y = \dfrac{1}{x}$, $y = x$, $y = 0$, and $x = 2$.

$$= \int_0^1 \int_0^x 1 \, dy \, dx + \int_1^2 \int_0^{1/x} 1 \, dy \, dx$$

$$= \int_0^1 x \, dx + \int_1^2 \frac{1}{x} \, dx$$

$$= \frac{1}{2} x^2 \Big|_0^1 + \ln x \Big|_1^2$$

$$= \frac{1}{2} + \ln 2$$

For practice, do the problem again, this time using horizontal cross sections. Notice that this solution also involves the use of two iterated integrals.

Selecting the order of integration

Before evaluating a double integral, it is a good idea to take a moment to decide which of the two possible orders of integration will lead to the simpler calculation. A wise choice of order can result in substantial savings of time and effort.

EXAMPLE 2.6

Set up the iterated integral you would prefer to use to evaluate $\iint_R y \, dA$, where R is the region bounded by $y = \ln x, y = 1, x = 0$, and $y = 0$.

SOLUTION

The region is sketched in Figure 2.8. To the left of $x = 1$, vertical cross sections are bounded below by the x axis ($y = 0$), while to the right of $x = 1$, they are bounded below by the curve $y = \ln x$. As a result, the use of vertical cross sections will involve the two iterated integrals

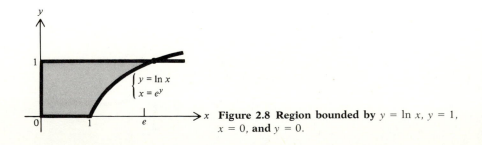

Figure 2.8 **Region bounded by** $y = \ln x, y = 1,$ $x = 0,$ **and** $y = 0.$

$$\iint\limits_{R} y \; dA = \int_{0}^{1} \int_{0}^{1} y \; dy \; dx + \int_{1}^{e} \int_{\ln x}^{1} y \; dy \; dx$$

the second of which will eventually require the integration of $(\ln x)^2$, which is not easy.

On the other hand, the horizontal cross sections are always bounded on the left by the y axis ($x = 0$) and on the right by the curve $y = \ln x$, whose equation can be rewritten as $x = e^y$. The inequalities that describe R in terms of its horizontal cross sections are

$$0 \leq y \leq 1 \qquad \text{and} \qquad 0 \leq x \leq e^y$$

and the corresponding integral is

$$\iint\limits_{R} y \; dA = \int_{0}^{1} \int_{0}^{e^y} y \; dx \; dy$$

Clearly this iterated integral, in which x is the first variable of integration, is preferable to the pair of iterated integrals that are needed if y is to be the first variable of integration.

For practice, evaluate $\iint\limits_{R} y \; dA$ using both methods. Each calculation will involve integration by parts. The answer is 1.

In the next example, the choice of the order of integration is critical. The wrong choice leads to an integral that is impossible to evaluate by elementary methods.

EXAMPLE 2.7

Evaluate $\iint\limits_{R} e^{y^2} \; dA$, where R is the triangle bounded by the lines $y = \dfrac{1}{2} x$, $y = 1$, and $x = 0$.

SOLUTION

The region is sketched in Figure 2.9. The two possible iterated integrals are

$$\int_{0}^{2} \int_{x/2}^{1} e^{y^2} \; dy \; dx \qquad \text{and} \qquad \int_{0}^{1} \int_{0}^{2y} e^{y^2} \; dx \; dy$$

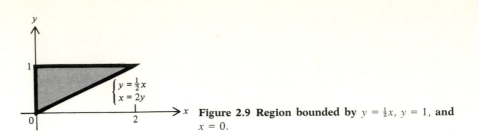

Figure 2.9 Region bounded by $y = \frac{1}{2}x$, $y = 1$, **and** $x = 0$.

The first integral, in which y is the first variable of integration, requires the evaluation of $\int e^{y^2}\, dy$, which cannot be done by elementary methods. The second, on the other hand, is easy,

$$\int_0^1 \int_0^{2y} e^{y^2}\, dx\, dy = \int_0^1 \left(xe^{y^2}\Big|_{x=0}^{x=2y} \right) dy = \int_0^1 2ye^{y^2}\, dy = e^{y^2}\Big|_0^1 = e - 1$$

Why double integrals can be evaluated by iteration

Double integrals can be evaluated as iterated integrals because, for nonnegative functions, both give the volume under the graph of the function. Since both the double integral and the corresponding iterated integral are equal to the volume, they must be equal to each other.

In Section 1 you saw why a double integral can be interpreted as the volume under a surface. Here is an argument that establishes the same geometric interpretation for iterated integrals.

Suppose $f(x, y) \geq 0$ on R, where R is a region described by the inequalities $a \leq x \leq b$ and $g_1(x) \leq y \leq g_2(x)$. Let S be the solid bounded above by the surface $z = f(x, y)$ and below by R. The goal is to show that

$$\text{Volume of } S = \int_a^b \int_{g_1(x)}^{g_2(x)} f(x, y)\, dy\, dx$$

The first step is to show that

$$\text{Volume of } S = \int_a^b A(x)\, dx$$

where $A(x)$ is the area of the cross section (Figure 2.10a) formed by slicing S with a plane parallel to the yz plane.

To see why the volume of S is the integral of the cross-sectional area, imagine that the interval $a \leq x \leq b$ is divided into n subintervals of width Δx, with x_j denoting the start of the jth subinterval. These subintervals determine n vertical slabs, each parallel to the yz plane. The jth slab is shown in Figure 2.10b. Then

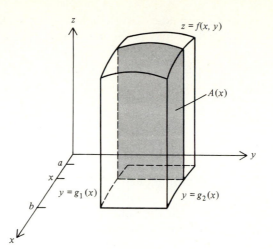

(a)

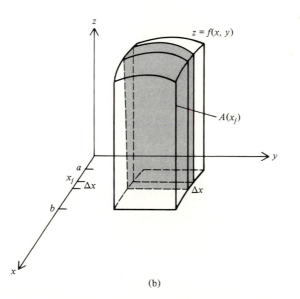

(b)

Figure 2.10 (a) Cross-sectional slice. (b) The jth slab.

$$\text{Volume of } j\text{th slab} \approx A(x_j)\,\Delta x$$

and

$$\text{Volume of } S \approx \sum_{j=1}^{n} A(x_j)\,\Delta x$$

The approximation improves as n increases, and it follows that

$$\text{Volume of } S = \lim_{n \to \infty} \sum_{j=1}^{n} A(x_j)\,\Delta x = \int_{a}^{b} A(x)\,dx$$

The second step in the argument is to observe that the cross-sectional area $A(x)$ is the area under the curve $z = f(x, y)$ for fixed x, between $y = g_1(x)$ and $y = g_2(x)$. (To see this, look at Figure 2.10a again.) Since the area under a curve is given by the definite integral, it follows that

$$A(x) = \int_{g_1(x)}^{g_2(x)} f(x, y) \, dy$$

Combining these two steps you get

$$\text{Volume of } S = \int_a^b A(x) \, dx = \int_a^b \int_{g_1(x)}^{g_2(x)} f(x, y) \, dy \, dx$$

which says that the volume under the surface is equal to the iterated integral. (A similar argument shows that the volume is also equal to the iterated integral with the order of integration reversed.)

Problems In Problems 1 through 6, use inequalities to describe R in terms of its vertical and horizontal cross sections.

1. R is the region bounded by $y = x^2$ and $y = 3x$.

2. R is the region bounded by $y = \sqrt{x}$ and $y = x^2$.

3. R is the rectangle with vertices $(-1, 1)$, $(2, 1)$, $(2, 2)$, and $(-1, 2)$.

4. R is the triangle with vertices $(1, 0)$, $(1, 1)$, and $(2, 0)$.

5. R is the region bounded by $y = \ln x$, $y = 0$, and $x = e$.

6. R is the region bounded by $y = e^x$, $y = 2$, and $x = 0$.

In Problems 7 through 22, evaluate the given double integral for the specified region R.

7. $\iint_R 3xy^2 \, dA$, where R is the rectangle bounded by the lines $x = -1$, $x = 2$, $y = -1$, and $y = 0$.

8. $\iint_R (x + 2y) \, dA$, where R is the triangle with vertices $(0, 0)$, $(1, 0)$, and $(0, 2)$.

9. $\iint_R xe^y \, dA$, where R is the triangle with vertices $(0, 0)$, $(1, 0)$, and $(1, 1)$.

10. $\iint\limits_{R} 48xy\, dA$, where R is the region bounded by $y = x^3$ and $y = \sqrt{x}$.

11. $\iint\limits_{R} (2y - x)\, dA$, where R is the region bounded by $y = x^2$ and $y = 2x$.

12. $\iint\limits_{R} 12x\, dA$, where R is the region bounded by $y = x^2$ and $y = 6 - x$.

13. $\iint\limits_{R} 2\, dA$, where R is the region bounded by $y = \dfrac{16}{x}$, $y = x$, and $x = 2$.

14. $\iint\limits_{R} y\, dA$, where R is the region bounded by $y = \sqrt{x}$, $y = 2 - x$, and $y = 0$.

15. $\iint\limits_{R} 4x\, dA$, where R is the region bounded by $y = 4 - x^2$, $y = 3x$, and $x = 0$.

16. $\iint\limits_{R} 4x\, dA$, where R is the region in the first quadrant bounded by $y = 4 - x^2$, $y = 3x$, and $y = 0$.

17. $\iint\limits_{R} (2x + 1)\, dA$, where R is the triangle with vertices $(-1, 0)$, $(1, 0)$, and $(0, 1)$.

18. $\iint\limits_{R} 2x\, dA$, where R is the region bounded by $y = \dfrac{1}{x^2}$, $y = x$, $x = 2$, and $y = 0$.

19. $\iint\limits_{R} \dfrac{1}{y^2 + 1}\, dA$, where R is the triangle bounded by the lines $y = \dfrac{1}{2}x$, $y = -x$, and $y = 2$.

20. $\iint\limits_{R} e^{y^3} \, dA$, where R is the region bounded by $y = \sqrt{x}$, $y = 1$, and $x = 0$.

21. $\iint\limits_{R} 12x^2 e^{y^2} \, dA$, where R is the region in the first quadrant bounded by $y = x^3$ and $y = x$.

22. $\iint\limits_{R} y \, dA$, where R is the region bounded by $y = \ln x$, $y = 0$, and $x = e$.

In Problems 23 through 34, sketch the region of integration for the given iterated integral, and set up an equivalent iterated integral with the order of integration reversed. (In some cases, two or more iterated integrals may be needed.)

23. $\displaystyle\int_0^2 \int_0^{4-x^2} f(x, y) \, dy \, dx$

24. $\displaystyle\int_0^1 \int_0^{2y} f(x, y) \, dx \, dy$

25. $\displaystyle\int_0^1 \int_{x^3}^{\sqrt{x}} f(x, y) \, dy \, dx$

26. $\displaystyle\int_0^4 \int_{y/2}^{\sqrt{y}} f(x, y) \, dx \, dy$

27. $\displaystyle\int_1^{e^2} \int_{\ln x}^2 f(x, y) \, dy \, dx$

28. $\displaystyle\int_0^{\ln 3} \int_{e^x}^3 f(x, y) \, dy \, dx$

29. $\displaystyle\int_{-1}^1 \int_{x^2+1}^2 f(x, y) \, dy \, dx$

30. $\displaystyle\int_{-1}^1 \int_{-\sqrt{y+1}}^{\sqrt{y+1}} f(x, y) \, dx \, dy$

31. $\displaystyle\int_0^1 \int_x^{2-x} f(x, y) \, dy \, dx$

32. $\displaystyle\int_0^3 \int_{y/3}^{\sqrt{4-y}} f(x, y) \, dx \, dy$

33. $\displaystyle\int_{-3}^2 \int_{x^2}^{6-x} f(x, y) \, dy \, dx$

34. $\displaystyle\int_2^3 \int_x^{16/x} f(x, y) \, dy \, dx$

3 APPLICATIONS OF THE DOUBLE INTEGRAL

In this section, you will see a sampling of applications of the double integral. Most are derived from the definition of the integral as the limit of a sum using a standard strategy. Here is a summary of that strategy.

How to apply the definition of the double integral

> **The following steps may lead to a double integral giving a numerical quantity associated with a plane region R.**
>
> **Step 1. Approximate R by n rectangles.**
> **Step 2. Estimate the quantity associated with one of the rectangles.**

> **Step 3.** **Add the estimates from Step 2 to estimate the quantity associated with R.**
> **Step 4.** **Take the limit as n increases without bound to pass from the estimate to the exact value.**

The area of a plane region

The area of a region R in the xy plane can be computed as the double integral over R of the constant function $f(x, y) = 1$. To see this, imagine that R is approximated by n rectangles of area ΔA, as shown in Figure 3.1. The area of R is approximately the sum of the areas of the

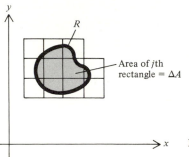

Figure 3.1 Approximation of R by rectangles.

rectangles, and the approximation improves as n increases. Since the area of each rectangle is ΔA, it follows that

$$\text{Area of } R = \lim_{n \to \infty} \sum_{j=1}^{n} \Delta A = \lim_{n \to \infty} \sum_{j=1}^{n} 1 \, \Delta A = \iint_R 1 \, dA$$

Area formula

> **The area of a region R in the xy plane is given by the formula**
> $$\text{Area of } R = \iint_R 1 \, dA$$

Here are two examples.

EXAMPLE 3.1

Find the area of the region R bounded by the curves $y = x^3$ and $y = x^2$.

SOLUTION

The region R is shown in Figure 3.2. Using the area formula you get

$$\text{Area of } R = \iint_R 1 \, dA = \int_0^1 \int_{x^3}^{x^2} 1 \, dy \, dx$$

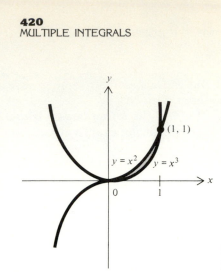

Figure 3.2 Region bounded by $y = x^3$ **and** $y = x^2$.

$$= \int_0^1 \left(y \Big|_{y=x^3}^{y=x^2} \right) dx$$

$$= \int_0^1 (x^2 - x^3) \, dx$$

$$= \left(\frac{1}{3} x^3 - \frac{1}{4} x^4 \right) \Big|_0^1$$

$$= \frac{1}{12}$$

The area calculated in Example 3.1 using a double integral is the same as that calculated in Example 2.5 of Chapter 6 using a single integral. Compare the two solutions. Observe, in particular, that after the integration with respect to y, the double integral in Example 3.1 reduces to the single integral that was used in Example 2.5.

EXAMPLE 3.2

Find the area of the region R in the first quadrant bounded by $y = x^3$, $y = x - 2$, $y = 0$, and $y = 1$.

SOLUTION

From the sketch in Figure 3.3 you see that the use of vertical cross sections to describe R will lead to three iterated integrals. To avoid this, use horizontal cross sections.

$$\text{Area of } R = \int_0^1 \int_{y^{1/3}}^{y+2} 1 \, dx \, dy$$

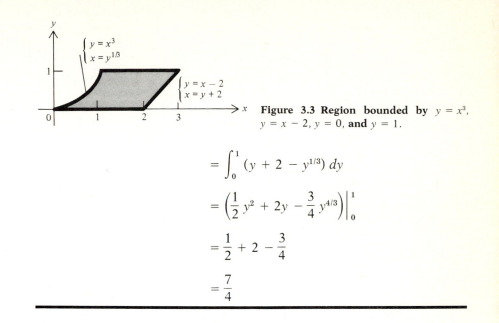

Figure 3.3 Region bounded by $y = x^3$, $y = x - 2$, $y = 0$, and $y = 1$.

$$= \int_0^1 (y + 2 - y^{1/3}) \, dy$$

$$= \left(\frac{1}{2} y^2 + 2y - \frac{3}{4} y^{4/3} \right) \Big|_0^1$$

$$= \frac{1}{2} + 2 - \frac{3}{4}$$

$$= \frac{7}{4}$$

The volume under a surface As you saw in Section 1, the double integral over R of a nonnegative function $f(x, y)$ gives the volume under the graph of f and above R.

Volume formula

> **If $f(x, y) \geq 0$ for all points (x, y) in R, and if S is the solid bounded above by the surface $z = f(x, y)$ and below by R, then**
>
> $$\text{Volume of } S = \iint\limits_R f(x, y) \, dA$$

Here is an example illustrating the use of this formula.

EXAMPLE 3.3

Find the volume under the surface $z = e^{-x} e^{-y}$ and above the triangle with vertices $(0, 0)$, $(1, 0)$, and $(0, 1)$.

SOLUTION

The triangle is shown in Figure 3.4a. The surface under which you are to find the volume is the graph of the function $f(x, y) = e^{-x} e^{-y}$, which is nonnegative for all values of x and y. Hence,

$$\text{Volume} = \int_0^1 \int_0^{1-x} e^{-x} e^{-y} \, dy \, dx$$

$$= \int_0^1 \left(-e^{-x}e^{-y} \Big|_{y=0}^{y=1-x} \right) dx$$

$$= \int_0^1 (-e^{-x}e^{x-1} + e^{-x}) \, dx$$

$$= \int_0^1 (-e^{-1} + e^{-x}) \, dx$$

$$= (-xe^{-1} - e^{-x}) \Big|_0^1$$

$$= 1 - 2e^{-1}$$

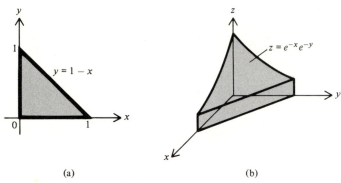

(a)   (b)

Figure 3.4 Triangle and solid for Example 3.3.

Notice that you do not have to draw the three-dimensional region to compute its volume (although you do have to check that the function is nonnegative for all points in the region of integration). The picture of the solid in Figure 3.4b is shown here simply to help you visualize the situation.

The average value of a function   In Chapter 6, Section 4, you saw that the average value of a continuous function $f(x)$ over an interval $a \le x \le b$ is given by the integral formula

$$\text{Average value} = \frac{1}{b-a} \int_a^b f(x) \, dx$$

This says that to find the average value of a function of one variable over an interval, you integrate the function over the interval and divide by the length of the interval. The two-variable procedure is similar. In particular, to find the average value of a function of two variables over a region, you integrate the function over the region and divide by the area of the region.

Average value formula

The average value of a continuous function $f(x, y)$ over a region R is given by the formula

$$\text{Average value} = \frac{1}{\text{area of } R} \iint\limits_{R} f(x, y) \, dA$$

EXAMPLE 3.4

Suppose R represents the surface of a lake and $f(x, y)$ the depth of the lake at the point (x, y). Find an expression for the average depth of the lake.

SOLUTION

According to the average value formula,

$$\text{Average depth} = \frac{1}{\text{area of } R} \iint\limits_{R} f(x, y) \, dA$$

Moreover, the area of R is itself a double integral,

$$\text{Area of } R = \iint\limits_{R} 1 \, dA$$

Hence,

$$\text{Average depth} = \frac{\iint\limits_{R} f(x, y) \, dA}{\iint\limits_{R} 1 \, dA}$$

EXAMPLE 3.5

Find the average value of the function $f(x, y) = xe^{y}$ on the triangle with vertices $(0, 0)$, $(1, 0)$, and $(1, 1)$.

SOLUTION

The triangle is shown in Figure 3.5. Its area can be calculated without integration using the familiar formula from plane geometry:

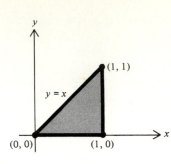

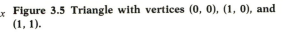

Figure 3.5 Triangle with vertices (0, 0), (1, 0), and (1, 1).

$$\text{Area} = \frac{1}{2}(\text{base})(\text{height}) = \frac{1}{2}(1)(1) = \frac{1}{2}$$

Hence,

$$\text{Average value} = \frac{1}{1/2}\int_0^1 \int_0^x xe^y \, dy \, dx$$

$$= 2\int_0^1 \left(xe^y \Big|_{y=0}^{y=x} \right) dx$$

$$= 2\int_0^1 (xe^x - x) \, dx$$

$$= 2\left(xe^x - e^x - \frac{1}{2}x^2 \right)\Big|_0^1$$

$$= 2\left[\left(e - e - \frac{1}{2} \right) - (-1) \right]$$

$$= 1$$

Why the average-value formula works

Imagine that the region R is approximated in the usual way by n rectangles of area ΔA and let (x_j, y_j) be a point in the jth rectangle. The numerical average of the n function values $f(x_1, y_1)$, $f(x_2, y_2)$, . . . , $f(x_n, y_n)$ is

$$\frac{1}{n}\sum_{j=1}^n f(x_j, y_j)$$

which can be rewritten as

$$\frac{1}{n\,\Delta A}\sum_{j=1}^n f(x_j, y_j)\,\Delta A$$

Since R is approximated by n rectangles of area ΔA, the term $n\,\Delta A$ is approximately the area of R. Hence the average of the n function values is approximately

$$\frac{1}{\text{Area of } R} \sum_{j=1}^{n} f(x_j, y_j)\ \Delta A$$

As n increases and the size of each rectangle decreases, the numerical average of the n function values becomes increasingly sensitive to fluctuations in f, thereby approximating with increasing accuracy the average value of f over the entire region R. Hence,

$$\begin{aligned}
\text{Average value of}\atop f(x, y) \text{ over } R &= \lim_{n \to \infty} \left[\frac{1}{\text{area of } R} \sum_{j=1}^{n} f(x_j, y_j)\ \Delta A \right] \\
&= \frac{1}{\text{area of } R} \iint\limits_{R} f(x, y)\ dA
\end{aligned}$$

Joint probability density functions One of the most important applications of integration in the social, managerial, and life sciences is the computation of probabilities. The technique of integrating probability density functions to find probabilities was introduced in Chapter 6, Section 2, and discussed in more detail in Chapter 7, Section 3. Recall that a probability density function for a single variable x is a nonnegative function $f(x)$ such that the probability that x is between a and b is given by the formula

$$P(a \le x \le b) = \int_a^b f(x)\ dx$$

In geometric terms, the probability $P(a \le x \le b)$ is the area under the graph of f from $x = a$ to $x = b$ (Figure 3.6a).

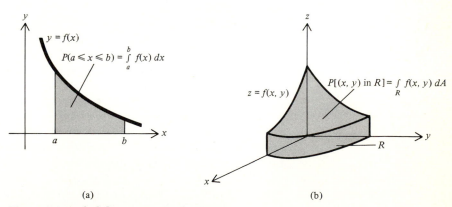

(a) $y = f(x)$; $P(a \le x \le b) = \int_a^b f(x)\ dx$

(b) $z = f(x, y)$; $P[(x, y) \text{ in } R] = \int_R f(x, y)\ dA$

Figure 3.6 Probability as area and volume.

In situations involving two variables, you compute probabilities by evaluating double integrals of a two-variable density function. In particular, you integrate a **joint probability density function,** which is a nonnegative function $f(x, y)$ such that the probability that x is between a and b and y is between c and d is given by the formula

$$P(a \leq x \leq b \text{ and } c \leq y \leq d) = \int_c^d \int_a^b f(x, y) \, dx \, dy$$

$$= \int_a^b \int_c^d f(x, y) \, dy \, dx$$

More generally, the probability that the ordered pair (x, y) lies in a region R is given by the formula

$$P[(x, y) \text{ in } R] = \iint_R f(x, y) \, dA$$

In geometric terms, the probability that (x, y) is in R is the volume under the graph of f above the region R (Figure 3.6b).

The techniques for constructing joint probability density functions from experimental data are beyond the scope of this book and are discussed in most probability and statistics texts. The use of double integrals to compute probabilities once the appropriate density functions are known is illustrated in the next two examples.

EXAMPLE 3.6

Smoke detectors manufactured by a certain firm contain two independent circuits, one manufactured at the firm's California plant and the other at the firm's plant in Ohio. Reliability studies suggest that if x denotes the life span (in years) of a randomly selected circuit from the California plant and y the life span (in years) of a randomly selected circuit from the Ohio plant, the joint probability density function for x and y is $f(x, y) = e^{-x}e^{-y}$. If the smoke detector will operate as long as either of its circuits is operating, find the probability that a randomly selected smoke detector will fail within 1 year.

SOLUTION

Since the smoke detector will operate as long as either of its circuits is operating, it will fail within 1 year if and only if *both* of its circuits fail within 1 year. The desired probability is therefore the probability that both $0 \leq x \leq 1$ and $0 \leq y \leq 1$. The points (x, y) for which both these inequalities hold form the square R shown in Figure 3.7. The corresponding probability is the double integral of the density function f over this region R. That is,

$$P(0 \leq x \leq 1 \text{ and } 0 \leq y \leq 1) = \int_0^1 \int_0^1 e^{-x}e^{-y} \, dx \, dy$$

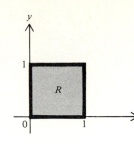

Figure 3.7 **Square consisting of all points** (x, y) **for which** $0 \le x \le 1$ **and** $0 \le y \le 1$.

$$= \int_0^1 \left(-e^{-x} e^{-y} \Big|_{x=0}^{x=1} \right) dy$$

$$= \int_0^1 -(e^{-1} - 1) e^{-y} \, dy$$

$$= (e^{-1} - 1) e^{-y} \Big|_0^1$$

$$= (e^{-1} - 1)^2$$

$$= 0.3996$$

EXAMPLE 3.7

Suppose x denotes the time (in minutes) that a person sits in the waiting room of a certain dentist and y the time (in minutes) that a person stands in line at a certain bank. You have an appointment with the dentist, after which you are planning to cash a check at the bank. If $f(x, y)$ is the joint probability density function for x and y, write down an iterated integral that gives the probability that your *total* waiting time will be no more than 20 minutes.

SOLUTION

The goal is to find the probability that $x + y \le 20$. The points (x, y) for which $x + y \le 20$ lie on or below the line $x + y = 20$. Moreover, since x and y stand for nonnegative quantities, only those points in the first quadrant are meaningful in this particular context. The problem, then, is to find the probability that a randomly selected point (x, y) lies in R, where R is the region in the first quadrant bounded by the line $x + y = 20$ and the coordinate axes (Figure 3.8). This probability is given by the double integral of the density function f over the region R. That is,

$$P[(x, y) \text{ in } R] = \iint_R f(x, y) \, dA = \int_0^{20} \int_0^{20-x} f(x, y) \, dy \, dx$$

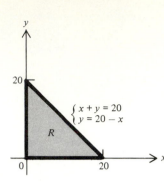

Figure 3.8 Triangle consisting of all points (x, y) **for which** $x + y \le 20$.

Fluid flow   The following example illustrates how you can use a double integral to compute the rate at which fluid flows through a plane region. This procedure generalizes the one-variable technique used in Example 4.6 of Chapter 6 to find the rate of flow of blood through an artery.

EXAMPLE 3.8

Suppose R represents a flat filter through which air is flowing. The speed of the air varies, depending on the portion of the filter through which it is flowing. Suppose $f(x, y)$ is the speed (in feet per minute) of the air that passes through the point (x, y) on the filter, and assume that the flow of air is always perpendicular to the filter. Find an expression for the rate (in cubic feet per minute) at which the air flows through the filter.

SOLUTION

Approximate the filter R by n rectangles of area ΔA, and let (x_j, y_j) be a point in the jth rectangle.

To estimate the volume (cubic feet) of air that passes through the jth rectangle in 1 minute, multiply the area of the rectangle (square feet) by the speed of the air (feet per minute). The situation is illustrated in Figure 3.9. Thus,

$$\text{Rate of flow through } j\text{th rectangle} \approx f(x_j, y_j) \, \Delta A \quad \text{cubic feet per minute}$$

and

$$\text{Rate of flow through } R \approx \sum_{j=1}^{n} f(x_j, y_j) \, \Delta A \quad \text{cubic feet per minute}$$

The approximation improves as n increases, and so

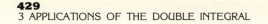

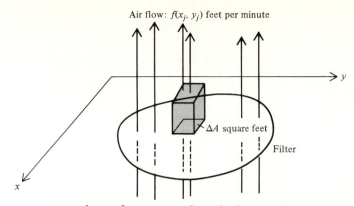

Figure 3.9 Volume of air passing through jth rectangle in 1 minute.

$$\text{Rate of flow through } R = \lim_{n \to \infty} \sum_{j=1}^{n} f(x_j, y_j) \, \Delta A$$

$$= \iint_R f(x, y) \, dA$$

That is, the rate at which volume passes through R is the double integral over R of the linear speed of the flow.

Problems   In Problems 1 through 10, use a double integral to find the area of R.

1. R is the triangle with vertices $(-4, 0)$, $(2, 0)$, and $(2, 6)$.

2. R is the triangle with vertices $(0, -1)$, $(-2, 1)$, and $(2, 1)$.

3. R is the region bounded by $y = \frac{1}{2}x^2$ and $y = 2x$.

4. R is the region bounded by $y = \sqrt{x}$ and $y = x^2$.

5. R is the region bounded by $y = x^2 - 4x + 3$ and the x axis.

6. R is the region bounded by $y = x^2 + 6x + 5$ and the x axis.

7. R is the region bounded by $y = \ln x$, $y = 0$, and $x = e$.

8. R is the region bounded by $y = x$, $y = \ln x$, $y = 0$, and $y = 1$.

9. R is the region in the first quadrant bounded by $y = 4 - x^2$, $y = 3x$, and $y = 0$.

10. R is the region in the first quadrant that lies under the curve $y = \dfrac{16}{x}$ and is bounded by this curve and the lines $y = x$, $y = 0$, and $x = 8$.

In Problems 11 through 16, find the volume of the solid bounded above by the surface $z = f(x, y)$ and below by the region R.

11. $f(x, y) = 6 - 2x - y$; R is the rectangle with vertices $(0, 0)$, $(1, 0)$, $(0, 2)$, and $(1, 2)$.

12. $f(x, y) = 2x + y$; R is the triangle bounded by $y = x$, $y = 2 - x$, and the x axis.

13. $f(x, y) = e^{y^2}$; R is the triangle bounded by $y = \frac{1}{2}x$, $x = 0$, and $y = 1$.

14. $f(x, y) = x + 1$; R is the region in the first quadrant bounded by $y = 8 - x^2$, $y = x^2$, and the y axis.

15. $f(x, y) = e^{2x+y}$; R is the triangle with vertices $(0, 0)$, $(1, 0)$, and $(0, 1)$.

16. $f(x, y) = 4xe^y$; R is the triangle bounded by $y = 2x$, $y = 2$, and $x = 0$.

Air pollution 17. Suppose R is the region within the boundary of Los Angeles County and $f(x, y)$ the number of units of carbon monoxide in the air at noon at the point (x, y) in R. Find an expression for the average level of carbon monoxide in the county at noon.

Elevation 18. Suppose the rectangle R in the first quadrant bounded by the coordinate axes and the lines $x = 3$ and $y = 1$ represents the region inside the boundary of a certain national park. Suppose the elevation above sea level at any point (x, y) in the park is $f(x, y) = 900(2x + y^2)$ feet. Find the average elevation in the park.

Property value 19. Suppose the triangle R with vertices $(0, 0)$, $(1, 0)$, and $(1, 1)$ represents the region inside the boundary of a certain rural congressional district. Suppose the property value at any point (x, y) in the district is $f(x, y) = 400xe^{-y}$ dollars per acre. Find the average property value in the district.

In Problems 20 through 25, find the average value of the given function over the specified region R.

20. $f(x, y) = 6xy$; R is the triangle with vertices $(0, 0)$, $(0, 1)$, and $(3, 1)$.

21. $f(x, y) = 3y$; R is the triangle with vertices $(0, 0)$, $(4, 0)$, and $(2, 2)$.

22. $f(x, y) = e^{x^2}$; R is the triangle with vertices $(0, 0)$, $(1, 0)$, and $(1, 1)$.

23. $f(x, y) = x$; R is the region bounded by $y = 4 - x^2$ and the x axis.

24. $f(x, y) = e^{x^3}$; R is the region in the first quadrant bounded by $y = x^2$, $y = 0$, and $x = 1$.

25. $f(x, y) = e^x y^{-1/2}$; R is the region in the first quadrant bounded by $y = x^2$, $x = 0$, and $y = 1$.

Probability   26. Suppose the joint probability density function for the nonnegative variables x and y is $f(x, y) = 2e^{-2x}e^{-y}$. Find the probability that $0 \le x \le 1$ and $1 \le y \le 2$.

Probability   27. Suppose the joint probability density function for the nonnegative variables x and y is $f(x, y) = xe^{-x}e^{-y}$. Find the probability that $0 \le x \le 1$ and $0 \le y \le 2$.

Probability   28. Suppose the joint probability density function for the nonnegative variables x and y is $f(x, y) = 2e^{-2x}e^{-y}$. Find the probability that $x + y \le 1$.

Probability   29. Suppose the joint probability density function for the nonnegative variables x and y is $f(x, y) = xe^{-x}e^{-y}$. Find the probability that $x + y \le 1$.

Health care   30. Suppose x is the length of time (in days) that a person stays in the hospital after abdominal surgery and y the length of time (in days) that a person stays in the hospital after orthopedic surgery. On Monday, the patient in bed 107A undergoes an emergency appendectomy (abdominal surgery), while the patient's roommate in bed 107B undergoes (orthopedic) surgery for the repair of torn knee cartilage. If the joint probability density function for x and y is $f(x, y) = \frac{1}{12}e^{-x/4}e^{-y/3}$, find the probability that both patients will be discharged from the hospital within 3 days.

Warranty protection   31. A certain appliance consisting of two independent electronic components will be usable as long as either one of its components is still operating. The appliance carries a warranty from the manufacturer guaranteeing replacement if the appliance becomes unusable within 1 year of the date of purchase. Let x denote the life span (in years) of the first component and y the life span (in years) of the second, and suppose that the joint probability density function for x and y is $f(x, y) = \frac{1}{4}e^{-x/2}e^{-y/2}$. You purchase one of these appliances, selected at random from the manufacturer's stock. Find the probability that the warranty will expire before your appliance becomes unusable. (*Hint:* You want the probability that the appliance does *not* fail during the first year, which you can compute as 1 minus the probability that it *does* fail during this period.)

Customer service   32. Suppose x denotes the time (in minutes) that a person stands in line at a certain bank and y the duration (in minutes) of a routine transaction at the teller's window. You arrive at the bank to deposit a check. If the joint probability density function for x and y is $f(x, y) = \frac{1}{8}e^{-x/4}e^{-y/2}$, find the probability that you will complete your business at the bank within 8 minutes.

Insurance sales

33. Suppose x denotes the time (in minutes) that a person spends with an agent choosing a life insurance policy and y the time (in minutes) that the agent spends doing the paperwork once the client has decided. You arrange to meet with an insurance agent to buy a life insurance policy. If the joint probability density function for x and y is $f(x, y) = \frac{1}{300}e^{-x/30}e^{-y/10}$, find the probability that the entire transaction will take more than half an hour. (*Hint:* The probability you want is 1 minus the probability that the transaction will take no more than 30 minutes.)

Circulation

34. Suppose R represents the cross section of an artery through which blood is flowing, and let $f(x, y)$ denote the speed (in centimeters per second) at which the blood passes through the point (x, y) in R. Find an expression for the rate (in cubic centimeters per minute) at which blood flows through the artery.

Air purification

35. Suppose R represents a flat filter that absorbs all the ozone from the air passing through it. Let $f(x, y)$ denote the speed (in feet per minute) of the air that flows through the filter at the point (x, y). If the air contains 50 units of ozone per cubic foot and the air flow is always perpendicular to the filter, find an expression for the number of units of ozone collected by the filter in 10 minutes.

CHAPTER SUMMARY AND PROFICIENCY TEST

Important terms, symbols, and formulas

Double integral: $\displaystyle\iint\limits_{R} f(x, y)\, dA = \lim_{n \to \infty} \sum_{j=1}^{n} f(x_j, y_j)\, \Delta A$

Iterated integrals: If R is described by vertical cross sections,

$$\iint\limits_{R} f(x, y)\, dA = \int_{a}^{b} \int_{g_1(x)}^{g_2(x)} f(x, y)\, dy\, dx$$

If R is described by horizontal cross sections,

$$\iint\limits_{R} f(x, y)\, dA = \int_{c}^{d} \int_{h_1(y)}^{h_2(y)} f(x, y)\, dx\, dy$$

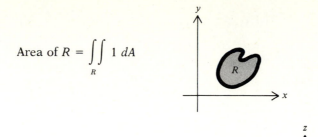

$$\text{Area of } R = \iint_R 1 \, dA$$

$$\text{Volume of } S = \iint_R f(x, y) \, dA$$

$$\text{Average value of } f \text{ over } R = \frac{1}{\text{area of } R} \iint_R f(x, y) \, dA$$

Joint probability density function:

$$P[(x, y) \text{ in } R] = \iint_R f(x, y) \, dA$$

Fluid flow: The rate at which volume passes through R is the double integral over R of the linear speed of the flow.

Proficiency test   In Problems 1 through 5, evaluate the given iterated integral.

1. $\displaystyle\int_0^1 \int_{-2}^0 (2x + 3y) \, dy \, dx$

2. $\displaystyle\int_0^1 \int_0^y x\sqrt{1 - y^3} \, dx \, dy$

3. $\displaystyle\int_0^1 \int_{-x}^x \frac{6xy^2}{x^5 + 1} \, dy \, dx$

4. $\displaystyle\int_0^1 \int_0^2 e^{-x-y} \, dy \, dx$

5. $\displaystyle\int_0^1 \int_0^x xe^{2y} \, dy \, dx$

In Problems 6 through 10, evaluate the given double integral for the specified region R.

6. $\displaystyle\iint_R 6x^2 y \, dA$, where R is the rectangle with vertices $(-1, 0)$, $(2, 0)$, $(2, 3)$, and $(-1, 3)$.

7. $\iint\limits_{R} (x + 2y) \, dA$, where R is the triangle with vertices $(0, 0)$, $(0, 3)$, and $(1, 1)$.

8. $\iint\limits_{R} 40x^2y \, dA$, where R is the region bounded by $y = \sqrt{x}$ and $y = \frac{1}{2}x$.

9. $\iint\limits_{R} 6y^2e^{x^2} \, dA$, where R is the region bounded by $y = \sqrt[3]{x}$ and $y = x$.

10. $\iint\limits_{R} 32xy^3 \, dA$, where R is the region bounded by $y = \dfrac{8}{x}$, $y = \sqrt{x}$, and $y = 1$.

In Problems 11 through 14, sketch the region of integration for the given iterated integral, and set up an equivalent iterated integral (or integrals) with the order of integration reversed.

11. $\displaystyle\int_{0}^{4} \int_{x^2/8}^{\sqrt{x}} f(x, y) \, dy \, dx$ 12. $\displaystyle\int_{0}^{2} \int_{1}^{e^y} f(x, y) \, dx \, dy$

13. $\displaystyle\int_{0}^{2} \int_{x^2}^{8-2x} f(x, y) \, dy \, dx$ 14. $\displaystyle\int_{-3}^{1} \int_{x^2}^{6-x} f(x, y) \, dy \, dx$

15. Find the area of the region bounded by $y = x^2 - 4$ and $y = x - 2$.

16. Find the area of the region bounded by $y = \ln x$, $y = 0$, $y = 1$, and $x = 0$.

17. Find the area of the region in the first quadrant bounded by $y = x^2$, $y = 2 - x$, and $y = 0$.

18. Find the volume under the surface $z = 2xy$ above the triangle with vertices $(0, 0)$, $(2, 0)$, and $(0, 1)$.

19. Find the volume under the surface $z = xe^y$ and above the region bounded by $y = x$ and $y = x^2$.

20. Find the average value of $f(x, y) = 2xy$ over the region bounded by $y = x^2$ and $y = x$.

21. Suppose R is the region inside the boundary of a certain county in the midwest. After a winter storm, the depth of snow at the point (x, y) in R was $f(x, y)$ meters.
    (a) Assuming x and y are measured in kilometers, find an expression for the number of cubic meters of snow that fell on the county. (Remember that 1 kilometer is 1,000 meters.)
    (b) Find an expression for average depth (in meters) of snow in the county.

22. Suppose the joint probability density function for the nonnegative variables x and y is $f(x, y) = 6e^{-2x}e^{-3y}$.

(a) Find the probability that $0 \le x \le 1$ and $0 \le y \le 2$.

(b) Find the probability that $x + y \le 2$.

23. Suppose x denotes the time (in minutes) that a person spends in a certain doctor's waiting room and y the duration (in minutes) of a complete physical examination. You arrive at the doctor's office for a physical 50 minutes before you are due to leave for a meeting. If the joint probability density function for x and y is $f(x, y) = \frac{1}{500}e^{-x/10}e^{-y/50}$, find the probability that you will be late leaving for your meeting.

24. Suppose R represents a flat filter that absorbes some of the ozone from the air passing through it. At the point (x, y) on the filter, the speed of the air is $f(x, y)$ feet per minute, and the fraction of ozone removed is $g(x, y)$. If the air contains 40 units of ozone per cubic foot before it flows through the filter and the air flow is always perpendicular to the filter, find an expression for the number of units of ozone collected by the filter in 30 minutes.

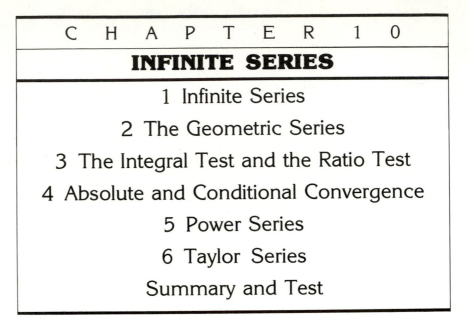

1 INFINITE SERIES The sum of infinitely many numbers may be finite. This statement, which may seem paradoxical at first, plays a central role in mathematics and has a variety of important applications. The purpose of this chapter is to explore its meaning and some of its consequences.

You are already familiar with the phenomenon of a finite-valued infinite sum. You know, for example, that the repeating decimal 0.333 . . . stands for the infinite sum $\frac{3}{10} + \frac{3}{100} + \frac{3}{1,000} + \ldots$ and that its value is the finite number $\frac{1}{3}$. What may not be familiar, however, is exactly what it means to say that this infinite sum "adds up to" $\frac{1}{3}$. The situation will be clarified in this introductory section.

Infinite series An expression of the form

$$a_1 + a_2 + \cdots + a_n + \cdots$$

is called an **infinite series.** It is customary to use summation notation to write series compactly as follows:

$$a_1 + a_2 + \cdots + a_n + \cdots = \sum_{n=1}^{\infty} a_n$$

The use of this notation is illustrated in the following example.

EXAMPLE 1.1

(a) Write out some representative terms of the series $\displaystyle\sum_{n=1}^{\infty} \frac{1}{2^n}$.

(b) Use summation notation to write the series $1 - \frac{1}{4} + \frac{1}{9} - \frac{1}{16} + \cdots$ in compact form.

SOLUTION

(a) $\displaystyle\sum_{n=1}^{\infty} \frac{1}{2^n} = \frac{1}{2} + \frac{1}{4} + \frac{1}{8} + \cdots + \frac{1}{2^n} + \cdots$

(b) The nth term of this series is

$$\frac{(-1)^{n+1}}{n^2}$$

where the factor $(-1)^{n+1}$ generates the alternating signs starting with a plus sign when $n = 1$. Thus,

$$1 - \frac{1}{4} + \frac{1}{9} - \frac{1}{16} + \cdots = \sum_{n=1}^{\infty} \frac{(-1)^{n+1}}{n^2}$$

Applications of infinite series Infinite series arise in a variety of practical situations. In Section 2 you will see applications of series to such fields as economics, medicine, and probability. In Section 6 you will see how infinite series are used in the approximation of functions that may be difficult or impossible to evaluate directly. As a preview of these applications, and to give you a more concrete feel for how series arise in practice, here is an investment problem that leads to an infinite series.

EXAMPLE 1.2

Find an expression for the amount of money you should invest today at an annual interest rate of 10 percent compounded continuously so that, starting next year, you can make annual withdrawals of $400 in perpetuity.

SOLUTION

The amount you should invest today to generate the desired sequence of withdrawals is the sum of the present values of the individual withdrawals. You compute the present value of each withdrawal using the formula $P = Be^{-rt}$ from Chapter 4, Section 5, with $B = 400$, $r = 0.1$, and t being the time (in years) at which the withdrawal is made. Thus,

P_1 = present value of 1st withdrawal = $400e^{-0.1(1)}$ = $400e^{-0.1}$

P_2 = present value of 2nd withdrawal = $400e^{-0.1(2)}$ = $400e^{-0.2}$

.
.
.

P_n = present value of nth withdrawal = $400e^{-0.1(n)}$

and so on. The situation is illustrated in Figure 1.1. The amount that

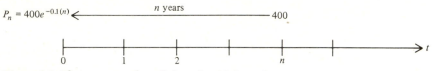

Figure 1.1 The present value of the nth withdrawal.

you should invest today is the sum of these (infinitely many) present values. That is,

$$\text{Amount to be invested} = \sum_{n=1}^{\infty} P_n = \sum_{n=1}^{\infty} 400e^{-0.1(n)}$$

In Section 2 you will learn how to show that this infinite series "adds up to" $3,808.33. As an experiment, use your calculator to add the first several terms of this series, and see what happens as you include more and more terms.

Convergence and divergence of infinite series

Roughly speaking, an infinite series is said to **converge** if it "adds up to" a finite number and to **diverge** if it does not. A more precise statement of the criterion for convergence involves the sum

$$S_n = a_1 + a_2 + \cdots + a_n$$

of the first n terms of the series. This (finite) sum is called the nth **partial sum** of the series, and its behavior as n approaches infinity determines the convergence or divergence of the series. Here is a summary of the criterion.

Convergence and divergence

An infinite series $\sum_{n=1}^{\infty} a_n$ with nth partial sum

$$S_n = a_1 + a_2 + \cdots + a_n$$

is said to converge if there is a (finite) number S such that

$$\lim_{n \to \infty} S_n = S$$

In this case, S is said to be the sum of the series, and one writes

$$\sum_{n=1}^{\infty} a_n = S$$

If S_n does not have a finite limit as n approaches infinity, the series is said to diverge.

According to this criterion, to determine the convergence or divergence of an infinite series, you start by finding an expression for the sum of the first n terms of the series and then take the limit of this finite sum as n approaches infinity. This procedure should be reminiscent of the one that was used in Chapter 7, Section 2, to determine the convergence or divergence of an improper integral. This is not the last time that similarities between infinite series and improper integrals will arise.

The notions of convergence and divergence of infinite series are illustrated in the following examples.

EXAMPLE 1.3

Investigate the possible convergence of the series $\sum_{n=1}^{\infty} \dfrac{1}{n(n+1)}$.

SOLUTION

To get a feel for the series, you might begin by computing a few of its partial sums.

$$S_1 = \frac{1}{1 \cdot 2} = \frac{1}{2}$$

$$S_2 = \frac{1}{1 \cdot 2} + \frac{1}{2 \cdot 3} = \frac{1}{2} + \frac{1}{6} = \frac{2}{3}$$

$$S_3 = \left(\frac{1}{1 \cdot 2} + \frac{1}{2 \cdot 3} \right) + \frac{1}{3 \cdot 4} = \frac{2}{3} + \frac{1}{12} = \frac{3}{4}$$

$$S_4 = \left(\frac{1}{1 \cdot 2} + \frac{1}{2 \cdot 3} + \frac{1}{3 \cdot 4} \right) + \frac{1}{4 \cdot 5} = \frac{3}{4} + \frac{1}{20} = \frac{4}{5}$$

$$S_5 = \left(\frac{1}{1 \cdot 2} + \frac{1}{2 \cdot 3} + \frac{1}{3 \cdot 4} + \frac{1}{4 \cdot 5} \right) + \frac{1}{5 \cdot 6} = \frac{4}{5} + \frac{1}{30} = \frac{5}{6}$$

These calculations *suggest* that $S_n = n/(n + 1)$ and that the sum of the series is 1. However, since the reason for the observed pattern among the partial sums is probably not clear, it is advisable to seek an algebraic solution to the problem.

The trick is to use the fact that

$$\frac{1}{n(n + 1)} = \frac{1}{n} - \frac{1}{n + 1}$$

to write the series as

$$\sum_{n=1}^{\infty} \frac{1}{n(n + 1)} = \sum_{n=1}^{\infty} \left(\frac{1}{n} - \frac{1}{n + 1}\right)$$

and the nth partial sum as

$$S_n = \left(\frac{1}{1} - \frac{1}{2}\right) + \left(\frac{1}{2} - \frac{1}{3}\right) + \left(\frac{1}{3} - \frac{1}{4}\right) + \cdots + \left(\frac{1}{n} - \frac{1}{n + 1}\right)$$

Since all but the first and last terms in this sum cancel, S_n can be rewritten as

$$S_n = 1 - \frac{1}{n + 1} = \frac{n}{n + 1}$$

Finally, since

$$\lim_{n \to \infty} S_n = \lim_{n \to \infty} \frac{n}{n + 1} = 1$$

it follows that the series converges to 1. That is,

$$\sum_{n=1}^{\infty} \frac{1}{n(n + 1)} = 1$$

The series in the preceding example is known as a **telescoping series** because of the cancellation that takes place within the nth partial sum.

The series in the next example is the one that corresponds to the decimal representation of the fraction $\frac{1}{3}$. It is an example of an important type of series that will be examined in more detail in Section 2.

EXAMPLE 1.4

Investigate the possible convergence of the series $\sum_{n=1}^{\infty} \frac{3}{10^n}$.

SOLUTION

Notice that each term of the series

$$\sum_{n=1}^{\infty} \frac{3}{10^n} = \frac{3}{10} + \frac{3}{10^2} + \frac{3}{10^3} + \cdots$$

is $\frac{1}{10}$ times the preceding term. This leads to the following trick for finding a compact formula for S_n.

Begin with the nth partial sum in the form

$$S_n = \frac{3}{10} + \frac{3}{10^2} + \frac{3}{10^3} + \cdots + \frac{3}{10^n}$$

and multiply both sides of this equation by $\frac{1}{10}$ to get

$$\frac{1}{10} S_n = \frac{3}{10^2} + \frac{3}{10^3} + \cdots + \frac{3}{10^n} + \frac{3}{10^{n+1}}$$

Now subtract the expression for $\frac{1}{10}S_n$ from the expression for S_n. Most of the terms cancel, and you are left with

$$\frac{9}{10} S_n = \frac{3}{10} - \frac{3}{10^{n+1}} = \frac{3}{10}\left(1 - \frac{1}{10^n}\right)$$

or
$$S_n = \frac{1}{3}\left(1 - \frac{1}{10^n}\right)$$

Then,
$$\lim_{n\to\infty} S_n = \lim_{n\to\infty} \frac{1}{3}\left(1 - \frac{1}{10^n}\right) = \frac{1}{3}$$

and so
$$\sum_{n=1}^{\infty} \frac{3}{10^n} = \frac{1}{3}$$

as expected.

A necessary condition for convergence

The individual terms a_n of a convergent series $\displaystyle\sum_{n=1}^{\infty} a_n$ must approach zero as n increases without bound. To see this, write a_n as the difference of two partial sums as follows:

$$a_n = (a_1 + a_2 + \cdots + a_n) - (a_1 + a_2 + \cdots + a_{n-1}) = S_n - S_{n-1}$$

Then, if $\displaystyle\sum_{n=1}^{\infty} a_n$ converges to some number S, both partial sums S_n and S_{n-1} approach S as n approaches infinity, and so,

$$\lim_{n\to\infty} a_n = \lim_{n\to\infty} S_n - \lim_{n\to\infty} S_{n-1} = S - S = 0$$

This observation can be rephrased as a test for *divergence* as follows:

Test for divergence

> **If** $\lim\limits_{n\to\infty} a_n \neq 0$, **then** $\sum\limits_{n=1}^{\infty} a_n$ **diverges.**

The use of this test is illustrated in the next example.

EXAMPLE 1.5

Investigate the possible convergence of the series $\sum\limits_{n=1}^{\infty} \dfrac{n+3}{2n+1}$.

SOLUTION

The limit of the nth term is

$$\lim_{n\to\infty} a_n = \lim_{n\to\infty} \frac{n+3}{2n+1} = \frac{1}{2}$$

which is not equal to zero. Hence the series cannot converge and must diverge.

Look again at the series in Example 1.5. Its individual terms a_n decrease as n increases. In particular,

$$a_1 = \frac{4}{3} = 1.3333$$

$$a_2 = \frac{5}{5} = 1$$

$$a_3 = \frac{6}{7} = 0.8571$$

$$a_4 = \frac{7}{9} = 0.7778$$

$$a_5 = \frac{8}{11} = 0.7273$$

$$\cdot$$
$$\cdot$$
$$\cdot$$

$$a_{50} = \frac{53}{101} = 0.5248$$

.
.
.

$$a_{100} = \frac{103}{201} = 0.5124$$

and so on. However, it is because the terms do not decrease *to zero* (but approach $\frac{1}{2}$ instead) that the series $\sum\limits_{n=1}^{\infty} a_n$ cannot converge.

Warning against misuse of the divergence test

The test for divergence states that a series whose individual terms do not approach zero must diverge. The converse of this test is *not* true. That is, the fact that the individual terms of a series approach zero does not guarantee that the series converges. Here is an example.

EXAMPLE 1.6

Investigate the possible convergence of the series $\sum\limits_{n=1}^{\infty} \dfrac{1}{n}$.

SOLUTION

Since

$$\lim_{n\to\infty} a_n = \lim_{n\to\infty} \frac{1}{n} = 0$$

it is possible, but not guaranteed, that the series converges.

It is natural to proceed by computing a few of the partial sums to get a better feel for the series.

$$S_1 = 1$$

$$S_2 = 1 + \frac{1}{2} = \frac{3}{2} = 1.5$$

$$S_3 = \left(1 + \frac{1}{2}\right) + \frac{1}{3} = \frac{3}{2} + \frac{1}{3} = \frac{11}{6} = 1.8333$$

$$S_4 = \left(1 + \frac{1}{2} + \frac{1}{3}\right) + \frac{1}{4} = \frac{11}{6} + \frac{1}{4} = \frac{25}{12} = 2.0833$$

$$S_5 = \left(1 + \frac{1}{2} + \frac{1}{3} + \frac{1}{4}\right) + \frac{1}{5} = \frac{25}{12} + \frac{1}{5} = \frac{137}{60} = 2.2833$$

$$S_6 = \left(1 + \frac{1}{2} + \frac{1}{3} + \frac{1}{4} + \frac{1}{5}\right) + \frac{1}{6} = \frac{137}{60} + \frac{1}{6} = \frac{147}{60} = 2.45$$

Unfortunately, there is no obvious pattern among these partial sums, nor is it even clear whether or not they are approaching a finite limit.

Actually, this series diverges. To see this, group its terms as follows:

$$1 + \underbrace{\frac{1}{2}} + \underbrace{\frac{1}{3} + \frac{1}{4}}_{2 \text{ terms}} + \underbrace{\frac{1}{5} + \frac{1}{6} + \frac{1}{7} + \frac{1}{8}}_{2^2 \text{ terms}} + \underbrace{\frac{1}{9} + \cdots + \frac{1}{16}}_{2^3 \text{ terms}} + \underbrace{\frac{1}{17}} + \cdots$$

The sum of the terms in each group is greater than or equal to $\frac{1}{2}$. For example,

$$\frac{1}{3} + \frac{1}{4} > \frac{1}{4} + \frac{1}{4} = \frac{1}{2}$$

and

$$\frac{1}{5} + \frac{1}{6} + \frac{1}{7} + \frac{1}{8} > \frac{1}{8} + \frac{1}{8} + \frac{1}{8} + \frac{1}{8} = \frac{1}{2}$$

By including sufficiently many of these groups, you can make the corresponding partial sum as large as you like, from which it follows that the infinite series must diverge.

The series $\sum_{n=1}^{\infty} \frac{1}{n}$ in Example 1.6 is known as the **harmonic series** and is one of the most well-known examples of a divergent series whose individual terms approach zero.

It is instructive to compare the convergent series $\sum_{n=1}^{\infty} \frac{1}{n(n + 1)}$ from Example 1.3 with the divergent harmonic series $\sum_{n=1}^{\infty} \frac{1}{n}$ from Example 1.6. Each is a series of positive terms that approach zero as n increases. Yet one converges and has a finite sum, while the other diverges and has an "infinite sum." The reason for the difference is that, as n increases, the terms $\frac{1}{n(n + 1)}$ approach zero more quickly than the terms $\frac{1}{n}$ do. The situation is illustrated in the following table, and should be reminiscent of the difference between convergent and divergent improper integrals. This relationship between infinite series and improper integrals will be explored further in Section 3.

n	1	2	3	4	5	10	50
$\dfrac{1}{n}$	1	0.5	0.3333	0.25	0.2	0.1	0.02
$\dfrac{1}{n(n+1)}$	0.5	0.1667	0.0833	0.05	0.0333	0.0091	0.0004

A useful property of convergent series

Like finite sums, convergent infinite series obey a distributive law with respect to multiplication. According to this law, if you multiply each term of a convergent series by a constant c, you get a new series whose sum is c times the sum of the original series.

The distributive law for series

For any constant c,

$$\sum_{n=1}^{\infty} ca_n = c \sum_{n=1}^{\infty} a_n$$

This law follows from the fact that the sum of an infinite series is determined by its nth partial sum, which, being a finite sum, obeys the familiar distributive law from elementary algebra. In particular,

$$\sum_{n=1}^{\infty} ca_n = \lim_{n \to \infty} (ca_1 + ca_2 + \cdots + ca_n)$$

$$= \lim_{n \to \infty} c(a_1 + a_2 + \cdots + a_n)$$

$$= c \lim_{n \to \infty} (a_1 + a_2 + \cdots + a_n)$$

$$= c \sum_{n=1}^{\infty} a_n$$

The distributive law also holds for divergent series in the sense that if $\sum_{n=1}^{\infty} a_n$ diverges, so does $\sum_{n=1}^{\infty} ca_n$ (provided, of course, that $c \neq 0$).

The distributive law allows you to factor constants out of infinite series. Such factorization will be used several times in the next section to simplify calculations.

Problems

In Problems 1 through 6, use summation notation to write the given series in compact form.

1. $\dfrac{1}{3} + \dfrac{1}{9} + \dfrac{1}{27} + \dfrac{1}{81} + \cdots$

2. $1 + \dfrac{1}{8} + \dfrac{1}{27} + \dfrac{1}{64} + \cdots$

3. $\dfrac{1}{2} + \dfrac{2}{3} + \dfrac{3}{4} + \dfrac{4}{5} + \cdots$

4. $\dfrac{1}{3} + \dfrac{2}{5} + \dfrac{3}{7} + \dfrac{4}{9} + \cdots$

5. $\dfrac{1}{2} - \dfrac{4}{3} + \dfrac{9}{4} - \dfrac{16}{5} + \cdots$

6. $-3 + \dfrac{9}{4} - \dfrac{27}{9} + \dfrac{81}{16} - \cdots$

In Problems 7 through 10, find the fourth partial sum S_4 of the given series.

7. $\displaystyle\sum_{n=1}^{\infty} \dfrac{1}{2^n}$

8. $\displaystyle\sum_{n=1}^{\infty} \dfrac{n}{n+1}$

9. $\displaystyle\sum_{n=1}^{\infty} \dfrac{(-1)^n}{n}$

10. $\displaystyle\sum_{n=1}^{\infty} \dfrac{(-1)^{n+1}}{n(n+1)}$

In Problems 11 through 18, find the sum of the given convergent series by taking the limit of a compact expression for the nth partial sum.

11. $\displaystyle\sum_{n=1}^{\infty} \left(\dfrac{1}{n+3} - \dfrac{1}{n+4} \right)$

12. $\displaystyle\sum_{n=1}^{\infty} \left(\dfrac{1}{n} - \dfrac{1}{n+2} \right)$

13. $\displaystyle\sum_{n=1}^{\infty} \dfrac{1}{(n+1)(n+2)}$

14. $\displaystyle\sum_{n=1}^{\infty} \dfrac{1}{(n+2)(n+3)}$

15. $\displaystyle\sum_{n=1}^{\infty} \dfrac{6}{10^n}$

16. $\displaystyle\sum_{n=1}^{\infty} \dfrac{1}{2^n}$

17. $\displaystyle\sum_{n=1}^{\infty} \dfrac{4}{3^n}$

18. $\displaystyle\sum_{n=1}^{\infty} \left(-\dfrac{1}{5} \right)^n$

In Problems 19 through 25, determine whether or not the given series converges, and if it does, find its sum.

19. $\displaystyle\sum_{n=1}^{\infty} \dfrac{n}{2n+1}$

20. $\displaystyle\sum_{n=1}^{\infty} \dfrac{1}{n+1}$

21. $\displaystyle\sum_{n=1}^{\infty} \left(\dfrac{1}{n+1} - \dfrac{1}{n} \right)$

22. $\displaystyle\sum_{n=1}^{\infty} \dfrac{n^2}{50n^2 + 1}$

23. $\displaystyle\sum_{n=1}^{\infty} \left(-\dfrac{2}{3} \right)^n$

24. $\displaystyle\sum_{n=1}^{\infty} \left(-\dfrac{3}{2} \right)^n$

25. $\sum\limits_{n=1}^{\infty} \dfrac{1}{\sqrt{n}}$ (*Hint:* Show that $S_n \geq \dfrac{n}{\sqrt{n}} = \sqrt{n}$.)

2 THE GEOMETRIC SERIES

A **geometric series** is an infinite series in which the ratio of successive terms is constant. For example, the series

$$3 + \frac{3}{2} + \frac{3}{4} + \frac{3}{8} + \cdots$$

is geometric because each term is one half of the preceding term. In general, a geometric series is a series of the form

$$\sum_{n=m}^{\infty} ar^n = ar^m + ar^{m+1} + ar^{m+2} + \cdots$$

The constant r, which is the ratio of the successive terms of a geometric series, is known as the **ratio** of the series. The ratio of a geometric series may be positive or negative. For example,

$$\sum_{n=0}^{\infty} \frac{2}{(-3)^n} = 2 - \frac{2}{3} + \frac{2}{9} - \frac{2}{27} + \cdots$$

is a geometric series with ratio $r = -\frac{1}{3}$.

Elementary geometric series

The most elementary geometric series are those for which $m = 0$ and $a = 1$, that is, series of the form

$$\sum_{n=0}^{\infty} r^n = 1 + r + r^2 + \cdots$$

Because of the distributive law, every geometric series can be written as a multiple of one of these elementary geometric series. In particular,

$$\sum_{n=m}^{\infty} ar^n = ar^m + ar^{m+1} + ar^{m+2} + \cdots$$

$$= ar^m(1 + r + r^2 + \cdots) \qquad \text{(distributive law)}$$

$$= ar^m \sum_{n=0}^{\infty} r^n$$

For this reason, it suffices to study the elementary geometric series.

Convergence of geometric series

A geometric series $\sum\limits_{n=0}^{\infty} r^n$ converges if the absolute value of its ratio r is less than 1 and diverges otherwise. The divergence when $|r| \geq 1$

follows from the fact that, in this case, the individual terms r^n do not approach zero as n increases. When $|r| < 1$, you can establish the convergence of the series and derive a formula for its sum by means of a calculation similar to the one you saw in Example 1.4 of the preceding section. In particular, begin with the nth partial sum in the form

$$S_n = 1 + r + r^2 + \cdots + r^{n-1}$$

(Notice that the power of r in the nth term is only $n - 1$ because the first term in the sum is $1 = r^0$.) Now multiply both sides of this equation by r to get

$$rS_n = r + r^2 + \cdots + r^{n-1} + r^n$$

and subtract the expression for rS_n from the expression for S_n to get

$$(1 - r)S_n = 1 - r^n$$

or
$$S_n = \frac{1 - r^n}{1 - r}$$

Since $|r| < 1$,
$$\lim_{n \to \infty} r^n = 0$$

and so
$$\lim_{n \to \infty} S_n = \lim_{n \to \infty} \frac{1 - r^n}{1 - r} = \frac{1}{1 - r}$$

Hence the series converges, and its sum is $\dfrac{1}{1 - r}$.

The sum of a geometric series

If $|r| < 1$,

$$\sum_{n=0}^{\infty} r^n = \frac{1}{1 - r}$$

Notice that this formula for the sum of a geometric series is valid only for elementary series of the form

$$\sum_{n=0}^{\infty} r^n = 1 + r + r^2 + \cdots$$

that start with the term $r^0 = 1$. To apply the formula to any other geometric series, you will first have to rewrite the series as a constant multiple of an elementary one. The use of the formula is illustrated in the following examples.

EXAMPLE 2.1

Find the sum of the series $\displaystyle\sum_{n=0}^{\infty} \left(-\frac{2}{3}\right)^n$.

SOLUTION

This is an elementary geometric series with $r = -\frac{2}{3}$. Hence,

$$\sum_{n=0}^{\infty} \left(-\frac{2}{3}\right)^n = \frac{1}{1 - (-2/3)} = \frac{1}{5/3} = \frac{3}{5}$$

EXAMPLE 2.2

Find the sum of the series $\displaystyle\sum_{n=0}^{\infty} \frac{3}{2^n}$.

SOLUTION

$$\sum_{n=0}^{\infty} \frac{3}{2^n} = 3 \sum_{n=0}^{\infty} \left(\frac{1}{2}\right)^n = 3 \left(\frac{1}{1 - 1/2}\right) = 3 \left(\frac{1}{1/2}\right) = 6$$

EXAMPLE 2.3

Find the sum of the series $\displaystyle\sum_{n=2}^{\infty} \frac{2}{5^n}$.

SOLUTION

$$\sum_{n=2}^{\infty} \frac{2}{5^n} = \frac{2}{5^2} + \frac{2}{5^3} + \frac{2}{5^4} + \cdots$$

$$= \frac{2}{5^2} \left(1 + \frac{1}{5} + \frac{1}{5^2} + \cdots\right)$$

$$= \frac{2}{25} \sum_{n=0}^{\infty} \left(\frac{1}{5}\right)^n$$

$$= \frac{2}{25} \left(\frac{1}{1 - 1/5}\right) = \frac{2}{25} \left(\frac{5}{4}\right) = \frac{1}{10}$$

EXAMPLE 2.4

Find the sum of the series $\displaystyle\sum_{n=1}^{\infty} \frac{(-2)^n}{3^{n+1}}$.

SOLUTION

$$\sum_{n=1}^{\infty} \frac{(-2)^n}{3^{n+1}} = \frac{-2}{3^2} + \frac{(-2)^2}{3^3} + \frac{(-2)^3}{3^4} + \cdots$$

$$= \frac{-2}{3^2}\left[1 + \left(-\frac{2}{3}\right) + \left(-\frac{2}{3}\right)^2 + \cdots\right]$$

$$= -\frac{2}{9}\sum_{n=0}^{\infty}\left(-\frac{2}{3}\right)^n$$

$$= -\frac{2}{9}\left(\frac{1}{1 + 2/3}\right) = -\frac{2}{9}\left(\frac{3}{5}\right) = -\frac{2}{15}$$

Applications of geometric series

Geometric series arise in many branches of mathematics and in a variety of applications. Here are four illustrations taken from number theory, economics, medicine, and probability.

Repeating decimals

A repeating decimal is a geometric series, and its value is the sum of the series. Here is an example.

EXAMPLE 2.5

Express the repeating decimal 0.232323 . . . as a fraction.

SOLUTION

Write the decimal as a geometric series as follows:

$$0.232323\ldots = \frac{23}{100} + \frac{23}{10,000} + \frac{23}{1,000,000} + \cdots$$

$$= \frac{23}{100}\left[1 + \frac{1}{100} + \left(\frac{1}{100}\right)^2 + \cdots\right]$$

$$= \frac{23}{100}\sum_{n=0}^{\infty}\left(\frac{1}{100}\right)^n$$

$$= \frac{23}{100}\left(\frac{1}{1 - 1/100}\right) = \frac{23}{100}\left(\frac{100}{99}\right) = \frac{23}{99}$$

The multiplier effect in economics

A tax rebate that returns a certain amount of money to taxpayers can result in spending that is many times this amount. This phenomenon is known in economics as the **multiplier effect.** It occurs because the portion of the rebate that is spent by one individual becomes income for one or more others who, in turn, spend some of it again, creating income for yet other individuals to spend. If the fraction of income that is saved remains constant as this process continues indefinitely, the total amount spent as a result of the rebate is the sum of a geometric series. Here is an example.

EXAMPLE 2.6

Suppose that nationwide, approximately 90 percent of all income is spent and 10 percent saved. How much additional spending will be generated by a 40 billion dollar tax rebate if savings habits do not change?

SOLUTION

The amount (in billions) spent by original recipients of the rebate is

$$0.9(40)$$

This becomes new income, of which 90 percent or

$$0.9[0.9(40)] = (0.9)^2(40)$$

is spent. This, in turn, generates additional spending of

$$0.9[(0.9)^2(40)] = (0.9)^3(40)$$

and so on. The total amount spent if this process continues indefinitely is

$$0.9(40) + (0.9)^2(40) + (0.9)^3(40) + \cdots$$

$$= 0.9(40)[1 + 0.9 + (0.9)^2 + \cdots]$$

$$= 36 \sum_{n=0}^{\infty} (0.9)^n = 36 \left(\frac{1}{1 - 0.9} \right) = 360 \text{ billion}$$

Accumulation of medication in the body

In Example 2.5 of Chapter 7, a nuclear power plant generated radioactive waste which decayed exponentially. The problem was to determine the long-run accumulation of waste from the plant. Because the waste was being generated *continuously*, the solution involved an (improper) integral. The situation in the next example is similar. A patient receives medication which is eliminated from the body exponentially, and the goal is to determine the long-run accumulation of medication in the patient's body. In this case, however, the medication is being administered in *discrete* doses, and the solution to the problem involves an infinite series rather than an integral.

EXAMPLE 2.7

A patient is given an injection of 10 units of a certain drug every 24 hours. The drug is eliminated exponentially so that the fraction that remains in the patient's body after t days is $f(t) = e^{-t/5}$. If the treatment is continued indefinitely, approximately how many units of the drug will eventually be in the patient's body just prior to an injection?

SOLUTION

Of the original dose of 10 units, only $10e^{-1/5}$ units are left in the patient's body after 1 day (just prior to the second injection). That is,

$$\text{Amount in body after 1 day} = S_1 = 10e^{-1/5}$$

The medication in the patient's body after 2 days consists of what remains from the first *two* doses. Of the original dose, only $10e^{-2/5}$ units are left (since 2 days have elapsed), and of the second dose, $10e^{-1/5}$ units remain. (See Figure 2.1.) Hence

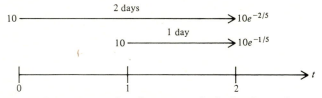

Figure 2.1 **Amount of medication in the body after 2 days.**

$$\text{Amount in body after 2 days} = S_2 = 10e^{-1/5} + 10e^{-2/5}$$

Similarly,

$$\text{Amount in body after } n \text{ days} = S_n = 10e^{-1/5} + 10e^{-2/5} + \cdots + 10e^{-n/5}$$

The amount S of medication in the patient's body in the long run is the limit of S_n as n approaches infinity. That is,

$$S = \lim_{n \to \infty} S_n = \sum_{n=1}^{\infty} 10e^{-n/5}$$

$$= \sum_{n=1}^{\infty} 10(e^{-1/5})^n = 10e^{-1/5} \sum_{n=0}^{\infty} (e^{-1/5})^n$$

$$= 10e^{-1/5}\left(\frac{1}{1 - e^{-1/5}}\right) = 45.17 \text{ units}$$

Geometric random variables

Suppose a random experiment with two possible outcomes (traditionally called "success" and "failure") is performed repeatedly until the first success occurs. The number x of repetitions required to produce the first success is known as a **geometric random variable.**

A geometric random variable is a discrete random variable that can take on any positive integer value. For any positive integer n, the probability that $x = n$ is the probability that the first $n - 1$ trials of the experiment result in failure and the nth in success:

$$\underbrace{\text{F} \quad \text{F} \quad \text{F} \quad \cdots \quad \text{F}}_{n - 1 \text{ failures}} \quad \underbrace{\text{S}}_{\text{1st success}}$$

If, on each trial, the probability of success is p, then the probability of failure must be $1 - p$, and the probability of $n - 1$ failures followed by 1 success is the product

$$(1 - p)(1 - p)(1 - p) \cdots (1 - p)p$$

That is,
$$P(x = n) = (1 - p)^{n-1}p$$

If $p \neq 0$, success will eventually occur if the experiment is repeated indefinitely. Hence, $P(x \geq 1)$ ought to be 1. To compute this probability, you find the sum of a certain geometric series as follows:

$$P(x \geq 1) = \sum_{n=1}^{\infty} P(x = n)$$

$$= \sum_{n=1}^{\infty} (1 - p)^{n-1}p$$

$$= p + (1 - p)p + (1 - p)^2p + \cdots$$

$$= p[1 + (1 - p) + (1 - p)^2 + \cdots]$$

$$= p \sum_{n=0}^{\infty} (1 - p)^n$$

$$= \frac{p}{1 - (1 - p)} = \frac{p}{p} = 1$$

A simple application of geometric random variables is given in the next example.

EXAMPLE 2.8

In a certain two-person board game played with 1 die, each player needs a six to start play. You and your opponent take turns rolling the die. Find the probability that you will get the first six if you roll first.

SOLUTION

In this context, "success" is getting a six and "failure" is getting any of the other 5 possible numbers. Hence, the probability of success is $p = \frac{1}{6}$ and the probability of failure is $1 - p = \frac{5}{6}$.

Let x denote the number of the roll on which the first six comes up. If x is odd, the first six will occur on one of your rolls, while if x is even, the first six will occur on one of your opponent's rolls. Hence, the probability that you get the first six is

$$P(x \text{ is odd}) = P(x = 1) + P(x = 3) + P(x = 5) + \cdots$$

$$= \frac{1}{6} + \left(\frac{5}{6}\right)^2 \left(\frac{1}{6}\right) + \left(\frac{5}{6}\right)^4 \left(\frac{1}{6}\right) + \cdots$$

$$= \frac{1}{6} + \left(\frac{25}{36}\right)\left(\frac{1}{6}\right) + \left(\frac{25}{36}\right)^2 \left(\frac{1}{6}\right) + \cdots$$

$$= \frac{1}{6} \sum_{n=0}^{\infty} \left(\frac{25}{36}\right)^n$$

$$= \frac{1}{6} \left(\frac{1}{1 - 25/36}\right) = \frac{1}{6} \left(\frac{36}{11}\right) = \frac{6}{11} = 0.5455$$

Problems In Problems 1 through 14, determine whether the given geometric series converges, and if so, find its sum.

1. $\displaystyle\sum_{n=0}^{\infty} \left(\frac{4}{5}\right)^n$

2. $\displaystyle\sum_{n=0}^{\infty} \left(-\frac{4}{5}\right)^n$

3. $\displaystyle\sum_{n=0}^{\infty} \frac{2}{3^n}$

4. $\displaystyle\sum_{n=0}^{\infty} \frac{2}{(-3)^n}$

5. $\displaystyle\sum_{n=1}^{\infty} \left(\frac{3}{2}\right)^n$

6. $\displaystyle\sum_{n=1}^{\infty} \frac{3}{2^n}$

7. $\displaystyle\sum_{n=2}^{\infty} \frac{3}{(-4)^n}$

8. $\displaystyle\sum_{n=2}^{\infty} \left(-\frac{4}{3}\right)^n$

9. $\displaystyle\sum_{n=1}^{\infty} 5(0.9)^n$

10. $\displaystyle\sum_{n=1}^{\infty} e^{-0.2n}$

11. $\displaystyle\sum_{n=1}^{\infty} \frac{3^n}{4^{n+2}}$ 12. $\displaystyle\sum_{n=2}^{\infty} \frac{(-2)^{n-1}}{3^{n+1}}$

13. $\displaystyle\sum_{n=0}^{\infty} \frac{4^{n+1}}{5^{n-1}}$ 14. $\displaystyle\sum_{n=2}^{\infty} (-1)^n \frac{2^{n+1}}{3^{n-3}}$

In Problems 15 through 18, express the given decimal as a fraction.

15. 0.3333 . . . 16. 0.5555 . . .

17. 0.252525 . . . 18. 1.405405405 . . .

Multiplier effect 19. Suppose that nationwide, approximately 92 percent of all income is spent and 8 percent saved. How much additional spending will be generated by a 50 billion dollar tax cut if savings habits do not change?

Present value 20. An investment guarantees annual payments of $1,000 in perpetuity, with the payments beginning immediately. Find the present value of this investment if the prevailing annual interest rate remains fixed at 12 percent compounded continuously. (*Hint:* The present value of the investment is the sum of the present values of the individual payments.)

Present value 21. How much should you invest today at an annual interest rate of 15 percent compounded continuously so that, starting next year, you can make annual withdrawals of $2,000 in perpetuity?

Accumulation of medication 22. A patient is given an injection of 20 units of a certain drug every 24 hours. The drug is eliminated exponentially so that the fraction that remains in the patient's body after t days is $f(t) = e^{-t/2}$. If the treatment is continued indefinitely, approximately how many units of the drug will eventually be in the patient's body just prior to an injection?

Group membership 23. Each January first, the administration of a certain private college adds 6 new members to its board of trustees. If the fraction of trustees that remain active for at least t years is $f(t) = e^{-0.2t}$, approximately how many active trustees will the college have on December thirty-first in the long run?

Games of chance 24. You and a friend take turns rolling a die until one of you wins by getting a three or a four. If your friend rolls first, find the probability that you will win.

Numismatics 25. In 1959, the design on the reverse side of United States pennies was changed from one depicting a pair of wheat stalks to one depicting the Lincoln Memorial. Today, very few of the old pennies are still in circulation. If only 0.2 percent of the pennies now in circulation were minted before 1959, find the probability that you will have to examine at least 100 pennies to find one with the old design. (Incidentally, there is an easy way to do this problem that does not involve infinite series. For practice, do it both ways.)

Quality control 26. Three inspectors take turns checking electronic components as they come off an assembly line. If 10 percent of all the components produced on the assembly line are defective, find the probability that the inspector who checks the first component will be the one who finds the first defective component.

3 THE INTEGRAL TEST AND THE RATIO TEST

The convergence or divergence of an infinite series is determined by the behavior of its nth partial sum S_n. In Section 1 you saw examples in which algebraic tricks were used to find compact formulas for the nth partial sums of series. Once these formulas for S_n were obtained, it was an easy matter to take the limit as n increased without bound to determine the convergence or divergence of the series. Unfortunately, it is often difficult or even impossible to find a compact formula for the nth partial sum of a series, and other techniques must be used to determine convergence or divergence.

This section contains two tests you can use to try to determine if a given series converges. The tests have the advantage that they do not require knowledge of a compact formula for S_n. They have the disadvantage that, at best, they tell you only if a series converges or diverges and cannot be used to find the actual sum of the series. However, once you know that a series converges, you can use numerical methods to approximate its sum.

The integral test The individual terms $a_1, a_2, \ldots, a_n, \ldots$ of an infinite series $\sum\limits_{n=1}^{\infty} a_n$ can be thought of as the values $f(1), f(2), \ldots, f(n), \ldots$ of the function $f(x)$ obtained by writing x in place of n in the expression for a_n. For example, the terms of the series $\sum\limits_{n=1}^{\infty} \frac{1}{n^2}$ are the values for $x = 1, 2,$ 3, . . . of the function $f(x) = \frac{1}{x^2}$. The **integral test** states that, under certain circumstances, the convergence or divergence of the series $\sum\limits_{n=1}^{\infty} a_n$ can be determined from the convergence or divergence of the improper integral of the corresponding function $f(x)$.

The integral test

> If $a_n = f(n)$ for $n = 1, 2, \ldots$, where $f(x)$ **is positive and decreasing for** $x \geq 1$, **then** $\sum\limits_{n=1}^{\infty} a_n$ **converges if and only if** $\int_{1}^{\infty} f(x)\, dx$
>
> **converges.**

This relationship between infinite series and improper integrals should not be surprising. Infinite series $\sum\limits_{n=1}^{\infty} a_n$ and improper integrals

$\int_{1}^{\infty} f(x)\, dx$ converge if their terms a_n or integrands $f(x)$ approach zero "fast enough." Roughly speaking, the integral test states that what is fast enough for integrals is fast enough for series, and conversely.

An outline of the proof of the integral test will be given later in this section. Here are some examples illustrating its use.

EXAMPLE 3.1

Use the integral test to determine whether the series $\sum\limits_{n=1}^{\infty} \dfrac{1}{n^2}$ converges

or diverges.

SOLUTION

Observe that

$$\frac{1}{n^2} = f(n), \qquad \text{where} \qquad f(x) = \frac{1}{x^2}$$

and that, as required, $f(x)$ is positive and decreasing for $x \geq 1$. Now evaluate the improper integral to get

$$\int_{1}^{\infty} \frac{1}{x^2}\, dx = \lim_{N \to \infty} \int_{1}^{N} \frac{1}{x^2}\, dx = \lim_{N \to \infty} \left(-\frac{1}{x}\Big|_{1}^{N} \right) = \lim_{N \to \infty} \left(-\frac{1}{N} + 1 \right) = 1$$

Since the improper integral converges, it follows by the integral test that the series also converges.

When applying the integral test, remember that the value of the improper integral is not necessarily equal to the sum of the corresponding series. In the preceding example, for instance, the value of the integral was 1, which is clearly not the sum of the series, since

$$\sum_{n=1}^{\infty} \frac{1}{n^2} = 1 + \frac{1}{4} + \frac{1}{9} + \cdots > 1$$

In fact, using advanced techniques, it can be shown that the sum of this series is actually $\dfrac{\pi^2}{6}$. (You might find it interesting to compute some partial sums of this series and compare them with the decimal representation of $\dfrac{\pi^2}{6}$.)

In the next example, the integral test is applied to the harmonic series, which was investigated in Example 1.6 of this chapter.

EXAMPLE 3.2

Use the integral test to show that the harmonic series $\displaystyle\sum_{n=1}^{\infty} \dfrac{1}{n}$ diverges.

SOLUTION

The corresponding function is $f(x) = \dfrac{1}{x}$, which is positive and decreasing for $x \geq 1$. Since

$$\int_1^{\infty} \frac{1}{x}\, dx = \lim_{N \to \infty} \int_1^N \frac{1}{x}\, dx = \lim_{N \to \infty} \left(\ln |x| \Big|_1^N \right) = \lim_{N \to \infty} (\ln N - \ln 1) = \infty$$

it follows that the harmonic series diverges.

EXAMPLE 3.3

Use the integral test to determine whether the series $\displaystyle\sum_{n=2}^{\infty} \dfrac{1}{n \ln n}$ converges or diverges.

SOLUTION

Since this series starts with $n = 2$ instead of $n = 1$, the corresponding improper integral will start with $x = 2$. The appropriate function is $f(x) = \dfrac{1}{x \ln x}$, which is positive and decreasing for $x \geq 2$. Since

$$\int_2^{\infty} \frac{1}{x \ln x}\, dx = \lim_{N \to \infty} \int_2^N \frac{1}{x \ln x}\, dx$$

$$= \lim_{N \to \infty} \left(\ln|\ln x| \Big|_2^N \right)$$

$$= \lim_{N \to \infty} (\ln|\ln N| - \ln|\ln 2|) = \infty$$

it follows that the series diverges.

Why the integral test works

Suppose $f(x)$ is positive and decreasing for $x \geq 1$ with $f(n) = a_n$ for $n = 1, 2, \ldots$ To see why the series $\sum\limits_{n=1}^{\infty} a_n$ must diverge if the integral $\int_{1}^{\infty} f(x)\, dx$ diverges, look at Figure 3.1a, in which rectangles are used to approximate the area under the graph of f. Since the length of the base of each rectangle is 1, the area of the rectangle is equal to its height. The area of the first rectangle is therefore $f(1) = a_1$, the area of the second $f(2) = a_2$, and so on. Since the area under the curve is less than (or equal to) the area of the rectangles, it follows that

$$\int_{1}^{n+1} f(x)\, dx \leq a_1 + a_2 + \cdots + a_n = S_n$$

(where S_n is the nth partial sum of the series).

If the improper integral $\int_{1}^{\infty} f(x)\, dx$ diverges to ∞, the left-hand side of

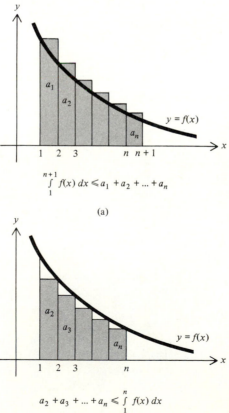

$$\int_{1}^{n+1} f(x)\, dx \leqslant a_1 + a_2 + \ldots + a_n$$

(a)

$$a_2 + a_3 + \ldots + a_n \leqslant \int_{1}^{n} f(x)\, dx$$

(b)

Figure 3.1 Rectangles that approximate the area under the curve from above and from below.

this inequality increases without bound as n increases, forcing S_n on the right-hand side to do the same. Since the partial sums increase without bound, the series itself must diverge.

To see why the series must converge if the integral converges, look at Figure 3.1b. In this case, the area of the rectangles is less than (or equal to) the area under the curve. Hence,

$$a_2 + a_3 + \cdots + a_n \le \int_1^n f(x)\, dx$$

or, equivalently,

$$S_n = a_1 + a_2 + a_3 + \cdots + a_n \le a_1 + \int_1^n f(x)\, dx$$

If the improper integral $\int_1^\infty f(x)\, dx$ converges to L, the right-hand side of this inequality increases and approaches $a_1 + L$ as n approaches infinity, from which it follows that

$$S_n \le a_1 + L \qquad \text{(for all } n)$$

Since the terms $a_1, a_2, \ldots$ are all positive, the partial sums S_n increase as n increases, but since they cannot exceed $a_1 + L$, they must approach some finite limit. That is, the series $\displaystyle\sum_{n=1}^{\infty} a_n$ must converge.

The ratio test The value of the integral test is limited by the fact that it requires the integration of the function obtained from the terms of the series. For many of the series you will encounter, it will not be possible to carry out this integration. Fortunately, there are several other convergence tests that do not involve integration. One of the most useful of these is the **ratio test,** which involves the ratio of successive terms of the series.

The ratio test

Suppose $\displaystyle\sum_{n=1}^{\infty} a_n$ is a series of positive terms and let

$$\rho = \lim_{n \to \infty} \frac{a_{n+1}}{a_n}$$

If $\rho > 1$, (including $\rho = \infty$), the series diverges.
If $\rho < 1$, the series converges.
If $\rho = 1$, the test is inconclusive.

For geometric series, the quotient $\dfrac{a_{n+1}}{a_n}$ in the ratio test is simply the (constant) ratio r of the series. Thus the ratio test is consistent with the fact that a geometric series converges if $|r| < 1$ and diverges if $|r| > 1$. In fact, as you will see later in this section, the proof of the ratio test involves the comparison of series with geometric series.

Here are four examples illustrating the use of the ratio test.

EXAMPLE 3.4

Use the ratio test to determine whether the series $\sum\limits_{n=1}^{\infty} \dfrac{n}{2^n}$ converges or diverges.

SOLUTION

$$\rho = \lim_{n \to \infty} \frac{a_{n+1}}{a_n} = \lim_{n \to \infty} \frac{(n+1)/2^{n+1}}{n/2^n} = \lim_{n \to \infty} \frac{n+1}{2^{n+1}} \cdot \frac{2^n}{n}$$

$$= \lim_{n \to \infty} \frac{n+1}{2n} = \frac{1}{2} < 1$$

and so the series converges.

Many of the series to which the ratio test can be applied involve products of the form $n(n-1)(n-2) \cdots 3 \cdot 2 \cdot 1$. It is customary to abbreviate such a product using the symbol $n!$, which is read "n factorial." For example, $3! = 3 \cdot 2 \cdot 1 = 6$. Moreover, $0!$ is *defined* to be 1 (because, as it turns out, this definition leads to compact representations of certain formulas involving factorials).

Factorial notation

If n is a positive integer,

$$n! = n(n-1)(n-2) \cdots 3 \cdot 2 \cdot 1$$

Moreover,

$$0! = 1$$

The nth term of the series in the next example involves n factorial.

EXAMPLE 3.5

Use the ratio test to determine whether the series $\sum\limits_{n=1}^{\infty} \dfrac{2^n}{n!}$ converges or diverges.

SOLUTION

$$\rho = \lim_{n \to \infty} \frac{2^{n+1}}{(n+1)!} \cdot \frac{n!}{2^n}$$

Since $(n+1)! = (n+1)(n)(n-1) \cdots 3 \cdot 2 \cdot 1 = (n+1)n!$

and $$2^{n+1} = 2(2^n)$$

it follows that

$$\rho = \lim_{n \to \infty} \frac{2(2^n)}{(n+1)n!} \cdot \frac{n!}{2^n} = \lim_{n \to \infty} \frac{2}{n+1} = 0 < 1$$

Hence the series converges.

EXAMPLE 3.6

Use the ratio test to determine whether the series $\sum\limits_{n=1}^{\infty} \dfrac{n!}{e^n}$ converges or diverges.

SOLUTION

$$\rho = \lim_{n \to \infty} \frac{(n+1)!}{e^{n+1}} \cdot \frac{e^n}{n!} = \lim_{n \to \infty} \frac{n+1}{e} = \infty$$

Hence the series diverges.

For the next example, recall from Chapter 4, Section 1, that

$$\lim_{n \to \infty} \left(1 + \frac{1}{n}\right)^n = e$$

EXAMPLE 3.7

Use the ratio test to determine whether the series $\sum\limits_{n=1}^{\infty} \dfrac{n^n}{n!}$ converges or diverges.

SOLUTION

$$\rho = \lim_{n \to \infty} \frac{(n+1)^{n+1}}{(n+1)!} \cdot \frac{n!}{n^n} = \lim_{n \to \infty} \frac{(n+1)^{n+1}}{(n+1)n^n} = \lim_{n \to \infty} \left(\frac{n+1}{n}\right)^n$$

$$= \lim_{n \to \infty} \left(1 + \frac{1}{n}\right)^n = e \approx 2.718 > 1$$

Hence the series diverges.

Why the ratio test works

Suppose $\sum\limits_{n=1}^{\infty} a_n$ is a series of positive terms and let

$$\rho = \lim_{n\to\infty} \frac{a_{n+1}}{a_n}$$

If $\rho > 1$, the ratio $\dfrac{a_{n+1}}{a_n}$ will eventually be greater than 1. That is, for large values of n,

$$\frac{a_{n+1}}{a_n} > 1 \qquad \text{or} \qquad a_{n+1} > a_n$$

This says that the individual terms of the series eventually increase as n increases and so cannot approach zero as required for convergence. Hence the series must diverge.

If $\rho < 1$, the series eventually grows no faster than a convergent geometric series and hence must converge. To see this, choose any number r such that $\rho < r < 1$. Since $\dfrac{a_{n+1}}{a_n}$ approaches ρ as n increases, this quotient will eventually be so close to ρ as to be less than r. (See Figure 3.2.) That is, there exists an integer N such that for all $n \geq N$,

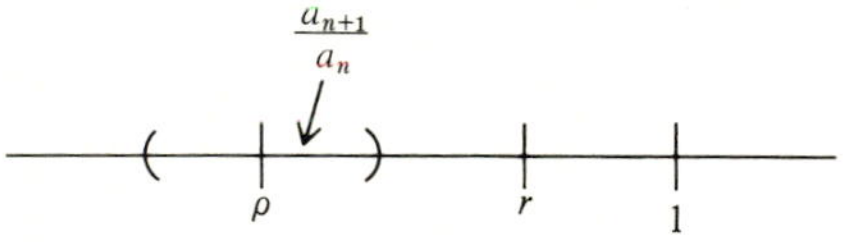

Figure 3.2 The ratio $\dfrac{a_{n+1}}{a_n}$ is close to ρ and hence less than r.

$$\frac{a_{n+1}}{a_n} < r \qquad \text{or} \qquad a_{n+1} < ra_n$$

In particular,

$$a_{N+1} < ra_N$$
$$a_{N+2} < ra_{N+1} < r^2 a_N$$
$$a_{N+3} < ra_{N+2} < r^3 a_N$$

and so on. Thus,

$$\sum_{n=N+1}^{\infty} a_n < \sum_{n=1}^{\infty} a_N r^n$$

But the geometric series on the right-hand side of this inequality converges since its ratio r is less than 1. It follows that the series on the left-hand side of the inequality must also converge, as must the original series $\sum\limits_{n=1}^{\infty} a_n$, which is obtained from this series by adding a finite number of terms.

Problems In each of the following problems, determine whether the given series converges or diverges.

1. $\displaystyle\sum_{n=1}^{\infty} \frac{1}{n^3}$

2. $\displaystyle\sum_{n=1}^{\infty} \frac{1}{n!}$

3. $\displaystyle\sum_{n=1}^{\infty} \frac{1}{\sqrt{n}}$

4. $\displaystyle\sum_{n=1}^{\infty} \frac{2^n}{n^2}$

5. $\displaystyle\sum_{n=1}^{\infty} \frac{n^2}{2^n}$

6. $\displaystyle\sum_{n=1}^{\infty} n^{-3/4}$

7. $\displaystyle\sum_{n=1}^{\infty} n^{-4/3}$

8. $\displaystyle\sum_{n=1}^{\infty} \frac{n}{n^2 + 1}$

9. $\displaystyle\sum_{n=1}^{\infty} \frac{3^n}{n!}$

10. $\displaystyle\sum_{n=1}^{\infty} \frac{n!}{2^n}$

11. $\displaystyle\sum_{n=1}^{\infty} \frac{3^n}{2^n(n!)}$

12. $\displaystyle\sum_{n=1}^{\infty} \frac{n^2}{(n^3 + 2)^2}$

13. $\displaystyle\sum_{n=1}^{\infty} \frac{n^2}{\sqrt{n^3 + 2}}$

14. $\displaystyle\sum_{n=1}^{\infty} e^{-n}$

15. $\displaystyle\sum_{n=1}^{\infty} \frac{3}{e^{2n}}$

16. $\displaystyle\sum_{n=1}^{\infty} ne^{-n^2}$

17. $\displaystyle\sum_{n=1}^{\infty} \frac{n!}{e^{n^2}}$

18. $\displaystyle\sum_{n=1}^{\infty} ne^{-n}$

19. $\displaystyle\sum_{n=2}^{\infty} \frac{3^n}{\ln n}$

20. $\displaystyle\sum_{n=2}^{\infty} \frac{1}{n(\ln n)^2}$

21. $\displaystyle\sum_{n=1}^{\infty} \frac{\ln n}{2^n}$

22. $\displaystyle\sum_{n=1}^{\infty} \frac{\ln n}{n^2}$

23. $\displaystyle\sum_{n=1}^{\infty} \frac{(n!)^2}{(2n)!}$

24. $\displaystyle\sum_{n=1}^{\infty} \frac{2^n}{3^{2n}(n + 1)}$

25. $\displaystyle\sum_{n=1}^{\infty} \frac{2^{2n}}{3^n(n + 1)}$

26. $\displaystyle\sum_{n=1}^{\infty} \frac{2^{2n}}{3^n(n + 1)!}$

27. $\displaystyle\sum_{n=1}^{\infty} \frac{n!}{n^n}$

4 ABSOLUTE AND CONDITIONAL CONVERGENCE

The two convergence tests introduced in the preceding section apply to series that contain only positive terms. In this section you will see some techniques you can use to test the convergence of series whose terms need not be positive. Such series will arise in connection with material to be covered later in this chapter.

Alternating series   A series whose terms alternate in sign is said to be an **alternating series.** For example, the series

$$\sum_{n=1}^{\infty} \left(-\frac{1}{2}\right)^n = -\frac{1}{2} + \frac{1}{4} - \frac{1}{8} + \frac{1}{16} - \cdots$$

is alternating, as is the series

$$\sum_{n=1}^{\infty} \frac{(-1)^{n+1}}{n} = 1 - \frac{1}{2} + \frac{1}{3} - \frac{1}{4} + \cdots$$

In general, an alternating series is a series of the form

$$\sum_{n=1}^{\infty} (-1)^n a_n = -a_1 + a_2 - a_3 + a_4 - \cdots$$

or

$$\sum_{n=1}^{\infty} (-1)^{n+1} a_n = a_1 - a_2 + a_3 - a_4 + \cdots$$

where the constants a_n are positive.

A simple convergence test states that an alternating series must converge if the absolute values of its terms decrease and approach zero as n increases. Here is a more precise statement of the test.

Alternating series test   **If $a_1, a_2, \ldots, a_n, \ldots$ are positive numbers such that**

(i) $a_1 \geq a_2 \geq a_3 \geq \cdots$

and

(ii) $\lim_{n \to \infty} a_n = 0$

then the alternating series

$$\sum_{n=1}^{\infty} (-1)^n a_n \quad \textbf{and} \quad \sum_{n=1}^{\infty} (-1)^{n+1} a_n$$

converge.

The formal proof of this test is not hard, but will be omitted in favor of the following geometric argument, which shows rather convincingly why the test should work.

Consider the alternating series

$$\sum_{n=1}^{\infty} (-1)^{n+1}a_n = a_1 - a_2 + a_3 - a_4 + \cdots$$

Its partial sums are

$$S_1 = a_1$$

$$S_2 = a_1 - a_2$$

$$S_3 = a_1 - a_2 + a_3$$

$$S_4 = a_1 - a_2 + a_3 - a_4$$

and so on. Figure 4.1 shows a few of these partial sums represented as

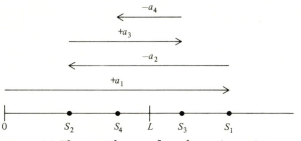

Figure 4.1 The partial sums of an alternating series.

points on a horizontal line. Notice that since the terms of the series alternate in sign, the partial sums $S_1, S_2, S_3, \ldots$ oscillate. Since the absolute values of the terms of the series decrease, each oscillation is smaller than the preceding one. Moreover, since the terms of the series approach zero as n increases, the amplitudes of the oscillations approach zero, forcing the partial sums to approach some limit L. Since the partial sums approach a finite limit L, the series converges and its sum is L.

The use of the alternating series test is illustrated in the next example.

EXAMPLE 4.1

Determine whether the series $\sum_{n=1}^{\infty} \dfrac{(-1)^{n+1}}{n}$ converges or diverges.

SOLUTION

This series,

$$\sum_{n=1}^{\infty} \frac{(-1)^{n+1}}{n} = 1 - \frac{1}{2} + \frac{1}{3} - \frac{1}{4} + \cdots$$

is an alternating series of the form $\sum_{n=1}^{\infty} (-1)^{n+1} a_n$, with $a_n = \dfrac{1}{n}$. Since

(i) $\quad 1 \geq \dfrac{1}{2} \geq \dfrac{1}{3} \geq \dfrac{1}{4} \cdots$

and

(ii) $\lim\limits_{n \to \infty} \dfrac{1}{n} = 0$

it follows from the alternating series test that the series converges.

The series in Example 4.1 is known as the **alternating harmonic series.** Notice that it converges even though the harmonic series

$$\sum_{n=1}^{\infty} \frac{1}{n} = 1 + \frac{1}{2} + \frac{1}{3} + \frac{1}{4} + \cdots$$

diverges. The convergence of the alternating harmonic series is due to the "cancellation" between its positive and negative terms.

Absolute and conditional convergence In general, a series $\sum_{n=1}^{\infty} a_n$ that contains a mixture of positive and negative terms may converge either because the absolute values of the terms approach zero "fast enough" or because of "cancellation" between its positive and negative terms. This distinction leads to the following definitions.

Absolute convergence

A series $\sum_{n=1}^{\infty} a_n$ is said to converge absolutely if both it and the corresponding series $\sum_{n=1}^{\infty} |a_n|$ of absolute values converge.

Conditional convergence

A series $\sum_{n=1}^{\infty} a_n$ is said to converge conditionally if it converges but the corresponding series $\sum_{n=1}^{\infty} |a_n|$ of absolute values diverges.

For example, the alternating series

$$\sum_{n=1}^{\infty} \frac{(-1)^{n+1}}{n^2} = 1 - \frac{1}{4} + \frac{1}{9} - \frac{1}{16} + \cdots$$

converges by the alternating series test. The corresponding series of absolute values

$$\sum_{n=1}^{\infty} \left| \frac{(-1)^{n+1}}{n^2} \right| = \sum_{n=1}^{\infty} \frac{1}{n^2} = 1 + \frac{1}{4} + \frac{1}{9} + \frac{1}{16} + \cdots$$

also converges (by the integral test). Hence the original series converges absolutely.

On the other hand, the alternating harmonic series

$$\sum_{n=1}^{\infty} \frac{(-1)^{n+1}}{n} = 1 - \frac{1}{2} + \frac{1}{3} - \frac{1}{4} + \cdots$$

converges, while the corresponding series of absolute values, the harmonic series

$$\sum_{n=1}^{\infty} \left| \frac{(-1)^{n+1}}{n} \right| = \sum_{n=1}^{\infty} \frac{1}{n} = 1 + \frac{1}{2} + \frac{1}{3} + \frac{1}{4} + \cdots$$

diverges. Hence the alternating harmonic series converges conditionally.

To show that a series converges absolutely, it suffices to show that the corresponding series of absolute values converges. Roughly speaking, this is because if $\sum_{n=1}^{\infty} |a_n|$ converges (without the aid of any cancellation of positive and negative terms), then $\sum_{n=1}^{\infty} a_n$ (in which cancellation is possible) must converge as well.

For easy reference, here is a summary of the convergence possibilities for a series $\sum_{n=1}^{\infty} a_n$ that may contain both positive and negative terms.

Hierarchy of convergence possibilities

Absolute convergence:

$$\sum_{n=1}^{\infty} a_n \text{ converges and } \sum_{n=1}^{\infty} |a_n| \text{ converges}$$

Conditional convergence:

$$\sum_{n=1}^{\infty} a_n \text{ converges but } \sum_{n=1}^{\infty} |a_n| \text{ diverges}$$

Divergence:

$$\sum_{n=1}^{\infty} a_n \text{ diverges and } \sum_{n=1}^{\infty} |a_n| \text{ diverges}$$

The two convergence tests introduced in the preceding section are for series whose terms are positive and cannot be used to test for (conditional) convergence of a series $\sum_{n=1}^{\infty} a_n$ containing both positive and negative terms. They can, however, be applied to the corresponding series $\sum_{n=1}^{\infty} |a_n|$ which contains only positive terms. In the case of the ratio test, this leads to the following generalization.

Generalized ratio test

> **Suppose $\sum_{n=1}^{\infty} a_n$ is a series of nonzero terms and let**
>
> $$\rho = \lim_{n \to \infty} \left| \frac{a_{n+1}}{a_n} \right|$$
>
> **If $\rho > 1$, (including $\rho = \infty$), the series diverges.**
> **If $\rho < 1$, the series converges absolutely.**
> **If $\rho = 1$, the test is inconclusive.**

Notice that if $\rho > 1$ in the generalized ratio test, you can conclude that the *original* series (and not just the series of absolute values) diverges. This is because $\rho > 1$ implies that neither the terms of the original series nor the terms of the series of absolute values can approach zero.

Here are two examples.

EXAMPLE 4.2

Determine whether the series $\sum_{n=1}^{\infty} (-1)^n \dfrac{3^n}{n!}$ converges absolutely, converges conditionally, or diverges.

SOLUTION

The presence of the terms 3^n and $n!$ suggests that the ratio test may be appropriate. Applying the generalized ratio test, you get

$$\rho = \lim_{n \to \infty} \left| \frac{(-1)^{n+1} 3^{n+1}}{(n+1)!} \cdot \frac{n!}{(-1)^n 3^n} \right| = \lim_{n \to \infty} \frac{3}{n+1} = 0 < 1$$

which implies that the series converges absolutely.

EXAMPLE 4.3

Determine whether the series $\displaystyle\sum_{n=1}^{\infty} (-1)^n \frac{1}{\sqrt{n}}$ converges absolutely, converges conditionally, or diverges.

SOLUTION

The series of absolute values is

$$\sum_{n=1}^{\infty} \left|(-1)^n \frac{1}{\sqrt{n}}\right| = \sum_{n=1}^{\infty} \frac{1}{\sqrt{n}}$$

which diverges by the integral test. However, the original series

$$\sum_{n=1}^{\infty} (-1)^n \frac{1}{\sqrt{n}} = -1 + \frac{1}{\sqrt{2}} - \frac{1}{\sqrt{3}} + \cdots$$

is an alternating series that converges by the alternating series test. Hence this series converges conditionally (and not absolutely).

Problems Determine whether the given series converges absolutely, converges conditionally, or diverges.

1. $\displaystyle\sum_{n=1}^{\infty} \frac{(-1)^n}{2n}$

2. $\displaystyle\sum_{n=1}^{\infty} \frac{(-1)^{n+1}}{n^3}$

3. $\displaystyle\sum_{n=1}^{\infty} \left(-\frac{1}{3}\right)^n$

4. $\displaystyle\sum_{n=1}^{\infty} \left(-\frac{3}{2}\right)^n$

5. $\displaystyle\sum_{n=1}^{\infty} \frac{3}{(-2)^n}$

6. $\displaystyle\sum_{n=1}^{\infty} \frac{(-1)^{n+1}}{2^n}$

7. $\displaystyle\sum_{n=1}^{\infty} (-1)^{n+1} \frac{n}{2n+1}$

8. $\displaystyle\sum_{n=1}^{\infty} (-1)^{n+1} n^{-3/4}$

9. $\displaystyle\sum_{n=1}^{\infty} -n^{-3/4}$

10. $\displaystyle\sum_{n=1}^{\infty} \frac{n}{(-2)^n}$

11. $\displaystyle\sum_{n=1}^{\infty} \frac{n!}{(-2)^n}$

12. $\displaystyle\sum_{n=1}^{\infty} \frac{(-2)^n}{n!}$

13. $\displaystyle\sum_{n=1}^{\infty} \frac{(-2)^n}{n^2}$

14. $\displaystyle\sum_{n=1}^{\infty} (-1)^n \left(\frac{1}{n} - \frac{1}{n+1}\right)$

15. $\displaystyle\sum_{n=1}^{\infty} \frac{(-3)^n}{(-2)^{n+1}(n!)}$

16. $\displaystyle\sum_{n=2}^{\infty} \frac{-1}{n \ln n}$

17. $\displaystyle\sum_{n=2}^{\infty} \frac{(-1)^n}{n \ln n}$

18. $\displaystyle\sum_{n=2}^{\infty} \frac{(-1)^n}{n(\ln n)^2}$

19. $\displaystyle\sum_{n=1}^{\infty} (-1)^n \frac{n^2}{(n^3 + 1)^2}$

20. $\displaystyle\sum_{n=1}^{\infty} \frac{(-1)^n \, n!}{e^{n^2}}$

21. $\displaystyle\sum_{n=1}^{\infty} (-1)^n \frac{\ln n}{3^n}$

22. $\displaystyle\sum_{n=1}^{\infty} \frac{2^{2n}}{(-3)^n(n + 1)}$

5 POWER SERIES

A power series is an infinite series of the form

$$\sum_{n=0}^{\infty} a_n x^n = a_0 + a_1 x + a_2 x^2 + \cdots$$

in which each term is a constant times a power of x. As you will see, power series are important because they lead to a method for approximating complicated functions by simpler ones.

Convergence of power series

If the variable x in a power series is given a specific numerical value, the series becomes a series of constant terms, which either converges or diverges. Thus a power series may converge for some values of x and diverge for other values of x. It can be shown that to each power series there corresponds a symmetric interval of the form $-R \le x \le R$ inside of which the series converges (absolutely) and outside of which it diverges (Figure 5.1). (At the endpoints $x = -R$

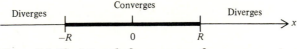

Figure 5.1 **The interval of convergence for a power series.**

and $x = R$, convergence or divergence is possible.) The number R is called the **radius of convergence** of the series, and the set of all points for which the series converges is called its **interval of convergence.** Thus the interval of convergence of a power series consists of the interval $-R < x < R$, where R is the radius of convergence, and possibly one or both of the endpoints $x = -R$ and $x = R$. If the series happens to converge for all values of x, the radius of convergence is said to be ∞. Every power series converges for $x = 0$. If this is the only value of x for which the series converges, the radius of convergence is said to be zero.

Radius of convergence

> **For each power series $\sum\limits_{n=0}^{\infty} a_n x^n$ there is a number R (possibly 0 or ∞) such that the series converges for $|x| < R$ and diverges for $|x| > R$.**

For many power series, you can find the radius of convergence by applying the generalized ratio test. Here are some examples.

EXAMPLE 5.1

Find the radius of convergence and the interval of convergence for the power series $\sum\limits_{n=1}^{\infty} \dfrac{x^n}{n(2^n)}$.

SOLUTION

As all power series do, this one converges if $x = 0$. If $x \neq 0$,

$$\rho = \lim_{n\to\infty} \left| \frac{x^{n+1}}{(n+1)(2^{n+1})} \cdot \frac{n(2^n)}{x^n} \right| = \lim_{n\to\infty} \frac{n}{2(n+1)} |x| = \frac{1}{2} |x|$$

By the ratio test, the series converges if $\rho < 1$, that is, if

$$\frac{1}{2} |x| < 1 \qquad \text{or, equivalently, if} \qquad |x| < 2$$

The series diverges if $\rho > 1$, that is, if

$$\frac{1}{2} |x| > 1 \qquad \text{or, equivalently, if} \qquad |x| > 2$$

Hence the radius of convergence is $R = 2$.

When $x = 2$, the series becomes the harmonic series

$$\sum_{n=1}^{\infty} \frac{2^n}{n(2^n)} = \sum_{n=1}^{\infty} \frac{1}{n}$$

which diverges. When $x = -2$, the series becomes the alternating harmonic series

$$\sum_{n=1}^{\infty} \frac{(-2)^n}{n(2^n)} = \sum_{n=1}^{\infty} \frac{(-1)^n 2^n}{n(2^n)} = \sum_{n=1}^{\infty} \frac{(-1)^n}{n}$$

which converges (conditionally) by the alternating series test. Hence the interval of convergence is $-2 \leq x < 2$.

EXAMPLE 5.2

Find the radius of convergence and the interval of convergence for the power series $\displaystyle\sum_{n=0}^{\infty} \frac{x^n}{n!}$.

SOLUTION

The series converges for $x = 0$. If $x \neq 0$,

$$\rho = \lim_{n \to \infty} \left| \frac{x^{n+1}}{(n+1)!} \cdot \frac{n!}{x^n} \right| = \lim_{n \to \infty} \frac{|x|}{n+1} = 0 < 1$$

In this case, $\rho < 1$ for all values of x, and so the series converges for all x. That is, the radius of convergence is $R = \infty$, and the interval of convergence is $-\infty < x < \infty$.

EXAMPLE 5.3

Find the radius of convergence and the interval of convergence for the power series $\displaystyle\sum_{n=0}^{\infty} \frac{n!x^n}{2^{n+1}}$.

SOLUTION

If $x \neq 0$,

$$\rho = \lim_{n \to \infty} \left| \frac{(n+1)!x^{n+1}}{2^{n+2}} \cdot \frac{2^{n+1}}{n!x^n} \right| = \lim_{n \to \infty} \left(\frac{n+1}{2} \right) |x| = \infty$$

Since $\rho > 1$ for all $x \neq 0$, the series converges only for $x = 0$. Hence the radius of convergence is $R = 0$, and the interval of convergence consists of the single point $x = 0$.

Power series in $x - a$　In some applications, you will encounter series of the form

$$\sum_{n=0}^{\infty} a_n(x - a)^n = a_0 + a_1(x - a) + a_2(x - a)^2 + \cdots$$

in which each term is a constant times a power of $x - a$. Series of this sort are called power series in $x - a$, and their intervals of convergence are symmetric intervals of the form $a - R < x < a + R$, including possibly one or both of the endpoints $x = a - R$ and $x = a + R$. Here is an example.

EXAMPLE 5.4

Find the radius of convergence and the interval of convergence for the power series $\sum_{n=0}^{\infty} n^2(x - 2)^n$.

SOLUTION

When $x = 2$, the series clearly converges. If $x \neq 2$,

$$\rho = \lim_{n \to \infty} \left| \frac{(n + 1)^2(x - 2)^{n+1}}{n^2(x - 2)^n} \right| = \lim_{n \to \infty} \left(\frac{n + 1}{n} \right)^2 |x - 2| = |x - 2|$$

By the ratio test, the series converges if $\rho < 1$, that is, if

$$|x - 2| < 1 \qquad \text{or, equivalently, if} \qquad 1 < x < 3$$

The series diverges if $\rho > 1$, that is, if

$$|x - 2| > 1 \qquad \text{or, equivalently, if} \qquad x < 1 \qquad \text{or} \qquad x > 3$$

Hence the radius of convergence is $R = 1$.
   If $x = 1$, the series becomes

$$\sum_{n=0}^{\infty} n^2(1 - 2)^n = \sum_{n=0}^{\infty} (-1)^n n^2$$

and if $x = 3$, the series becomes

$$\sum_{n=0}^{\infty} n^2(3 - 2)^n = \sum_{n=0}^{\infty} n^2$$

both of which diverge. Hence the interval of convergence is $1 < x < 3$.

Functions defined by power series   A power series can be thought of as a function $f(x) = \sum_{n=0}^{\infty} a_n x^n$ whose domain is the interval of convergence of the series. For any x in this interval, the value $f(x)$ of the function is the corresponding sum of the series.

   Consider, for example, the power series $\sum_{n=0}^{\infty} x^n$. This is a geometric series with ratio $r = x$, and so it converges for $|x| < 1$. According to the formula derived in Section 2, the sum of this series for $|x| < 1$ is $\frac{1}{1 - x}$. Thus,

$$\frac{1}{1 - x} = \sum_{n=0}^{\infty} x^n \qquad \text{for } |x| < 1$$

(Of course, the function $\dfrac{1}{1 - x}$ is defined for other values of x as well, but it is only on the interval $|x| < 1$ that it is equal to the sum of the power series.)

In the preceding illustration involving the geometric series, an algebraic formula was found for the sum of a given power series. In most applications, the situation is reversed. In particular, a function is given and the goal is to find a power series that converges to it on some interval. If such a series can be found, its partial sums can be used to approximate the original function. Even if the original function is difficult or impossible to evaluate directly, the partial sums of the corresponding power series are just polynomials and can be evaluated with ease. It is precisely this sort of approximation that is used to program electronic computers and calculators to generate values of functions such as e^x, $\ln x$, and the trigonometric functions.

In Section 6 you will see a general procedure you can use to find power series that converge to given functions. The remainder of this section is devoted to some special techniques by which you can sometimes modify the power series for a given function to get power series that converge to related functions.

Substitution in power series

If $f(x) = \sum\limits_{n=0}^{\infty} a_n x^n$ for $|x| < R$ and $g(x)$ is any function of x, you can get the power series for $f(g(x))$ by replacing x by $g(x)$ in the series for $f(x)$. The new series will converge for any value of x for which $|g(x)| < R$.

Substitution in power series

If

$$f(x) = \sum_{n=0}^{\infty} a_n x^n \qquad \textbf{for } |x| < R$$

then

$$f(g(x)) = \sum_{n=0}^{\infty} a_n [g(x)]^n \qquad \textbf{for } |g(x)| < R$$

Here is an example.

EXAMPLE 5.5

Starting with an appropriate geometric series, find a power series for the function $\dfrac{1}{1 + 3x^2}$ and specify the interval of convergence.

SOLUTION

Start with the fact that

$$\frac{1}{1 - x} = \sum_{n=0}^{\infty} x^n \qquad \text{for } |x| < 1$$

and replace x by $-3x^2$ to get

$$\frac{1}{1 + 3x^2} = \sum_{n=0}^{\infty} (-3x^2)^n = \sum_{n=0}^{\infty} (-3)^n x^{2n}$$

The new interval of convergence is $|-3x^2| < 1$, which can be rewritten as $|x^2| < \dfrac{1}{3}$ or $|x| < \dfrac{1}{\sqrt{3}}$.

Differentiation of power series

If you know the power series that converges to a function $f(x)$ for $|x| < R$, you can find the power series that converges to its derivative $f'(x)$ by differentiating this series term by term. Moreover, the differentiation will not affect the radius of convergence (although it may affect the convergence at the endpoints of the interval of convergence).

Differentiation of power series

If

$$f(x) = \sum_{n=0}^{\infty} a_n x^n \qquad \textbf{for } |x| < R$$

then

$$f'(x) = \sum_{n=0}^{\infty} \frac{d}{dx} (a_n x^n) = \sum_{n=1}^{\infty} n a_n x^{n-1} \qquad \textbf{for } |x| < R$$

(Notice that the series for $f'(x)$ starts with $n = 1$ since the derivative of the original constant term a_0 is zero.)

The proof of this result is complicated and will be omitted. Here are two applications.

EXAMPLE 5.6

Starting with a geometric series, find a power series for the function $\dfrac{1}{(1-x)^2}$ and specify an interval on which the series converges to the function.

SOLUTION

Start with

$$\frac{1}{1-x} = \sum_{n=0}^{\infty} x^n \quad \text{for } |x| < 1$$

and differentiate both sides of this equation to get

$$\frac{1}{(1-x)^2} = \sum_{n=1}^{\infty} nx^{n-1} \quad \text{for } |x| < 1$$

The next example involves both substitution and differentiation.

EXAMPLE 5.7

Starting with a geometric series, find a power series for the function $\dfrac{x}{(1+3x)^2}$ and specify an interval on which the series converges to the function.

SOLUTION

Start with

$$\frac{1}{1-x} = \sum_{n=0}^{\infty} x^n \quad \text{for } |x| < 1$$

and differentiate to get

$$\frac{1}{(1-x)^2} = \sum_{n=1}^{\infty} nx^{n-1} \quad \text{for } |x| < 1$$

Now replace x by $-3x$ to get

$$\frac{1}{(1+3x)^2} = \sum_{n=1}^{\infty} n(-3x)^{n-1} = \sum_{n=1}^{\infty} (-3)^{n-1}nx^{n-1} \quad \text{for } |x| < \frac{1}{3}$$

and finally multiply by x to get

$$\frac{x}{(1+3x)^2} = \sum_{n=1}^{\infty} (-3)^{n-1}nx^n \quad \text{for } |x| < \frac{1}{3}$$

Integration of power series

If you know the power series that converges to a given function, you can find the power series that converges to its antiderivative by integrating the series term by term. The integration will not affect the radius of convergence. However, it will result in the introduction of a constant of integration that will have to be evaluated.

Integration of power series

If

$$f(x) = \sum_{n=0}^{\infty} a_n x^n \qquad \textbf{for } |x| < R$$

then

$$\int f(x) \, dx = \sum_{n=0}^{\infty} \int a_n x^n \, dx + C = \sum_{n=0}^{\infty} \frac{a_n x^{n+1}}{n+1} + C \qquad \textbf{for } |x| < R$$

The proof of this result is omitted. Here is an application.

EXAMPLE 5.8

Starting with a geometric series, find a power series for the function $\ln(1 + x)$ and specify an interval on which the series converges to the function.

SOLUTION

Start with

$$\frac{1}{1-x} = \sum_{n=0}^{\infty} x^n \qquad \text{for } |x| < 1$$

Replace x by $-x$ to get

$$\frac{1}{1+x} = \sum_{n=0}^{\infty} (-1)^n x^n \qquad \text{for } |x| < 1$$

and integrate to get

$$\ln(1 + x) = \sum_{n=0}^{\infty} (-1)^n \frac{x^{n+1}}{n+1} + C \qquad \text{for } |x| < 1$$

To find the constant C, evaluate both sides of this equation for $x = 0$. Since $\ln 1 = 0$ and each term of the series is zero when $x = 0$, it follows that $C = 0$. Hence,

$$\ln(1 + x) = \sum_{n=0}^{\infty} (-1)^n \frac{x^{n+1}}{n+1} \qquad \text{for } |x| < 1$$

The power series for e^x　Here is an interesting argument that combines the differentiation of a power series and the solution of a separable differential equation to derive the power series for e^x.

Consider the series $\sum\limits_{n=0}^{\infty} \dfrac{x^n}{n!}$ which, as you saw in Example 5.2, converges for all x. Let

$$f(x) = \sum_{n=0}^{\infty} \frac{x^n}{n!} \qquad \text{for all } x$$

Then, for all x,

$$f'(x) = \sum_{n=1}^{\infty} \frac{nx^{n-1}}{n!} = \sum_{n=1}^{\infty} \frac{x^{n-1}}{(n-1)!}$$

$$= 1 + x + \frac{x^2}{2!} + \cdots$$

$$= \sum_{n=0}^{\infty} \frac{x^n}{n!} = f(x)$$

By solving the separable differential equation

$$\frac{dy}{dx} = y$$

it is easy to show that the only functions that are equal to their own derivatives are those of the form

$$f(x) = Ae^x$$

Hence,　　　　　　$$Ae^x = \sum_{n=0}^{\infty} \frac{x^n}{n!} \qquad \text{for all } x$$

When $x = 0$, the left-hand side reduces to A, while the right-hand side reduces to 1 (since $0! = 1$). Hence $A = 1$ and

$$e^x = \sum_{n=0}^{\infty} \frac{x^n}{n!} \qquad \text{for all } x$$

Approximation by power series　The use of the partial sums of the series for e^x to approximate powers of e is illustrated in the next example. Approximation by partial sums of power series will be considered further in Section 6.

EXAMPLE 5.9

Use the 7th partial sum of the series for e^x to approximate e.

SOLUTION

Start with the fact that

$$e^x = \sum_{n=0}^{\infty} \frac{x^n}{n!} \qquad \text{for all } x$$

and take $x = 1$ to get

$$e = \sum_{n=0}^{\infty} \frac{1}{n!}$$

Then

$$e \approx S_7 = \frac{1}{1} + \frac{1}{1} + \frac{1}{2} + \frac{1}{3 \cdot 2} + \frac{1}{4 \cdot 3 \cdot 2} + \frac{1}{5 \cdot 4 \cdot 3 \cdot 2} + \frac{1}{6 \cdot 5 \cdot 4 \cdot 3 \cdot 2}$$

$$= 1 + 1 + \frac{1}{2} + \frac{1}{6} + \frac{1}{24} + \frac{1}{120} + \frac{1}{720} = 2.71806$$

By analyzing the error that results from this sort of approximation, it can be shown that the actual value of e, rounded off to five decimal places, is 2.71828.

Problems In Problems 1 through 18, find the radius of convergence and the interval of convergence for the given power series.

1. $\displaystyle\sum_{n=0}^{\infty} x^n$

2. $\displaystyle\sum_{n=1}^{\infty} \frac{x^n}{2n}$

3. $\displaystyle\sum_{n=0}^{\infty} nx^n$

4. $\displaystyle\sum_{n=0}^{\infty} \frac{(n+1)x^n}{3^n}$

5. $\displaystyle\sum_{n=1}^{\infty} \frac{x^n}{n(2^{n+1})}$

6. $\displaystyle\sum_{n=0}^{\infty} n!x^n$

7. $\displaystyle\sum_{n=0}^{\infty} \frac{x^n}{n!(2^{n+1})}$

8. $\displaystyle\sum_{n=2}^{\infty} \frac{x^n}{\ln n}$

9. $\displaystyle\sum_{n=1}^{\infty} (\ln n)x^n$

10. $\displaystyle\sum_{n=0}^{\infty} \frac{n!x^n}{(2n)!}$

11. $\displaystyle\sum_{n=0}^{\infty} (-1)^n \frac{x^{2n+1}}{(2n+1)!}$

12. $\displaystyle\sum_{n=0}^{\infty} (-1)^n \frac{x^{2n+1}}{2n+1}$

13. $\displaystyle\sum_{n=0}^{\infty} 9^n n^2 x^{2n}$

14. $\displaystyle\sum_{n=0}^{\infty} \frac{(-3x)^{2n}}{4^n(n+1)}$

15. $\displaystyle\sum_{n=0}^{\infty} n(x + 2)^n$

16. $\displaystyle\sum_{n=0}^{\infty} \frac{2^{n+1}(x - 2)^n}{n!}$

17. $\displaystyle\sum_{n=0}^{\infty} \frac{n!(x + 5)^n}{2^n}$

18. $\displaystyle\sum_{n=0}^{\infty} \frac{n(x - 1)^{2n}}{4^n}$

In Problems 19 through 26, modify a geometric series to find a power series for the given function and specify an interval on which the series converges to the function.

19. $\dfrac{x}{1 - 5x}$

20. $\dfrac{1}{1 + 8x^3}$

21. $\dfrac{1}{(1 + \frac{1}{2}x)^2}$

22. $\dfrac{2x}{(1 - 9x^2)^2}$

23. $\dfrac{x}{(1 - x)^3}$

24. $\ln(1 + 2x)$

25. $\ln(1 - 3x)$

26. $x \ln(1 + x^2)$

27. Starting with a geometric series, show that

$$\ln x = \sum_{n=0}^{\infty} (-1)^n \frac{(x - 1)^{n+1}}{n + 1} \qquad \text{for } 0 < x < 2$$

In Problems 28 through 31, modify the series for e^x to find a power series for the given function.

28. e^{-x^2}

29. $e^{-x/2}$

30. $x^2 e^{2x}$

31. $2x e^{x^2/2}$

32. Use the 8th partial sum of the series for e^x to approximate $\dfrac{1}{e}$.

33. Use the 8th partial sum of the series for e^x to approximate e^2.

34. Use the 6th partial sum of the series for e^x to approximate $\sqrt{e}$.

6 TAYLOR SERIES

The purpose of this section is to develop a general procedure for finding power series that converge to given functions and to illustrate how the partial sums of these series can be used to approximate the functions. Even if the original functions are difficult or impossible to evaluate directly, the partial sums of their power series are just polynomials and can be evaluated with ease. The approximation of functions by the partial sums of their power series is how electronic computers and calculators are programmed to generate values of functions such as e^x, $\ln x$, and the trigonometric functions.

Derivation of the Taylor coefficients

Imagine that a function $f(x)$ is given and that you would like to find the corresponding coefficients a_n such that the power series $\sum\limits_{n=0}^{\infty} a_n x^n$ converges to $f(x)$ on some interval. To discover what these coefficients might be, suppose that

$$f(x) = \sum_{n=0}^{\infty} a_n x^n = a_0 + a_1 x + a_2 x^2 + a_3 x^3 + \cdots + a_n x^n + \cdots$$

If $x = 0$, only the first term in the sum is nonzero and so

$$a_0 = f(0)$$

Now differentiate the series term by term to get

$$f'(x) = a_1 + 2a_2 x + 3a_3 x^2 + \cdots + na_n x^{n-1} + \cdots$$

and let $x = 0$ to conclude that

$$a_1 = f'(0)$$

Differentiate again to get

$$f''(x) = 2a_2 + 3 \cdot 2a_3 x + \cdots + n(n-1)a_n x^{n-2} + \cdots$$

and let $x = 0$ to conclude that

$$f''(0) = 2a_2 \qquad \text{or} \qquad a_2 = \frac{f''(0)}{2}$$

Similarly, the third derivative of f is

$$f^{(3)}(x) = 3 \cdot 2a_3 + \cdots + n(n-1)(n-2)x^{n-3} + \cdots$$

and $\qquad f^{(3)}(0) = 3 \cdot 2a_3 \qquad \text{or} \qquad a_3 = \frac{f^{(3)}(0)}{3!}$

and so on. In general,

$$f^{(n)}(0) = n!a_n \qquad \text{or} \qquad a_n = \frac{f^{(n)}(0)}{n!}$$

where $f^{(n)}(0)$ denotes the nth derivative of f evaluated at $x = 0$, and $f^{(0)}(0) = f(0)$.

The preceding argument shows that *if* there is any power series that converges to $f(x)$, it must be the one whose coefficients are obtained from the derivatives of f by the formula $a_n = \dfrac{f^{(n)}(0)}{n!}$. This series is known as the **Taylor series** of f (about $x = 0$), and the corresponding coefficients a_n are called the **Taylor coefficients** of f.

Taylor series

> **The Taylor series of $f(x)$ about $x = 0$ is the power series $\displaystyle\sum_{n=0}^{\infty} a_n x^n$,**
>
> **where**
>
> $$a_n = \frac{f^{(n)}(0)}{n!}$$

Convergence of Taylor series

Although the Taylor series of a function is the only power series that can possibly converge to the function, it need not actually do so. There are, in fact, some functions whose Taylor series converge, but not to the values of the function. Fortunately, such functions rarely arise in practice. Most of the functions you are likely to encounter are equal to their Taylor series wherever the series converge.

Calculation of Taylor series

The use of the formula for the Taylor coefficients is illustrated in the following examples.

EXAMPLE 6.1

Find the Taylor series of the function $f(x) = e^x$ about $x = 0$.

SOLUTION

Compute the Taylor coefficients as follows:

$$f(x) = e^x \qquad f(0) = 1 \qquad a_0 = \frac{f(0)}{0!} = \frac{1}{0!}$$

$$f'(x) = e^x \qquad f'(0) = 1 \qquad a_1 = \frac{f'(0)}{1!} = \frac{1}{1!}$$

$$f''(x) = e^x \qquad f''(0) = 1 \qquad a_2 = \frac{f''(0)}{2!} = \frac{1}{2!}$$

$$f^{(3)}(x) = e^x \qquad f^{(3)}(0) = 1 \qquad a_3 = \frac{f^{(3)}(0)}{3!} = \frac{1}{3!}$$

$$\cdot \qquad\qquad \cdot \qquad\qquad \cdot$$
$$\cdot \qquad\qquad \cdot \qquad\qquad \cdot$$
$$\cdot \qquad\qquad \cdot \qquad\qquad \cdot$$

$$f^{(n)}(x) = e^x \qquad f^{(n)}(0) = 1 \qquad a_n = \frac{f^{(n)}(0)}{n!} = \frac{1}{n!}$$

The corresponding Taylor series is

$$\sum_{n=0}^{\infty} \frac{x^n}{n!}$$

which converges (by the ratio test) for all x. Moreover, as you saw in the preceding section, the sum of this series is indeed e^x. That is,

$$e^x = \sum_{n=0}^{\infty} \frac{x^n}{n!} = 1 + x + \frac{x^2}{2!} + \frac{x^3}{3!} + \cdots \qquad \text{for all } x$$

It should come as no surprise that the Taylor series of the function $f(x) = \dfrac{1}{1-x}$ is the geometric series. Here is the calculation.

EXAMPLE 6.2

Find the Taylor series of the function $f(x) = \dfrac{1}{1-x}$ about $x = 0$.

SOLUTION

Compute the Taylor coefficients as follows:

$$f(x) = \frac{1}{1-x} \qquad f(0) = 1 \qquad a_0 = \frac{f(0)}{0!} = \frac{1}{0!} = 1$$

$$f'(x) = \frac{1}{(1-x)^2} \qquad f'(0) = 1 \qquad a_1 = \frac{f'(0)}{1!} = \frac{1}{1!} = 1$$

$$f''(x) = \frac{2 \cdot 1}{(1-x)^3} \qquad f''(0) = 2! \qquad a_2 = \frac{f''(0)}{2!} = \frac{2!}{2!} = 1$$

$$f^{(3)}(x) = \frac{3 \cdot 2 \cdot 1}{(1-x)^4} \qquad f^{(3)}(0) = 3! \qquad a_3 = \frac{f^{(3)}(0)}{3!} = \frac{3!}{3!} = 1$$

$$\cdot \qquad\qquad\qquad \cdot \qquad\qquad\qquad \cdot$$
$$\cdot \qquad\qquad\qquad \cdot \qquad\qquad\qquad \cdot$$
$$\cdot \qquad\qquad\qquad \cdot \qquad\qquad\qquad \cdot$$

$$f^{(n)}(x) = \frac{n!}{(1-x)^{n+1}} \qquad f^{(n)}(0) = n! \qquad a_n = \frac{f^{(n)}(0)}{n!} = \frac{n!}{n!} = 1$$

The corresponding Taylor series is

$$\sum_{n=0}^{\infty} x^n$$

which, as you already know, converges to $f(x)$ for $|x| < 1$. Thus,

$$\frac{1}{1-x} = \sum_{n=0}^{\infty} x^n = 1 + x + x^2 + \cdots \qquad \text{for } |x| < 1$$

**Taylor series
about** $x = a$

In certain situations it will be inappropriate (or impossible) to represent a given function by a series of the form $\sum\limits_{n=0}^{\infty} a_n x^n$, and you will have to use a series of the form $\sum\limits_{n=0}^{\infty} a_n(x - a)^n$ instead. This is the case, for example, for the function $f(x) = \ln x$, which is undefined for $x \leq 0$ and so cannot possibly be the sum of a series of the form $\sum\limits_{n=0}^{\infty} a_n x^n$ on a symmetric interval about $x = 0$.

Using an argument similar to the one at the beginning of this section, you can show that, if $f(x) = \sum\limits_{n=0}^{\infty} a_n(x - a)^n$, the coefficients of the series are related to the derivatives of f by the formula

$$a_n = \frac{f^{(n)}(a)}{n!}$$

The power series in $x - a$ with these coefficients is called the Taylor series of $f(x)$ about $x = a$. Notice that the Taylor series about $x = 0$ is simply a special case of this more general type of Taylor series.

**Taylor series
about** $x = a$

The Taylor series of $f(x)$ **about** $x = a$ **is the power series** $\sum\limits_{n=0}^{\infty} a_n(x - a)^n$, **where**

$$a_n = \frac{f^{(n)}(a)}{n!}$$

Here is an example.

EXAMPLE 6.3

Find the Taylor series of the function $f(x) = \ln x$ about $x = 1$.

SOLUTION

Compute the Taylor coefficients as follows:

$$f(x) = \ln x \qquad\qquad f(1) = 0 \qquad\qquad a_0 = \frac{f(1)}{0!} = 0$$

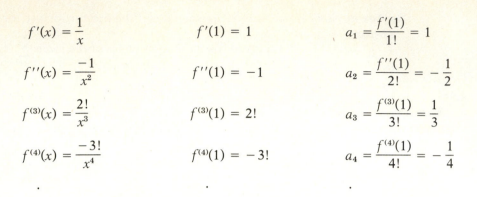

$$f'(x) = \frac{1}{x} \qquad f'(1) = 1 \qquad a_1 = \frac{f'(1)}{1!} = 1$$

$$f''(x) = \frac{-1}{x^2} \qquad f''(1) = -1 \qquad a_2 = \frac{f''(1)}{2!} = -\frac{1}{2}$$

$$f^{(3)}(x) = \frac{2!}{x^3} \qquad f^{(3)}(1) = 2! \qquad a_3 = \frac{f^{(3)}(1)}{3!} = \frac{1}{3}$$

$$f^{(4)}(x) = \frac{-3!}{x^4} \qquad f^{(4)}(1) = -3! \qquad a_4 = \frac{f^{(4)}(1)}{4!} = -\frac{1}{4}$$

$$f^{(n)}(x) = (-1)^{n+1}\frac{(n-1)!}{x^n} \quad f^{(n)}(1) = (-1)^{n+1}(n-1)! \quad a_n = \frac{(-1)^{n+1}}{n}$$

The corresponding Taylor series is

$$\sum_{n=1}^{\infty} \frac{(-1)^{n+1}}{n}(x-1)^n$$

To check the convergence of this series, apply the generalized ratio test. Since

$$\rho = \lim_{n\to\infty}\left|\frac{(-1)^{n+2}(x-1)^{n+1}}{n+1} \cdot \frac{n}{(-1)^{n+1}(x-1)^n}\right|$$

$$= \lim_{n\to\infty}\left(\frac{n}{n+1}\right)|x-1| = |x-1|$$

it follows that the series converges if $|x-1| < 1$, or, equivalently, if $0 < x < 2$. In fact, it can be shown that the series also converges at the endpoint $x = 2$ and that the sum of the series is $\ln 2$. Thus,

$$\ln x = \sum_{n=1}^{\infty} \frac{(-1)^{n+1}}{n}(x-1)^n$$

$$= (x-1) - \frac{1}{2}(x-1)^2 + \frac{1}{3}(x-1)^3 - \cdots \qquad \text{for } 0 < x \le 2$$

Taylor polynomials The partial sums of a Taylor series of a function are polynomials that can be used to approximate the function. In general, the more terms there are in the partial sum, the better the approximation will be. It is customary to let $P_n(x)$ denote the $(n+1)$st partial sum, which is a polynomial of degree (at most) n. In particular,

$$P_n(x) = f(a) + f'(a)(x - a) + \frac{f''(a)}{2!}(x - a)^2 + \cdots + \frac{f^{(n)}(a)}{n!}(x - a)^n$$

The polynomial $P_n(x)$ is known as the **Taylor polynomial** of degree n of $f(x)$ about $x = a$.

Approximation by Taylor polynomials

The approximation of $f(x)$ by its Taylor polynomials $P_n(x)$ is most accurate near $x = a$ and for large values of n. Here is a geometric argument that should give you some additional insight into the situation.

Observe first that *at* $x = a$, f and P_n are equal, as are, respectively, their first n derivatives. For example,

$$P_2(x) = f(a) + f'(a)(x - a) + \frac{f''(a)}{2!}(x - a)^2 \qquad \text{and} \qquad P_2(a) = f(a)$$

$$P_2'(x) = f'(a) + f''(a)(x - a) \qquad\qquad\qquad \text{and} \qquad P_2'(a) = f'(a)$$

$$P_2''(x) = f''(a) \qquad\qquad\qquad\qquad\qquad \text{and} \qquad P_2''(a) = f''(a)$$

The fact that $P_n(a) = f(a)$ implies that the graphs of P_n and f intersect at $x = a$. The fact that $P_n'(a) = f'(a)$ implies further that the graphs have the same slope at $x = a$. The fact that $P_n''(a) = f''(a)$ imposes the further restriction that the graphs have the same concavity at $x = a$. In general, as n increases, the number of matching derivatives increases, and the graph of P_n approximates more closely that of f near $x = a$. The situation is illustrated in Figure 6.1, which shows the graphs of e^x and its first three Taylor polynomials.

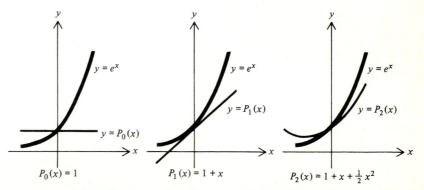

Figure 6.1 The graphs of e^x and its first three Taylor polynomials.

The use of Taylor polynomials to approximate functions is illustrated in the following examples.

EXAMPLE 6.4

Use an appropriate Taylor polynomial of degree 3 to approximate $\sqrt{4.1}$.

SOLUTION

The goal is to estimate $f(x) = \sqrt{x}$ when $x = 4.1$. Since 4.1 is close to 4 and since the values of f and its derivatives at $x = 4$ are easy to compute, it is natural to use a Taylor polynomial about $x = 4$.

Compute the Taylor coefficients about $x = 4$ as follows:

$$f(x) = \sqrt{x} \qquad f(4) = 2 \qquad a_0 = \frac{2}{0!} = 2$$

$$f'(x) = \frac{1}{2} x^{-1/2} \qquad f'(4) = \frac{1}{4} \qquad a_1 = \frac{1/4}{1!} = \frac{1}{4}$$

$$f''(x) = -\frac{1}{4} x^{-3/2} \qquad f''(4) = -\frac{1}{32} \qquad a_2 = \frac{-1/32}{2!} = -\frac{1}{64}$$

$$f^{(3)}(x) = \frac{3}{8} x^{-5/2} \qquad f^{(3)}(4) = \frac{3}{256} \qquad a_3 = \frac{3/256}{3!} = \frac{1}{512}$$

The corresponding Taylor polynomial of degree 3 is

$$P_3(x) = 2 + \frac{1}{4}(x - 4) - \frac{1}{64}(x - 4)^2 + \frac{1}{512}(x - 4)^3$$

and so

$$\sqrt{4.1} \approx P_3(4.1) = 2 + \frac{1}{4}(0.1) - \frac{1}{64}(0.1)^2 + \frac{1}{512}(0.1)^3 = 2.02485$$

Incidentally, all five decimal places of this estimate are correct. Rounded off to seven decimal places, the true value of $\sqrt{4.1}$ is 2.0248457.

In the next example, a Taylor polynomial is used to estimate a definite integral whose exact value cannot be calculated by elementary methods.

EXAMPLE 6.5

Use a Taylor polynomial of degree 8 to approximate $\int_0^1 e^{-x^2}\, dx$.

SOLUTION

The easiest way to get the Taylor series of e^{-x^2} is to start with the series

$$e^x = \sum_{n=0}^{\infty} \frac{x^n}{n!}$$

for e^x and replace x by $-x^2$ to get

$$e^{-x^2} = \sum_{n=0}^{\infty} \frac{(-1)^n x^{2n}}{n!}$$

Thus, $\qquad e^{-x^2} \approx 1 - x^2 + \frac{1}{2} x^4 - \frac{1}{6} x^6 + \frac{1}{24} x^8$

and so $\displaystyle\int_0^1 e^{-x^2}\, dx \approx \int_0^1 \left(1 - x^2 + \frac{1}{2} x^4 - \frac{1}{6} x^6 + \frac{1}{24} x^8\right) dx$

$$= \left(x - \frac{1}{3} x^3 + \frac{1}{10} x^5 - \frac{1}{42} x^7 + \frac{1}{216} x^9\right)\Big|_0^1$$

$$= 1 - \frac{1}{3} + \frac{1}{10} - \frac{1}{42} + \frac{1}{216} = 0.7475$$

The accuracy of Taylor approximations

The difference between a function $f(x)$ and its nth-degree Taylor polynomial $P_n(x)$ is called the nth **remainder** and is denoted by $R_n(x)$. Thus,

$$R_n(x) = f(x) - P_n(x)$$

It is proved in more advanced courses that the nth remainder is given by the following formula, which resembles the formula for the $(n + 1)$st term of the Taylor series.

The remainder formula

If $P_n(x)$ is the nth degree Taylor polynomial of $f(x)$ about $x = a$, then

$$f(x) - P_n(x) = R_n(x)$$

where

$$R_n(x) = \frac{f^{n+1}(c)}{(n + 1)!} (x - a)^{n+1}$$

for some number c between a and x.

Notice that the number c in the formula for $R_n(x)$ is not specified. Its value depends on n and x and usually cannot be determined. Nevertheless, the knowledge that c must lie between a and x is often sufficient for the use of the formula.

One way the remainder formula is used is to determine whether or not the Taylor series of a given function actually converges to the function. In particular, since

$$f(x) = P_n(x) + R_n(x)$$

the Taylor series of f converges to $f(x)$ if and only if

$$\lim_{n\to\infty} R_n(x) = 0$$

Another important application of the remainder formula is the estimation of the accuracy of Taylor approximations. This is illustrated in the following two examples.

EXAMPLE 6.6

Use a Taylor polynomial of degree 3 to approximate ln 1.2, and estimate the accuracy of the approximation.

SOLUTION

From Example 6.3 you know that

$$\ln x = \sum_{n=1}^{\infty} \frac{(-1)^{n+1}}{n} (x-1)^n \qquad \text{for } 0 < x \leq 2$$

Hence,

$$P_3(x) = (x-1) - \frac{1}{2}(x-1)^2 + \frac{1}{3}(x-1)^3$$

and

$$\ln 1.2 \approx P_3(1.2) = 0.2 - \frac{1}{2}(0.2)^2 + \frac{1}{3}(0.2)^3 = 0.1827$$

To estimate the accuracy of this approximation, look at the absolute value of the corresponding remainder $R_3(1.2)$, which is given by the formula

$$|R_3(1.2)| = \left| \frac{f^{(4)}(c)}{4!}(1.2-1)^4 \right| = \left| \frac{f^{(4)}(c)}{4!}(0.2)^4 \right|$$

for some number c with $1 \leq c \leq 1.2$.

From Example 6.3 you know that

$$f^{(4)}(x) = -\frac{3!}{x^4}$$

and so

$$|R_3(1.2)| = \left| -\frac{3!}{c^4}\left(\frac{1}{4!}\right)(0.2)^4 \right| = \frac{(0.2)^4}{4c^4}$$

Moreover, since $1 \leq c \leq 1.2$,

$$\frac{1}{c^4} \leq \frac{1}{1^4} = 1$$

Hence, $$|R_3(1.2)| \leq \frac{(0.2)^4}{4} = 0.0004$$

which says that the approximation error is no greater than 0.0004. (In fact, the true value of ln 1.2, rounded off to four decimal places, is 0.1823, which differs from the approximation 0.1827 by 0.0004.)

EXAMPLE 6.7

Use a Taylor polynomial to approximate $e^{0.5}$ with an error of less than 0.00005.

SOLUTION

The strategy is to use the remainder formula to determine the smallest value of n that will guarantee the desired accuracy, and then to use the corresponding Taylor polynomial P_n for the approximation.

Let $f(x) = e^x$. Then $f^{(n+1)}(x) = e^x$, and so the absolute value of the nth remainder (about $x = 0$) is

$$|R_n(0.5)| = \left| \frac{f^{(n+1)}(c)}{(n+1)!} (0.5)^{n+1} \right| = \frac{e^c}{(n+1)!} (0.5)^{n+1}$$

for some number c with $0 \leq c \leq 0.5$.

Since $c \leq 0.5$, it follows that

$$|R_n(0.5)| \leq \frac{e^{0.5}}{(n+1)!} (0.5)^{n+1}$$

Unfortunately, this estimate involves $e^{0.5}$, which is the number you are trying to approximate. However, since $e \approx 2.718 < 4$, it follows that

$$e^{0.5} = \sqrt{e} < \sqrt{4} = 2$$

and so $$|R_n(0.5)| < \frac{2}{(n+1)!} (0.5)^{n+1}$$

Now, using your calculator, compute $\frac{2(0.5)^{n+1}}{(n+1)!}$ for $n = 1, 2, 3, \ldots$ until you get an answer that is less than 0.00005. You should find that for $n = 4$,

$$\frac{2(0.5)^5}{5!} = 0.00052 > 0.00005$$

and for $n = 5$,

$$\frac{2(0.5)^6}{6!} = 0.000043 < 0.00005$$

Thus $n = 5$ guarantees the desired accuracy, and you use $P_5(0.5)$ for the approximation.

From Example 6.1 you know that the Taylor series for e^x is

$$e^x = \sum_{n=0}^{\infty} \frac{x^n}{n!}$$

so that
$$P_5(x) = 1 + x + \frac{x^2}{2!} + \frac{x^3}{3!} + \frac{x^4}{4!} + \frac{x^5}{5!}$$

Hence,

$$e^{0.5} \approx P_5(0.5) = 1 + 0.5 + \frac{(0.5)^2}{2!} + \frac{(0.5)^3}{3!} + \frac{(0.5)^4}{4!} + \frac{(0.5)^5}{5!} = 1.648698$$

where the approximation error is less than 0.00005.

Problems   In Problems 1 through 8, use the formula for the Taylor coefficients to find the Taylor series of f about $x = 0$.

1. $f(x) = e^{3x}$

2. $f(x) = e^{-2x}$

3. $f(x) = \ln(1 + x)$

4. $f(x) = \dfrac{1}{2 - x}$

5. $f(x) = \dfrac{e^x + e^{-x}}{2}$

6. $f(x) = \dfrac{e^x - e^{-x}}{2}$

7. $f(x) = (1 + x)e^x$

8. $f(x) = (3x + 2)e^x$

In Problems 9 through 14, use the formula for the Taylor coefficients to find the Taylor series of f about $x = a$.

9. $f(x) = e^{2x};\ a = 1$

10. $f(x) = e^{-3x};\ a = -1$

11. $f(x) = \dfrac{1}{x};\ a = 1$

12. $f(x) = \ln 2x;\ a = \dfrac{1}{2}$

13. $f(x) = \dfrac{1}{2 - x};\ a = 1$

14. $f(x) = \dfrac{1}{1 + x};\ a = 2$

In Problems 15 through 20, use a Taylor polynomial of degree n to approximate the given number.

15. $\sqrt{3.8};\ n = 3$

16. $\sqrt{1.2};\ n = 3$

17. $\ln 1.1;\ n = 5$

18. $\ln 0.7;\ n = 5$

19. $e^{0.3};\ n = 4$

20. $\dfrac{1}{\sqrt{e}};\ n = 4$

In Problems 21 through 24, use a Taylor polynomial of degree n to approximate the given integral.

21. $\int_0^{1/2} e^{-x^2} \, dx; \ n = 6$

22. $\int_{-0.2}^{0.1} e^{-x^2} \, dx; \ n = 4$

23. $\int_0^{0.1} \dfrac{1}{1 + x^2} \, dx; \ n = 4$

24. $\int_{-1/2}^{0} \dfrac{1}{1 - x^3} \, dx; \ n = 9$

In Problems 25 through 30, use a Taylor polynomial of degree n to approximate the given number, and estimate the accuracy of the approximation. (In Problem 27, use the fact that $e < 3$ in your estimate of the accuracy.)

25. $\ln 1.3; \ n = 5$

26. $\ln 0.9; \ n = 3$

27. $e^{0.2}; \ n = 3$

28. $\dfrac{1}{e}; \ n = 7$

29. $\sqrt{1.02}; \ n = 2$

30. $\sqrt{4.5}; \ n = 2$

In Problems 31 through 36, use an appropriate Taylor polynomial to approximate the given number with an error less than 0.00005. (In Problem 33, use the fact that $e < 3$ in your estimate of the accuracy.)

31. $\ln 1.2$

32. $\ln 0.8$

33. e

34. $e^{-0.2}$

35. $\sqrt{1.1}$

36. $\sqrt{4.05}$

Important terms, symbols, and formulas

CHAPTER SUMMARY AND PROFICIENCY TEST

Infinite series: $\displaystyle\sum_{n=1}^{\infty} a_n = a_1 + a_2 + a_3 + \cdots$

nth partial sum: $S_n = a_1 + a_2 + \cdots + a_n$

Convergence and divergence: $\displaystyle\sum_{n=1}^{\infty} a_n$ converges to S if and only if $\lim\limits_{n \to \infty} S_n = S$.

Test for divergence: If $\lim\limits_{n \to \infty} a_n \neq 0$, then $\displaystyle\sum_{n=1}^{\infty} a_n$ diverges.

Distributive law: $\displaystyle\sum_{n=1}^{\infty} ca_n = c \sum_{n=1}^{\infty} a_n$

Harmonic series: $\displaystyle\sum_{n=1}^{\infty} \dfrac{1}{n}$ diverges.

Geometric series: If $|r| < 1$, $\sum\limits_{n=0}^{\infty} r^n = \dfrac{1}{1-r}$.

Integral test: If $a_n = f(n)$, where $f(x)$ is positive and decreasing for $x \geq 1$, then

$\sum\limits_{n=1}^{\infty} a_n$ converges if and only if $\int_{1}^{\infty} f(x)\, dx$ converges.

Ratio test: Suppose $\sum\limits_{n=1}^{\infty} a_n$ is a series of positive terms and $\rho = \lim\limits_{n\to\infty} \dfrac{a_{n+1}}{a_n}$

If $\rho > 1$, the series diverges.   If $\rho < 1$, the series converges.

Factorial notation: $n! = n(n-1)(n-2)\ \ldots\ 3 \cdot 2 \cdot 1;\ 0! = 1$

Conditional convergence; absolute convergence

Alternating series test: If

(i) $a_1 \geq a_2 \geq a_3 \geq \cdots$ and   (ii) $\lim\limits_{n\to\infty} a_n = 0$

then the alternating series $\sum\limits_{n=1}^{\infty} (-1)^n a_n$ and $\sum\limits_{n=1}^{\infty} (-1)^{n+1} a_n$ converge.

Generalized ratio test: Suppose $\sum\limits_{n=1}^{\infty} a_n$ is a series of nonzero terms and

$$\rho = \lim\limits_{n\to\infty} \left| \dfrac{a_{n+1}}{a_n} \right|$$

If $\rho > 1$, the series diverges.   If $\rho < 1$, the series converges absolutely.

Power series: $\sum\limits_{n=0}^{\infty} a_n x^n$; $\sum\limits_{n=0}^{\infty} a_n(x-a)^n$

Radius of convergence; interval of convergence

Substitution in power series

Differentiation of power series term by term

Integration of power series term by term

Taylor series of $f(x)$ about $x = a$: $\sum\limits_{n=0}^{\infty} a_n(x-a)^n$, where $a_n = \dfrac{f^{(n)}(a)}{n!}$

Important Taylor series:

$$e^x = \sum\limits_{n=0}^{\infty} \dfrac{x^n}{n!} \qquad \text{for all } x$$

$$\dfrac{1}{1-x} = \sum\limits_{n=0}^{\infty} x^n \qquad \text{for } |x| < 1$$

$$\ln x = \sum\limits_{n=1}^{\infty} \dfrac{(-1)^{n+1}}{n}(x-1)^n \qquad \text{for } 0 < x \leq 2$$

Taylor polynomial of degree n: $P_n(x)$

Taylor approximation: $f(x) = P_n(x) + R_n(x)$, where

$$R_n(x) = \dfrac{f^{n+1}(c)}{(n+1)!}(x-a)^{n+1} \qquad \text{for some number } c \text{ between } a \text{ and } x.$$

Proficiency test   In Problems 1 through 3, find the sum of the given convergent series by taking the limit of a compact expression for the nth partial sum.

1. $\displaystyle\sum_{n=1}^{\infty} \left(\frac{1}{n+1} - \frac{1}{n+3} \right)$

2. $\displaystyle\sum_{n=2}^{\infty} \frac{1}{n(n-1)}$

3. $\displaystyle\sum_{n=1}^{\infty} \frac{2}{(-3)^n}$

In Problems 4 through 7, determine whether the given geometric series converges, and if so, find its sum.

4. $\displaystyle\sum_{n=1}^{\infty} \frac{3}{(-5)^n}$

5. $\displaystyle\sum_{n=0}^{\infty} \left(-\frac{3}{2} \right)^n$

6. $\displaystyle\sum_{n=1}^{\infty} e^{-0.5n}$

7. $\displaystyle\sum_{n=2}^{\infty} \frac{2^{n+1}}{3^{n-3}}$

8. Express the repeating decimal 1.545454 . . . as a fraction.

9. A ball is dropped from a height of 6 feet and allowed to bounce indefinitely. How far will the ball travel if it always rebounds to 80 percent of its previous height?

10. How much should you invest today at an annual interest rate of 12 percent compounded continuously so that, starting next year, you can make annual withdrawals of \$500 in perpetuity?

11. A patient is given an injection of 10 units of a certain drug every 24 hours. The drug is eliminated exponentially so that the fraction that remains in the patient's body after t days is $f(t) = e^{-0.8t}$. If the treatment is continued indefinitely, approximately how many units of the drug will eventually be in the patient's body immediately following an injection?

12. You and two friends take turns rolling a die until one of you wins by getting a six. Find the probability that you will win if you roll second.

In Problems 13 through 20, determine whether the given series converges or diverges.

13. $\displaystyle\sum_{n=1}^{\infty} \frac{3^n}{n^2}$

14. $\displaystyle\sum_{n=1}^{\infty} n^{-3/2}$

15. $\displaystyle\sum_{n=1}^{\infty} \frac{2^n}{(2n)!}$

16. $\displaystyle\sum_{n=1}^{\infty} \frac{\ln n}{3^n}$

17. $\displaystyle\sum_{n=2}^{\infty} \frac{1}{n\sqrt{\ln n}}$

18. $\displaystyle\sum_{n=1}^{\infty} \frac{e^{-\sqrt{n}}}{\sqrt{n}}$

19. $\displaystyle\sum_{n=1}^{\infty} \frac{5^n}{3^{2n}(n+1)}$

20. $\displaystyle\sum_{n=1}^{\infty} \frac{(2n)!}{(n!)^2}$

In Problems 21 through 23, determine whether the given series converges absolutely, converges conditionally, or diverges.

21. $\displaystyle\sum_{n=1}^{\infty} \frac{(-1)^n}{3\sqrt{n}}$

22. $\displaystyle\sum_{n=1}^{\infty} (-1)^n \frac{n!}{3^n}$

23. $\displaystyle\sum_{n=1}^{\infty} \frac{(-2)^{n+1}}{(-3)^n(n!)}$

In Problems 24 through 27, find the radius of convergence and the interval of convergence for the given power series.

24. $\displaystyle\sum_{n=1}^{\infty} \frac{3^{n+1}x^n}{n}$

25. $\displaystyle\sum_{n=0}^{\infty} \frac{nx^{2n}}{n!}$

26. $\displaystyle\sum_{n=1}^{\infty} \frac{3^{2n+1}(2x)^{2n}}{n}$

27. $\displaystyle\sum_{n=1}^{\infty} \frac{4^n(x+1)^{2n+1}}{n^2}$

In Problems 28 and 29, modify a geometric series to find a power series for the given function, and specify an interval on which the series converges to the function.

28. $\ln(1 + 3x^2)$

29. $\dfrac{-2x^2}{(1+2x)^2}$

In Problems 30 through 33, use the formula for the Taylor coefficients to find the Taylor series of f about $x = a$.

30. $f(x) = \dfrac{1}{x+3};\ a = 0$

31. $f(x) = e^{-3x};\ a = 0$

32. $f(x) = (1 + 2x)e^x;\ a = 0$

33. $f(x) = \dfrac{1}{(1+x)^2};\ a = -2$

34. Use a Taylor polynomial of degree 3 to approximate $\sqrt{0.9}$.

35. Use a Taylor polynomial of degree 10 to approximate

$$\int_0^{1/2} \frac{x}{1+x^3}\, dx$$

36. Use a Taylor polymonial of degree 5 to approximate $1/\sqrt{e}$, and estimate the accuracy of the approximation.

37. Use an appropriate Taylor polynomial to approximate $\ln 0.9$ with an error less than 0.00005.

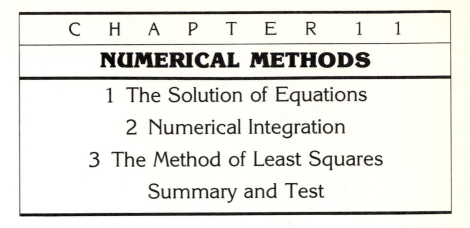

C H A P T E R   1 1

NUMERICAL METHODS

1 The Solution of Equations

2 Numerical Integration

3 The Method of Least Squares

Summary and Test

1 THE SOLUTION OF EQUATIONS

Functions and data that arise in practical situations are often much more unruly than those in the simplified examples found in calculus books. Direct calculation is frequently difficult or impossible, and approximation must be used. In Section 6 of the preceding chapter, you saw how to use Taylor polynomials to approximate differentiable functions. In this chapter, you will be introduced to some additional approximation techniques. In particular, Section 1 will deal with techniques for approximating the solutions of equations, Section 2 with techniques for estimating definite integrals, and Section 3 with a method for fitting the "best" straight line through experimental data. All these techniques belong to the branch of mathematics known as **numerical analysis,** which has grown in importance as calculators and computers have reduced the time and effort required to perform the numerical calculations.

The roots of an equation

A value of a variable that satisfies an equation is said to be a **solution** or **root** of the equation. In this section, you will see two techniques that can be used to approximate roots of equations of the form

$f(x) = 0$. In geometric terms, a root of such an equation is an x intercept of the graph of f.

Most techniques for approximating a root of an equation require that you start with a rough estimate of the location of the root. The following property of continuous functions is often used to obtain this preliminary estimate.

The intermediate value theorem

If $f(x)$ **is continuous on the interval** $a \le x \le b$, **and if** $f(a)$ **and** $f(b)$ **have opposite signs, then the equation** $f(x) = 0$ **has at least one root between** $x = a$ **and** $x = b$.

In geometric terms, the intermediate value theorem says that if f is continuous and if the points $(a, f(a))$ and $(b, f(b))$ lie on opposite sides of the x axis, then the graph of f (which is an unbroken curve) must cross the x axis somewhere between $x = a$ and $x = b$. The situation is illustrated in Figure 1.1.

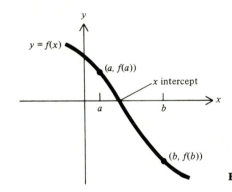

Figure 1.1 The intermediate value theorem.

The use of the intermediate value theorem to estimate the location of roots is illustrated in the following example.

EXAMPLE 1.1

Make a rough sketch of the function $f(x) = x^3 - x^2 - 1$, and use the intermediate value theorem to estimate the location of the roots of the equation $f(x) = 0$.

SOLUTION

The graph (obtained with the aid of the first derivative) is sketched in Figure 1.2. It indicates that the equation $f(x) = 0$ has only one root. Moreover, since

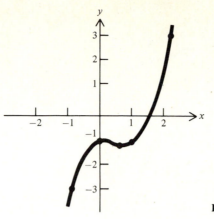

Figure 1.2 The graph of $y = x^3 - x^2 - 1$.

$$f(1) = -1 < 0 \quad \text{and} \quad f(2) = 3 > 0$$

it follows from the intermediate value theorem that the root lies between $x = 1$ and $x = 2$.

The bisection method

The intermediate value theorem is the basis of the following elementary procedure that can be used to approximate roots of equations to any desired degree of accuracy.

Suppose f is a continuous function on $a \leq x \leq b$ with, say, $f(a) > 0$ and $f(b) < 0$. Then, by the intermediate value theorem, the equation $f(x) = 0$ has a root somewhere between $x = a$ and $x = b$. To estimate the location of this root with more precision, evaluate f at the midpoint, $x_1 = \dfrac{a + b}{2}$, of the interval $a \leq x \leq b$. If it happens that $f(x_1) = 0$, you have found the root you were looking for. If $f(x_1) < 0$, the root must lie between $x = a$ (where f is positive) and $x = x_1$ (where f is negative). On the other hand, if $f(x_1) > 0$, the root must lie between $x = x_1$ (where f is positive) and $x = b$ (where f is negative). In either case, you have found a new interval, only half as wide as the original one, which must contain the desired root.

If you now apply this procedure to the new interval, you will get another interval, half again as wide, which must contain the root. Continuing in this manner, you can pinpoint the location of the root to any desired degree of accuracy. The situation is illustrated in Figure 1.3.

The use of the bisection method is illustrated in the following two examples. Notice that at each stage of the procedure, the midpoint x_n

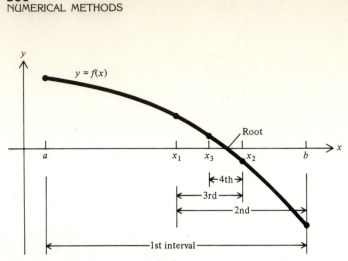

Figure 1.3 The bisection method.

of the interval approximates the root with a maximum error E_n, which is no greater than the width of the next interval.

EXAMPLE 1.2

Use the bisection method to approximate the roots of the equation $x^3 - x^2 - 1 = 0$ with an error no greater than $\frac{1}{16}$.

SOLUTION

The goal is to find the roots of the equation $f(x) = 0$, where

$$f(x) = x^3 - x^2 - 1$$

As you saw in Example 1.1, there is only one such root, and since $f(1) = -1 < 0$ and $f(2) = 3 > 0$, this root must lie in the interval $1 \le x \le 2$. Starting with this interval, apply the bisection method as follows:

The first estimate (see Figure 1.4):

$$x_1 = \frac{1+2}{2} = \frac{3}{2}; \qquad E_1 = \frac{1}{2}; \qquad f\left(\frac{3}{2}\right) = 0.125 > 0$$

Next interval: $1 \le x \le \frac{3}{2}$ $\quad \left[\text{since } f(1) < 0 \text{ and } f\left(\frac{3}{2}\right) > 0\right]$

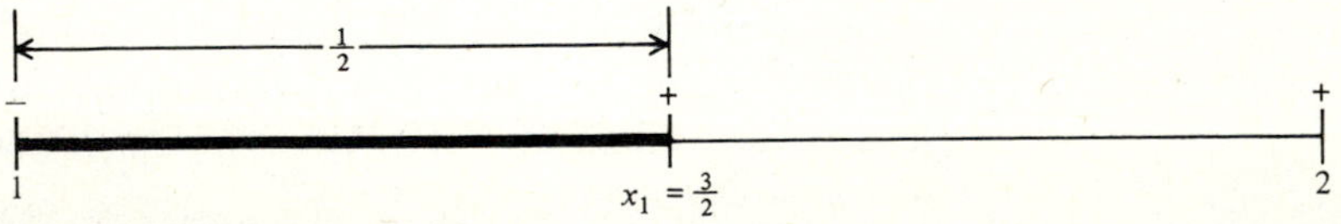

Figure 1.4 The first estimate.

The second estimate (see Figure 1.5):

$$x_2 = \frac{1}{2}\left(1 + \frac{3}{2}\right) = \frac{5}{4}; \qquad E_2 = \frac{1}{4}; \qquad f\left(\frac{5}{4}\right) = -0.6094 < 0$$

Next interval: $\dfrac{5}{4} \le x \le \dfrac{3}{2}$

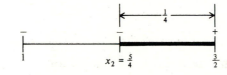

$x_2 = \frac{5}{4}$ **Figure 1.5 The second estimate.**

The third estimate (see Figure 1.6):

$$x_3 = \frac{1}{2}\left(\frac{5}{4} + \frac{3}{2}\right) = \frac{11}{8}; \qquad E_3 = \frac{1}{8}; \qquad f\left(\frac{11}{8}\right) = -0.2910 < 0$$

Next interval: $\dfrac{11}{8} \le x \le \dfrac{3}{2}$

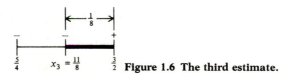

$x_3 = \frac{11}{8}$ **Figure 1.6 The third estimate.**

The fourth estimate (see Figure 1.7):

$$x_4 = \frac{1}{2}\left(\frac{11}{8} + \frac{3}{2}\right) = \frac{23}{16}; \qquad E_4 = \frac{1}{16}$$

$x_4 = \frac{23}{16}$ **Figure 1.7 The fourth estimate.**

Thus the desired root is approximately $x_4 = \frac{23}{16} = 1.4375$, and the error is no greater than $E_4 = \frac{1}{16} = 0.0625$.

EXAMPLE 1.3

Use the bisection method to approximate $\sqrt{5}$ with an error no greater than $\frac{1}{16}$.

SOLUTION

Observe that $\sqrt{5}$ is the root of the equation $f(x) = 0$, where

$$f(x) = x^2 - 5$$

Since $f(2) < 0$ and $f(3) > 0$, apply the bisection method starting with the interval $2 \leq x \leq 3$.

The first estimate:

$$x_1 = \frac{2 + 3}{2} = \frac{5}{2}; \qquad E_1 = \frac{1}{2}; \qquad f\left(\frac{5}{2}\right) = 1.25 > 0$$

Next interval: $2 \leq x \leq \dfrac{5}{2}$

The second estimate:

$$x_2 = \frac{1}{2}\left(2 + \frac{5}{2}\right) = \frac{9}{4}; \qquad E_2 = \frac{1}{4}; \qquad f\left(\frac{9}{4}\right) = 0.0625 > 0$$

Next interval: $2 \leq x \leq \dfrac{9}{4}$

The third estimate:

$$x_3 = \frac{1}{2}\left(2 + \frac{9}{4}\right) = \frac{17}{8}; \qquad E_3 = \frac{1}{8}; \qquad f\left(\frac{17}{8}\right) = -0.4844 < 0$$

Next interval: $\dfrac{17}{8} \leq x \leq \dfrac{9}{4}$

The fourth estimate:

$$x_4 = \frac{1}{2}\left(\frac{17}{8} + \frac{9}{4}\right) = \frac{35}{16}; \qquad E_4 = \frac{1}{16}$$

Hence $\sqrt{5}$ is approximately $x_4 = \frac{35}{16} = 2.1875$ with an error no greater than $E_4 = \frac{1}{16} = 0.0625$.

Convergence of the bisection method

The approximations $x_1, x_2, \ldots$ generated by the bisection method converge to the true value of the desired root. However, the rate of this convergence is fairly slow. After four steps in Example 1.3, for instance, the maximum possible error was $E_4 = \frac{1}{16} = 0.0625$. After n steps, the maximum error would be $\dfrac{1}{2^n}$, and more than 14 steps would be required to guarantee four-decimal-place accuracy.

Because of the slow rate of convergence of the bisection method, other methods of approximation are usually preferred. One, known as **Newton's method,** uses the derivative of the function to increase the rate of convergence.

Newton's method The basic idea behind Newton's method is illustrated in Figure 1.8a, in which r is a root of the equation $f(x) = 0$, x_0 is an approximation to r, and x_1 is a better approximation obtained by taking the x intercept of the line that is tangent to the graph of f at $(x_0, f(x_0))$.

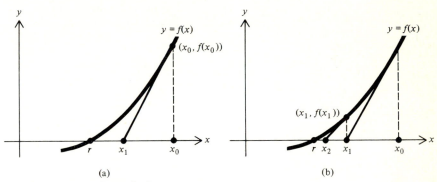

(a)   (b)

Figure 1.8 Newton's method.

To find a formula for the improved approximation x_1, recall that the slope of the tangent line through $(x_0, f(x_0))$ is the derivative $f'(x_0)$. Hence,

$$f'(x_0) = \text{slope} = \frac{\Delta y}{\Delta x} = \frac{f(x_0) - 0}{x_0 - x_1}$$

or, equivalently,

$$x_1 = x_0 - \frac{f(x_0)}{f'(x_0)}$$

If the procedure is repeated using x_1 as the initial approximation, an even better approximation is obtained (Figure 1.8b). This approximation, x_2, is related to x_1 as x_1 was related to x_0. That is,

$$x_2 = x_1 - \frac{f(x_1)}{f'(x_1)}$$

The process can be continued until the desired degree of accuracy is obtained. In general, the nth approximation x_n is related to the $(n - 1)$st by the formula

$$x_n = x_{n-1} - \frac{f(x_{n-1})}{f'(x_{n-1})}$$

Newton's method

To approximate a root of the equation $f(x) = 0$, start with a preliminary estimate x_0 and generate a sequence of increasingly accurate approximations $x_1, x_2, x_3, \ldots$ using the formula

$$x_n = x_{n-1} - \frac{f(x_{n-1})}{f'(x_{n-1})}$$

The use of Newton's method is illustrated in the following example.

EXAMPLE 1.4

Use three repetitions of Newton's method to approximate $\sqrt{5}$.

SOLUTION

As in Example 1.3, the goal is to find the root of the equation $f(x) = 0$, where

$$f(x) = x^2 - 5$$

The derivative of f is $f'(x) = 2x$, and so

$$x - \frac{f(x)}{f'(x)} = x - \frac{x^2 - 5}{2x} = \frac{x^2 + 5}{2x}$$

Thus, for $n = 1, 2, \ldots$

$$x_n = \frac{x_{n-1}^2 + 5}{2x_{n-1}}$$

A convenient choice for the preliminary estimate is $x_0 = 2$ (since $2^2 = 4$, which is close to 5). Then,

$$x_1 = \frac{x_0^2 + 5}{2x_0} = 2.250 \qquad \text{(using } x_0 = 2\text{)}$$

$$x_2 = \frac{x_1^2 + 5}{2x_1} = 2.2361111 \qquad \text{(using } x_1 = 2.250\text{)}$$

$$x_3 = \frac{x_2^2 + 5}{2x_2} = 2.2360680 \qquad \text{(using } x_2 = 2.2361111\text{)}$$

Thus, $\sqrt{5}$ is approximately 2.236. (Actually, rounded off to seven decimal places, $\sqrt{5} = 2.2360680$, which is exactly the value obtained by Newton's method after only three repetitions!

Estimating the accuracy of Newton's method

A systematic analysis of the error involved in approximation by Newton's method is more complicated than that for the bisection method. However, there is a simple rule of thumb, which states that, to use Newton's method to approximate a root to a given number of decimal places, round off all calculations to one more decimal place and stop when there is no change from one approximation to the next. This is illustrated in the following examples.

EXAMPLE 1.5

Use Newton's method to approximate the root of the equation $x^3 - x^2 - 1 = 0$ to three decimal places.

SOLUTION

As in Example 1.2, let

$$f(x) = x^3 - x^2 - 1$$

Then,

$$x - \frac{f(x)}{f'(x)} = x - \frac{x^3 - x^2 - 1}{3x^2 - 2x} = \frac{2x^3 - x^2 + 1}{3x^2 - 2x}$$

and so, for $n = 1, 2, 3, \ldots$,

$$x_n = \frac{2x_{n-1}^3 - x_{n-1}^2 + 1}{3x_{n-1}^2 - 2x_{n-1}}$$

Since $f(1) = -1 < 0$ and $f(2) = 3 > 0$, there is a root of the equation $f(x) = 0$ between 1 and 2. Either of these two values is a reasonable choice for the initial estimate x_0. If you take $x_0 = 1$, you get

$$x_1 = \frac{2x_0^3 - x_0^2 + 1}{3x_0^2 - 2x_0} = 2 \qquad \text{(using } x_0 = 1)$$

$$x_2 = \frac{2x_1^3 - x_1^2 + 1}{3x_1^2 - 2x_1} = 1.6250 \qquad \text{(using } x_1 = 2)$$

$$x_3 = \frac{2x_2^3 - x_2^2 + 1}{3x_2^2 - 2x_2} = 1.4858 \qquad \text{(using } x_2 = 1.6250)$$

$$x_4 = \frac{2x_3^3 - x_3^2 + 1}{3x_3^2 - 2x_3} = 1.4660 \qquad \text{(using } x_3 = 1.4858)$$

$$x_5 = \frac{2x_4^3 - x_4^2 + 1}{3x_4^2 - 2x_4} = 1.4656 \qquad \text{(using } x_4 = 1.4660)$$

$$x_6 = \frac{2x_5^3 - x_5^2 + 1}{3x_5^2 - 2x_5} = 1.4656 \qquad \text{(using } x_5 = 1.4656)$$

Since x_5 and x_6 are identical (to four decimal places), no further computation is necessary. You now simply round off the number 1.4656 to conclude that, to three-decimal-place accuracy, the desired root is 1.466.

EXAMPLE 1.6

Use Newton's method to approximate the solution of the equation $e^x = -x$ to three decimal places.

SOLUTION

From the graphs of $y = e^x$ and $y = -x$ in Figure 1.9, you see that the equation has a solution somewhere between $x = -1$ and $x = 0$.

Let

$$f(x) = e^x + x$$

so that the desired solution is the root of the equation $f(x) = 0$. Then,

$$x - \frac{f(x)}{f'(x)} = x - \frac{e^x + x}{e^x + 1} = \frac{(x - 1)e^x}{e^x + 1}$$

and so, for $n = 1, 2, 3, \ldots$,

$$x_n = \frac{(x_{n-1} - 1)e^{x_{n-1}}}{e^{x_{n-1}} + 1}$$

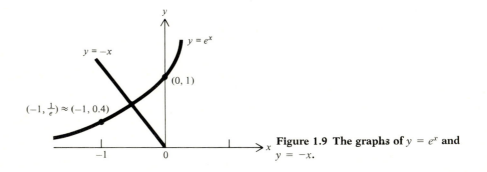

Figure 1.9 The graphs of $y = e^x$ and $y = -x$.

Taking $x_0 = -0.5$ (as suggested by the graph), you get

$$x_1 = \frac{(x_0 - 1)e^{x_0}}{e^{x_0} + 1} = -0.5663$$

$$x_2 = \frac{(x_1 - 1)e^{x_1}}{e^{x_1} + 1} = -0.5671$$

$$x_3 = \frac{(x_2 - 1)e^{x_2}}{e^{x_2} + 1} = -0.5671$$

Hence, to three-decimal-place accuracy, the solution of the equation $e^x = -x$ is -0.567.

Convergence of Newton's method

It can be shown that if the approximations $x_1, x_2, x_3, \ldots$ generated by Newton's method converge to some (finite) number, this number must be a root of the equation $f(x) = 0$. However, there are cases in which the approximations do not converge, even though the equation has a root. Roughly speaking, this occurs if the initial estimate x_0 is not "close enough" to the root. This phenomenon is illustrated in Figure 1.10, which shows successive "approximations" that fail to converge to the root but instead increase without bound. In such cases, it is usually possible to achieve convergence by reworking the problem using a better initial estimate.

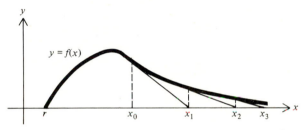

Figure 1.10 Newton's method fails because x_0 is not close enough to r.

Problems

In Problems 1 through 6, use the bisection method to approximate, with an error no greater than $\frac{1}{16}$, the indicated root of the given equation.

1. The root between 1 and 2 of $x^2 - x - 1 = 0$.

2. The positive root of $x^2 - 3x + 1 = 0$.

3. The negative root of $x^2 - 2x - 1 = 0$.

4. The root of $x^3 + x^2 - 1 = 0$.

5. The root of $x^3 + 2x^2 + x - 5 = 0$.

6. The root of $x^3 + 2x^2 - x + 1 = 0$.

In Problems 7 through 10, use the bisection method to approximate the given number with an error no greater than $\frac{1}{16}$.

7. $\sqrt{2}$

8. $\sqrt{7}$

9. $\sqrt[3]{9}$

10. $\sqrt[3]{-20}$

In Problems 11 through 18, use Newton's method to find all the roots of the given equation to three-decimal-place accuracy.

11. $x^2 - 12 = 0$　　　　　　　　　12. $x^3 + 49 = 0$

13. $x^3 + x^2 - 1 = 0$　　　　　　　14. $x^3 + 2x^2 - x + 1 = 0$

15. $e^x = -2x$　　　　　　　　　　16. $e^{-x} = x - 1$

17. $x^2 - 5x + 1 = 0$ (*Hint:* There are two roots.)

18. $x^4 - 4x^3 + 10 = 0$ (*Hint:* There are two roots.)

In Problems 19 through 22, use Newton's method to compute the given number to five decimal places.

19. $\sqrt{2}$　　　　　　　　　　　20. $\sqrt{7}$

21. $\sqrt[3]{9}$　　　　　　　　　　21. $\sqrt[3]{-20}$

2 NUMERICAL INTEGRATION

In this section you will see some techniques you can use to approximate definite integrals. Numerical methods such as these are needed when the function to be integrated does not have an elementary antiderivative.

Approximation by rectangles

If $f(x)$ is positive on the interval $a \leq x \leq b$, the definite integral $\int_a^b f(x)\,dx$ is equal to the area under the graph of f between $x = a$ and $x = b$. As you saw in Chapter 6, Section 3, one way to approximate this area is to use n rectangles, as shown in Figure 2.1. In particular, you divide the interval $a \leq x \leq b$ into n equal subintervals of width $\Delta x = \dfrac{b - a}{n}$ and let x_j denote the beginning of the jth subinterval. The base of the jth rectangle is the jth subinterval, and its height is $f(x_j)$. Hence the area of the jth rectangle is $f(x_j)\,\Delta x$. The sum of the areas of all n rectangles is an approximation to the area under the curve and

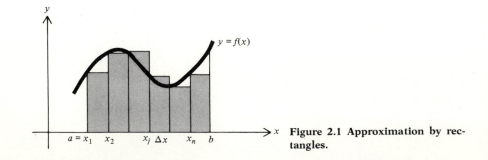

Figure 2.1 Approximation by rectangles.

hence an approximation to the corresponding definite integral. Thus,

$$\int_a^b f(x)\, dx \approx f(x_1)\, \Delta x + f(x_2)\, \Delta x + \cdots + f(x_n)\, \Delta x$$

This approximation improves as the number of rectangles increases, and you can estimate the integral to any desired degree of accuracy by taking n large enough. However, since fairly large values of n are usually required to achieve reasonable accuracy, approximation by rectangles is rarely used in practice.

Approximation by trapezoids The accuracy of the approximation improves significantly if trapezoids are used instead of rectangles. Figure 2.2 shows the area from Figure 2.1 approximated by n trapezoids. Notice how much better the approximation is in this case.

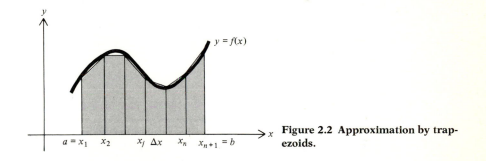

Figure 2.2 Approximation by trapezoids.

The jth trapezoid is shown in greater detail in Figure 2.3. Notice that it consists of a rectangle with a right triangle on top of it. Since,

$$\text{Area of rectangle} = f(x_{j+1})\, \Delta x$$

and

$$\text{Area of triangle} = \tfrac{1}{2}[f(x_j) - f(x_{j+1})]\, \Delta x$$

it follows that

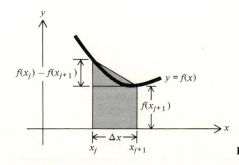

Figure 2.3 The jth trapezoid.

$$\text{Area of trapezoid} = f(x_{j+1})\,\Delta x + \tfrac{1}{2}[f(x_j) - f(x_{j+1})]\,\Delta x$$
$$= \tfrac{1}{2}[f(x_j) + f(x_{j+1})]\,\Delta x$$

The sum of the areas of all n trapezoids is an approximation to the area under the curve and hence an approximation to the corresponding definite integral. Thus,

$$\int_a^b f(x)\,dx \approx \tfrac{1}{2}[f(x_1) + f(x_2)]\,\Delta x + \tfrac{1}{2}[f(x_2) + f(x_3)]\,\Delta x + \cdots$$

$$+ \tfrac{1}{2}[f(x_n) + f(x_{n+1})]\,\Delta x$$

$$= \frac{\Delta x}{2}[f(x_1) + 2f(x_2) + \cdots + 2f(x_n) + f(x_{n+1})]$$

This approximation formula is known as the **trapezoidal rule** and applies even if the function f is not positive.

The trapezoidal rule

$$\int_a^b f(x)\,dx \approx \frac{\Delta x}{2}[f(x_1) + 2f(x_2) + \cdots + 2f(x_n) + f(x_{n+1})]$$

The use of the trapezoidal rule is illustrated in the following example.

EXAMPLE 2.1

Use the trapezoidal rule with $n = 10$ to approximate $\displaystyle\int_1^2 \frac{1}{x}\,dx$.

SOLUTION

Since
$$\Delta x = \frac{2 - 1}{10} = 0.1$$

the interval $1 \le x \le 2$ is divided into 10 subintervals, as shown in Figure 2.4, by

$$x_1 = 1,\ x_2 = 1.1,\ x_3 = 1.2,\ \ldots,\ x_{10} = 1.9,\ x_{11} = 2$$

Then, by the trapezoidal rule,

$$\int_1^2 \frac{1}{x}\,dx \approx \frac{0.1}{2}\left(\frac{1}{1} + \frac{2}{1.1} + \frac{2}{1.2} + \frac{2}{1.3} + \frac{2}{1.4} + \frac{2}{1.5} + \frac{2}{1.6} + \frac{2}{1.7} + \frac{2}{1.8}\right.$$

$$\left. + \frac{2}{1.9} + \frac{1}{2}\right)$$

$$= 0.693771$$

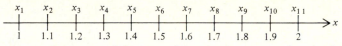

Figure 2.4 Division of the interval $1 \leq x \leq 2$ into ten subintervals.

The definite integral in Example 2.1 can be evaluated directly. In particular,

$$\int_1^2 \frac{1}{x} \, dx = \ln |x| \Big|_1^2 = \ln 2 = 0.693147$$

Thus the approximation of this particular integral by the trapezoidal rule with $n = 10$ is accurate (after round off) to two decimal places.

The accuracy of the trapezoidal rule

The difference between the true value of the integral $\int_a^b f(x) \, dx$ and the approximation generated by the trapezoidal rule when n subintervals are used is denoted by E_n. The following estimate for the absolute value of E_n is proved in more advanced courses.

Error estimate for the trapezoidal rule

If M is the maximum value of $|f''(x)|$ on the interval $a \leq x \leq b$, then

$$|E_n| \leq \frac{M(b - a)^3}{12n^2}$$

The use of this formula is illustrated in the next example.

EXAMPLE 2.2

Estimate the accuracy of the approximation of $\int_1^2 \frac{1}{x} \, dx$ by the trapezoidal rule with $n = 10$.

SOLUTION

Starting with $f(x) = \frac{1}{x}$, compute the derivatives

$$f'(x) = -\frac{1}{x^2} \quad \text{and} \quad f''(x) = \frac{2}{x^3}$$

and observe that the largest value of $|f''(x)|$ for $1 \leq x \leq 2$ is $|f''(1)| = 2$. Apply the error formula with

$$M = 2, \quad a = 1, \quad b = 2, \quad \text{and} \quad n = 10$$

to get
$$|E_{10}| \leq \frac{2(2-1)^3}{12(10)^2} = 0.00167$$

That is, the error in the approximation in Example 2.1 is guaranteed to be no greater than 0.00167. (In fact, to five decimal places, the error is 0.00062, as you can see by comparing the approximation obtained in Example 2.1 with the decimal representation of ln 2.)

With the aid of the error estimate you can decide in advance how many subintervals to use to achieve a desired degree of accuracy. Here is an example.

EXAMPLE 2.3

How many subintervals are required to guarantee that the error will be less than 0.00005 in the approximation of $\int_1^2 \frac{1}{x} \, dx$ using the trapezoidal rule?

SOLUTION

From Example 2.2 you know that $M = 2$, $a = 1$, and $b = 2$, so that
$$|E_n| \leq \frac{2(2-1)^3}{12n^2} = \frac{1}{6n^2}$$

The goal is to find the smallest positive integer n for which
$$\frac{1}{6n^2} < 0.00005$$

or, equivalently,
$$n^2 > \frac{1}{6(0.00005)}$$

or
$$n > \sqrt{\frac{1}{6(0.00005)}} = 57.74$$

The smallest such integer is $n = 58$, and so 58 subintervals are required to ensure the desired accuracy.

The relatively large number of subintervals required in Example 2.3 to ensure accuracy to within 0.00005 suggests that approximation by trapezoids may not be sufficiently accurate for some applications. There is another approximation technique that is no harder to use than the trapezoidal rule, but that is substantially more accurate. Like the trapezoidal rule, it is based on the approximation of the area under a curve by columns, but unlike the trapezoidal rule, it uses curves rather than lines as the tops of the columns.

Approximation using parabolas The approximation of a definite integral using parabolas is based on the following construction (which is illustrated in Figure 2.5 for $n = 6$). Divide the interval $a \leq x \leq b$ into an even number of subin-

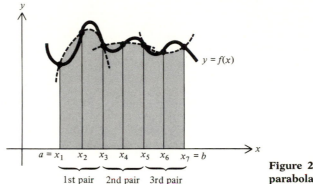

Figure 2.5 **Approximation using parabolas.**

tervals so that adjacent subintervals can be paired with none left over. Approximate the portion of the graph that lies above the first pair of subintervals by the (unique) parabola that passes through the three points $(x_1, f(x_1))$, $(x_2, f(x_2))$, and $(x_3, f(x_3))$, and use the area under this parabola between x_1 and x_3 to approximate the corresponding area under the curve. Do the same for the remaining pairs of subintervals and use the sum of the resulting areas to approximate the total area under the graph. It can be shown that this construction leads to the following approximation scheme known as **Simpson's rule.**

Simpson's rule

$$\int_a^b f(x)\, dx \approx \frac{\Delta x}{3} \left[f(x_1) + 4f(x_2) + 2f(x_3) + 4f(x_4) + 2f(x_5) + \cdots \right.$$
$$\left. + 2f(x_{n-1}) + 4f(x_n) + f(x_{n+1}) \right]$$

Notice that the first and last function values in the approximating sum in Simpson's rule are multiplied by 1, while the others are multiplied alternately by 4 and 2.

The proof of Simpson's rule is based on the fact that the equation of a parabola is a polynomial of the form $y = Ax^2 + Bx + C$. For each pair of subintervals, the three given points are used to find the coefficients A, B, and C, and the resulting polynomial is then integrated to get the corresponding area. The details of the proof are straightforward but tedious and will be omitted.

The use of Simpson's rule is illustrated in the following example.

EXAMPLE 2.4

Use Simpson's rule with $n = 10$ to approximate $\int_1^2 \frac{1}{x} \, dx$.

SOLUTION

As in Example 2.1, $\Delta x = 0.1$, and hence the interval $1 \leq x \leq 2$ is divided into the 10 subintervals by

$$x_1 = 1, x_2 = 1.1, x_3 = 1.2, \ldots, x_{10} = 1.9, x_{11} = 2$$

Then, by Simpson's rule,

$$\int_1^2 \frac{1}{x} \, dx \approx \frac{0.1}{3} \left(\frac{1}{1} + \frac{4}{1.1} + \frac{2}{1.2} + \frac{4}{1.3} + \frac{2}{1.4} + \frac{4}{1.5} + \frac{2}{1.6} + \frac{4}{1.7} + \frac{2}{1.8} \right.$$
$$\left. + \frac{4}{1.9} + \frac{1}{2} \right)$$
$$= 0.693150$$

Notice that this is an excellent approximation to the true value $\ln 2 = 0.693147$.

The accuracy of Simpson's rule

The error estimate for Simpson's rule turns out to involve the fourth derivative $f^{(4)}(x)$.

Error estimate for Simpson's rule

If M is the maximum value of $|f^{(4)}(x)|$ on the interval $a \leq x \leq b$, then

$$|E_n| \leq \frac{M(b - a)^5}{180n^4}$$

Here is an application of this formula.

EXAMPLE 2.5

Estimate the accuracy of the approximation of $\int_1^2 \frac{1}{x} \, dx$ by Simpson's rule with $n = 10$.

SOLUTION

Starting with $f(x) = \dfrac{1}{x}$, compute the derivatives

$$f'(x) = -\frac{1}{x^2} \qquad f''(x) = \frac{2}{x^3} \qquad f^{(3)}(x) = -\frac{6}{x^4} \qquad f^{(4)}(x) = \frac{24}{x^5}$$

and observe that the largest value of $|f^{(4)}(x)|$ on the interval $1 \leq x \leq 2$ is $|f^{(4)}(1)| = 24$.

Now apply the error formula with $M = 24$, $a = 1$, $b = 2$, and $n = 10$ to get

$$|E_{10}| \leq \frac{24(2-1)^5}{180(10)^4} = 0.000013$$

That is, the error in the approximation in Example 2.4 is guaranteed to be no greater than 0.000013.

In the next example, the error estimate is used to determine the number of subintervals that are required to ensure a specified degree of accuracy.

EXAMPLE 2.6

How many subintervals are required to ensure accuracy to within 0.00005 in the approximation of $\displaystyle\int_1^2 \frac{1}{x}\,dx$ by Simpson's rule?

SOLUTION

From Example 2.5 you know that $M = 24$, $a = 1$, and $b = 2$. Hence,

$$|E_n| \leq \frac{24(2-1)^5}{180n^4} = \frac{2}{15n^4}$$

The goal is to find the smallest positive (even) integer n for which

$$\frac{2}{15n^4} < 0.00005$$

or, equivalently,

$$n^4 > \frac{2}{15(0.00005)}$$

or

$$n > \left(\frac{2}{15(0.00005)}\right)^{1/4} = 7.19$$

The smallest such (even) integer is $n = 8$, and so eight subintervals are required to ensure the desired accuracy.

For a striking illustration of the superiority of Simpson's rule, compare the results of Examples 2.3 and 2.6. With the trapezoidal rule, 58 subintervals are required to ensure accuracy to within 0.00005 in the approximation of $\int_1^2 \frac{1}{x}\, dx$, while with Simpson's rule, the number of required subintervals is only eight.

Areas under the standard normal curve

In Chapter 7, Section 4, you learned how to compute probabilities using Table III (at the back of the book), which gives areas under the standard normal curve. The table itself was generated using numerical methods to approximate definite integrals of the standard normal density function $f(z) = \dfrac{1}{\sqrt{2\pi}}\, e^{-z^2/2}$. In the next example, Simpson's rule is used to obtain one of the entries in the table.

EXAMPLE 2.7

Use Simpson's rule with $n = 10$ to approximate the probability $P(z \le 1)$, where z is a standard normal random variable.

SOLUTION

By the symmetry of the standard normal curve, and because the total area under the curve is 1, the desired probability can be written (see Figure 2.6) as

$$P(z \le 1) = P(z \le 0) + P(0 \le z \le 1) = 0.5 + P(0 \le z \le 1)$$

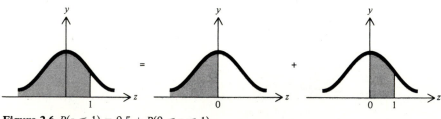

Figure 2.6 $P(z \le 1) = 0.5 + P(0 \le z \le 1)$.

The probability $P(0 \le z \le 1)$ is the definite integral $\int_0^1 f(z)\, dz$, where $f(z)$ is the standard normal density function

$$f(z) = \frac{1}{\sqrt{2\pi}} e^{-z^2/2}$$

To approximate this integral using 10 subintervals, observe that

$$\Delta z = \frac{1 - 0}{10} = 0.1$$

and that

$$z_1 = 0, z_2 = 0.1, z_3 = 0.2, \ldots, z_{10} = 0.9, z_{11} = 1$$

Then, by Simpson's rule,

$$P(0 \leq z \leq 1) = \int_0^1 \frac{1}{\sqrt{2\pi}} e^{-z^2/2} \, dz = \frac{1}{\sqrt{2\pi}} \int_0^1 e^{-z^2/2} \, dz$$

$$\approx \frac{1}{\sqrt{2\pi}} \left(\frac{0.1}{3}\right) [e^0 + 4e^{-(0.1)^2/2} + 2e^{-(0.2)^2/2}$$

$$+ 4e^{-(0.3)^2/2} + 2e^{-(0.4)^2/2} + 4e^{-(0.5)^2/2} + 2e^{-(0.6)^2/2}$$

$$+ 4e^{-(0.7)^2/2} + 2e^{-(0.8)^2/2} + 4e^{-(0.9)^2/2} + e^{-1/2}]$$

$$= \frac{1}{\sqrt{2\pi}} \left(\frac{0.1}{3}\right) (25.66875)$$

$$= 0.3413$$

Hence, $P(z \leq 1) = 0.5 + P(0 \leq z \leq 1) \approx 0.5 + 0.3413 = 0.8413$

which is identical to the entry in Table III at the back of the book.

Problems In Problems 1 through 8, approximate the given integral using
(a) the trapezoidal rule, and
(b) Simpson's rule
with the specified number of subintervals.

1. $\int_1^2 x^2 \, dx$; $n = 4$

2. $\int_4^6 \frac{1}{\sqrt{x}} \, dx$; $n = 10$

3. $\int_0^1 \frac{1}{1 + x^2} \, dx$; $n = 4$

4. $\int_2^3 \frac{1}{x^2 - 1} \, dx$; $n = 4$

5. $\int_{-1}^0 \sqrt{1 + x^2} \, dx$; $n = 4$

6. $\int_0^3 \sqrt{9 - x^2} \, dx$; $n = 6$

7. $\int_0^1 e^{-x^2} \, dx$; $n = 4$

8. $\int_0^2 e^{x^2} \, dx$; $n = 10$

In Problems 9 through 14, approximate the given integral and esti-
mate the error $|E_n|$ using
  (a) the trapezoidal rule, and
  (b) Simpson's rule
with the specified number of subintervals.

9. $\int_1^2 \frac{1}{x^2} \, dx$; $n = 4$ 10. $\int_0^2 x^3 \, dx$; $n = 8$

11. $\int_1^3 \sqrt{x} \, dx$; $n = 10$ 12. $\int_1^2 \ln x \, dx$; $n = 4$

13. $\int_0^1 e^{x^2} \, dx$; $n = 4$ 14. $\int_0^{0.6} e^{x^3} \, dx$; $n = 6$

Normal distribution 15. Use Simpson's rule with $n = 8$ to approximate the probability
$P(z \leq 0.8)$, where z is a standard normal random variable. Compare your answer with the corresponding entry in Table III at the
back of the book.

Normal distribution 16. Use Simpson's rule with $n = 8$ to approximate the probability
$P(z \leq 1.6)$, where z is a standard normal random variable. Compare your answer with the corresponding entry in Table III at the
back of the book.

In Problems 17 through 22, determine how many subintervals are required to guarantee accuracy to within 0.00005 in the approximation
of the given integral by
  (a) the trapezoidal rule, and
  (b) Simpson's rule.

17. $\int_1^3 \frac{1}{x} \, dx$ 18. $\int_0^4 (x^4 + 2x^2 + 1) \, dx$

19. $\int_1^2 \frac{1}{\sqrt{x}} \, dx$ 20. $\int_1^2 \ln (1 + x) \, dx$

21. $\int_{1.2}^{2.4} e^x \, dx$ 22. $\int_0^2 e^{x^2} \, dx$

3 THE METHOD
OF LEAST
SQUARES

Throughout this text, you have seen examples in which functions relating two or more variables were differentiated or integrated to obtain useful information about practical situations. In this section, you
will see one of the techniques that researchers use to determine such
functions from observed data.

Suppose data consisting of n points $(x_1, y_1), (x_2, y_2), \ldots, (x_n, y_n)$ are
known and the goal is to find a function $y = f(x)$ that fits the data rea-
sonably well. The first step is to decide what type of function to try.

Sometimes this can be done by a theoretical analysis of the under-lying practical situation and sometimes by inspection of the graph of the n points. Two sets of data are plotted in Figure 3.1. In Figure 3.1a,

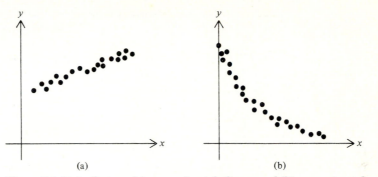

(a)   (b)

Figure 3.1 Data that are (a) approximately linear and (b) approximately exponential.

the points lie roughly along a straight line, so a linear function, $y = mx + b$, would be an appropriate choice in this case. In Figure 3.1b, the points appear to follow an exponential curve, and a function of the form $y = Ae^{-kx}$ would be reasonable.

The least-squares criterion

Once the type of function has been chosen, the next step is to deter-mine the particular function of this type whose graph is "closest" to the given set of points. A convenient way to measure how close a curve is to a set of points is to compute the sum of the squares of the vertical distances from the points to the curve. In Figure 3.2, for ex-

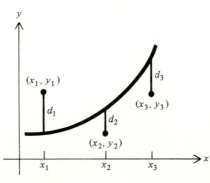

Figure 3.2 Sum of the squares of the ver-tical distances: $d_1^2 + d_2^2 + d_3^2$.

ample, this is the sum $d_1^2 + d_2^2 + d_3^2$. The closer the curve is to the points, the smaller this sum will be, and the curve for which this sum is smallest is said to be **closest** to the set of points **according to the least-squares criterion.**

The use of the least-squares criterion to fit a linear function to a set of points is illustrated in the following example. The computation involves the technique from Chapter 8, Section 4, for minimizing a function of two variables.

EXAMPLE 3.1

Use the least-squares criterion to find the equation of the line that is closest to the three points $(1, 1)$, $(2, 3)$, and $(4, 3)$.

SOLUTION

As indicated in Figure 3.3, the sum of the squares of the vertical distances from the three given points to the line $y = mx + b$ is

$$d_1^2 + d_2^2 + d_3^2 = (m + b - 1)^2 + (2m + b - 3)^2 + (4m + b - 3)^2$$

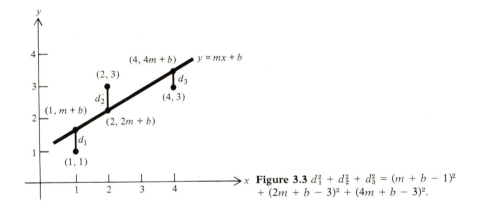

Figure 3.3 $d_1^2 + d_2^2 + d_3^2 = (m + b - 1)^2 + (2m + b - 3)^2 + (4m + b - 3)^2$.

This sum depends on the coefficients m and b that define the line, and so the sum can be thought of as a function $S(m, b)$ of the two variables m and b. The goal, therefore, is to find the values of m and b that minimize the function

$$S(m, b) = (m + b - 1)^2 + (2m + b - 3)^2 + (4m + b - 3)^2$$

You do this by setting the partial derivatives $\dfrac{\partial S}{\partial m}$ and $\dfrac{\partial S}{\partial b}$ equal to zero to get

$$\frac{\partial S}{\partial m} = 2(m + b - 1) + 4(2m + b - 3) + 8(4m + b - 3)$$

$$= 42m + 14b - 38 = 0$$

and $\dfrac{\partial S}{\partial b} = 2(m + b - 1) + 2(2m + b - 3) + 2(4m + b - 3)$

$$= 14m + 6b - 14 = 0$$

and solving the resulting simplified equations

$$21m + 7b = 19$$
$$7m + 3b = 7$$

simultaneously for m and b to conclude that

$$m = \frac{4}{7} \quad \text{and} \quad b = 1$$

It can be shown that the critical point $(m, b) = (\frac{4}{7}, 1)$ does indeed minimize the function $S(m, b)$, and so it follows that

$$y = \frac{4}{7} x + 1$$

is the equation of the line that is closest to the three given points.

The least-squares line

The line that is closest to a set of points according to the least-squares criterion is called the **least-squares line** for the points. (The term **regression line** is also used, especially in statistical work.) The procedure used in Example 3.1 can be generalized to give the following formulas for the slope m and the y intercept b of the least-squares line for an arbitrary set of n points $(x_1, y_1), (x_2, y_2), \ldots, (x_n, y_n)$. The formulas involve sums of the x and y values. All the sums run from $j = 1$ to $j = n$. To simplify the notation, the indices are omitted and, for example, Σx is used instead of $\displaystyle\sum_{j=1}^{n} x_j$.

The least-squares lines

The equation of the least-squares line for the n points (x_1, y_1), $(x_2, y_2), \ldots, (x_n, y_n)$ is $y = mx + b$, where

$$m = \frac{n\Sigma xy - \Sigma x \Sigma y}{n\Sigma x^2 - (\Sigma x)^2} \quad \textbf{and} \quad b = \frac{\Sigma x^2 \Sigma y - \Sigma x \Sigma xy}{n\Sigma x^2 - (\Sigma x)^2}$$

EXAMPLE 3.2

Use the formulas to find the least-squares line for the points $(1, 1)$, $(2, 3)$, and $(4, 3)$ from Example 3.1.

SOLUTION

Arrange your calculations as follows:

x	y	xy	x^2
1	1	1	1
2	3	6	4
4	3	12	16
$\Sigma x = 7$	$\Sigma y = 7$	$\Sigma xy = 19$	$\Sigma x^2 = 21$

Then use the formulas with $n = 3$ to get

$$m = \frac{3(19) - 7(7)}{3(21) - (7)^2} = \frac{4}{7} \quad \text{and} \quad b = \frac{21(7) - 7(19)}{3(21) - (7)^2} = 1$$

from which it follows that the equation of the least-squares line is

$$y = \frac{4}{7}x + 1$$

Least-squares prediction The least-squares line (or curve) that fits data collected in the past can be used to make rough predictions about the future. This is illustrated in Figure 3.4, which shows the least-squares line for a com-

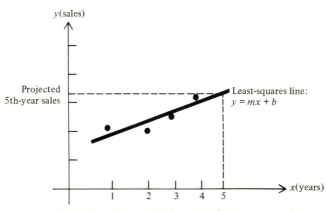

Figure 3.4 **Projected 5th-year sales:** $y = 5m + b$.

pany's annual sales for its first four years of operation. A reasonable estimate of fifth-year sales is the value of y obtained from the equation of the line when $x = 5$.

Least-squares prediction is illustrated further in the next example. To keep the calculations relatively simple, an unrealistically small set of data is used.

EXAMPLE 3.3

A college admissions officer has compiled the following data relating students' high-school and college grade-point averages.

High-school GPA	2.0	2.5	3.0	3.0	3.5	3.5	4.0	4.0
College GPA	1.5	2.0	2.5	3.5	2.5	3.0	3.0	3.5

Find the equation of the least-squares line for these data, and use it to predict the college GPA of a student whose high-school GPA is 3.75.

SOLUTION

Let x denote the high-school GPA and y the college GPA, and arrange the calculations as follows:

x	y	xy	x^2
2.0	1.5	3.0	4.0
2.5	2.0	5.0	6.25
3.0	2.5	7.5	9.0
3.0	3.5	10.5	9.0
3.5	2.5	8.75	12.25
3.5	3.0	10.5	12.25
4.0	3.0	12.0	16.0
4.0	3.5	14.0	16.0
$\Sigma x = 25.5$	$\Sigma y = 21.5$	$\Sigma xy = 71.25$	$\Sigma x^2 = 84.75$

Use the least-squares formulas with $n = 8$ to get

$$m = \frac{8(71.25) - 25.5(21.5)}{8(84.75) - (25.5)^2} = 0.78$$

and

$$b = \frac{84.75(21.5) - 25.5(71.25)}{8(84.75) - (25.5)^2} = 0.19$$

The equation of the least-squares line is therefore

$$y = 0.78x + 0.19$$

To predict the college GPA y of a student whose high-school GPA x is 3.75, substitute $x = 3.75$ into the equation of the least-squares line. This gives

$$y = 0.78(3.75) + 0.19 = 3.12$$

which suggests that the student's college GPA might be about 3.1.

A graph of the original data and of the corresponding least-squares line is shown in Figure 3.5. Actually, in practice, it is a good idea to plot the data *before* proceeding with the calculations. By looking at the graph you will usually be able to tell whether approximation by a

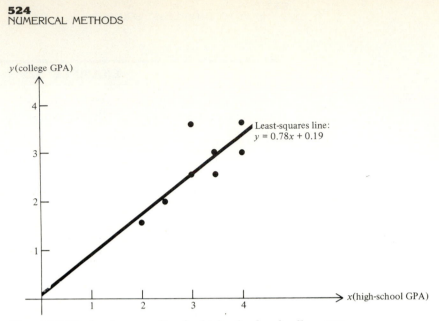

Figure 3.5 The least-squares line for high-school and college GPAs.

straight line is appropriate or whether a curve of some other shape should be used instead.

Nonlinear curve fitting

In each of the preceding examples, the least-squares criterion was used to fit a linear function to a set of data. With appropriate modifications, the procedure can also be used to fit nonlinear functions to data. For example, to find the quadratic function $y = Ax^2 + Bx + C$ whose graph is closest to a set of points, you proceed as in Example 3.1 and form the sum of the squares of the vertical distances from the given points to the graph. In this case, the sum is a function $S(A, B, C)$ of the coefficients $A, B,$ and C. To minimize this sum, you set all three of its first-order partial derivatives equal to zero and solve the resulting system of three equations in three unknowns. A computer is often used to perform the necessary calculations.

Exponential curve fitting

There is a special technique that is often used to fit exponential functions to data. It is based on the fact that if y is an exponential function of x, then $\ln y$ is a linear function of x. In particular, if

$$y = Ae^{kx}$$

then $\qquad \ln y = \ln(Ae^{kx}) = \ln A + \ln(e^{kx}) = \ln A + kx$

which is a linear function of x with slope k and vertical intercept $\ln A$.

To fit an exponential function to the points $(x_1, y_1), (x_2, y_2), \ldots,$

(x_n, y_n), you first fit a linear function

$$\ln y = mx + b$$

to the points $(x_1, \ln y_1)$, $(x_2, \ln y_2)$, . . . , $(x_n, \ln y_n)$ and then solve for y to get the desired exponential function

$$y = e^{mx+b} = e^b e^{mx}$$

The technique is illustrated in the following example.

EXAMPLE 3.4

A medical researcher studying the growth of bacteria in a certain culture has compiled the following data:

Number of minutes	0	20	40
Number of bacteria (in units of 1,000)	6	9	14

Theoretical considerations suggest that the bacteria should be growing exponentially. Find an exponential function that fits these data.

SOLUTION

Let x denote the number of minutes and y the number of bacteria (in units of 1,000). Begin by finding the least-squares line for the data

x	0	20	40
$\ln y$	$\ln 6 = 1.7918$	$\ln 9 = 2.1972$	$\ln 14 = 2.6391$

Arrange your calculations as follows:

x	$\ln y$	$x \ln y$	x^2
0	1.7918	0	0
20	2.1972	43.944	400
40	2.6391	105.564	1,600
$\Sigma x = 60$	$\Sigma \ln y = 6.6281$	$\Sigma x \ln y = 149.508$	$\Sigma x^2 = 2,000$

Then apply the least-squares formulas with $n = 3$ to get

$$m = \frac{3(149.508) - 60(6.6281)}{3(2,000) - (60)^2} = 0.02$$

and

$$b = \frac{2,000(6.6281) - 60(149.508)}{3(2,000) - (60)^2} = 1.786$$

The corresponding least-squares line is

$$\ln y = 0.02x + 1.786$$

from which it follows that

$$y = e^{0.02x+1.786} = e^{1.786}e^{0.02x}$$

or
$$y = 5.97e^{0.02x}$$

That is, after x minutes, approximately $5.97e^{0.02x}$ thousand bacteria should be present in the culture.

Problems In Problems 1 through 4, plot the given points, and use the method of Example 3.1 to find the corresponding least-squares line.

1. (0, 1), (2, 3), (4, 2)

2. (1, 1), (2, 2), (6, 0)

3. (1, 2), (2, 4), (4, 4), (5, 2)

4. (1, 5), (2, 4), (3, 2), (6, 0)

In Problems 5 through 8, plot the given points, and use the formulas to find the corresponding least-squares line.

5. (1, 2), (2, 2), (2, 3), (5, 5)

6. (−4, −1), (−3, 0), (−1, 0), (0, 1), (1, 2)

7. (−2, 5), (0, 4), (2, 3), (4, 2), (6, 1)

8. (−6, 2), (−3, 1), (0, 0), (0, −3), (1, −1), (3, −2)

College admissions
9. Over the past four years, a college admissions officer has compiled the following data (measured in units of 1,000) relating the number of college catalogs requested by high-school students by December 1 to the number of completed applications received by March 1.

Catalogs requested	4.5	3.5	4.0	5.0
Applications received	1.0	0.8	1.0	1.5

(a) Plot these data on a graph.
(b) Find the equation of the least-squares line.
(c) Use the least-squares line to predict the number of completed applications that will be received by March 1 if 4,800 catalogs are requested by December 1.

Sales
10. A company's annual sales (in units of 1 billion dollars) for its first five years of operation are shown in the following table:

Year	1	2	3	4	5
Sales	0.9	1.5	1.9	2.4	3.0

(a) Plot these data on a graph.

(b) Find the equation of the least-squares line.

(c) Use the least-squares line to predict the company's 6th-year sales.

Voter turnout 11. On election day, the polls in a certain state open at 8:00 A.M. Every two hours after that, an election official determines what percentage of the registered voters have already cast their ballots. The data through 6:00 P.M. are shown below.

Time	10:00	12:00	2:00	4:00	6:00
Percentage turnout	12	19	24	30	37

(a) Plot these data on a graph.

(b) Find the equation of the least-squares line. (Let x denote the number of hours after 8:00 A.M.

(c) Use the least-squares line to predict what percentage of the registered voters will have cast their ballots by the time the polls close at 8:00 P.M.

Public health 12. In a study of five industrial areas, a researcher obtained the following data relating the average number of units of a certain pollutant in the air and the incidence (per 100,000 people) of a certain disease.

Units of pollutant	3.4	4.6	5.2	8.0	10.7
Incidence of disease	48	52	58	76	96

(a) Plot these data on a graph.

(b) Find the equation of the least-squares line.

(c) Use the least-squares line to estimate the incidence of the disease in an area with an average pollution level of 7.3 units.

In Problems 13 and 14, plot the given points and use the method of Example 3.4 to find an exponential function that fits them.

13. (0, 5), (2, 7), (3, 9)

14. (1, 18), (2, 7), (3, 2), (4, 1)

Population growth 15. Population figures for a certain country are recorded below:

Year	1963	1973	1983
Population (in millions)	50	61	75

(a) Plot these data on a graph.

(b) Find an exponential function that fits the data. (Let x denote the number of years after 1963.)

(c) Use the exponential function from part (b) to predict the population in 1990.

Energy consumption   16. Data on the annual consumption (in billion kilowatthours) of electricity in a certain country are recorded below:

Year	1967	1972	1977	1982
Consumption	200	315	490	770

(a) Plot these data on a graph.

(b) Find an exponential function that fits the data.

(c) Use the exponential function from part (b) to predict the consumption of electricity in 1987.

CHAPTER SUMMARY AND PROFICIENCY TEST

Important terms, symbols, and formulas

Root of an equation

Intermediate value theorem

Bisection method

Newton's method: $x_n = x_{n-1} - \dfrac{f(x_{n-1})}{f'(x_{n-1})}$

Trapezoidal rule:

$$\int_a^b f(x)\, dx \approx \frac{\Delta x}{2} [f(x_1) + 2f(x_2) + \cdots + 2f(x_n) + f(x_{n+1})]$$

Error estimate: $|E_n| \le \dfrac{M(b-a)^3}{12n^2}$, where M is the maximum value of $|f''(x)|$ for $a \le x \le b$.

Simpson's rule:

$$\int_a^b f(x)\, dx \approx \frac{\Delta x}{3} [f(x_1) + 4f(x_2) + 2f(x_3) + 4f(x_4) + 2f(x_5) + \cdots + 2f(x_{n-1})$$

$$+ 4f(x_n) + f(x_{n+1})]$$

Error estimate: $|E_n| \le \dfrac{M(b-a)^5}{180n^4}$, where M is the maximum value of $|f^{(4)}(x)|$ for $a \le x \le b$.

Least-squares criterion

Least-squares line: $y = mx + b$, where

$$m = \frac{n\Sigma xy - \Sigma x \Sigma y}{n\Sigma x^2 - (\Sigma x)^2} \quad \text{and} \quad b = \frac{\Sigma x^2 \Sigma y - \Sigma x \Sigma xy}{n\Sigma x^2 - (\Sigma x)^2}$$

Least-squares prediction

Exponential curve fitting

Proficiency test   In Problems 1 and 2, use the bisection method to approximate, with an error no greater than $\frac{1}{16}$, the root of the given equation.

1. $x^3 + 2x - 6 = 0$ 2. $x^3 - 3x^2 + 3x + 1 = 0$

3. Use the bisection method to approximate $\sqrt{55}$ with an error no greater than $\frac{1}{16}$.

4. Use Newton's method to find $\sqrt{55}$ to five decimal places.

In Problems 5 and 6, use Newton's method to find all the roots of the given equation to three decimal places.

5. $x^3 + 3x^2 + 1 = 0$ 6. $e^{-x} = 3x$

In Problems 7 and 8, approximate the given integral and estimate the error $|E_n|$ using
   (a) the trapezoidal rule, and
   (b) Simpson's rule
with the specified number of subintervals.

7. $\int_1^3 \frac{1}{x} dx;\ n = 10$ 8. $\int_0^2 e^{x^2} dx;\ n = 8$

In Problems 9 and 10, determine how many subintervals are required to guarantee accuracy to within 0.00005 in the approximation of the given integral by
   (a) the trapezoidal rule, and
   (b) Simpson's rule.

9. $\int_1^3 \sqrt{x}\ dx$ 10. $\int_{0.5}^1 e^{x^2}\ dx$

11. Plot the points (1, 1), (1, 2), (3, 2), and (4, 3), and use partial derivatives to find the corresponding least-squares line.

12. The marketing manager for a certain company has compiled the following data relating monthly advertising expenditure and monthly sales (both measured in units of $1,000):

Advertising	3	4	7	9	10
Sales	78	86	138	145	156

   (a) Plot these data on a graph.
   (b) Find the least-squares line.
   (c) Use the least-squares line to predict monthly sales if the monthly advertising expenditure is $5,000.

13. Air pollution figures for a certain community are recorded below:

Year	1960	1970	1980
Units of pollution	4.1	5.4	7.3

(a) Plot these data on a graph.

(b) Find an exponential function that fits the data. (Let x denote the number of years after 1960.)

(c) Use the exponential function from part (b) to predict the pollution level in 1990.

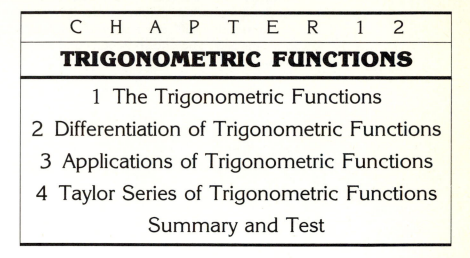

C H A P T E R   1 2

TRIGONOMETRIC FUNCTIONS

1 The Trigonometric Functions

2 Differentiation of Trigonometric Functions

3 Applications of Trigonometric Functions

4 Taylor Series of Trigonometric Functions

Summary and Test

1 THE TRIGONOMETRIC FUNCTIONS

In this chapter you will be introduced to some functions that are widely used in the natural sciences to study periodic or rhythmic phenomena such as oscillations, the periodic motion of planets, and the respiratory cycle and heartbeat of animals. These functions are also related to the measurement of angles and hence play an important role in such fields as architecture, navigation, and surveying.

Angles

An **angle** is formed when one line segment in the plane is rotated into another about their common endpoint. The resulting angle is said to be a **positive angle** if the rotation is in a counterclockwise direction and a **negative angle** if the rotation is in a clockwise direction. The situation is illustrated in Figure 1.1.

Positive angle   Negative angle

Figure 1.1 Positive and negative angles.

531

Measurement of angles

You are probably already familiar with the use of degrees to measure angles. A **degree** is the amount by which a line segment must be rotated so that its free endpoint traces out $\frac{1}{360}$ of a circle. Thus, for example, a complete counterclockwise rotation generating an entire circle contains 360°, one-half of a complete counterclockwise rotation contains 180°, and one-sixth of a complete counterclockwise rotation contains 60°. Some important angles are shown in Figure 1.2.

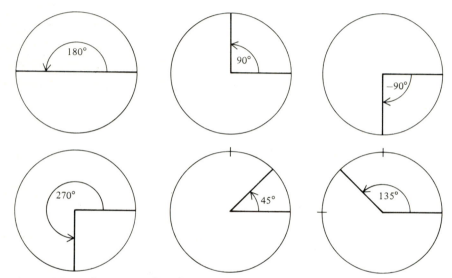

Figure 1.2 Angles measured in degrees.

Although measurement of angles in degrees is convenient for many geometric applications, there is another unit of angle measurement called a **radian** that leads to simpler rules for the differentiation and integration of trigonometric functions. One radian is defined to be the amount by which a line segment of length 1 must be rotated so that its free endpoint traces out a circular arc of length 1. The situation is illustrated in Figure 1.3.

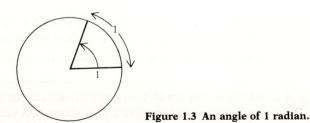

Figure 1.3 An angle of 1 radian.

The total circumference of a circle of radius 1 is 2π. Hence 2π radians is equal to 360°, π radians is equal to 180°, and, in general, the relationship between radians and degrees is given by the following proportion.

Conversion formula

$$\frac{\text{Degrees}}{180} = \frac{\text{radians}}{\pi}$$

Here is an example.

EXAMPLE 1.1

(a) Convert 45° to radians.

(b) Convert $\dfrac{\pi}{6}$ radians to degrees.

SOLUTION

(a) From the proportion

$$\frac{45}{180} = \frac{\text{radians}}{\pi}$$

it follows that

$$\text{Radians} = \frac{\pi}{4}$$

That is, 45° equals $\dfrac{\pi}{4}$ radians.

(b) From the proportion

$$\frac{\text{Degrees}}{180} = \frac{\pi/6}{\pi}$$

it follows that

$$\text{Degrees} = 30$$

That is, $\dfrac{\pi}{6}$ radians equals 30°.

For reference, six of the most important angles are shown in Figure 1.4 along with their measurements in degrees and in radians.

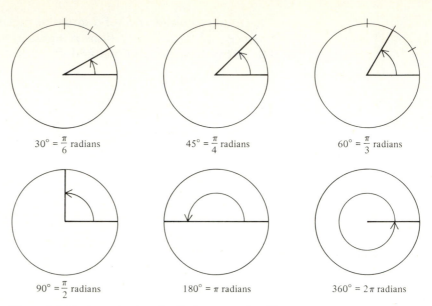

$30° = \dfrac{\pi}{6}$ radians $\qquad$ $45° = \dfrac{\pi}{4}$ radians $\qquad$ $60° = \dfrac{\pi}{3}$ radians

$90° = \dfrac{\pi}{2}$ radians $\qquad$ $180° = \pi$ radians $\qquad$ $360° = 2\pi$ radians

Figure 1.4 Six important angles in degrees and radians.

The sine and cosine Suppose the line segment joining the points (0, 0) and (1, 0) on the x axis is rotated through an angle of θ radians so that the free endpoint of the segment moves from (1, 0) to a point (x, y) as in Figure 1.5. The

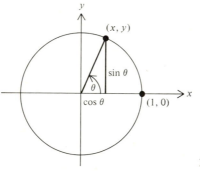

Figure 1.5 The sine and cosine of an angle.

x and y coordinates of the point (x, y) are known, respectively, as the **cosine** and **sine** of the angle θ. The symbol $\cos \theta$ is used to denote the cosine of θ, and $\sin \theta$ is used to denote its sine.

The sine and cosine

For any angle θ,

$$\cos \theta = x \qquad \textbf{and} \qquad \sin \theta = y$$

where (x, y) is the point to which (1, 0) is carried by a rotation of θ radians about the origin.

The values of the sine and cosine for multiples of $\frac{\pi}{2}$ can be read from Figure 1.6 and are summarized in the following table:

θ	0	$\frac{\pi}{2}$	π	$\frac{3\pi}{2}$	2π
$\cos \theta$	1	0	-1	0	1
$\sin \theta$	0	1	0	-1	0

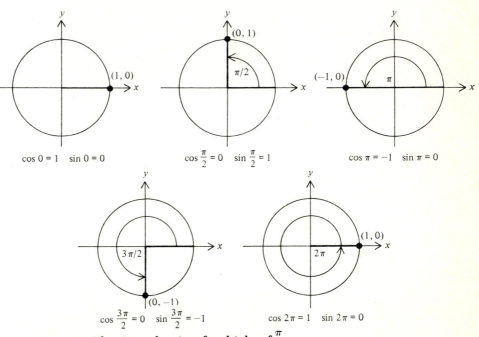

Figure 1.6 **The sine and cosine of multiples of** $\frac{\pi}{2}$.

The sine and cosine of a few other important angles are also easy to obtain geometrically, as you will see shortly. For other angles, you can use Table IV at the back of the book or your calculator to find the sine and cosine. (In Section 4, you will see the approximation techniques that are used to program computers and calculators to generate these values.)

Elementary properties of sine and cosine

Since there are 2π radians in a complete rotation, it follows that

$$\sin (\theta + 2\pi) = \sin \theta \quad \text{and} \quad \cos (\theta + 2\pi) = \cos \theta$$

That is, the sine and cosine functions are **periodic** with period 2π. The situation is illustrated in Figure 1.7.

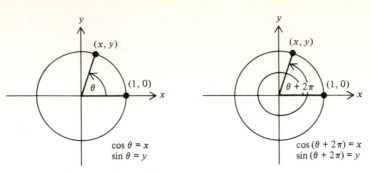

Figure 1.7 The periodicity of $\sin \theta$ and $\cos \theta$.

Since negative angles correspond to clockwise rotations, it follows that

$$\sin(-\theta) = -\sin\theta \quad \text{and} \quad \cos(-\theta) = \cos\theta$$

This is illustrated in Figure 1.8.

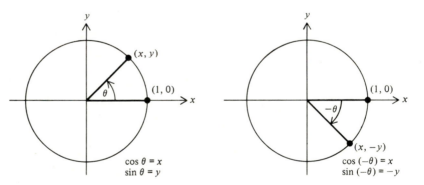

Figure 1.8 The sine and cosine of negative angles.

Properties of the sine and cosine

$$\sin(\theta + 2\pi) = \sin\theta \quad \textbf{and} \quad \cos(\theta + 2\pi) = \cos\theta$$

$$\sin(-\theta) = -\sin\theta \quad \textbf{and} \quad \cos(-\theta) = \cos\theta$$

The use of these properties is illustrated in the following example.

EXAMPLE 1.2

Evaluate the following expressions.

(a) $\cos(-\pi)$ (b) $\sin\left(-\dfrac{\pi}{2}\right)$ (c) $\cos 3\pi$ (d) $\sin\dfrac{5\pi}{2}$

SOLUTION

(a) Since $\cos \pi = -1$, it follows that

$$\cos(-\pi) = \cos \pi = -1$$

(b) Since $\sin \dfrac{\pi}{2} = 1$, it follows that

$$\sin\left(-\frac{\pi}{2}\right) = -\sin\frac{\pi}{2} = -1$$

(c) Since $3\pi = \pi + 2\pi$ and $\cos \pi = -1$, it follows that

$$\cos 3\pi = \cos(\pi + 2\pi) = \cos \pi = -1$$

(d) Since $\dfrac{5\pi}{2} = \dfrac{\pi}{2} + 2\pi$ and $\sin\dfrac{\pi}{2} = 1$, it follows that

$$\sin\frac{5\pi}{2} = \sin\left(\frac{\pi}{2} + 2\pi\right) = \sin\frac{\pi}{2} = 1$$

The graphs of sin θ and cos θ

It is obvious from the definitions of the sine and cosine that as θ goes from 0 to 2π, the function $\sin \theta$ oscillates between 1 and -1, starting with $\sin 0 = 0$, and the function $\cos \theta$ oscillates between 1 and -1, starting with $\cos 0 = 1$. This observation, together with the elementary properties previously derived, suggests that the graphs of the functions $\sin \theta$ and $\cos \theta$ resemble the curves in Figures 1.9 and 1.10, respectively.

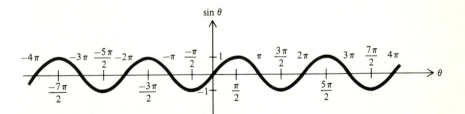

Figure 1.9 The graph of sin θ.

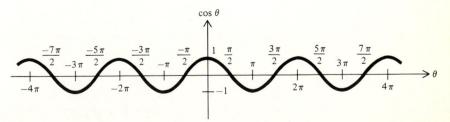

Figure 1.10 The graph of cos θ.

Other trigonometric functions

The tangent, cotangent, secant, and cosecant

Four other useful trigonometric functions can be defined in terms of the sine and cosine as follows.

For any angle θ,

$$\tan \theta = \frac{\sin \theta}{\cos \theta} \qquad \cot \theta = \frac{1}{\tan \theta} = \frac{\cos \theta}{\sin \theta}$$

$$\sec \theta = \frac{1}{\cos \theta} \qquad \csc \theta = \frac{1}{\sin \theta}$$

provided the denominators are not zero.

EXAMPLE 1.3

Evaluate the following expressions.

(a) $\tan \pi$ (b) $\cot \dfrac{\pi}{2}$ (c) $\sec(-\pi)$ (d) $\csc\left(-\dfrac{5\pi}{2}\right)$

SOLUTION

(a) Since $\sin \pi = 0$ and $\cos \pi = -1$, it follows that

$$\tan \pi = \frac{\sin \pi}{\cos \pi} = \frac{0}{-1} = 0$$

(b) Since $\sin \dfrac{\pi}{2} = 1$ and $\cos \dfrac{\pi}{2} = 0$, it follows that

$$\cot \frac{\pi}{2} = \frac{\cos(\pi/2)}{\sin(\pi/2)} = \frac{0}{1} = 0$$

(c) Since $\cos \pi = -1$, it follows that

$$\sec(-\pi) = \frac{1}{\cos(-\pi)} = \frac{1}{\cos \pi} = -1$$

(d) Since $\dfrac{5\pi}{2} = \dfrac{\pi}{2} + 2\pi$ and $\sin \dfrac{\pi}{2} = 1$, it follows that

$$\csc\left(-\frac{5\pi}{2}\right) = \frac{1}{\sin(-5\pi/2)} = \frac{1}{-\sin(5\pi/2)} = \frac{1}{-\sin(\pi/2)} = -1$$

Right triangles

If you had a high-school course in trigonometry, you may remember the following definitions of the sine, cosine, and tangent involving the sides of a right triangle like the one in Figure 1.11.

The trigonometry of right triangles

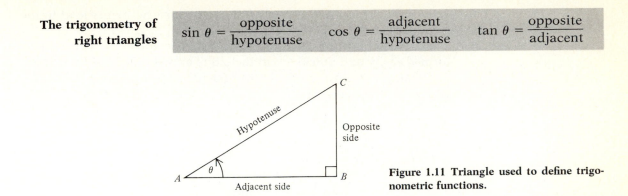

$$\sin \theta = \frac{\text{opposite}}{\text{hypotenuse}} \qquad \cos \theta = \frac{\text{adjacent}}{\text{hypotenuse}} \qquad \tan \theta = \frac{\text{opposite}}{\text{adjacent}}$$

Figure 1.11 Triangle used to define trigonometric functions.

The definitions that you have seen in this section involving the coordinates of points on a circle of radius 1 are equivalent to the definitions from high-school trigonometry. To see this, superimpose an xy coordinate system over the triangle ABC as shown in Figure 1.12, and draw the circle of radius 1 that is centered at the origin.

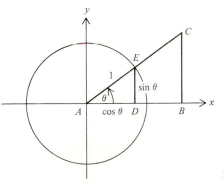

Figure 1.12 Similar triangles.

Since ABC and ADE are similar triangles, it follows that

$$\frac{AD}{AE} = \frac{AB}{AC}$$

But the length AD is the x coordinate of the point E on the circle, and so, by the definition of the cosine,

$$AD = \cos \theta$$

Moreover,

$$AE = 1 \qquad AB = \text{adjacent side} \qquad \text{and} \qquad AC = \text{hypotenuse}$$

Hence, $\qquad \dfrac{\cos \theta}{1} = \dfrac{\text{adjacent}}{\text{hypotenuse}} \qquad$ or $\qquad \cos \theta = \dfrac{\text{adjacent}}{\text{hypotenuse}}$

For practice, convince yourself that similar arguments lead to the formulas for the sine and tangent.

Many calculations involving trigonometric functions can be performed easily and quickly with the aid of appropriate right triangles. For example, from the well-known 30-60-90° triangle in Figure 1.13, you see immediately that

$$\sin 30° = \frac{1}{2} \quad \text{and} \quad \cos 30° = \frac{\sqrt{3}}{2}$$

or, using radian measure,

$$\sin \frac{\pi}{6} = \frac{1}{2} \quad \text{and} \quad \cos \frac{\pi}{6} = \frac{\sqrt{3}}{2}$$

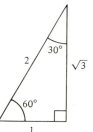

Figure 1.13 A 30-60-90° triangle.

Here is an example that further illustrates the use of right triangles.

EXAMPLE 1.4

Find $\tan \theta$ if $\sec \theta = \frac{3}{2}$.

SOLUTION

Since

$$\sec \theta = \frac{1}{\cos \theta} = \frac{\text{hypotenuse}}{\text{adjacent}}$$

begin by drawing a right triangle (Figure 1.14) in which the hypotenuse has length 3 and the side adjacent to the angle θ has length 2.

By the pythagorean theorem,

$$\text{Length of } BC = \sqrt{3^2 - 2^2} = \sqrt{5}$$

and so

$$\tan \theta = \frac{\text{opposite}}{\text{adjacent}} = \frac{\sqrt{5}}{2}$$

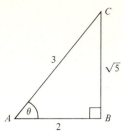

Figure 1.14 Right triangle for Example 1.4.

Important angles Using appropriate right triangles and the definition and elementary properties of the sine and cosine, you should now be able to find the sine and cosine of the important angles listed in the following table. Before proceeding, take a few minutes to fill in the table so that you will have this information summarized for easy reference in the future.

A table of important values

θ	0	$\dfrac{\pi}{6}$	$\dfrac{\pi}{4}$	$\dfrac{\pi}{3}$	$\dfrac{\pi}{2}$	$\dfrac{2\pi}{3}$	$\dfrac{3\pi}{4}$	$\dfrac{5\pi}{6}$	π	
$\sin \theta$										
$\cos \theta$										

Trigonometric identities An **identity** is an equation that holds for all values of its variable or variables. Identities involving trigonometric functions are called **trigonometric identities** and can be used to simplify trigonometric expressions. You have already seen some elementary trigonometric identities in this section. For example, the formulas

$$\cos(-\theta) = \cos\theta \quad \text{and} \quad \sin(-\theta) = -\sin\theta$$

are identities because they hold for all values of θ. Most trigonometry texts list dozens of trigonometric identities. Fortunately, only a few of them will be needed in this book.

One of the most important and well-known trigonometric identities is a simple consequence of the pythagorean theorem. It states that

$$\sin^2 \theta + \cos^2 \theta = 1$$

where $\sin^2 \theta$ stands for $(\sin\theta)^2$ and $\cos^2 \theta$ stands for $(\cos\theta)^2$. To see why this identity holds, look at Figure 1.15, which shows a point (x, y) on a circle of radius 1. By definition,

$$y = \sin\theta \quad \text{and} \quad x = \cos\theta$$

and, by the pythagorean theorem,

$$y^2 + x^2 = 1$$

from which the identity follows immediately.

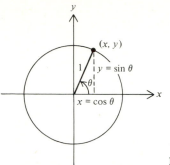

Figure 1.15 $\sin^2 \theta + \cos^2 \theta = 1$.

The pythagorean identity

For any angle θ,

$$\sin^2 \theta + \cos^2 \theta = 1$$

The following formulas for the cosine and sine of the sum of two angles are particularly useful. Proofs of these identities can be found in most trigonometry texts and will be omitted here.

The addition formulas

For any angles a and b,

$$\cos (a + b) = \cos a \cos b - \sin a \sin b$$

and

$$\sin (a + b) = \sin a \cos b + \cos a \sin b$$

If the angles a and b in the addition formulas are equal, the identities can be rewritten as follows.

The double-angle formulas

For any angle θ,

$$\cos 2\theta = \cos^2 \theta - \sin^2 \theta$$

and

$$\sin 2\theta = 2 \sin \theta \cos \theta$$

Some of these trigonometric identities will be used in Section 2 to derive the formulas for the derivatives of the sine and cosine functions. They can also be used in the solution of certain trigonometric equations.

Trigonometric equations
The following examples illustrate the use of trigonometric identities to solve trigonometric equations.

EXAMPLE 1.5

Find all the values of θ in the interval $0 \le \theta \le \pi$ that satisfy the equation $\sin 2\theta = \sin \theta$.

SOLUTION

Use the double-angle formula

$$\sin 2\theta = 2 \sin \theta \cos \theta$$

to rewrite the given equation as

$$2 \sin \theta \cos \theta = \sin \theta$$

or $\qquad 2 \sin \theta \cos \theta - \sin \theta = 0$

and factor to get

$$\sin \theta \, (2 \cos \theta - 1) = 0$$

which is satisfied if

$$\sin \theta = 0 \qquad \text{or} \qquad 2 \cos \theta - 1 = 0$$

From the graphs of the sine and cosine functions and from the table of important values, which you completed earlier in this section, it should be clear that the only solutions of

$$\sin \theta = 0$$

in the interval $0 \le \theta \le \pi$ are

$$\theta = 0 \qquad \text{and} \qquad \theta = \pi$$

and that the only solution of

$$2 \cos \theta - 1 = 0 \qquad \text{or} \qquad \cos \theta = \frac{1}{2}$$

in the interval $0 \le \theta \le \pi$ is

$$\theta = \frac{\pi}{3}$$

Hence the only solutions of the original equation in the specified interval are $\theta = 0$, $\theta = \dfrac{\pi}{3}$, and $\theta = \pi$.

EXAMPLE 1.6

Find all the values of θ in the interval $0 \le \theta \le 2\pi$ that satisfy the equation $\sin^2 \theta - \cos^2 \theta + \sin \theta = 0$.

SOLUTION

Use the pythagorean identity

$$\sin^2 \theta + \cos^2 \theta = 1 \quad \text{or} \quad \cos^2 \theta = 1 - \sin^2 \theta$$

to rewrite the original equation without any cosine terms as

$$\sin^2 \theta - (1 - \sin^2 \theta) + \sin \theta = 0$$

or

$$2 \sin^2 \theta + \sin \theta - 1 = 0$$

and factor to get

$$(2 \sin \theta - 1)(\sin \theta + 1) = 0$$

The only solutions of

$$2 \sin \theta - 1 = 0 \quad \text{or} \quad \sin \theta = \frac{1}{2}$$

in the interval $0 \le \theta \le 2\pi$ are

$$\theta = \frac{\pi}{6} \quad \text{and} \quad \theta = \frac{5\pi}{6}$$

and the only solution of

$$\sin \theta + 1 = 0 \quad \text{or} \quad \sin \theta = -1$$

in the interval is

$$\theta = \frac{3\pi}{2}$$

Hence the solutions of the original equation in the specified interval are $\theta = \dfrac{\pi}{6}$, $\theta = \dfrac{5\pi}{6}$, and $\theta = \dfrac{3\pi}{2}$.

Problems   In Problems 1 through 6, specify the number of degrees in the given angle.

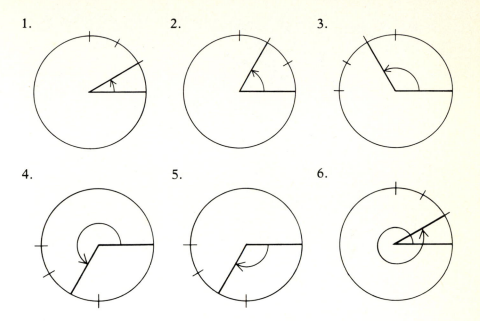

In Problems 7 through 12, show the given angle graphically.

7. 60°

8. −45°

9. 240°

10. 120°

11. −150°

12. 405°

In Problems 13 through 18, convert from degrees to radians.

13. 15°

14. 270°

15. −150°

16. −240°

17. 540°

18. 1°

In Problems 19 through 24, convert from radians to degrees.

19. $\dfrac{5\pi}{6}$

20. $\dfrac{2\pi}{3}$

21. $\dfrac{3\pi}{2}$

22. $-\dfrac{\pi}{12}$

23. $-\dfrac{3\pi}{4}$

24. 1

In Problems 25 through 30, specify the number of radians in the given angle.

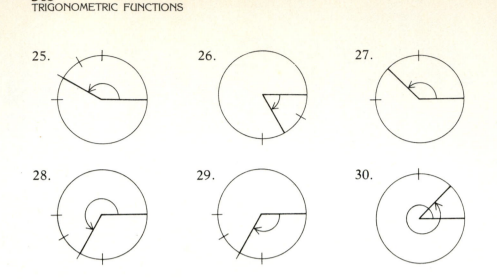

In Problems 31 through 36, show graphically the angles with the given radian measure.

31. $\dfrac{3\pi}{4}$ 32. $-\dfrac{\pi}{6}$

33. $\dfrac{2\pi}{3}$ 34. $\dfrac{5\pi}{6}$

35. $-\dfrac{5\pi}{6}$ 36. $\dfrac{3\pi}{2}$

In Problems 37 through 44, evaluate the given expression without using Table IV or your calculator.

37. $\cos \dfrac{7\pi}{2}$ 38. $\sin 5\pi$

39. $\cos\left(-\dfrac{5\pi}{2}\right)$ 40. $\sin\left(-\dfrac{7\pi}{2}\right)$

41. $\cot \dfrac{5\pi}{2}$ 42. $\sec 3\pi$

43. $\tan(-\pi)$ 44. $\csc\left(-\dfrac{7\pi}{2}\right)$

45. Use a right triangle to find $\cos \dfrac{\pi}{3}$ and $\sin \dfrac{\pi}{3}$.

46. Use a right triangle to find $\cos \dfrac{\pi}{4}$ and $\sin \dfrac{\pi}{4}$.

47. Complete the following table:

θ	$\dfrac{7\pi}{6}$	$\dfrac{5\pi}{4}$	$\dfrac{4\pi}{3}$	$\dfrac{3\pi}{2}$	$\dfrac{5\pi}{3}$	$\dfrac{7\pi}{4}$	$\dfrac{11\pi}{6}$	2π
$\sin\theta$								
$\cos\theta$								

In Problems 48 through 58, evaluate the given expression without using Table IV or your calculator.

48. $\cos\left(-\dfrac{2\pi}{3}\right)$

49. $\sin\left(-\dfrac{\pi}{6}\right)$

50. $\sin\left(-\dfrac{3\pi}{4}\right)$

51. $\cos\dfrac{7\pi}{3}$

52. $\sin\left(-\dfrac{7\pi}{6}\right)$

53. $\tan\dfrac{\pi}{6}$

54. $\cot\dfrac{\pi}{3}$

55. $\sec\dfrac{\pi}{3}$

56. $\csc\dfrac{2\pi}{3}$

57. $\tan\left(-\dfrac{\pi}{4}\right)$

58. $\cot\left(-\dfrac{5\pi}{3}\right)$

59. Find $\tan\theta$ if $\sec\theta = \frac{5}{4}$.

60. Find $\tan\theta$ if $\cos\theta = \frac{3}{5}$.

61. Find $\tan\theta$ if $\csc\theta = \frac{5}{3}$.

62. Find $\sin\theta$ if $\tan\theta = \frac{2}{3}$.

63. Find $\cos\theta$ if $\tan\theta = \frac{3}{4}$.

64. Find $\sec\theta$ if $\cot\theta = \frac{4}{3}$.

In Problems 65 through 73, find all the values of θ in the specified interval that satisfy the given equation.

65. $\cos\theta - \sin 2\theta = 0; \ 0 \le \theta \le 2\pi$

66. $3\sin^2\theta + \cos 2\theta = 2; \ 0 \le \theta \le 2\pi$

67. $\sin 2\theta = \sqrt{3}\cos\theta; \ 0 \le \theta \le \pi$

68. $\cos 2\theta = \cos\theta; \ 0 \le \theta \le \pi$

69. $2\cos^2\theta = \sin 2\theta; \ 0 \le \theta \le \pi$

70. $3\cos^2\theta - \sin^2\theta = 2; \ 0 \le \theta \le \pi$

71. $2 \cos^2 \theta + \sin \theta = 1$; $0 \le \theta \le \pi$

72. $\cos^2 \theta - \sin^2 \theta + \cos \theta = 0$; $0 \le \theta \le \pi$

73. $\sin^2 \theta - \cos^2 \theta + 3 \sin \theta = 1$; $0 \le \theta \le \pi$

74. Starting with the identity $\sin^2 \theta + \cos^2 \theta = 1$, derive the identity $1 + \tan^2 \theta = \sec^2 \theta$.

75. Starting with the addition formula

$$\sin (a + b) = \sin a \cos b + \cos a \sin b$$

derive the subtraction formula

$$\sin (a - b) = \sin a \cos b - \cos a \sin b$$

76. Starting with the addition formula

$$\cos (a + b) = \cos a \cos b - \sin a \sin b$$

derive the subtraction formula

$$\cos (a - b) = \cos a \cos b + \sin a \sin b$$

77. Starting with the addition formulas for the sine and cosine, derive the addition formula

$$\tan (a + b) = \frac{\tan a + \tan b}{1 - \tan a \tan b}$$

for the tangent.

78. Use the addition formula for the sine to derive the identity

$$\sin \left(\frac{\pi}{2} - \theta \right) = \cos \theta$$

79. Use the addition formula for the cosine to derive the identity

$$\cos \left(\frac{\pi}{2} - \theta \right) = \sin \theta$$

80. Give geometric arguments involving the coordinates of points on a circle of radius 1 to show why the identities in Problems 78 and 79 are valid.

2 DIFFERENTIATION OF TRIGONOMETRIC FUNCTIONS

The trigonometric functions are relatively easy to differentiate. In this section you will learn how. In the following section you will see a variety of rate-of-change and optimization problems that can be solved using derivatives of trigonometric functions.

The derivatives of the sine and cosine

Here are the formulas for the derivatives of the sine and cosine functions. Their proofs will be discussed later in this section, after you have had a chance to practice using the formulas.

The derivatives of $\sin \theta$ and $\cos \theta$

> **If θ is measured in radians,**
>
> $$\frac{d}{d\theta} \sin \theta = \cos \theta \qquad \textbf{and} \qquad \frac{d}{d\theta} \cos \theta = -\sin \theta$$

These formulas are usually used in conjunction with the chain rule in the following forms.

Chain rules for sine and cosine

> **If h is a differentiable function of θ and θ is measured in radians,**
>
> $$\frac{d}{d\theta} \sin h(\theta) = h'(\theta) \cos h(\theta)$$
>
> **and**
>
> $$\frac{d}{d\theta} \cos h(\theta) = -h'(\theta) \sin h(\theta)$$

Here are some examples.

EXAMPLE 2.1

Differentiate the function $f(\theta) = \sin (3\theta + 1)$.

SOLUTION

Using the chain rule for the sine function with $h(\theta) = 3\theta + 1$, you get

$$f'(\theta) = 3 \cos (3\theta + 1)$$

EXAMPLE 2.2

Differentiate the function $f(\theta) = \cos^2 \theta$.

SOLUTION

Since
$$\cos^2 \theta = (\cos \theta)^2$$

you use the chain rule for powers and the formula for the derivative of the cosine to get

$$f'(\theta) = 2 \cos \theta \, (-\sin \theta) = -2 \cos \theta \sin \theta$$

The derivatives of other trigonometric functions

The differentiation formulas for the sine and cosine can be used to obtain differentiation formulas for the other trigonometric functions. The procedure is illustrated in the next example.

EXAMPLE 2.3

Show that $\dfrac{d}{d\theta} \tan \theta = \sec^2 \theta$.

SOLUTION

Write
$$\tan \theta = \frac{\sin \theta}{\cos \theta}$$

and use the quotient rule to get

$$\frac{d}{d\theta} \tan \theta = \frac{\cos \theta \, (\cos \theta) - \sin \theta \, (-\sin \theta)}{\cos^2 \theta}$$

$$= \frac{\cos^2 \theta + \sin^2 \theta}{\cos^2 \theta}$$

$$= \frac{1}{\cos^2 \theta} \qquad (\text{since } \sin^2 \theta + \cos^2 \theta = 1)$$

$$= \sec^2 \theta \qquad \left(\text{since } \sec \theta = \frac{1}{\cos \theta}\right)$$

The derivative of $\tan \theta$

If θ is measured in radians,

$$\frac{d}{d\theta} \tan \theta = \sec^2 \theta$$

Chain rule for the tangent

If h is a differentiable function of θ and θ is measured in radians,

$$\frac{d}{d\theta} \tan h(\theta) = h'(\theta) \sec^2 h(\theta)$$

Here is an example illustrating the use of the chain rule for the tangent.

EXAMPLE 2.4

Differentiate the function $f(\theta) = \tan \theta^2$.

SOLUTION

Using the chain rule for tangents, you get

$$f'(\theta) = 2\theta \sec^2 \theta^2$$

The derivations of the differentiation formulas for the other trigonometric functions are left as exercises for you to do. (See Problems 21, 22, and 23 at the end of this section.) Only the formulas for the derivatives of the sine, cosine, and tangent will be needed for the applications in Section 3.

The proof that

$$\frac{d}{d\theta} \sin \theta = \cos \theta$$

Since the function $f(\theta) = \sin \theta$ is unrelated to any of the functions whose derivatives were obtained in previous chapters, it will be necessary to return to the *definition* of the derivative (from Chapter 2, Section 1) to find the derivative in this case. In terms of θ, the definition states that

$$\frac{f(\theta + \Delta\theta) - f(\theta)}{\Delta\theta} \to f'(\theta) \qquad \text{as} \qquad \Delta\theta \to 0$$

In performing the required calculations when $f(\theta) = \sin \theta$, you will need to use the addition formula

$$\sin (a + b) = \sin a \cos b + \cos a \sin b$$

from the preceding section, as well as the following two facts about the behavior of $\sin \theta$ and $\cos \theta$ as θ approaches zero.

Two important limits

> **If θ is measured in radians,**
>
> **(1)** $\dfrac{\sin \theta}{\theta} \to 1 \qquad$ as $\qquad \theta \to 0$
>
> **and**
>
> **(2)** $\dfrac{\cos \theta - 1}{\theta} \to 0 \qquad$ as $\qquad \theta \to 0$

To see why the first of these limits holds, look at Figure 2.1. From the definition of the sine of an angle it follows that

$$\sin \theta = y = \text{length of segment } PQ$$

and since θ is measured in radians it follows that

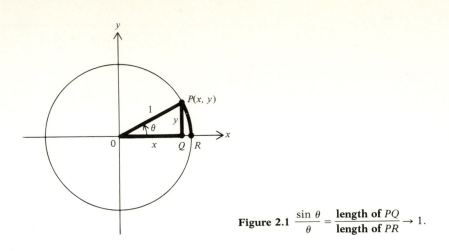

Figure 2.1 $\dfrac{\sin \theta}{\theta} = \dfrac{\textbf{length of } PQ}{\textbf{length of } PR} \to 1.$

$$\theta = \text{length of arc } PR$$

(Recall that 1 radian is the amount by which a line segment of length 1 must be rotated so that its free endpoint traces out a circular arc of length 1. Hence θ radians is the amount by which a line segment of length 1 must be rotated so that its free endpoint traces out a circular arc of length θ.)

Thus,

$$\frac{\sin \theta}{\theta} = \frac{\text{length of segment } PQ}{\text{length of arc } PR}$$

The picture suggests that as θ approaches zero, the length of the segment PQ and that of the arc PR will get closer and closer to one another so that their ratio will approach 1. (A formal proof of this fact can be found in more advanced texts.) Hence,

$$\frac{\sin \theta}{\theta} \to 1 \qquad \text{as} \qquad \theta \to 0$$

which establishes the first limit.

The second limit,

$$\frac{\cos \theta - 1}{\theta} \to 0 \qquad \text{as} \qquad \theta \to 0$$

can be derived algebraically from the first one. (See Problem 24 at the end of this section.) It also has the following simple geometric interpretation, which should give you additional insight into why it holds.

Figure 2.2 shows a portion of the graph of the function $\cos \theta$ near $\theta = 0$. The graph has a relative maximum when $\theta = 0$ and so, assuming that $\cos \theta$ is differentiable at $\theta = 0$ (which has not been proved), it

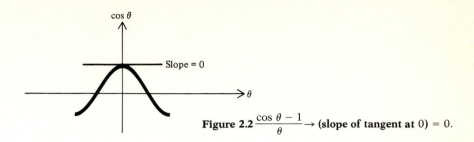

Figure 2.2 $\dfrac{\cos \theta - 1}{\theta} \rightarrow$ **(slope of tangent at** 0**)** $= 0.$

follows that the derivative of $\cos \theta$ at $\theta = 0$ is zero. But the difference quotient you would use to find this derivative is

$$\frac{\cos (0 + \Delta\theta) - \cos 0}{\Delta\theta} = \frac{\cos \Delta\theta - 1}{\Delta\theta}$$

Hence, $\qquad \dfrac{\cos \Delta\theta - 1}{\Delta\theta} \rightarrow 0 \qquad$ as $\qquad \Delta\theta \rightarrow 0$

which, except for the symbol used to denote the variable ($\Delta\theta$ instead of θ), is precisely the limit under consideration.

With these preliminaries out of the way, the proof of the formula for the derivative of $\sin \theta$ is not hard. Here are the details.

Let $f(\theta) = \sin \theta$. Then,

$$\frac{f(\theta + \Delta\theta) - f(\theta)}{\Delta\theta} = \frac{\sin (\theta + \Delta\theta) - \sin \theta}{\Delta\theta}$$

$$= \frac{\sin \theta \cos \Delta\theta + \cos \theta \sin \Delta\theta - \sin \theta}{\Delta\theta}$$

$$\text{(by the addition}$$
$$\text{formula for the sine)}$$

$$= \frac{\sin \theta \cos \Delta\theta - \sin \theta}{\Delta\theta} + \frac{\cos \theta \sin \Delta\theta}{\Delta\theta}$$

$$= \sin \theta \frac{\cos \Delta\theta - 1}{\Delta\theta} + \cos \theta \frac{\sin \Delta\theta}{\Delta\theta}$$

Now let $\Delta\theta$ approach zero to get the derivative. As $\Delta\theta$ approaches zero,

$$\frac{\cos \Delta\theta - 1}{\Delta\theta} \rightarrow 0 \qquad \text{[by limit (2)]}$$

and $\qquad \dfrac{\sin \Delta\theta}{\Delta\theta} \rightarrow 1 \qquad$ [by limit (1)]

and so $\qquad \dfrac{f(\theta + \Delta\theta) - f(\theta)}{\Delta\theta} \rightarrow (\sin \theta)(0) + (\cos \theta)(1) = \cos \theta$

That is, $\qquad$ $f'(\theta) = \cos \theta \qquad$ or $\qquad \dfrac{d}{d\theta} \sin \theta = \cos \theta$

The proof that
$\dfrac{d}{d\theta} \cos \theta = -\sin \theta$

The formula for the derivative of $\cos \theta$ can be obtained from the definition of the derivative by a calculation similar to the one just given for $\sin \theta$. (Problem 25 at the end of this section will suggest that you try it.) Here is an attractive alternative derivation that takes advantage of the formula for the derivative of the sine function and uses the identities

$$\cos \theta = \sin \left(\frac{\pi}{2} - \theta\right) \qquad \text{and} \qquad \sin \theta = \cos \left(\frac{\pi}{2} - \theta\right)$$

that relate the sine and cosine functions. (These identities are simple consequences of the addition formulas for the sine and cosine. They can also be obtained geometrically, directly from the definitions of the sine and cosine. See Problems 78, 79, and 80 of the preceding section.)

To obtain the derivative of $\cos \theta$, start with the identity

$$\cos \theta = \sin \left(\frac{\pi}{2} - \theta\right)$$

and differentiate using the chain rule for the sine to get

$$\frac{d}{d\theta} \cos \theta = (-1) \cos \left(\frac{\pi}{2} - \theta\right)$$

Then apply the identity

$$\sin \theta = \cos \left(\frac{\pi}{2} - \theta\right)$$

to the right-hand side to conclude that

$$\frac{d}{d\theta} \cos \theta = -\sin \theta$$

Problems
In Problems 1 through 20, differentiate the given function.

1. $f(\theta) = \sin 3\theta$
2. $f(\theta) = \cos 2\theta$
3. $f(\theta) = \sin (1 - 2\theta)$
4. $f(\theta) = \sin \theta^2$
5. $f(\theta) = \cos (\theta^3 + 1)$
6. $f(\theta) = \sin^2 \theta$
7. $f(\theta) = \cos^2 \left(\frac{\pi}{2} - \theta\right)$
8. $f(\theta) = \sin (2\theta + 1)^2$

9. $f(\theta) = \cos (1 + 3\theta)^2$

10. $f(\theta) = e^{-\theta} \sin \theta$

11. $f(\theta) = e^{-\theta/2} \cos 2\pi\theta$

12. $f(\theta) = \dfrac{\cos \theta}{1 - \cos \theta}$

13. $f(\theta) = \dfrac{\sin \theta}{1 + \sin \theta}$

14. $f(\theta) = \tan (5\theta + 2)$

15. $f(\theta) = \tan (1 - \theta^5)$

16. $f(\theta) = \tan^2 \theta$

17. $f(\theta) = \tan^2 \left(\dfrac{\pi}{2} - 2\pi\theta\right)$

18. $f(\theta) = \tan (\pi - 4\theta)^2$

19. $f(\theta) = \ln \sin^2 \theta$

20. $f(\theta) = \ln \tan^2 \theta$

21. Show that $\dfrac{d}{d\theta} \sec \theta = \sec \theta \tan \theta$.

22. Show that $\dfrac{d}{d\theta} \csc \theta = -\csc \theta \cot \theta$.

23. Show that $\dfrac{d}{d\theta} \cot \theta = -\csc^2 \theta$.

24. Use the fact that

$$\frac{\sin \theta}{\theta} \to 1 \qquad \text{as} \qquad \theta \to 0$$

to prove that

$$\frac{\cos \theta - 1}{\theta} \to 0 \qquad \text{as} \qquad \theta \to 0$$

Hint: Write

$$\frac{\cos \theta - 1}{\theta} = \frac{\cos \theta - 1}{\theta} \cdot \frac{\cos \theta + 1}{\cos \theta + 1}$$

and multiply out the terms on the right-hand side. Simplify the resulting expression using an appropriate trigonometric identity, and then let θ approach zero, keeping in mind that

$$\sin \theta \to 0 \qquad \text{and} \qquad \cos \theta \to 1 \qquad \text{as} \qquad \theta \to 0$$

25. Use the definition of the derivative to prove that $\dfrac{d}{d\theta} \cos \theta = -\sin \theta$.

3 APPLICATIONS OF TRIGONOMETRIC FUNCTIONS

In this section you will see a sampling of applications involving trigonometric functions and their derivatives. The first example is a related-rates problem, and the subsequent examples are optimization problems.

Related rates

In Chapter 3, Section 3, you learned how to use implicit differentiation to solve related-rates problems. Here is a related-rates problem involving trigonometric functions.

EXAMPLE 3.1

An observer watches a plane approach at a speed of 400 miles per hour and at an altitude of 4 miles. At what rate is the angle of elevation of the observer's line of sight changing with respect to time when the horizontal distance between the plane and the observer is 3 miles?

SOLUTION

Let x denote the horizontal distance between the plane and the observer, let t denote time (in hours), and draw a diagram representing the situation as in Figure 3.1.

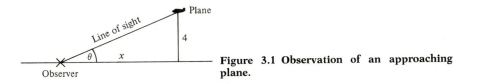

Figure 3.1 Observation of an approaching plane.

You know that $\dfrac{dx}{dt} = -400$ (the minus sign indicating that the distance x is decreasing), and your goal is to find $\dfrac{d\theta}{dt}$ when $x = 3$. From the right triangle in Figure 3.1 you see that

$$\tan \theta = \frac{4}{x}$$

Differentiate both sides of this equation with respect to t (remembering to use the chain rule) to get

$$\sec^2 \theta \frac{d\theta}{dt} = -\frac{4}{x^2} \frac{dx}{dt}$$

or

$$\frac{d\theta}{dt} = -\frac{4}{x^2} \cos^2 \theta \frac{dx}{dt}$$

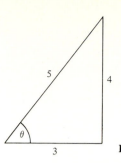

Figure 3.2 Triangle for the computation of cos θ when $x = 3$.

The right triangle in Figure 3.2 tells you that when $x = 3$,

$$\cos^2 \theta = \left(\frac{3}{5}\right)^2 = \frac{9}{25}$$

Substituting $x = 3$, $\dfrac{dx}{dt} = -400$, and $\cos^2 \theta = \dfrac{9}{25}$ into the formula for

$\dfrac{d\theta}{dt}$, you get

$$\frac{d\theta}{dt} = -\frac{4}{9}\left(\frac{9}{25}\right)(-400) = 64 \qquad \text{radians per hour}$$

Optimization problems   The next two examples are optimization problems involving trigonometric functions. To solve problems of this type, proceed as in Chapter 3, Section 1. In particular: (1) construct a function representing the quantity to be optimized in terms of a convenient variable; (2) identify an interval on which the function has a practical interpretation; (3) differentiate the function, and set the derivative equal to zero to find the critical points; and (4) compare the values of the function at the critical points in the interval and at the endpoints.

EXAMPLE 3.2

Two sides of a triangle are 4 inches long. What should the angle between these sides be to make the area of the triangle as large as possible?

SOLUTION

The triangle is shown in Figure 3.3. In general, the area of a triangle is given by the formula

$$\text{Area} = \frac{1}{2}(\text{base})(\text{height})$$

In this case,

$$\text{Base} = 4$$

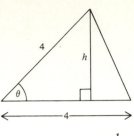

Figure 3.3 Triangle for Example 3.2.

and since $\sin \theta = \dfrac{h}{4}$,

$$\text{Height} = h = 4 \sin \theta$$

Hence the area of the triangle is given by the function

$$A(\theta) = \frac{1}{2}(4)(4 \sin \theta) = 8 \sin \theta$$

Since only values of θ between $\theta = 0$ and $\theta = \pi$ radians are meaningful in the context of this problem, the goal is to find the absolute maximum of the function $A(\theta)$ on the interval $0 \le \theta \le \pi$.

The derivative of $A(\theta)$ is

$$A'(\theta) = 8 \cos \theta$$

which is zero on the interval $0 \le \theta \le \pi$ only when $\theta = \dfrac{\pi}{2}$. Comparing

$$A(0) = 8 \sin 0 = 0$$

$$A\left(\frac{\pi}{2}\right) = 8 \sin \frac{\pi}{2} = 8(1) = 8$$

and

$$A(\pi) = 8 \sin \pi = 0$$

you can conclude that the area is maximized when the angle θ measures $\dfrac{\pi}{2}$ radians (or 90°), that is, when the triangle is a right triangle.

EXAMPLE 3.3

Illumination from a light source

A lamp with adjustable height hangs directly above the center of a circular kitchen table that is 8 feet in diameter. The illumination at the edge of the table is directly proportional to the cosine of the angle θ and inversely proportional to the square of the distance d, where θ and d are as shown in Figure 3.4. How close to the table should the lamp be pulled to maximize the illumination at the edge of the table?

SOLUTION

Let I denote the illumination at the edge of the table. Then,

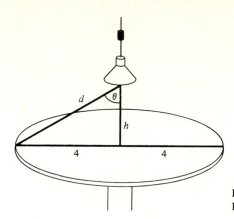

Figure 3.4 Hanging lamp and table for Example 3.3.

$$I = \frac{k \cos \theta}{d^2}$$

where k is a (positive) constant of proportionality. Moreover, from the right triangle in Figure 3.4,

$$\sin \theta = \frac{4}{d} \quad \text{or} \quad d = \frac{4}{\sin \theta}$$

Hence, $\qquad I(\theta) = \frac{k \cos \theta}{(4/\sin \theta)^2} = \frac{k}{16} \cos \theta \sin^2 \theta$

Only values of θ between 0 and $\frac{\pi}{2}$ radians are meaningful in the context of this problem. Hence, the goal is to find the absolute maximum of the function $I(\theta)$ on the interval $0 \le \theta \le \frac{\pi}{2}$.

Using the product rule and chain rule to differentiate $I(\theta)$, you get

$$I'(\theta) = \frac{k}{16} \left[\cos \theta \, (2 \sin \theta \cos \theta) + \sin^2 \theta \, (- \sin \theta) \right]$$

$$= \frac{k}{16} (2 \cos^2 \theta \sin \theta - \sin^3 \theta)$$

$$= \frac{k}{16} \sin \theta \, (2 \cos^2 \theta - \sin^2 \theta)$$

To find the critical points, set $I'(\theta) = 0$. That is,

$$\sin \theta = 0 \quad \text{or} \quad 2 \cos^2 \theta - \sin^2 \theta = 0$$

The only value of θ in the interval $0 \le \theta \le \frac{\pi}{2}$ for which $\sin \theta = 0$ is the endpoint $\theta = 0$. To solve the equation

$$2\cos^2\theta - \sin^2\theta = 0$$

rewrite it as

$$2\cos^2\theta = \sin^2\theta$$

divide both sides by $\cos^2\theta$

$$2 = \frac{\sin^2\theta}{\cos^2\theta} = \left(\frac{\sin\theta}{\cos\theta}\right)^2$$

and take the square root of each side to get

$$\pm\sqrt{2} = \frac{\sin\theta}{\cos\theta} = \tan\theta$$

Since both $\sin\theta$ and $\cos\theta$ are nonnegative on the interval $0 \le \theta \le \frac{\pi}{2}$, you may discard the negative square root and conclude that the critical point occurs when

$$\tan\theta = \sqrt{2}$$

Although it would not be hard to use your calculator (or Table IV at the back of the book) to find (or, at least, estimate) the angle θ for which $\tan\theta = \sqrt{2}$, it is not necessary to do so. Instead, look at the right triangle in Figure 3.5 in which

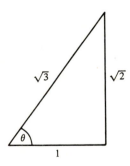

Figure 3.5 Right triangle with tan $\theta = \sqrt{2}$.

$$\tan\theta = \frac{\text{opposite}}{\text{adjacent}} = \frac{\sqrt{2}}{1} = \sqrt{2}$$

and observe that

$$\sin\theta = \frac{\text{opposite}}{\text{hypotenuse}} = \frac{\sqrt{2}}{\sqrt{3}} \quad \text{and} \quad \cos\theta = \frac{\text{adjacent}}{\text{hypotenuse}} = \frac{1}{\sqrt{3}}$$

Substitute these values into the equation for $I(\theta)$ to conclude that, at the critical point,

$$I(\theta) = \frac{k}{16}\cos\theta\sin^2\theta = \frac{k}{16}\left(\frac{1}{\sqrt{3}}\right)\left(\frac{\sqrt{2}}{\sqrt{3}}\right)^2 = \frac{k}{24\sqrt{3}}$$

Compare this value with the values of $I(\theta)$ at the endpoints $\theta = 0$ and $\theta = \dfrac{\pi}{2}$, which are

$$I(0) = \frac{k}{16} \cos (0) \sin^2 (0) = 0 \qquad (\text{since } \sin 0 = 0)$$

and
$$I\left(\frac{\pi}{2}\right) = \frac{k}{16} \cos \left(\frac{\pi}{2}\right) \sin^2 \left(\frac{\pi}{2}\right) = 0 \qquad \left(\text{since } \cos \frac{\pi}{2} = 0\right)$$

The largest of these possible maximum values is $k/24\sqrt{3}$. Hence, the absolute maximum of $I(\theta)$ on the interval is $k/24\sqrt{3}$, which occurs when $\tan \theta = \sqrt{2}$.

Finally, to find the height h that maximizes the illumination I, observe from Figure 3.4 that

$$\tan \theta = \frac{4}{h}$$

Since $\tan \theta = \sqrt{2}$, it follows that

$$h = \frac{4}{\tan \theta} = \frac{4}{\sqrt{2}} = 2.83 \text{ feet}$$

Minimization of travel time   In the next example, calculus is used to establish a general principle about the minimization of travel time. As you will see subsequently, the principle can be rephrased to give a well-known law about the reflection of light.

EXAMPLE 3.4

Two off-shore oil wells are, respectively, a and b miles out to sea. A motorboat that travels at a constant speed s transports workers from the first well to the shore and then proceeds to the second well. Show that the total travel time is minimized if the angle α between the motorboat's path of arrival and the shoreline is equal to the angle β between the shoreline and the motorboat's path of departure.

SOLUTION

Even though the goal is to prove that two angles are equal, it turns out that the easiest way to solve this problem is to let the variable x represent a convenient distance and to introduce trigonometry only at the end.

Begin with a sketch of the situation as in Figure 3.6, and define the variable x and the constant distance d as indicated. Then, by the pythagorean theorem,

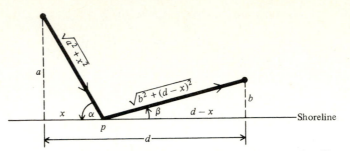

Figure 3.6 Path of motorboat from oil well to shore to oil well.

Distance from first well to $P = \sqrt{a^2 + x^2}$

and   Distance from second well to $P = \sqrt{b^2 + (d - x)^2}$

Since (for constant speeds),

$$\text{Time} = \frac{\text{distance}}{\text{speed}}$$

the total travel time is given by the function

$$T(x) = \frac{\sqrt{a^2 + x^2}}{s} + \frac{\sqrt{b^2 + (d - x)^2}}{s}$$

where $0 \le x \le d$.
  Then,

$$T'(x) = \frac{x}{s\sqrt{a^2 + x^2}} - \frac{d - x}{s\sqrt{b^2 + (d - x)^2}}$$

If you set $T'(x)$ equal to zero and do a little algebra, you get

$$\frac{x}{\sqrt{a^2 + x^2}} = \frac{d - x}{\sqrt{b^2 + (d - x)^2}}$$

Now look again at the right triangles in Figure 3.6 and observe that

$$\frac{x}{\sqrt{a^2 + x^2}} = \cos \alpha \quad \text{and} \quad \frac{d - x}{\sqrt{b^2 + (d - x)^2}} = \cos \beta$$

Hence, $T'(x) = 0$ implies that

$$\cos \alpha = \cos \beta$$

In the context of this problem, both α and β must measure between 0 and $\frac{\pi}{2}$ radians, and, for angles in this range, equality of the cosines implies equality of the angles themselves. (See, for example, the graph of the cosine function in Figure 1.10.) Hence, $T'(x) = 0$ implies that

$$\alpha = \beta$$

Moreover, it should be clear from the geometry of Figure 3.6 that no matter what the relative sizes of $a, b,$ and d may be, there is a (unique) point P for which $\alpha = \beta$. It follows that the function $T(x)$ has a critical point in the interval $0 \leq x \leq d$ and that at this critical point, $\alpha = \beta$.

To verify that this critical point is really the absolute minimum, you can use the second derivative

$$T''(x) = \frac{a^2}{s(a^2 + x^2)^{3/2}} + \frac{b^2}{s[b^2 + (d - x)^2]^{3/2}}$$

(For practice, check this calculation.) Since $T''(x)$ is positive for all values of x, the graph of $T(x)$ is always concave upward. It follows that the unique critical point on the interval $0 \leq x \leq \frac{\pi}{2}$, which occurs when $\alpha = \beta$, does indeed correspond to the absolute minimum of the total time T on this interval.

Reflection of light   According to **Fermat's principle** in optics, light traveling from one point to another takes the path that requires the least amount of time. Suppose (as illustrated in Figure 3.7) that a ray of light is transmitted from a source at a point A, strikes a reflecting surface (such as a mirror) at point P, and is subsequently received by an observer at point B.

Since, by Fermat's principle, the path from A to P to B minimizes time, it follows from the calculation in Example 3.4 that the angles α and β are equal. This, in turn, implies the **law of reflection,** which

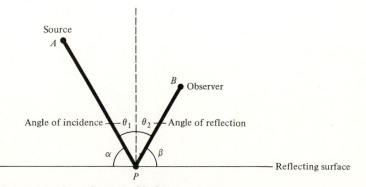

Figure 3.7 The reflection of light: $\theta_1 = \theta_2$.

states that the **angle of incidence** (θ_1 in Figure 3.7) must be equal to the **angle of reflection** (θ_2 in Figure 3.7).

Problems

Related rates

1. An observer watches a plane approach at a speed of 500 miles per hour and at an altitude of 3 miles. At what rate is the angle of elevation of the observer's line of sight changing with respect to time when the horizontal distance between the plane and the observer is 4 miles?

Related rates

2. A person 6 feet tall is watching a streetlight 18 feet high as he walks toward it at a speed of 5 feet per second. At what rate is the angle of elevation of the person's line of sight changing with respect to time when he is 9 feet from the base of the lamp?

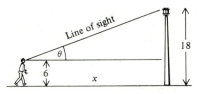

Related rates

3. An attendant is standing at the end of a pier 12 feet above the water and is pulling a rope attached to a rowboat at the rate of 4 feet of rope per minute. At what rate is the angle that the rope makes with the surface of the water changing with respect to time when the boat is 16 feet from the pier?

Related rates

4. A revolving searchlight in a lighthouse 2 miles off shore is following a jogger running along the shore. When the jogger is 1 mile from the point on the shore that is closest to the lighthouse, the searchlight is turning at the rate of 0.25 revolution per hour. How fast is the jogger running at that moment? (*Hint:* Since 0.25 revolution per hour is $\dfrac{\pi}{2}$ radians per hour, the problem is to find $\dfrac{dx}{dt}$ for the accompanying picture when $x = 1$ and $\dfrac{d\theta}{dt} = \dfrac{\pi}{2}$.)

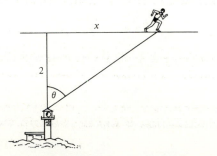

Maximization of
volume

5. You have a piece of metal that is 20 meters long and 6 meters wide, which you are going to bend to form a trough as indicated in the following diagram. At what angle should the sides meet so that the volume of the trough is as large as possible? (*Hint:* The volume is the length of the trough times the area of its triangular cross section.)

Maximization of area

6. Prove that of all isosceles triangles whose equal sides are of a specified length, the triangle of greatest area is the right triangle.

Maximization of area

7. The two sides and the base of an isosceles trapezoid are each 5 inches long. At what angle should the sides meet the horizontal top to maximize the area of the trapezoid? [*Hints:* (1) As indicated in the accompanying picture, the area of the trapezoid is the sum of the areas of two right triangles and the area of a rectangle. (2) You should be able to factor the derivative of the area function if you first simplify it using an appropriate trigonometric identity.]

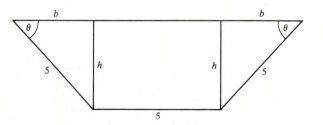

Illumination from a
light source

8. In Example 3.3, an alternative characterization of the illumination is that it is directly proportional to the sine of the angle ϕ and inversely proportional to the square of the distance d, where ϕ is the angle at which the ray of light meets the table. (That is, ϕ is the angle in the left-hand corner of the triangle in Figure 3.4 between the hypotenuse labeled d and the base labeled 4.) For practice, do Example 3.3 again, this time using this alternative characterization of the illumination. (Your final answer, of course, should be the same as before.)

Minimization of
horizontal clearance

9. Find the length of the longest pipe that can be carried horizontally around a corner joining two corridors that are $2\sqrt{2}$ feet wide. (*Hint:* Show that the horizontal clearance $C = x + y$ can be written as

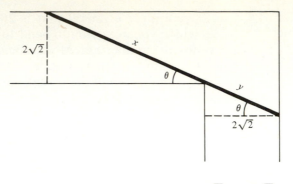

$$C(\theta) = \frac{2\sqrt{2}}{\sin \theta} + \frac{2\sqrt{2}}{\cos \theta}$$

and find the absolute *minimum* of $C(\theta)$ on an appropriate interval.)

Refraction of light 10. The accompanying figure shows a ray of light emitted from a source at point A under water and subsequently received by an observer at point B above the surface of the water. If v_1 is the speed of light in water and v_2 the speed of light in air, show that

$$\frac{\sin \theta_1}{\sin \theta_2} = \frac{v_1}{v_2}$$

where θ_1 is the **angle of incidence** and θ_2 the **angle of refraction.** (*Hint:* Apply Fermat's principle, which states that light takes the path requiring the least amount of time, and use the solution to Example 3.4 as a guide. You may assume without proof that the total time is minimized when its derivative is equal to zero.)

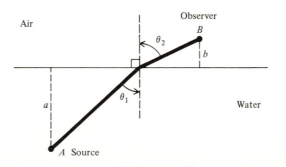

4 TAYLOR SERIES OF TRIGONOMETRIC FUNCTIONS In this section you will see how to find the Taylor series of the function $\sin x$ and how to use the partial sums of this series to generate values of $\sin x$ to any desired degree of accuracy. Similar calculations for the function $\cos x$ will be suggested in the problems at the end of the section.

Recall from Chapter 10, Section 6, that the Taylor series of a function $f(x)$ about $x = 0$ is the power series $\sum\limits_{n=0}^{\infty} a_n x^n$, where

$$a_n = \frac{f^{(n)}(0)}{n!}$$

The $(n + 1)$st partial sum of the Taylor series is the polynomial

$$P_n(x) = f(0) + f'(0)x + \frac{f''(0)}{2!} x^2 + \cdots + \frac{f^{(n)}(0)}{n!} x^n$$

This polynomial, known as the Taylor polynomial of degree n of $f(x)$ about $x = 0$, may be a good approximation to the original function $f(x)$ if x is near zero and n is large. The difference

$$R_n(x) = f(x) - P_n(x)$$

between $f(x)$ and its nth Taylor polynomial is known as the nth remainder and can be estimated using the formula

$$R_n(x) = \frac{f^{(n+1)}(c)}{(n + 1)!} x^{n+1}$$

where c is some (unspecified) number between 0 and x.

In the following example, the Taylor series of $\sin x$ about $x = 0$ is found. In subsequent examples, the remainder is used to show that the series converges to $\sin x$ for all x, and the partial sums of the series are used to approximate values of $\sin x$.

EXAMPLE 4.1

Find the Taylor series of the function $f(x) = \sin x$ about $x = 0$.

SOLUTION

Compute the Taylor coefficients as follows:

$$f(x) = \sin x \qquad f(0) = 0 \qquad a_0 = \frac{f(0)}{0!} = 0$$

$$f'(x) = \cos x \qquad f'(0) = 1 \qquad a_1 = \frac{f'(0)}{1!} = \frac{1}{1!}$$

$$f''(x) = -\sin x \qquad f''(0) = 0 \qquad a_2 = \frac{f''(0)}{2!} = 0$$

$$f^{(3)}(x) = -\cos x \qquad f^{(3)}(0) = -1 \qquad a_3 = \frac{f^{(3)}(0)}{3!} = -\frac{1}{3!}$$

$$f^{(4)}(x) = \sin x \qquad f^{(4)}(0) = 0 \qquad a_4 = \frac{f^{(4)}(0)}{4!} = 0$$

$$f^{(5)}(x) = \cos x \qquad f^{(5)}(0) = 1 \qquad a_5 = \frac{f^{(5)}(0)}{5!} = \frac{1}{5!}$$

and so on. In general, if n is even, $a_n = 0$, and if n is odd, $a_n = \pm \frac{1}{n!}$, where the signs alternate starting with $a_1 = +1$. Thus, the Taylor series of $\sin x$ about $x = 0$ is

$$x - \frac{x^3}{3!} + \frac{x^5}{5!} - \frac{x^7}{7!} + \cdots = \sum_{n=0}^{\infty} \frac{(-1)^n x^{2n+1}}{(2n+1)!}$$

The next example establishes that the Taylor series in Example 4.1 converges to $\sin x$ for all values of x. The technique will be to show that the remainder approaches zero as n increases without bound. The following limit will be needed in the calculation.

A useful limit

For all values of x,

$$\lim_{n \to \infty} \frac{x^n}{n!} = 0$$

Roughly speaking, this limit states that (for any fixed x), as n increases, the rate at which the denominator $n!$ increases is significantly greater than the rate at which the numerator x^n increases. Although this may seem plausible, it is not easy to prove directly. There is, however, a strikingly simple indirect proof based on the fact that the terms of a convergent series must approach zero. Recall (from Examples 5.2 and 6.1 in Chapter 10) that the Taylor series

$$\sum_{n=0}^{\infty} \frac{x^n}{n!}$$

of e^x converges for all x. (This involved a routine application of the ratio test.) It follows that the terms of this series must approach zero. That is,

$$\lim_{n \to \infty} \frac{x^n}{n!} = 0$$

EXAMPLE 4.2

Show that the Taylor series of $\sin x$ about $x = 0$ converges to $\sin x$ for all x.

SOLUTION

The goal is to show that the remainder $R_n(x)$ approaches zero as n increases without bound. Consider the remainder formula

$$R_n(x) = \frac{f^{(n+1)}(c)}{(n+1)!} x^{n+1}$$

where c is some number between 0 and x. Since $f(x) = \sin x$, the derivative $f^{(n+1)}(x)$ is either $\pm \sin x$ or $\pm \cos x$, all of which oscillate between -1 and 1. Hence for any c,

$$\left| f^{(n+1)}(c) \right| \leq 1$$

and it follows that

$$0 \leq |R_n(x)| = \frac{\left| f^{(n+1)}(c) \right|}{(n+1)!} |x|^{n+1} \leq \frac{1}{(n+1)!} |x|^{n+1}$$

Moreover, by the limit just established,

$$\lim_{n \to \infty} \frac{1}{(n+1)!} |x|^{n+1} = 0$$

Hence, $$0 \leq \lim_{n \to \infty} |R_n(x)| \leq \lim_{n \to \infty} \frac{1}{(n+1)!} |x|^{n+1} = 0$$

which implies that

$$\lim_{n \to \infty} |R_n(x)| = 0 \quad \text{or, equivalently,} \quad \lim_{n \to \infty} R_n(x) = 0$$

Since the remainder (which is the difference between $\sin x$ and the $(n+1)$st partial sum of its Taylor series) approaches zero for all x, it follows that the series converges to $\sin x$ for all x.

The approximation by partial sums of power series is used to program electronic computers and calculators to generate values of the trigonometric functions. This approximation technique is illustrated in the following two examples.

EXAMPLE 4.3

Use a Taylor polynomial of degree 5 to approximate $\sin\left(\frac{\pi}{20}\right)$, and estimate the accuracy of the approximation.

SOLUTION

For the approximation, use the Taylor polynomial

$$P_5(x) = x - \frac{x^3}{3!} + \frac{x^5}{5!}$$

obtained from the series in Example 4.1. In particular,

$$\sin\left(\frac{\pi}{20}\right) \approx P_5\left(\frac{\pi}{20}\right) = \frac{\pi}{20} - \frac{1}{3!}\left(\frac{\pi}{20}\right)^3 + \frac{1}{5!}\left(\frac{\pi}{20}\right)^5 = 0.1564345$$

To estimate the accuracy of this approximation, look at the absolute value of the corresponding remainder $R_5\left(\frac{\pi}{20}\right)$, which is given by the formula

$$\left|R_5\left(\frac{\pi}{20}\right)\right| = \frac{|f^{(6)}(c)|}{6!}\left(\frac{\pi}{20}\right)^6$$

where $0 \le c \le \frac{\pi}{20}$. Since all the derivatives of $f(x) = \sin x$ are either $\pm\sin x$ or $\pm\cos x$, it follows that

$$|f^{(6)}(c)| \le 1$$

Hence, $$\left|R_5\left(\frac{\pi}{20}\right)\right| \le \frac{(\pi/20)^6}{6!} = 0.00000002$$

which says that the approximation error is no greater than 0.00000002.

(Using the conversion formula from Section 1, you can show that $\frac{\pi}{20}$ radians is equal to 9°. Compare the estimate of sin (9°) obtained in this example with the value in Table IV at the back of the book and with the value generated by your calculator.)

EXAMPLE 4.4

Use a Taylor polynomial to approximate $\sin\left(\frac{\pi}{9}\right)$ with an error of less than 0.000005.

SOLUTION

The strategy is to use the remainder formula to determine the smallest value of n that will ensure the desired accuracy, and then to use the corresponding Taylor polynomial P_n for the approximation.

The absolute value of the nth remainder is

$$\left|R_n\left(\frac{\pi}{9}\right)\right| = \frac{|f^{(n+1)}(c)|}{(n+1)!}\left(\frac{\pi}{9}\right)^{n+1}$$

where $0 \le c \le \frac{\pi}{9}$. Since $f(x) = \sin x$,

$$|f^{(n+1)}(c)| \le 1$$

and so
$$\left| R_n \left(\frac{\pi}{9} \right) \right| \le \frac{(\pi/9)^{n+1}}{(n+1)!}$$

Now use your calculator to compute $\dfrac{(\pi/9)^{n+1}}{(n+1)!}$ for $n = 1, 2, 3, \ldots$ until you get an answer that is less than 0.000005. You should find that $n = 4$ gives

$$\frac{(\pi/9)^5}{5!} = 0.00004$$

and $n = 5$ gives

$$\frac{(\pi/9)^6}{6!} = 0.0000025$$

Thus $n = 5$ is the smallest integer that guarantees the desired accuracy, and you use $P_5 \left(\dfrac{\pi}{9} \right)$ for the approximation. That is,

$$\sin \left(\frac{\pi}{9} \right) \approx P_5 \left(\frac{\pi}{9} \right) = \frac{\pi}{9} - \frac{1}{3!} \left(\frac{\pi}{9} \right)^3 + \frac{1}{5!} \left(\frac{\pi}{9} \right)^5 = 0.3420203$$

where the approximation error is guaranteed to be less than 0.000005.

(You may be interested in converting $\dfrac{\pi}{9}$ radians to degrees and comparing the estimate obtained in this example with the value in Table IV and with the value you get using your calculator.)

The Taylor series of the function $\cos x$ about $x = 0$ can be obtained by a calculation similar to the one in Example 4.1, or by differentiation of the Taylor series for $\sin x$. The details are left as problems for you to do. (See Problems 1, 2, and 3 at the end of this section.) Here, for reference, is a summary of the results.

The Taylor series for the sine and cosine

For all x,

$$\sin x = x - \frac{x^3}{3!} + \frac{x^5}{5!} - \frac{x^7}{7!} + \cdots = \sum_{n=0}^{\infty} \frac{(-1)^n x^{2n+1}}{(2n+1)!}$$

and

$$\cos x = 1 - \frac{x^2}{2!} + \frac{x^4}{4!} - \frac{x^6}{6!} + \cdots = \sum_{n=0}^{\infty} \frac{(-1)^n x^{2n}}{(2n)!}$$

Problems

1. Use the formula for the Taylor coefficients to find the Taylor series of $f(x) = \cos x$ about $x = 0$.

2. Use the remainder $R_n(x)$ to show that the Taylor series of $\cos x$ about $x = 0$ converges to $\cos x$ for all x.

3. Get the Taylor series of $\cos x$ about $x = 0$ by differentiating the series for $\sin x$.

In Problems 4 through 11, use a Taylor polynomial of degree n to approximate the specified number, and estimate the accuracy of the approximation. (Don't forget that the differentiation formulas for trigonometric functions hold only if θ is measured in radians. In Problems 8 through 11, therefore, you will have to convert to radians before proceeding.)

4. $\sin \dfrac{\pi}{6}$; $n = 3$

5. $\cos \dfrac{\pi}{4}$; $n = 4$

6. $\cos \dfrac{\pi}{18}$; $n = 4$

7. $\sin \dfrac{\pi}{36}$; $n = 1$

8. $\sin 70°$; $n = 5$

9. $\cos 50°$; $n = 2$

10. $\cos 8°$; $n = 2$

11. $\sin 12°$; $n = 5$

In Problems 12 through 17, use an appropriate Taylor polynomial to approximate the specified number with an error of less than 0.000005. (In Problems 14 through 17, don't forget to convert to radians first.)

12. $\sin \dfrac{\pi}{18}$

13. $\cos \dfrac{\pi}{9}$

14. $\cos 65°$

15. $\sin 75°$

16. $\sin 3°$

17. $\cos 4°$

CHAPTER SUMMARY AND PROFICIENCY TEST

Important terms, symbols, and formulas

Angle measurement: $\dfrac{\text{Degrees}}{180} = \dfrac{\text{radians}}{\pi}$

Sine and cosine:

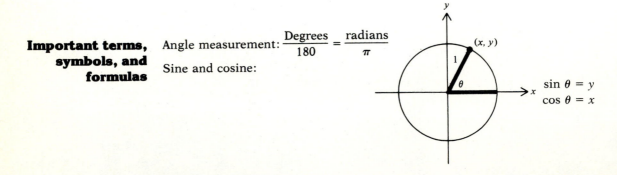

Important values:

θ	0	$\dfrac{\pi}{6}$	$\dfrac{\pi}{4}$	$\dfrac{\pi}{3}$	$\dfrac{\pi}{2}$	$\dfrac{2\pi}{3}$	$\dfrac{3\pi}{4}$	$\dfrac{5\pi}{6}$	π
$\sin\theta$	0	$\dfrac{1}{2}$	$\dfrac{\sqrt{2}}{2}$	$\dfrac{\sqrt{3}}{2}$	1	$\dfrac{\sqrt{3}}{2}$	$\dfrac{\sqrt{2}}{2}$	$\dfrac{1}{2}$	0
$\cos\theta$	1	$\dfrac{\sqrt{3}}{2}$	$\dfrac{\sqrt{2}}{2}$	$\dfrac{1}{2}$	0	$-\dfrac{1}{2}$	$-\dfrac{\sqrt{2}}{2}$	$-\dfrac{\sqrt{3}}{2}$	-1

Other trigonometric functions:

$$\tan\theta = \frac{\sin\theta}{\cos\theta}, \quad \cot\theta = \frac{\cos\theta}{\sin\theta}, \quad \sec\theta = \frac{1}{\cos\theta}, \quad \csc\theta = \frac{1}{\sin\theta}$$

Trigonometric identities:
  Periodicity: $\sin(\theta + 2\pi) = \sin\theta, \quad \cos(\theta + 2\pi) = \cos\theta$
  Negative angles: $\sin(-\theta) = -\sin\theta, \quad \cos(-\theta) = \cos\theta$
  Pythagorean identity: $\sin^2\theta + \cos^2\theta = 1$
  Addition formulas:

$$\cos(a + b) = \cos a \cos b - \sin a \sin b$$

$$\sin(a + b) = \sin a \cos b + \cos a \sin b$$

  Double-angle formulas: $\cos 2\theta = \cos^2\theta - \sin^2\theta$, $\sin 2\theta = 2\sin\theta\cos\theta$
Right triangles:

$$\sin\theta = \frac{\text{opposite}}{\text{hypotenuse}}$$

$$\cos\theta = \frac{\text{adjacent}}{\text{hypotenuse}}$$

$$\tan\theta = \frac{\text{opposite}}{\text{adjacent}}$$

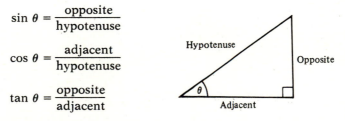

Differentiation formulas:

$$\frac{d}{d\theta}\sin\theta = \cos\theta, \quad \frac{d}{d\theta}\cos\theta = -\sin\theta, \quad \frac{d}{d\theta}\tan\theta = \sec^2\theta$$

Chain rules:

$$\frac{d}{d\theta}\sin h(\theta) = h'(\theta)\cos h(\theta)$$

$$\frac{d}{d\theta}\cos h(\theta) = -h'(\theta)\sin h(\theta)$$

$$\frac{d}{d\theta}\tan h(\theta) = h'(\theta)\sec^2 h(\theta)$$

Taylor series:

$$\sin x = x - \frac{x^3}{3!} + \frac{x^5}{5!} - \frac{x^7}{7!} + \cdots$$

$$\cos x = 1 - \frac{x^2}{2!} + \frac{x^4}{4!} - \frac{x^6}{6!} + \cdots$$

Proficiency test

1. Specify the radian measurement and degree measurement for each of the following angles.

   (a)   (b)

   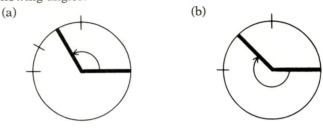

2. Convert 50° to radians.

3. Convert 0.25 radian to degrees.

4. Evaluate the given expression without using **Table IV or your calculator.**

   (a) $\sin\left(-\dfrac{5\pi}{3}\right)$ (b) $\cos\dfrac{15\pi}{4}$

   (c) $\sec\dfrac{7\pi}{3}$ (d) $\cot\dfrac{2\pi}{3}$

5. Find $\tan\theta$ if $\sin\theta = \frac{4}{5}$.

6. Find $\csc\theta$ if $\cot\theta = \dfrac{\sqrt{5}}{2}$.

In Problems 7 through 10, find all the values of θ in the specified interval that satisfy the given equation.

7. $2\cos\theta + \sin 2\theta = 0;\ 0 \le \theta \le 2\pi$

8. $3\sin^2\theta - \cos^2\theta = 2;\ 0 \le \theta \le \pi$

9. $2\sin^2\theta = \cos 2\theta;\ 0 \le \theta \le \pi$

10. $5\sin\theta - 2\cos^2\theta = 1;\ 0 \le \theta \le 2\pi$

11. Starting with the identity $\sin^2\theta + \cos^2\theta = 1$, derive the identity $1 + \cot^2\theta = \csc^2\theta$.

12. Starting with the double-angle formulas for the sine and cosine, derive the double-angle formula

$$\tan 2\theta = \frac{2 \tan \theta}{1 - \tan^2 \theta}$$

for the tangent.

13. (a) Starting with the addition formulas for the sine and cosine, derive the identities

$$\cos\left(\frac{\pi}{2} + \theta\right) = -\sin \theta \quad \text{and} \quad \sin\left(\frac{\pi}{2} + \theta\right) = \cos \theta$$

(b) Give geometric arguments to establish the identities in part (a).

In Problems 14 through 19, differentiate the given function.

14. $f(\theta) = \sin (3\theta + 1)^2$

15. $f(\theta) = \cos^2 (3\theta + 1)$

16. $f(\theta) = \tan (3\theta^2 + 1)$

17. $f(\theta) = \tan^2 (3\theta^2 + 1)$

18. $f(\theta) = \dfrac{\sin \theta}{1 - \cos \theta}$

19. $f(\theta) = \ln \cos^2 \theta$

20. Show that $\dfrac{d}{d\theta} \sin \theta \cos \theta = \cos 2\theta$.

21. On New Year's Eve, a holiday reveler is watching the descent of a lighted ball from atop a tall building that is 600 feet away. The ball is falling at the rate of 20 feet per minute. At what rate is the angle of elevation of the reveler's line of sight changing with respect to time when the ball is 800 feet from the ground?

22. A trough 9 meters long is to have a cross section consisting of an isosceles trapezoid in which the base and two sides are all 4 meters long. At what angle should the sides of the trapezoid meet the horizontal top to maximize the capacity of the trough?

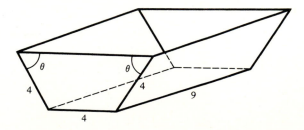

23. Find the length of the longest pipe that can be carried horizontally around a corner joining a corridor that is 8 feet wide and one that is $5\sqrt{5}$ feet wide.

24. A cable is run in a straight line from a power plant on one side of a river 900 meters wide to a point P on the other side and then along the river

bank to a factory, 3,000 meters downstream from the power plant. The cost of running the cable under the water is $5 per meter, while the cost over land is $4 per meter. If θ is the (smaller) angle between the segment of cable under the river and the opposite bank, show that $\cos \theta = \frac{4}{5}$ (the ratio of the per-meter costs) for the route that minimizes the total installation cost. (You may assume without proof that the absolute minimum occurs when the derivative of the cost function is zero.)

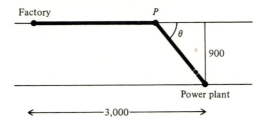

In Problems 25 and 26, use a Taylor polynomial of degree n to approximate the specified number, and estimate the accuracy of your approximation.

25. $\sin \dfrac{\pi}{10}$; $n = 3$ 26. $\cos 36°$; $n = 4$

In Problems 27 and 28, use an appropriate Taylor polynomial to approximate the specified numbers with an error of less than 0.000005.

27. $\cos \dfrac{\pi}{18}$ 28. $\sin 50°$

APPENDIX

A Algebra Review

B Limits and Continuity

A ALGEBRA REVIEW

This section contains a review of certain algebraic topics that are used throughout the book. A set of practice problems appears at the end of the section, and the answers to the odd-numbered problems are given at the back of the book.

Factoring

To factor an expression is to write it as a product of two or more terms, called **factors.** Factoring is used to simplify complicated expressions and is based on the distributive law for addition and multiplication.

The distributive law

> **For any real numbers** a, b, **and** c,
>
> $$ab + ac = a(b + c)$$

The factoring techniques you will need in this book are illustrated in the following examples.

EXAMPLE A.1

Factor the expression $3x^4 - 6x^3$.

SOLUTION

Since $3x^3$ is a factor of each of the terms in this expression, you can use the distributive law to "factor out" $3x^3$ and write

$$3x^4 - 6x^3 = 3x^3(x - 2)$$

EXAMPLE A.2

Factor the expression $x^2 - 9$.

SOLUTION

Your goal is to write $x^2 - 9$ as a product of the form

$$x^2 - 9 = (x + a)(x + b)$$

The distributive law implies that

$$(x + a)(x + b) = x^2 + (a + b)x + ab$$

Hence, your goal is to find integers a and b so that

$$x^2 - 9 = x^2 + (a + b)x + ab$$

In particular, you wish to find integers a and b for which

$$a + b = 0 \quad \text{and} \quad ab = -9$$

From the list $\qquad -1, 9 \qquad 1, -9 \qquad 3, -3$

of pairs of integers whose product is -9, choose $a = 3$ and $b = -3$ as the only pair whose sum is zero. It follows that

$$x^2 - 9 = (x + 3)(x - 3)$$

The expression $x^2 - 9$ in the preceding example is the difference of two perfect squares. Here is a well-known formula you can use to factor expressions of this form.

The difference of two squares

> **For any real numbers a and b,**
> $$a^2 - b^2 = (a + b)(a - b)$$

EXAMPLE A.3

Factor the expression $x^2 - 2x - 3$.

SOLUTION

Your goal is to find integers a and b such that

$$x^2 - 2x - 3 = (x + a)(x + b) = x^2 + (a + b)x + ab$$

That is, you wish to find integers a and b for which

$$a + b = -2 \quad \text{and} \quad ab = -3$$

From the list $\qquad 1, -3 \qquad -1, 3$

of pairs of integers whose product is -3, choose $a = -3$ and $b = 1$ as the only pair whose sum is -2. It follows that

$$x^2 - 2x - 3 = (x - 3)(x + 1)$$

EXAMPLE A.4

Factor the expression $x^3 - 8$.

SOLUTION

The fact that $2^3 = 8$ tells you that $x - 2$ must be a factor of this expression. That is, there are integers a and b for which

$$x^3 - 8 = (x - 2)(x^2 + ax + b)$$

Since $(x - 2)(x^2 + ax + b) = x^3 + (a - 2)x^2 + (b - 2a)x - 2b$

your goal is to find integers a and b for which

$$a - 2 = 0 \quad b - 2a = 0 \quad \text{and} \quad 2b = 8$$

The only such integers are $a = 2$ and $b = 4$ and hence

$$x^3 - 8 = (x - 2)(x^2 + 2x + 4)$$

Convince yourself by examining pairs of integers whose product is 4 that the expression $x^2 + 2x + 4$ cannot be factored further.

EXAMPLE A.5

Factor the expression $x^5 - 4x^3$.

SOLUTION

First factor out x^3 to get

$$x^5 - 4x^3 = x^3(x^2 - 4)$$

and then factor $x^2 - 4$ (which is the difference of two squares) to get

$$x^5 - 4x^3 = x^3(x + 2)(x - 2)$$

The solution of quadratic equations

An equation of the form

$$ax^2 + bx + c = 0$$

is said to be a **quadratic equation**. A **solution** of a quadratic equation (or of any equation) is a value of x for which the equation is true. A quadratic equation can have at most two solutions. One way to find the solutions of a quadratic equation is to factor it. Here are two examples.

EXAMPLE A.6

Solve the equation $x^2 - x - 6 = 0$.

SOLUTION

Factor the equation $x^2 - x - 6 = 0$

as $(x + 2)(x - 3) = 0$

Since the product $(x + 2)(x - 3)$ can be zero only if one (or both) of its factors is zero, it follows that the solutions are $x = -2$ (which makes the first factor zero) and $x = 3$ (which makes the second factor zero).

EXAMPLE A.7

Solve the equation $x^2 + 10x + 25 = 0$.

SOLUTION

Factoring you get $(x + 5)^2 = 0$

which implies that the only solution is $x = -5$.

There is a special formula you can use to solve quadratic equations. This is especially useful when the factors of the equation are not obvious or when the equation cannot be factored at all.

The quadratic formula

The solutions of the quadratic equation

$$ax^2 + bx + c = 0 \qquad (a \neq 0)$$

are

$$x = \frac{-b + \sqrt{b^2 - 4ac}}{2a} \quad \text{and} \quad x = \frac{-b - \sqrt{b^2 - 4ac}}{2a}$$

The term $b^2 - 4ac$ in the quadratic formula is called the **discriminant** of the quadratic equation. If the discriminant is positive, the equation has two solutions, if it is zero, the equation has only one solution, and if it is negative, the equation has no real solutions. (Do you see why?)

The use of the quadratic formula is illustrated in the following examples.

EXAMPLE A.8

Solve the equation $x^2 + 3x + 1 = 0$.

SOLUTION

This is a quadratic equation with $a = 1$, $b = 3$, and $c = 1$. Using the quadratic formula, you get

$$x = \frac{-3 + \sqrt{5}}{2} \quad \text{and} \quad x = \frac{-3 - \sqrt{5}}{2}$$

EXAMPLE A.9

Solve the equation $x^2 + 18x + 81 = 0$.

SOLUTION

This is a quadratic equation with $a = 1$, $b = 18$, and $c = 81$. Using the quadratic formula you find that the discriminant is zero and that the formulas for x give

$$x = \frac{-18 + \sqrt{0}}{2} = -9 \quad \text{and} \quad x = \frac{-18 - \sqrt{0}}{2} = -9$$

It follows that $x = -9$ is the only solution of this equation. (Notice that this equation could have been factored as $(x + 9)^2 = 0$ and the solution $x = -9$ obtained immediately.)

EXAMPLE A.10

Solve the equation $x^2 + x + 1 = 0$.

SOLUTION

This is a quadratic equation with $a = 1$, $b = 1$, and $c = 1$. Using the quadratic formula, you get

$$x = \frac{-1 + \sqrt{-3}}{2} \quad \text{and} \quad x = \frac{-1 - \sqrt{-3}}{2}$$

Since there is no real number that is the square root of -3, it follows that this equation has no real solution.

Exponential notation   The following rules define the expression a^x for $a > 0$ and all rational values of x.

Definition of a^x
$(a > 0)$

Integer powers: If n is a positive integer,

$$a^n = a \cdot a \cdots a$$

where the product $a \cdot a \cdots a$ contains n factors.

Fractional powers: If n and m are positive integers,

$$a^{n/m} = (\sqrt[m]{a})^n$$

where $\sqrt[m]{\ }$ denotes the positive mth root.

Negative powers: $\qquad\qquad a^{-x} = \dfrac{1}{a^x}$

Zero power: $\qquad\qquad\qquad a^0 = 1$

EXAMPLE A.11

Evaluate the following expressions.

(a) $9^{1/2}$ (b) $27^{2/3}$ (c) $8^{-1/3}$ (d) $(\tfrac{1}{100})^{-3/2}$ (e) 5^0

SOLUTION

(a) $9^{1/2} = \sqrt{9} = 3$

(b) $27^{2/3} = (\sqrt[3]{27})^2 = 3^2 = 9$

(c) $8^{-1/3} = \dfrac{1}{8^{1/3}} = \dfrac{1}{\sqrt[3]{8}} = \dfrac{1}{2}$

(d) $(\tfrac{1}{100})^{-3/2} = 100^{3/2} = (\sqrt{100})^3 = 10^3 = 1{,}000$

(e) $5^0 = 1$

Exponents obey the following useful laws.

Laws of exponents

The product law: $\qquad a^r a^s = a^{r+s}$

The quotient law: $\qquad \dfrac{a^r}{a^s} = a^{r-s}$

The power law: $\qquad\quad (a^r)^s = a^{rs}$

The use of these laws is illustrated in the next two examples.

EXAMPLE A.12

Evaluate the following expressions (without using a calculator).

(a) $(2^{-2})^3$ (b) $\dfrac{3^3}{3^{1/3}(3^{2/3})}$ (c) $2^{7/4}(8^{-1/4})$

SOLUTION

(a) $(2^{-2})^3 = 2^{-6} = \dfrac{1}{2^6} = \dfrac{1}{64}$

(b) $\dfrac{3^3}{3^{1/3}(3^{2/3})} = \dfrac{3^3}{3^{(1/3+2/3)}} = \dfrac{3^3}{3^1} = 3^2 = 9$

(c) $2^{7/4}(8^{-1/4}) = 2^{7/4}(2^3)^{-1/4} = 2^{7/4}(2^{-3/4}) = 2^{(7/4-3/4)} = 2$

EXAMPLE A.13

Solve each of the following equations for n.

(a) $\dfrac{a^5}{a^2} = a^n$ (b) $(a^n)^5 = a^{20}$

SOLUTION

(a) Since $\dfrac{a^5}{a^2} = a^{5-2} = a^3$, it follows that $n = 3$.

(b) Since $(a^n)^5 = a^{5n}$, it follows that $5n = 20$ or $n = 4$.

Summation notation

To describe a sum $a_1 + a_2 + \cdots + a_n$

it suffices to specify the general term a_i and to indicate that n terms of this form are to be added, starting with the term in which $i = 1$ and ending with the term in which $i = n$. It is customary to use the Greek letter Σ (sigma) to denote summation and to express the sum compactly as follows.

Summation notation

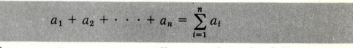

$$a_1 + a_2 + \cdots + a_n = \sum_{i=1}^{n} a_i$$

The use of summation notation is illustrated in the following examples.

EXAMPLE A.14

Use summation notation to express the following sums.

(a) $1 + 4 + 9 + 16 + 25 + 36 + 49 + 64$
(b) $(1 - x_1)^2 \, \Delta x + (1 - x_2)^2 \, \Delta x + \cdots + (1 - x_{15})^2 \, \Delta x$

SOLUTION

(a) This is a sum of 8 terms of the form i^2, starting with $i = 1$ and ending with $i = 8$. Hence,

$$1 + 4 + 9 + 16 + 25 + 36 + 49 + 64 = \sum_{i=1}^{8} i^2$$

(b) The ith term of this sum is $(1 - x_i)^2 \, \Delta x$. Hence,

$$(1 - x_1)^2 \, \Delta x + (1 - x_2)^2 \, \Delta x + \cdots + (1 - x_{15})^2 \, \Delta x = \sum_{i=1}^{15} (1 - x_i)^2 \, \Delta x$$

EXAMPLE A.15

Evaluate the following sums.

(a) $\displaystyle\sum_{i=1}^{4} (i^2 + 1)$ (b) $\displaystyle\sum_{i=1}^{3} (-2)^i$

SOLUTION

(a) $\displaystyle\sum_{i=1}^{4} (i^2 + 1) = (1^2 + 1) + (2^2 + 1) + (3^2 + 1) + (4^2 + 1)$

$$= 2 + 5 + 10 + 17 = 34$$

(b) $\displaystyle\sum_{i=1}^{3} (-2)^i = (-2)^1 + (-2)^2 + (-2)^3 = -2 + 4 - 8 = -6$

Problems   In Problems 1 through 16, factor the given expression.

1. $x^5 - 4x^4$ 2. $3x^3 - 12x^4$

3. $x^2 - 4$ 4. $64 - x^2$

5. $x^2 + x - 2$ 6. $x^2 + 3x - 10$

7. $x^2 - 7x + 12$ 8. $x^2 + 8x + 12$

9. $x^2 - 2x + 1$ 10. $x^2 + 6x + 9$

11. $x^3 - 1$ 12. $x^3 - 27$

13. $x^7 - x^5$ 14. $x^3 + 2x^2 + x$

15. $2x^3 - 8x^2 - 10x$ 16. $x^4 + 5x^3 - 14x^2$

In Problems 17 through 28, find all the solutions of the given quadratic equation.

17. $x^2 - 2x - 8 = 0$ 18. $x^2 - 4x + 3 = 0$

19. $x^2 + 10x + 25 = 0$ 20. $x^2 + 8x + 16 = 0$

21. $x^2 - 16 = 0$ 22. $x^2 - 25 = 0$

23. $2x^2 + 3x + 1 = 0$ 24. $-x^2 + 3x - 1 = 0$

25. $x^2 - 2x + 3 = 0$ 26. $x^2 - 2x + 1 = 0$

27. $4x^2 + 12x + 9 = 0$ 28. $x^2 + 12 = 0$

In Problems 29 through 37, evaluate the given expression without using a calculator.

29. 5^3 30. 2^{-3}

31. $16^{1/2}$ 32. $36^{-1/2}$

33. $8^{2/3}$ 34. $27^{-4/3}$

35. $(\frac{1}{4})^{1/2}$ 36. $(\frac{1}{4})^{-3/2}$

37. $\left(\dfrac{1}{\sqrt{2}}\right)^0$

In Problems 38 through 43, evaluate the given expression without using a calculator.

38. $\dfrac{3^4(3^3)}{(3^2)^3}$ 39. $\dfrac{2^{4/3}(2^{5/3})}{2^5}$

40. $\dfrac{5^{-3}(5^2)}{(5^{-2})^3}$ 41. $\dfrac{2(16^{3/4})}{2^3}$

42. $\dfrac{\sqrt{27}(\sqrt{3})^3}{9}$ 43. $[\sqrt{8}(2^{5/2})]^{-1/2}$

In Problems 44 through 51, solve the given equation for n. (Assume $a > 0$ and $a \neq 1$.)

44. $a^3a^7 = a^n$ 45. $\dfrac{a^5}{a^2} = a^n$

46. $a^4a^{-3} = a^n$ 47. $a^2a^n = \dfrac{1}{a}$

48. $(a^3)^n = a^{12}$ 49. $(a^n)^5 = \dfrac{1}{a^{10}}$

50. $a^{3/5}a^{-n} = \dfrac{1}{a^2}$ 51. $(a^n)^3 = \dfrac{1}{\sqrt{a}}$

In Problems 52 through 55, evaluate the given sum.

52. $\displaystyle\sum_{i=1}^{5} i^2$ 53. $\displaystyle\sum_{i=1}^{4} (3i + 1)$

54. $\displaystyle\sum_{i=1}^{5} 2^i$ 55. $\displaystyle\sum_{i=1}^{10} (-1)^i$

In Problems 56 through 61, use summation notation to express the given sum.

56. $1 + \frac{1}{2} + \frac{1}{3} + \frac{1}{4} + \frac{1}{5} + \frac{1}{6}$

57. $3 + 6 + 9 + 12 + 15 + 18 + 21 + 24 + 27 + 30$

58. $2 + 4 + 8 + 16 + 32 + 64$

59. $2x_1 + 2x_2 + 2x_3 + 2x_4 + 2x_5 + 2x_6$

60. $1 - 1 + 1 - 1 + 1 - 1$

61. $1 - 2 + 3 - 4 + 5 - 6 + 7 - 8$

B LIMITS AND CONTINUITY

In calculus, one infers what is or "should be" happening *at* a particular point from knowledge about what is happening at other points that approach it. Indeed, the development of the derivative itself, involving the approximation of tangents by secants, is based on such an inference. And indirectly, so are all the subsequent rules and applications developed in this book.

Central to this process of inference is the mathematical concept of **limit.** The purpose of this section is to give you a better feel for this important concept. The approach will be intuitive rather than formal. The ideas outlined here form the basis for a more rigorous development of the laws and procedures of calculus, and are at the heart of many of the branches of modern mathematics.

Here is a slightly imprecise definition of limit that will be sufficient for our purposes.

Limit

If the function values $f(x)$ get closer and closer to some number L whenever x gets closer and closer to some number a, we say that L is the limit of $f(x)$ as x approaches a, and we write

$$\lim_{x \to a} f(x) = L$$

In geometric terms, $\lim_{x \to a} f(x) = L$ means that the height of the graph of $y = f(x)$ approaches L as x approaches a.

Limits describe the behavior of a function *near* a particular point and not necessarily *at* the point itself. This is illustrated in Figure B.1. For all three of the functions graphed in this figure, the limit of $f(x)$ as x approaches a is equal to L. Yet the functions behave differently at $x = a$ itself. In Figure B.1a, $f(a)$ is equal to the limit L; in Figure B.1b, $f(a)$ has been defined (artificially) to be different from L; and in Figure B.1c, $f(a)$ is not defined at all.

Figure B.2 shows the graphs of two functions that do not have a limit as x approaches a. The function in Figure B.2a does not have a limit as x approaches a because $f(x)$ approaches 5 as x approaches a from the right, and it approaches a different value, 3, as x approaches a from the left. The function in Figure B.2b has no limit as x approaches

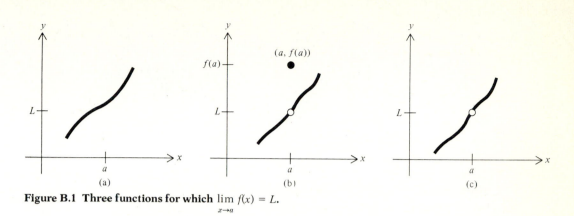

Figure B.1 Three functions for which $\lim\limits_{x \to a} f(x) = L$.

a because the values of $f(x)$ increase without bound and do not approach any (finite) number.

Properties of limits Limits obey the following algebraic laws. These laws, which should seem plausible on the basis of our informal definition, are proved formally in more theoretical courses. They are important because they simplify the calculation of limits of algebraic functions.

The limit of a sum, difference, or product

If $\lim\limits_{x \to a} f(x)$ **and** $\lim\limits_{x \to a} g(x)$ **exist, then**

$$\lim_{x \to a} [f(x) + g(x)] = \lim_{x \to a} f(x) + \lim_{x \to a} g(x)$$

$$\lim_{x \to a} [f(x) - g(x)] = \lim_{x \to a} f(x) - \lim_{x \to a} g(x)$$

$$\lim_{x \to a} [f(x)g(x)] = \lim_{x \to a} f(x) \cdot \lim_{x \to a} g(x)$$

That is, the limit of a sum (or difference or product) is the sum (or difference or product) of the individual limits.

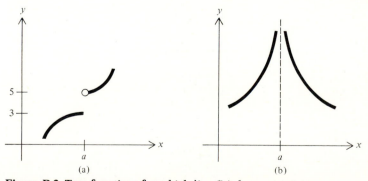

Figure B.2 Two functions for which $\lim\limits_{x \to a} f(x)$ **does not exist.**

The limit of a quotient

If $\lim\limits_{x \to a} f(x)$ and $\lim\limits_{x \to a} g(x)$ exist and if $\lim\limits_{x \to a} g(x) \neq 0$, then

$$\lim_{x \to a} \frac{f(x)}{g(x)} = \frac{\lim\limits_{x \to a} f(x)}{\lim\limits_{x \to a} g(x)}$$

That is, if the limit of the denominator is not zero, the limit of a quotient is the quotient of the individual limits.

The limit of a power

If $\lim\limits_{x \to a} f(x)$ exists, then

$$\lim_{x \to a} [f(x)]^p = \left[\lim_{x \to a} f(x)\right]^p$$

for any real number p. That is, the limit of a power is the power of the limit.

The next two properties deal with the limits of two elementary linear functions from which all other algebraic functions can be built.

The limit of a constant

If k is a constant, $\qquad\qquad \lim\limits_{x \to a} k = k$

That is, the limit of a constant is the constant itself.

In geometric terms, this says that the height of the graph of the constant function $f(x) = k$ approaches k as x approaches a. The situation is illustrated in Figure B.3a. Of course, in this case, the height of the graph is actually equal to k when $x = a$ and for all other values of x as well.

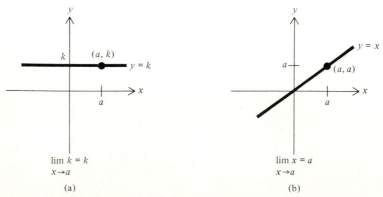

Figure B.3 The limits of two linear functions.

The next property says that the height of the linear function $f(x) = x$ approaches a as x approaches a. The situation is illustrated in Figure B.3b.

The limit of x

$$\lim_{x \to a} x = a$$

Computation of limits

The following examples illustrate how the properties of limits can be used to calculate limits of algebraic functions. In the first example, you will see how to find the limit of a polynomial.

EXAMPLE B.1

Find $\lim_{x \to 2} (x^2 - 3x + 1)$.

SOLUTION

Using the properties of limits, you get

$$\lim_{x \to 2} (x^2 - 3x + 1) = \left(\lim_{x \to 2} x\right)^2 - \lim_{x \to 2} 3 \cdot \lim_{x \to 2} x + \lim_{x \to 2} 1$$
$$= 2^2 - 3(2) + 1$$
$$= -1$$

In the next example, you will see how to find the limit of a rational function whose denominator does not approach zero.

EXAMPLE B.2

Find $\lim_{x \to 0} \dfrac{3x^2 - 8}{x - 2}$.

SOLUTION

Since

$$\lim_{x \to 0} (x - 2) = -2 \neq 0$$

you can use the rule for the limit of a quotient to get

$$\lim_{x \to 0} \frac{3x^2 - 8}{x - 2} = \frac{\lim_{x \to 0} (3x^2 - 8)}{\lim_{x \to 0} (x - 2)} = \frac{-8}{-2} = 4$$

In the next example, the denominator of the given rational function approaches zero while the numerator does not. When this happens you can conclude that the limit does not exist. The absolute value of such a quotient increases without bound and so does not approach any (finite) number.

EXAMPLE B.3

Find $\lim\limits_{x \to 2} \dfrac{x + 1}{x - 2}$.

SOLUTION

The rule for the limit of a quotient does not apply in this case since the limit of the denominator is

$$\lim_{x \to 2} (x - 2) = 0$$

Since the limit of the numerator is

$$\lim_{x \to 2} (x + 1) = 3$$

which is not equal to zero, you can conclude that the limit of the quotient does not exist.

The graph of the function $f(x) = \dfrac{x + 1}{x - 2}$ has been sketched in Figure B.4 to give you a better idea of what is actually happening here.

Notice that $f(x)$ increases without bound as x approaches 2 from the right and decreases without bound as x approaches 2 from the left.

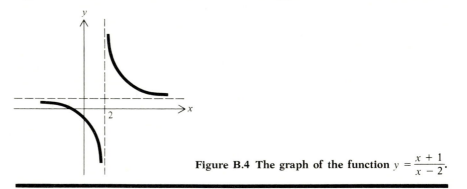

Figure B.4 The graph of the function $y = \dfrac{x + 1}{x - 2}$.

In the next example, both the numerator and denominator of the given rational function approach zero. When this happens, you have to simplify the function algebraically in order to find the desired limit.

EXAMPLE B.4

Find $\lim\limits_{x \to 1} \dfrac{x^2 + x - 2}{x - 1}$.

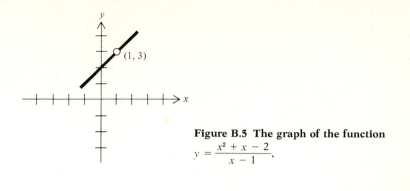

Figure B.5 The graph of the function
$$y = \frac{x^2 + x - 2}{x - 1}.$$

SOLUTION

As x approaches 1, both the numerator and denominator approach zero, from which you can draw no conclusion about the size of the quotient.

To proceed, observe that the given function is not defined when $x = 1$, but that for all other values of x, you can simplify it by dividing numerator and denominator by $x - 1$ to get

$$\frac{x^2 + x - 2}{x - 1} = \frac{(x - 1)(x + 2)}{x - 1} = x + 2$$

(Since $x \neq 1$, you are not dividing by zero.) Now take the limit as x approaches (but is not equal to) 1 to get

$$\lim_{x \to 1} \frac{x^2 + x - 2}{x - 1} = \lim_{x \to 1} (x + 2) = 3$$

The graph of the function $f(x) = \dfrac{x^2 + x - 2}{x - 1}$ is sketched in Figure B.5. It is the straight line $y = x + 2$ with a hole at the point $(1, 3)$.

In general, when both the numerator and denominator of a quotient approach zero as x approaches a, your strategy will be to simplify the quotient algebraically (as you did in Example B.4 by canceling $x - 1$). In most cases, the simplified form of the quotient will be valid for all values of x except $x = a$. Since you are interested in the behavior of the quotient *near $x = a$* and not *at $x = a$*, you may use the simplified form of the quotient to calculate the limit. Here is another example illustrating the technique.

EXAMPLE B.5

Find $\displaystyle\lim_{x \to 1} \frac{\sqrt{x} - 1}{x - 1}$.

SOLUTION

Both the numerator and denominator approach zero as x approaches 1. To simplify the quotient, multiply the numerator and denominator by $\sqrt{x} + 1$ to get

$$\frac{\sqrt{x} - 1}{x - 1} = \frac{(\sqrt{x} - 1)(\sqrt{x} + 1)}{(x - 1)(\sqrt{x} + 1)} = \frac{x - 1}{(x - 1)(\sqrt{x} + 1)} = \frac{1}{\sqrt{x} + 1}$$

and then take the limit to get

$$\lim_{x \to 1} \frac{\sqrt{x} - 1}{x - 1} = \lim_{x \to 1} \frac{1}{\sqrt{x} + 1} = \frac{1}{2}$$

The definition of the derivative

Using limit notation, you can write the definition of the derivative compactly as follows.

The derivative

$$f'(x) = \lim_{\Delta x \to 0} \frac{f(x + \Delta x) - f(x)}{\Delta x}$$

The calculations you did in Chapter 2, Section 1, to compute derivatives from the definition can be made more rigorous using limits. Here is an example.

EXAMPLE B.6

Use the definition to find the derivative of the function $f(x) = x^2$.

SOLUTION

Since both the numerator and denominator of the difference quotient $\dfrac{f(x + \Delta x) - f(x)}{\Delta x}$ approach zero as Δx approaches zero, begin by simplifying the difference quotient algebraically as follows.

$$\frac{f(x + \Delta x) - f(x)}{\Delta x} = \frac{(x + \Delta x)^2 - x^2}{\Delta x}$$

$$= \frac{x^2 + 2x \, \Delta x + (\Delta x)^2 - x^2}{\Delta x} = 2x + \Delta x$$

Then take the limit as Δx approaches zero to get

$$f'(x) = \lim_{\Delta x \to 0} \frac{f(x + \Delta x) - f(x)}{\Delta x} = \lim_{\Delta x \to 0} (2x + \Delta x) = 2x$$

Continuity The notion of continuity that was introduced informally in Chapter 2 can be defined more precisely in terms of limits. Recall that a function was said to be **continuous** if its graph was an unbroken curve. Here is the more formal definition.

Continuity

> **A function f is continuous at $x = a$ if**
>
> **(a)** $f(a)$ is defined
> **(b)** $\lim\limits_{x \to a} f(x)$ exists
>
> **and**
>
> **(c)** $\lim\limits_{x \to a} f(x) = f(a)$

Figure B.6 shows the graphs of three functions that are *not* continuous at $x = a$. The function in Figure B.6a is not continuous at $x = a$ because $f(a)$ is not defined. The function in Figure B.6b is not continuous at $x = a$ because $\lim\limits_{x \to a} f(x)$ does not exist. And the function in Figure B.6c is not continuous at $x = a$ because $\lim\limits_{x \to a} f(x) \neq f(a)$, even though $f(a)$ is defined and the limit exists.

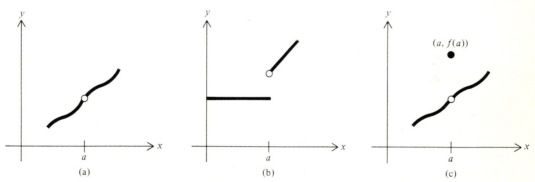

Figure B.6 Three functions with discontinuities at $x = a$.

A function whose graph is an unbroken curve near $x = a$ is continuous at $x = a$ because all three conditions in the definition of continuity are satisfied. Two such functions are sketched in Figure B.7.

You can use the properties of limits to show that a polynomial is continuous for every value of x and a rational function is continuous for every value of x for which its denominator is not zero. Here are some examples.

EXAMPLE B.7

Show that the polynomial $f(x) = 3x^2 - x + 5$ is continuous at $x = 1$.

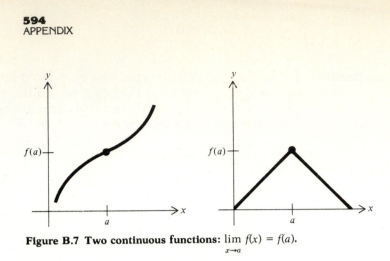

Figure B.7 Two continuous functions: $\lim\limits_{x \to a} f(x) = f(a)$.

SOLUTION

Verify that the three criteria for continuity are satisfied. Clearly, $f(1)$ is defined. In fact, $f(1) = 7$. Moreover,

$$\lim_{x \to 1} f(x) = 3 \left[\lim_{x \to 1} x\right]^2 - \lim_{x \to 1} x + \lim_{x \to 1} 5 = 3 - 1 + 5 = 7 = f(1)$$

Hence, f is continuous at $x = 1$.

EXAMPLE B.8

Show that the rational function $f(x) = \dfrac{x + 1}{x - 2}$ is continuous at $x = 3$.

SOLUTION

$$f(3) = \frac{3 + 1}{3 - 2} = 4$$

Moreover, since $\lim\limits_{x \to 3} (x - 2) = 1 \neq 0$, you can apply the rule for the limit of a quotient to get

$$\lim_{x \to 3} f(x) = \lim_{x \to 3} \frac{x + 1}{x - 2} = \frac{\lim\limits_{x \to 3} (x + 1)}{\lim\limits_{x \to 3} (x - 2)} = \frac{4}{1} = 4 = f(3)$$

Hence, f is continuous at $x = 3$.

EXAMPLE B.9

Show that the rational function $f(x) = \dfrac{x + 1}{x - 2}$ is not continuous at $x = 2$.

SOLUTION

Since division by zero is impossible, $f(2)$ is undefined, violating the first criterion for continuity.

Continuous extension of functions

The function shown in Figure B.8a has a discontinuity at $x = 1$ where it is not defined. By filling in the hole, you can extend this function to one that is continuous at $x = 1$ as shown in Figure B.8b.

The next example illustrates how you can perform this extension algebraically using limits.

EXAMPLE B.10

If

$$f(x) = \begin{cases} 2x & \text{if } 0 \le x < 1 \\ 2 & \text{if } x > 1 \end{cases}$$

define $f(1)$ so that the resulting function will be continuous at $x = 1$.

SOLUTION

According to the definition of continuity, $f(1)$ should be defined to be equal to the limit of $f(x)$ as x approaches 1. Since the function $f(x)$ is defined by one formula to the left of $x = 1$ and by a different formula to the right of $x = 1$, you will have to compute the limit of each of these separately to determine what the limit of f is (or if it even exists).

The limit of $f(x)$ as x approaches 1 from the left is

$$\lim_{x \to 1} 2x = 2$$

and the limit as x approaches 1 from the right is

$$\lim_{x \to 1} 2 = 2$$

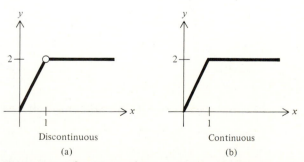

Discontinuous   Continuous

(a)   (b)

Figure B.8 A function and its continuous extension.

Since these are equal, you can conclude that the limit of $f(x)$ as x approaches 1 exists and

$$\lim_{x \to 1} f(x) = 2$$

To make f continuous at $x = 1$, define $f(1)$ to be 2.

Problems   In Problems 1 through 6, find $\lim_{x \to a} f(x)$ if it exists.

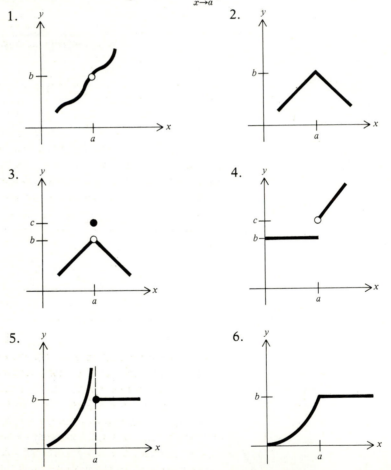

In Problems 7 through 28, find the indicated limit if it exists.

7. $\lim_{x \to 2} (3x^2 - 5x + 2)$

8. $\lim_{x \to -1} (x^3 - 2x^2 + x - 3)$

9. $\lim_{x \to 0} (x^5 - 6x^4 + 7)$

10. $\lim_{x \to 0} (1 - 5x^3)$

11. $\lim\limits_{x\to 3} (x - 1)^2(x + 1)$

12. $\lim\limits_{x\to -1} (x^2 + 1)(1 - 2x)^2$

13. $\lim\limits_{x\to 2} \dfrac{x + 1}{x + 2}$

14. $\lim\limits_{x\to 1} \dfrac{2x + 3}{x + 1}$

15. $\lim\limits_{x\to 5} \dfrac{x + 3}{5 - x}$

16. $\lim\limits_{x\to 3} \dfrac{2x + 3}{x - 3}$

17. $\lim\limits_{x\to 1} \dfrac{x^2 - 1}{x - 1}$

18. $\lim\limits_{x\to 3} \dfrac{9 - x^2}{x - 3}$

19. $\lim\limits_{x\to 5} \dfrac{x^2 - 3x - 10}{x - 5}$

20. $\lim\limits_{x\to 2} \dfrac{x^2 + x - 6}{x - 2}$

21. $\lim\limits_{x\to 4} \dfrac{(x + 1)(x - 4)}{(x - 1)(x - 4)}$

22. $\lim\limits_{x\to 0} \dfrac{x(x^2 - 1)}{x^2}$

23. $\lim\limits_{x\to -2} \dfrac{x^2 - x - 6}{x^2 + 3x + 2}$

24. $\lim\limits_{x\to 1} \dfrac{x^2 + 4x - 5}{x^2 - 1}$

25. $\lim\limits_{x\to 4} \dfrac{\sqrt{x} - 2}{x - 4}$

26. $\lim\limits_{x\to 9} \dfrac{\sqrt{x} - 3}{x - 9}$

27. $\lim\limits_{x\to 1} \dfrac{x - 1}{\sqrt{x} - 1}$

28. $\lim\limits_{x\to 9} \dfrac{x - 9}{\sqrt{x} - 3}$

In Problems 29 through 41, decide if the given function is continuous at the specified value of x.

29. $f(x) = 5x^2 - 6x + 1; x = 2$

30. $f(x) = x^3 - 2x^2 + x - 5; x = 0$

31. $f(x) = \dfrac{x + 2}{x + 1}; x = 1$

32. $f(x) = \dfrac{2x - 4}{3x - 2}; x = 2$

33. $f(x) = \dfrac{x + 1}{x - 1}; x = 1$

34. $f(x) = \dfrac{2x + 1}{3x - 6}; x = 2$

35. $f(x) = \dfrac{\sqrt{x} - 2}{x - 4}; x = 4$

36. $f(x) = \dfrac{\sqrt{x} - 2}{x - 4}; x = 2$

37. $f(x) = \begin{cases} x + 1 & \text{if } x \le 2 \\ 2 & \text{if } x > 2 \end{cases}; x = 2$

38. $f(x) = \begin{cases} 0 & \text{if } x < 1 \\ x - 1 & \text{if } x \ge 1 \end{cases}; x = 1$

39. $f(x) = \begin{cases} x + 1 & \text{if } x < 0 \\ x - 1 & \text{if } x \ge 0 \end{cases}; x = 0$

40. $f(x) = \begin{cases} x^2 + 1 & \text{if } x \le 3 \\ 2x + 4 & \text{if } x > 3 \end{cases}; x = 3$

41. $f(x) = \begin{cases} \dfrac{x^2 - 1}{x + 1} & \text{if } x < -1 \\ x^2 - 3 & \text{if } x \geq -1 \end{cases}$; $x = -1$

In Problems 42 through 54, list all the values of x for which the given function is *not* continuous.

42. $f(x) = 3x^2 - 6x + 9$

43. $f(x) = x^5 - x^3$

44. $f(x) = \dfrac{x + 1}{x - 2}$

45. $f(x) = \dfrac{3x - 1}{2x - 6}$

46. $f(x) = \dfrac{3x + 3}{x + 1}$

47. $f(x) = \dfrac{x^2 - 1}{x + 1}$

48. $f(x) = \dfrac{3x - 2}{(x + 3)(x - 6)}$

49. $f(x) = \dfrac{x}{(x + 5)(x - 1)}$

50. $f(x) = \dfrac{x}{x^2 - x}$

51. $f(x) = \dfrac{x^2 - 2x + 1}{x^2 - x - 2}$

52. $f(x) = \begin{cases} 2x + 3 & \text{if } x \leq 1 \\ 6x - 1 & \text{if } x > 1 \end{cases}$

53. $f(x) = \begin{cases} x^2 & \text{if } x \leq 2 \\ 9 & \text{if } x > 2 \end{cases}$

54. $f(x) = \begin{cases} x - 1 & \text{if } x < 1 \\ 1 & \text{if } x = 1 \\ 1 - x & \text{if } x > 1 \end{cases}$

In Problems 55 through 58, define $f(a)$ (if possible) so that the resulting function will be continuous at $x = a$.

55. $f(x) = \begin{cases} 3x + 1 & \text{if } x < 1 \\ 5x - 1 & \text{if } x > 1 \end{cases}$; $a = 1$

56. $f(x) = \begin{cases} x & \text{if } x < 0 \\ -x & \text{if } x > 0 \end{cases}$; $a = 0$

57. $f(x) = \begin{cases} \dfrac{2}{4 - x} & \text{if } x < 2 \\ \dfrac{1}{x} & \text{if } x > 2 \end{cases}$; $a = 2$

58. $f(x) = \begin{cases} \dfrac{\sqrt{x} - 1}{x - 1} & \text{if } x < 1 \\ \dfrac{x^2 + x - 2}{3x^2 - 3} & \text{if } x > 1 \end{cases}$; $a = 1$

TABLES

Table I Powers of e

x	e^x	e^{-x}	x	e^x	e^{-x}	x	e^x	e^{-x}
0.00	1.0000	1.00000	0.50	1.6487	.60653	1.00	2.7183	.36788
0.01	1.0101	0.99005	0.51	1.6653	.60050	1.20	3.3201	.30119
0.02	1.0202	.98020	0.52	1.6820	.59452	1.30	3.6693	.27253
0.03	1.0305	.97045	0.53	1.6989	.58860	1.40	4.0552	.24660
0.04	1.0408	.96079	0.54	1.7160	.58275	1.50	4.4817	.22313
0.05	1.0513	.95123	0.55	1.7333	.57695	1.60	4.9530	.20190
0.06	1.0618	.94176	0.56	1.7507	.57121	1.70	5.4739	.18268
0.07	1.0725	.93239	0.57	1.7683	.56553	1.80	6.0496	.16530
0.08	1.0833	.92312	0.58	1.7860	.55990	1.90	6.6859	.14957
0.09	1.0942	.91393	0.59	1.8040	.55433	2.00	7.3891	.13534
0.10	1.1052	.90484	0.60	1.8221	.54881	3.00	20.086	.04979
0.11	1.1163	.89583	0.61	1.8404	.54335	4.00	54.598	.01832
0.12	1.1275	.88692	0.62	1.8589	.53794	5.00	148.41	.00674
0.13	1.1388	.87809	0.63	1.8776	.53259	6.00	403.43	.00248
0.14	1.1503	.86936	0.64	1.8965	.52729	7.00	1096.6	.00091
0.15	1.1618	.86071	0.65	1.9155	.52205	8.00	2981.0	.00034
0.16	1.1735	.85214	0.66	1.9348	.51685	9.00	8103.1	.00012
0.17	1.1853	.84366	0.67	1.9542	.51171	10.00	22026.5	.00005
0.18	1.1972	.83527	0.68	1.9739	.50662			
0.19	1.2092	.82696	0.69	1.9937	.50158			
0.20	1.2214	.81873	0.70	2.0138	.49659			
0.21	1.2337	.81058	0.71	2.0340	.49164			
0.22	1.2461	.80252	0.72	2.0544	.48675			
0.23	1.2586	.79453	0.73	2.0751	.48191			
0.24	1.2712	.78663	0.74	2.0959	.47711			
0.25	1.2840	.77880	0.75	2.1170	.47237			
0.26	1.2969	.77105	0.76	2.1383	.46767			
0.27	1.3100	.76338	0.77	2.1598	.46301			
0.28	1.3231	.75578	0.78	2.1815	.45841			
0.29	1.3364	.74826	0.79	2.2034	.45384			
0.30	1.3499	.74082	0.80	2.2255	.44933			
0.31	1.3634	.73345	0.81	2.2479	.44486			
0.32	1.3771	.72615	0.82	2.2705	.44043			
0.33	1.3910	.71892	0.83	2.2933	.43605			
0.34	1.4049	.71177	0.84	2.3164	.43171			
0.35	1.4191	.70469	0.85	2.3396	.42741			
0.36	1.4333	.69768	0.86	2.3632	.42316			
0.37	1.4477	.69073	0.87	2.3869	.41895			
0.38	1.4623	.68386	0.88	2.4109	.41478			
0.39	1.4770	.67706	0.89	2.4351	.41066			
0.40	1.4918	.67032	0.90	2.4596	.40657			
0.41	1.5068	.66365	0.91	2.4843	.40252			
0.42	1.5220	.65705	0.92	2.5093	.39852			
0.43	1.5373	.65051	0.93	2.5345	.39455			
0.44	1.5527	.64404	0.94	2.5600	.39063			
0.45	1.5683	.63763	0.95	2.5857	.38674			
0.46	1.5841	.63128	0.96	2.6117	.38298			
0.47	1.6000	.62500	0.97	2.6379	.37908			
0.48	1.6161	.61878	0.98	2.6645	.37531			
0.49	1.6323	.61263	0.99	2.6912	.37158			

Excerpted from *Handbook of Mathematical Tables and Formulas,* 5th ed., by R. S. Burington. Copyright © 1973 by McGraw-Hill, Inc. Used with permission of McGraw-Hill Book Company.

Table II   The natural logarithm (base e)

x	$\ln x$	x	$\ln x$	x	$\ln x$	x	$\ln x$
.01	−4.60517	0.50	−0.69315	1.00	0.00000	1.5	0.40547
.02	−3.91202	.51	.67334	1.01	.00995	1.6	7000
.03	.50656	.52	.65393	1.02	.01980	1.7	0.53063
.04	.21888	.53	.63488	1.03	.02956	1.8	8779
		.54	.61619	1.04	.03922	1.9	0.64185
.05	−2.99573	.55	.59784	1.05	.04879	2.0	9315
.06	.81341	.56	.57982	1.06	.05827	2.1	0.74194
.07	.65926	.57	.56212	1.07	.06766	2.2	8846
.08	.52573	.58	.54473	1.08	.07696	2.3	0.83291
.09	.40795	.59	.52763	1.09	.08618	2.4	7547
0.10	−2.30259	0.60	−0.51083	1.10	.09531	2.5	0.91629
.11	.20727	.61	.49430	1.11	.10436	2.6	5551
.12	.12026	.62	.47804	1.12	.11333	2.7	9325
.13	.04022	.63	.46204	1.13	.12222	2.8	1.02962
.14	−1.96611	.64	.44629	1.14	.13103	2.9	6471
.15	.89712	.65	.43078	1.15	.13976	3.0	9861
.16	.83258	.66	.41552	1.16	.14842	4.0	1.38629
.17	.77196	.67	.40048	1.17	.15700	5.0	1.60944
.18	.71480	.68	.38566	1.18	.16551	10.0	2.30258
.19	.66073	.69	.37106	1.19	.17395		
0.20	−1.60944	0.70	−0.35667	1.20	.18232		
.21	.56065	.71	.34249	1.21	.19062		
.22	.51413	.72	.32850	1.22	.19885		
.23	.46968	.73	.31471	1.23	.20701		
.24	.42712	.74	.30111	1.24	.21511		
.25	.38629	.75	.28768	1.25	.22314		
.26	.34707	.76	.27444	1.26	.23111		
.27	.30933	.77	.26136	1.27	.23902		
.28	.27297	.78	.24846	1.28	.24686		
.29	.23787	.79	.23572	1.29	.25464		
0.30	−1.20397	0.80	−0.22314	1.30	.26236		
.31	.17118	.81	.21072	1.31	.27003		
.32	.13943	.82	.19845	1.32	.27763		
.33	.10866	.83	.18633	1.33	.28518		
.34	.07881	.84	.17435	1.34	.29267		
.35	−1.04982	.85	−0.16252	1.35	.30010		
.36	.02165	.86	.15032	1.36	.30748		
.37	−0.99425	.87	.13926	1.37	.31481		
.38	.96758	.88	.12783	1.38	.32208		
.39	.94161	.89	.11653	1.39	.32930		
0.40	−0.91629	0.90	−0.10536	1.40	.33647		
.41	.89160	.91	.09431	1.41	.34359		
.42	.86750	.92	.08338	1.42	.35066		
.43	.84397	.93	.07257	1.43	.35767		
.44	.82098	.94	.06188	1.44	.36464		
.45	.79851	.95	.05129	1.45	.37156		
.46	.77653	.96	.04082	1.46	.37844		
.47	.75502	.97	.03046	1.47	.38526		
.48	.73397	.98	.02020	1.48	.39204		
.49	.71335	.99	.01005	1.49	.39878		

From *Calculus and Analytic Geometry* by S. K. Stein. Copyright © 1973 by McGraw-Hill, Inc. Used with permission of McGraw-Hill Book Company.

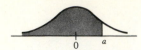

a	.00	.01	.02	.03	.04	.05	.06	.07	.08	.09
.0	.5000	.5040	.5080	.5120	.5160	.5199	.5239	.5279	.5319	.5359
.1	.5398	.5438	.5478	.5517	.5557	.5596	.5636	.5675	.5714	.5753
.2	.5793	.5832	.5871	.5910	.5948	.5987	.6026	.6064	.6103	.6141
.3	.6179	.6217	.6255	.6293	.6331	.6368	.6406	.6443	.6480	.6517
.4	.6554	.6591	.6628	.6664	.6700	.6736	.6772	.6808	.6844	.6879
.5	.6915	.6950	.6985	.7019	.7054	.7088	.7123	.7157	.7190	.7224
.6	.7257	.7291	.7324	.7357	.7389	.7422	.7454	.7486	.7517	.7549
.7	.7580	.7611	.7642	.7673	.7704	.7734	.7764	.7794	.7823	.7825
.8	.7881	.7910	.7939	.7967	.7995	.8023	.8051	.8078	.8106	.8133
.9	.8159	.8186	.8212	.8238	.8264	.8289	.8315	.8340	.8365	.8389
1.0	.8413	.8438	.8461	.8485	.8508	.8531	.8554	.8577	.8599	.8621
1.1	.8643	.8665	.8686	.8708	.8729	.8749	.8770	.8790	.8810	.8830
1.2	.8849	.8869	.8888	.8907	.8925	.8944	.8962	.8980	.8997	.9015
1.3	.9032	.9049	.9066	.9082	.9099	.9115	.9131	.9147	.9162	.9177
1.4	.9192	.9207	.9222	.9236	.9251	.9265	.9279	.9292	.9306	.9319
1.5	.9332	.9345	.9357	.9370	.9382	.9394	.9406	.9418	.9429	.9441
1.6	.9452	.9463	.9474	.9484	.9495	.9505	.9515	.9525	.9535	.9545
1.7	.9554	.9564	.9573	.9582	.9591	.9599	.9608	.9616	.9625	.9633
1.8	.9641	.9649	.9656	.9664	.9671	.9678	.9686	.9693	.9699	.9706
1.9	.9713	.9719	.9726	.9732	.9738	.9744	.9750	.9756	.9761	.9767
2.0	.9772	.9778	.9783	.9788	.9793	.9798	.9803	.9808	.9812	.9817
2.1	.9821	.9826	.9830	.9834	.9838	.9842	.9846	.9850	.9854	.9857
2.2	.9861	.9864	.9868	.9871	.9875	.9878	.9881	.9884	.9887	.9890
2.3	.9893	.9896	.9898	.9901	.9904	.9906	.9909	.9911	.9913	.9916
2.4	.9918	.9920	.9922	.9925	.9927	.9929	.9931	.9932	.9934	.9936
2.5	.9938	.9940	.9941	.9943	.9945	.9946	.9948	.9949	.9951	.9952
2.6	.9953	.9955	.9956	.9957	.9959	.9960	.9961	.9962	.9963	.9964
2.7	.9965	.9966	.9967	.9968	.9969	.9970	.9971	.9972	.9973	.9974
2.8	.9974	.9975	.9976	.9977	.9977	.9978	.9979	.9979	.9980	.9981
2.9	.9981	.9982	.9982	.9983	.9984	.9984	.9985	.9985	.9986	.9986
3.0	.9987	.9987	.9987	.9988	.9988	.9989	.9989	.9989	.9990	.9990
3.1	.9990	.9991	.9991	.9991	.9992	.9992	.9992	.9992	.9993	.9993
3.2	.9993	.9993	.9994	.9994	.9994	.9994	.9994	.9995	.9995	.9995
3.3	.9995	.9995	.9995	.9996	.9996	.9996	.9996	.9996	.9996	.9997
3.4	.9997	.9997	.9997	.9997	.9997	.9997	.9997	.9997	.9997	.9998

Table IV   Trigonometric functions

Degrees	Radians	sin	cos	tan	Degrees	Radians	sin	cos	tan
0	0.0000	0.0000	1.000	0.0000	45	0.7854	0.7071	0.7071	1.000
1	0.01745	0.01745	0.9998	0.01746	46	0.8028	0.7193	0.6947	1.036
2	0.03491	0.03490	0.9994	0.03492	47	0.8203	0.7314	0.6820	1.072
3	0.05236	0.05234	0.9986	0.05241	48	0.8378	0.7431	0.6691	1.111
4	0.06981	0.06976	0.9976	0.06993	49	0.8552	0.7547	0.6561	1.150
5	0.08727	0.08716	0.9962	0.08749	50	0.8727	0.7660	0.6428	1.192
6	0.1047	0.1045	0.9945	0.1051	51	0.8901	0.7772	0.6293	1.235
7	0.1222	0.1219	0.9926	0.1228	52	0.9076	0.7880	0.6157	1.280
8	0.1396	0.1392	0.9903	0.1405	53	0.9250	0.7986	0.6018	1.327
9	0.1571	0.1564	0.9877	0.1584	54	0.9425	0.8090	0.5878	1.376
10	0.1745	0.1736	0.9848	0.1763	55	0.9599	0.8192	0.5736	1.428
11	0.1920	0.1908	0.9816	0.1944	56	0.9774	0.8290	0.5592	1.483
12	0.2094	0.2079	0.9782	0.2126	57	0.9948	0.8387	0.5446	1.540
13	0.2269	0.2250	0.9744	0.2309	58	1.012	0.8480	0.5299	1.600
14	0.2444	0.2419	0.9703	0.2493	59	1.030	0.8572	0.5150	1.664
15	0.2618	0.2588	0.9659	0.2680	60	1.047	0.8660	0.5000	1.732
16	0.2792	0.2756	0.9613	0.2868	61	1.065	0.8746	0.4848	1.804
17	0.2967	0.2924	0.9563	0.3057	62	1.082	0.8830	0.4695	1.881
18	0.3142	0.3090	0.9511	0.3249	63	1.100	0.8910	0.4540	1.963
19	0.3316	0.3256	0.9455	0.3443	64	1.117	0.8988	0.4384	2.050
20	0.3491	0.3420	0.9397	0.3640	65	1.134	0.9063	0.4226	2.144
21	0.3665	0.3584	0.9336	0.3839	66	1.152	0.9136	0.4067	2.246
22	0.3840	0.3746	0.9272	0.4040	67	1.169	0.9205	0.3907	2.356
23	0.4014	0.3907	0.9205	0.4245	68	1.187	0.9272	0.3746	2.475
24	0.4189	0.4067	0.9136	0.4452	69	1.204	0.9336	0.3584	2.605
25	0.4363	0.4226	0.9063	0.4663	70	1.222	0.9397	0.3420	2.748
26	0.4538	0.4384	0.8988	0.4877	71	1.239	0.9455	0.3256	2.904
27	0.4712	0.4540	0.8910	0.5095	72	1.257	0.9511	0.3090	3.078
28	0.4887	0.4695	0.8830	0.5317	73	1.274	0.9563	0.2924	3.271
29	0.5062	0.4848	0.8746	0.5543	74	1.292	0.9613	0.2756	3.487
30	0.5236	0.5000	0.8660	0.5774	75	1.309	0.9659	0.2588	3.732
31	0.5410	0.5150	0.8572	0.6009	76	1.326	0.9703	0.2419	4.011
32	0.5585	0.5299	0.8480	0.6249	77	1.344	0.9744	0.2250	4.332
33	0.5760	0.5446	0.8387	0.6494	78	1.361	0.9782	0.2079	4.705
34	0.5934	0.5592	0.8290	0.6745	79	1.379	0.9816	0.1908	5.145
35	0.6109	0.5736	0.8192	0.7002	80	1.396	0.9848	0.1736	5.671
36	0.6283	0.5878	0.8090	0.7265	81	1.414	0.9877	0.1564	6.314
37	0.6458	0.6018	0.7986	0.7536	82	1.431	0.9903	0.1392	7.115
38	0.6632	0.6157	0.7880	0.7813	83	1.449	0.9926	0.1219	8.144
39	0.6807	0.6293	0.7772	0.8098	84	1.466	0.9945	0.1045	9.514
40	0.6981	0.6428	0.7660	0.8391	85	1.484	0.9962	0.08716	11.43
41	0.7156	0.6561	0.7547	0.8693	86	1.501	0.9976	0.06976	14.30
42	0.7330	0.6691	0.7431	0.9004	87	1.518	0.9986	0.05234	19.08
43	0.7505	0.6820	0.7314	0.9325	88	1.536	0.9994	0.03490	28.64
44	0.7679	0.6947	0.7193	0.9657	89	1.553	0.9998	0.01745	57.29
45	0.7854	0.7071	0.7071	1.000	90	1.571	1.000	0.0000	———

ANSWERS TO ODD-NUMBERED PROBLEMS AND PROFICIENCY TESTS

Chapter 1, Section 1 (page 6)

1. $f(1) = 6$, $f(0) = -2$, $f(-2) = 0$

3. $g(-1) = -2$, $g(1) = 2$, $g(2) = \frac{5}{2}$

5. $h(2) = 2\sqrt{3}$, $h(0) = 2$, $h(-4) = 2\sqrt{3}$

7. $f(1) = 1$, $f(5) = \frac{1}{27}$, $f(13) = \frac{1}{125}$

9. $f(1) = 0$, $f(2) = 2$, $f(3) = 2$

11. All real numbers x except $x = -2$.

13. All real numbers x for which $x \geq 5$.

15. All real numbers t.

17. All real numbers t for which $t \geq 2$.

19. All real numbers x for which $|x| > 3$.

21. All real numbers t except $t = 1$.

23. (a) \$4,500   (b) \$371

25. (a) $33\frac{1}{3}°$ Celsius   (b) Decreased by $7.5°$ Celsius

27. (a) All real numbers n except $n = 0$.
    (b) All positive integers n.
    (c) 7 minutes   (d) 12th trial
    (e) The time required will approach but never exceed 3 minutes.

29. (a) All real numbers x except $x = 300$.
    (b) All real numbers x for which $0 \leq x \leq 100$.
    (c) 120   (d) 300   (e) 60

605

31. (a) 192 feet   (b) 80 feet   (c) 256 feet
    (d) After 4 seconds

33. $g[h(x)] = 3x^2 + 14x + 10$

35. $g[h(x)] = x^3 + 2x^2 + 4x + 2$

37. $g[h(x)] = \dfrac{1}{(x - 1)^2}$

39. $g[h(x)] = |x|$

41. $f(x - 2) = 2x^2 - 11x + 15$

43. $f(x - 1) = x^5 - 3x^2 + 6x - 3$

45. $f(x^2 + 3x - 1) = \sqrt{x^2 + 3x - 1}$

47. $f(x + 1) = \dfrac{x}{x + 1}$

49. $h(x) = 3x - 5,\ g(u) = \sqrt{u}$

51. $h(x) = x^2 + 1,\ g(u) = \dfrac{1}{u}$

53. $h(x) = x + 3,\ g(u) = \sqrt{u} - \dfrac{1}{(u + 1)^3}$

55. (a) $C[q(t)] = 625t^2 + 25t + 900$ (b) \$6,600

**Chapter 1,
Section 2
(page 16)**

1.   3.

5.   7.

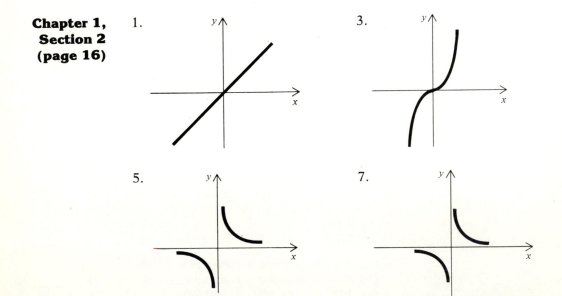

9.

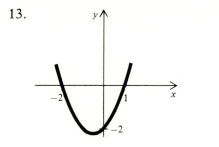

11.

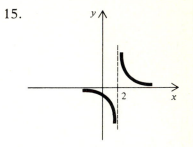

13.

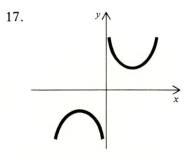

15.

17.

19. $P(x) = (x - 20)(120 - x)$
Optimal price = \$70 per recorder

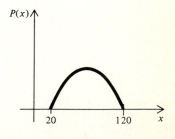

21. (a)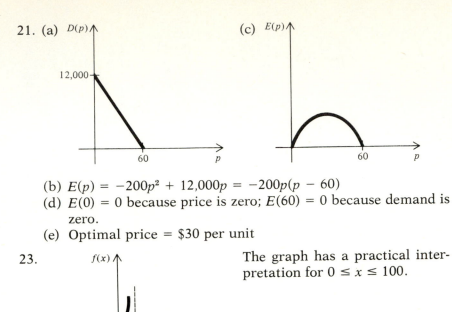

(c)

(b) $E(p) = -200p^2 + 12,000p = -200p(p - 60)$
(d) $E(0) = 0$ because price is zero; $E(60) = 0$ because demand is zero.
(e) Optimal price = $30 per unit

23.

The graph has a practical interpretation for $0 \leq x \leq 100$.

25. (a)

(b) The graph has a practical interpretation for $n = 1, 2, 3, \ldots$

(c) As n increases without bound, the height of the graph decreases and approaches 3. That is, as the number of trials increases, the time required for the rat to traverse the maze decreases, approaching a lower bound of 3 minutes.

27.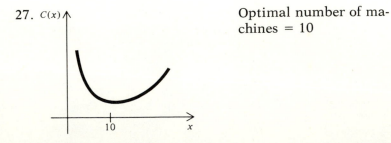

Optimal number of machines = 10

29. $A(x) = x + 4 + \dfrac{16}{x}$

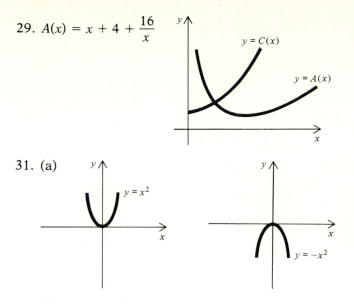

31. (a)

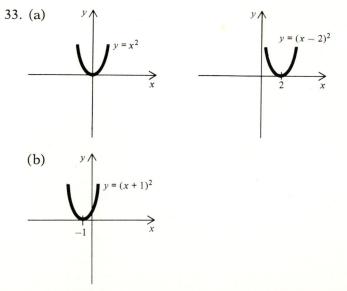

(b) The graph of g is obtained by reflecting the graph of f across the x axis.

33. (a)

(b)

(c) The graph of g is obtained by translating the graph of f horizontally by c units.

**Chapter 1,
Section 3
(page 28)**

1. $-\frac{7}{2}$ 3. -1 5. Undefined

7. $m = 5, b = 2$ 9. $m = -1, b = 2$ 11. $m = \frac{1}{2}, b = -3$

13. $m = 2, b = -3$

15. $m = 0, b = 2$

17. $y = x - 2$

19. $y = -\frac{1}{2}x + \frac{1}{2}$

21. $y = 5$

23. $y = -x + 1$

25. $y = x + 5$

27. $x = 1$

29. $C(x) = 25x + 600$

31. (a) $F(x) = -12.5x + 150$ (b) $87.50

33. (a) $V(x) = -1,900x + 20,000$ (b) $12,400

35. (a) $N(x) = 3x + 157$, where x is the number of days since the start of the program.
    (b) 289

37. (a) $24; $48
    (b) The function is not linear because the rate of change is not constant.

Chapter 1, Section 4 (page 39)

1. $(-\frac{1}{2}, \frac{7}{2})$

3. None

5. $(1, 0)$

7. $(1, 1)$ and $(0, 0)$

9. None

11. $(-1, 2)$

13. $(\frac{1}{2}, 4)$ and $(-\frac{1}{2}, 4)$

15. None

17. 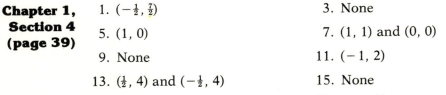 $\left(\dfrac{3 + \sqrt{5}}{2}, \dfrac{1 + \sqrt{5}}{2}\right)$ and $\left(\dfrac{3 - \sqrt{5}}{2}, \dfrac{1 - \sqrt{5}}{2}\right)$

19. (a) 4   (b) 7

21. Join the second club if fewer than 80 hours of tennis will be played and the first if more than 80 hours will be played.

23. $p = $40, q = 360$ units

25. (a) $p = $80, q = 70$ units   (b)
    (c) $S(10) = 0.$ Manufacturers will not supply any units unless the market price exceeds $10.

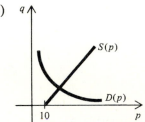

27. Of course!

**Chapter 1,
Section 5
(page 47)**

1. $P(x) = 20(25 - x)(x - 3)$; optimal price = $14

3. $R(x) = -\frac{1}{2}x(x - 155)$; optimal size = 77 or 78

5. $R(x) = 2(100 - x)(80 + x)$, where x is the number of days after July first; optimal harvest date = July 11

7. $f(x) = 2x + \dfrac{7,200}{x}$; optimal dimensions: 60 meters by 60 meters

9. $C(x) = 4x^2 + \dfrac{1,000}{x}$ 11. $V(x) = 4x(9 - x)^2$

13. $C(r) = 0.08\pi \left(r^2 + \dfrac{2}{r}\right)$

15. (a) $C(x) = \begin{cases} 1.5x & \text{if } 0 < x < 50 \\ x & \text{if } x \geq 50 \end{cases}$ (b) $23.50

17. $C(x) = \begin{cases} 13 & \text{if } 0 < x \leq 1 \\ 24 & \text{if } 1 < x \leq 2 \\ 35 & \text{if } 2 < x \leq 3 \\ 46 & \text{if } 3 < x \leq 4 \\ 57 & \text{if } 4 < x \leq 5 \\ 68 & \text{if } 5 < x \leq 6 \\ 79 & \text{if } 6 < x \leq 7 \end{cases}$

19. (a) $f(x) = \begin{cases} 0.25x - 410 & \text{if } 8,000 < x \leq 10,000 \\ 0.27x - 610 & \text{if } 10,000 < x \leq 12,000 \\ 0.29x - 850 & \text{if } 12,000 < x \leq 14,000 \\ 0.31x - 1,130 & \text{if } 14,000 < x \leq 16,000 \end{cases}$

    (b) $m_1 = 0.25$, $m_2 = 0.27$, $m_3 = 0.29$, $m_4 = 0.31$

21. $R(p) = kp$

23. $R(t) = k(M - t)$, where M is the temperature of the surrounding medium.

25. $R(x) = kx(n - x)$, where n is the total number of people involved.

27. $C(s) = \dfrac{k_1}{s} + k_2 s$

29. $D(t) = \sqrt{(90t)^2 + (975 - 60t)^2}$

31. $A(x) = 8x + \dfrac{100}{x} + 57$

**Chapter 1,
Proficiency test
(page 54)**

1. (a) All real numbers x.
    (b) All real numbers x except $x = 1$ and $x = -2$.
    (c) All real numbers x for which $|x| \geq 3$.

2. (a) \$45   (b) \$1   (c) 9 months from now
   (d) The price will approach \$40.

3. (a) $g[h(x)] = x^2 - 4x + 4$

   (b) $g[h(x)] = \dfrac{1}{2x + 5}$

   (c) $g[h(x)] = \sqrt{-2x - 3}$

4. (a) $f(x - 2) = x^2 - 5x + 10$

   (b) $f(x^2 + 1) = \sqrt{x^2 + 1} + \dfrac{2}{x^2}$

   (c) $f(x + 1) - f(x) = 2x + 1$

5. $c = -4$

6. (a)   (b)

   (c)

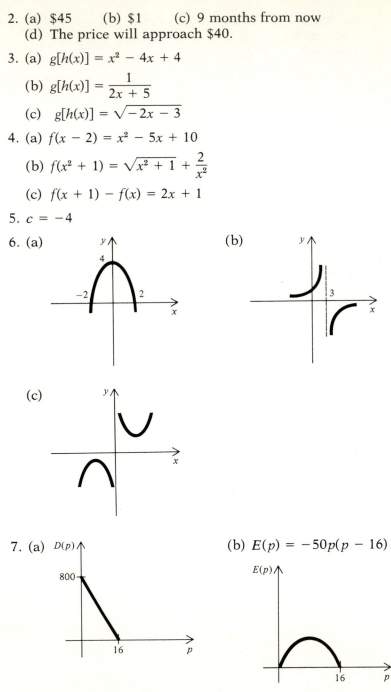

7. (a) $D(p)$ (b) $E(p) = -50p(p - 16)$

   (c) Optimal price = \$8 per unit

8. (a)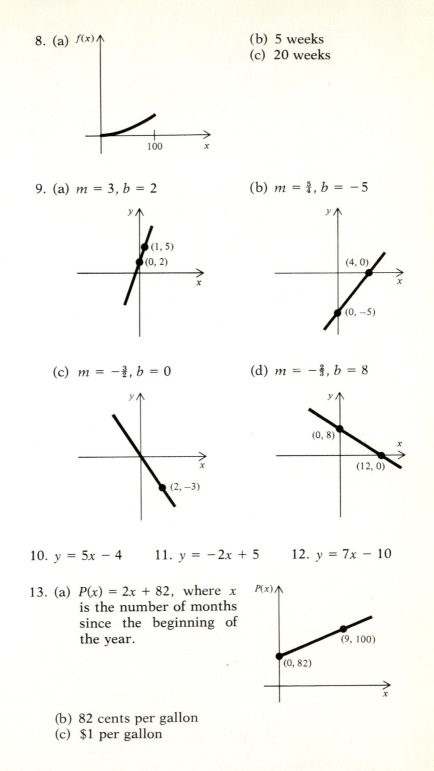

(b) 5 weeks
(c) 20 weeks

9. (a) $m = 3, b = 2$

(b) $m = \frac{5}{4}, b = -5$

(c) $m = -\frac{3}{2}, b = 0$

(d) $m = -\frac{2}{3}, b = 8$

10. $y = 5x - 4$ 11. $y = -2x + 5$ 12. $y = 7x - 10$

13. (a) $P(x) = 2x + 82$, where x is the number of months since the beginning of the year.

(b) 82 cents per gallon
(c) $1 per gallon

14. (a) $C(x) = 400x + 3,200,$
where x is the number of
months since the paper
first appeared.

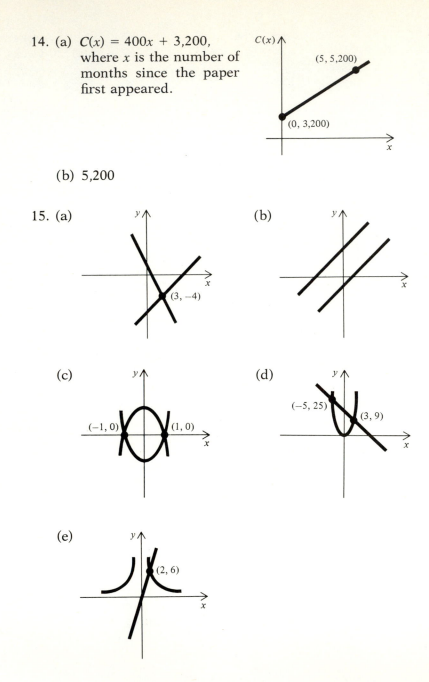

(b) 5,200

15. (a)

(b)

(c)

(d)

(e)

16. Call the first plumber if the work will take less than $1\frac{1}{2}$ hours and
the second if the work will take more than $1\frac{1}{2}$ hours.

17. (a) 150   (b) $1,500 profit   (c) 180

18. $P(x) = (50 - x)(x - 10)$
Optimal price = \$30

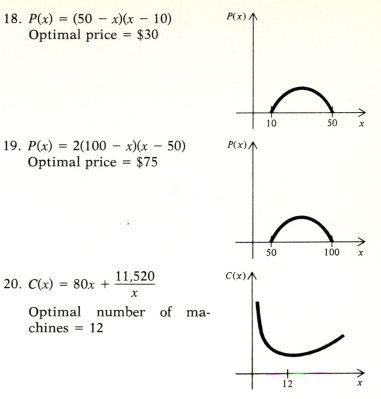

19. $P(x) = 2(100 - x)(x - 50)$
Optimal price = \$75

20. $C(x) = 80x + \dfrac{11{,}520}{x}$

Optimal number of machines = 12

21. $R(x) = k(n - x)$, where n is the total number of relevant facts in the subject's memory.

**Chapter 2,
Section 1
(page 67)**

1. $f'(x) = 5$, $m = 5$

3. $\dfrac{dy}{dx} = 4x - 3$, $m = -3$

5. $f'(x) = -\dfrac{2}{x^2}$, $m = -8$

7. $\dfrac{dy}{dx} = \dfrac{1}{2\sqrt{x}}$, $m = \frac{1}{6}$

9. $y = 11x + 16$

11. $y = \frac{1}{2}x + 2$

13. (a) 3.31

(b) 3

15.

17.

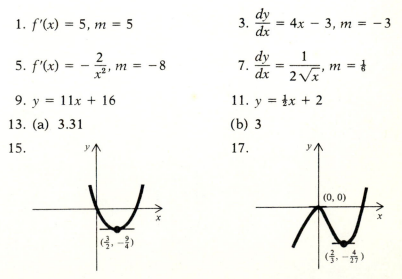

19. The graph of the function is rising for $a \le x \le b$.

21. (a) $\dfrac{d}{dx}(x^2) = 2x$, $\dfrac{d}{dx}(x^2 - 3) = 2x$. The graphs are "parallel."

    (b) $\dfrac{d}{dx}(x^2 + 5) = 2x$

23. (a) $\dfrac{d}{dx}(x^2) = 2x$, $\dfrac{d}{dx}(x^3) = 3x^2$

    (b) $\dfrac{d}{dx}(x^4) = 4x^3$, $\dfrac{d}{dx}(x^{27}) = 27x^{26}$

Chapter 2,
Section 2
(page 74)

1. $\dfrac{dy}{dx} = 2x + 2$

3. $f'(x) = 9x^8 - 40x^7 + 1$

5. $\dfrac{dy}{dx} = -\dfrac{1}{x^2} - \dfrac{2}{x^3}$

7. $f'(x) = \tfrac{1}{2}x^{-1/2} - \tfrac{1}{2}x^{-3/2}$

9. $\dfrac{dy}{dx} = -16 - \dfrac{2}{x^2} - \dfrac{3}{2}\sqrt{x} - \dfrac{1}{3x^2} + \dfrac{1}{3}$

11. $f'(x) = 12x - 1$

13. $\dfrac{dy}{dx} = -300x - 20$

15. $f'(x) = \tfrac{1}{3}(5x^4 - 6x^2)$

17. $\dfrac{dy}{dx} = \dfrac{-3}{(x - 2)^2}$

19. $f'(x) = \dfrac{-x^2 - 2}{(x^2 - 2)^2}$

21. $\dfrac{dy}{dx} = \dfrac{-3}{(x + 5)^2}$

23. $f'(x) = \dfrac{11x^2 - 10x - 7}{(2x^2 + 5x - 1)^2}$

25. $\dfrac{dy}{dx} = -24x^2 + 44x + 7$

27. $y = -6x + 6$

29. $y = -\tfrac{1}{16}x + 2$

31. $y = 3x - 3$

33. $y = 6x - 2$

35. (a) $\dfrac{dy}{dx} = \dfrac{-4x + 9}{x^4}$ (b) $\dfrac{dy}{dx} = -3x^{-4}(2x - 3) + 2x^{-3}$

37.

39. $a = \frac{8}{9}, b = -\frac{16}{3}$ 41. $y = 6x, y = -14x$

43. (a) $E(p) = -200p^2 + 12,000p$ (b) $30 per unit

Chapter 2,   1. (a) $C'(t) = 200t + 400$
Section 3   (b) Increasing at the rate of 1,400 per year.
(page 82)   (c) 1,500

3. (a) $f'(x) = -3x^2 + 12x + 15$
   (b) 24 radios per hour   (c) 26

5. (a) $P'(t) = \dfrac{6}{(t + 1)^2}$ thousand per year

   (b) 1,500 per year   (c) 1,000   (d) 60 per year
   (e) The rate of growth will approach zero.

9. (a) $241   (b) $244

11. (a) $248   (b) $248.05

13. Daily output will increase by approximately 10 units.

15. (a) 20 people per month   (b) 0.39 percent per month

17. (a) $280 per year   (b) 17.95 percent per year

19. (a) $P(x) = \dfrac{100}{12 + x}$

   (b) 7.69 percent per year
   (c) The percentage rate will approach zero.

21. (a) After 3 seconds   (b) 96 feet per second

23. (a) 32 feet per second   (b) 128 feet

    (c) 32 feet per second   (d) 96 feet per second

25. (a) Rate of change of cost with respect to output: dollars per unit.
    (b) Rate of change of output with respect to time: units per hour.
    (c) Rate of change of cost with respect to time: dollars per hour.

Chapter 2,   1. $\dfrac{dy}{dx} = 6(3x - 2)$ 3. $\dfrac{dy}{dx} = \dfrac{x + 1}{\sqrt{x^2 + 2x - 3}}$
Section 4
(page 91)   5. $\dfrac{dy}{dx} = \dfrac{-4x}{(x^2 + 1)^3}$ 7. $\dfrac{dy}{dx} = -x(x^2 - 9)^{-3/2}$

9. $\dfrac{dy}{dx} = -\dfrac{2x}{(x^2 - 1)^2}$

11. -160

13. $\frac{2}{3}$

15. -16

17. $f'(x) = 8(2x + 1)^3$

19. $f'(x) = \dfrac{-5}{(5x - 6)^2}$

21. $f'(x) = -2(4x - 1)^{-3/2}$

23. $f'(x) = 8x^2(x^5 - 4x^3 - 7)^7(5x^2 - 12)$

25. $f'(x) = 24x(1 - x^2)^{-5}$

27. $f'(x) = -\frac{5}{2}(3x + 1)^{-1/2}(2x - 1)^{-3/2}$

29. $f'(x) = (x + 2)^2(2x - 1)^4(16x + 17)$

31. $f'(x) = \dfrac{(x + 1)^4(9 - x)}{(1 - x)^5}$

33. $y = 594x - 1{,}161$

35. $y = -6x + 26$

39. (a) \$2,025 per year

(b) 10.125 percent per year

41. 0.31 parts per million per year

43. Decreasing at the rate of 6 pounds per week.

Chapter 2, Section 5 (page 102)

1. $f'(x) > 0$ for $-2 < x < 2$; $f'(x) < 0$ for $x < -2$ and $x > 2$.

3. $f'(x) > 0$ for $x < -4$ and $0 < x < 2$; $f'(x) < 0$ for $-4 < x < -2$, $-2 < x < 0$, and $x > 2$.

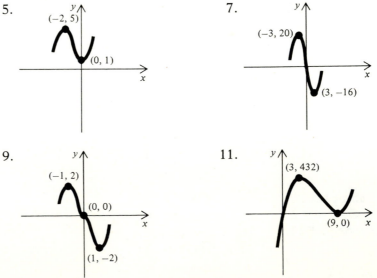

5. $(-2, 5)$ $(0, 1)$

7. $(-3, 20)$ $(3, -16)$

9. $(-1, 2)$ $(0, 0)$ $(1, -2)$

11. $(3, 432)$ $(9, 0)$

13.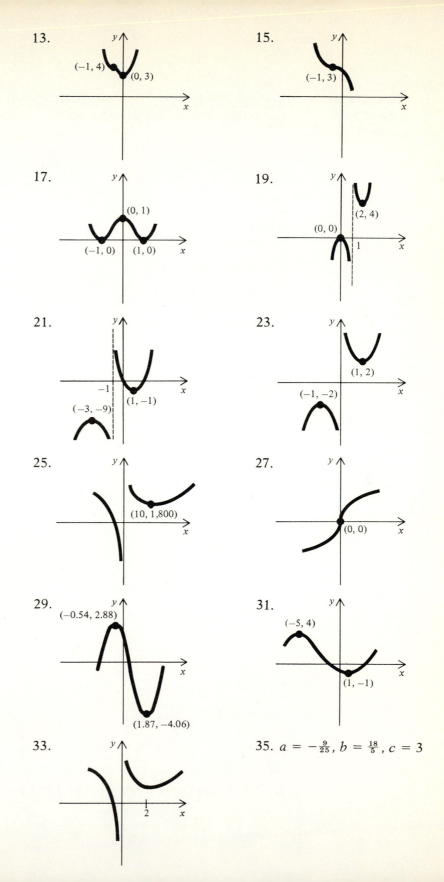
(−1, 4) (0, 3)

15.
(−1, 3)

17.
(0, 1)
(−1, 0) (1, 0)

19.
(2, 4)
(0, 0)
1

21.
−1
(1, −1)
(−3, −9)

23.
(1, 2)
(−1, −2)

25.
(10, 1,800)

27.
(0, 0)

29.
(−0.54, 2.88)
(1.87, −4.06)

31.
(−5, 4)
(1, −1)

33.
2

35. $a = -\frac{9}{25}$, $b = \frac{18}{5}$, $c = 3$

	Absolute maximum	Absolute minimum
Chapter 2, Section 6 (page 112)	1. $f(1) = 10$	$f(-2) = 1$
	3. $f(0) = 2$	$f(2) = -\frac{40}{3}$
	5. $f(-1) = 2$	$f(-2) = -56$
	7. $f(-3) = 3{,}125$	$f(0) = -1{,}024$
	9. $f(3) = \frac{10}{3}$	$f(1) = 2$
	11. none	$f(1) = 2$
	13. none	none
	15. $f(0) = 1$	none

17. $12.50 per radio

19. (a) Membership = 46,400 in 1964
    (b) Membership = 12,100 in 1971

21. At the central axis.

27. (a) $A(q) = 3q + 1 + \dfrac{48}{q}$

    (b) $q = 4$ (c) $q = 4$

Chapter 2, Proficiency test (page 116)

1. (a) $f'(x) = 2x - 3$ (b) $f'(x) = -\dfrac{1}{(x - 2)^2}$

2. (a) $f'(x) = 24x^3 - 21x^2 + 2$

   (b) $f'(x) = 3x^2 + \dfrac{2}{x^3} + \dfrac{1}{\sqrt{x}} + \dfrac{3}{x^2}$

   (c) $\dfrac{dy}{dx} = \dfrac{-5}{(2x + 1)^2}$

   (d) $\dfrac{dy}{dx} = 2(x + 1)(2x + 5)^2(5x + 8)$

   (e) $f'(x) = 20(5x^4 - 3x^2 + 2x + 1)^9(10x^3 - 3x + 1)$

   (f) $f'(x) = \dfrac{x}{\sqrt{x^2 + 1}}$

   (g) $\dfrac{dy}{dx} = \dfrac{4(x + 1)}{(1 - x)^3}$

   (h) $\dfrac{dy}{dx} = \dfrac{3(3x + 1)}{\sqrt{6x + 5}} + 3\sqrt{6x + 5}$

3. (a) $y = -x + 1$ (b) $y = -x - 1$
   (c) $y = x$ (d) $y = -\frac{2}{3}x + \frac{5}{3}$

4. (a) 1,652 people per week   (b) 1,514 people

5. (a) Output will increase by approximately 12,000 units.
   (b) Output will increase by 12,050 units.

6. 0.30 percent per month

7. (a) $\dfrac{dy}{dx} = 3(30x + 11)$ (b) $\dfrac{dy}{dx} = \dfrac{-4}{(2x + 3)^3}$

8. (a) 2   (b) $\frac{3}{2}$

9. $1,663.20 per hour

10. (a)   (b)

    (c)   (d)

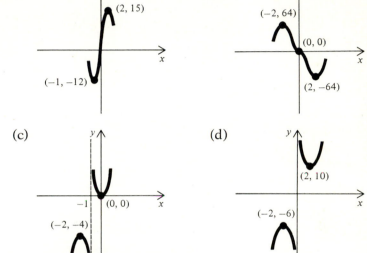

11.

	Absolute maximum	**Absolute minimum**
(a)	$f(-3) = 40$	$f(-1) = -12$
(b)	$f(2) = 6$	$f(3) = -37$
(c)	$f(-\frac{1}{2}) = f(1) = \frac{1}{2}$	$f(0) = 0$
(d)	none	$f(2) = 10$

**Chapter 3,
Section 1
(page 128)**

1. The field should be a square, 80 meters by 80 meters.

3. The playground should be a square, 60 meters by 60 meters.

5. $14   7. 80

9. 10 days from now

11. 2 meters by 2 meters by $\frac{4}{3}$ meters

13. 12 inches by 12 inches by 3 inches

15. The mathematician.

17. $8 + 5\sqrt{2}$ centimeters by $4 + \dfrac{5\sqrt{2}}{2}$ centimeters

19. Radius = 1 inch, height = 4 inches

21. (a) 8   (b) $160   (c) $160

23. 11:00 A.M.   25. 17

27. 3 times per year   29. 4,000

31. (a) Number = $\sqrt{\dfrac{pQ}{ns}}$

Chapter 3, Section 2 (page 142)

1. $f''(x) = 450x^8 - 120x^3$

3. $\dfrac{d^2y}{dx^2} = 2 - \dfrac{6}{x^4}$

5. $f''(x) = 10(x^2 + 1)^3(9x^2 + 1)$

7. $\dfrac{d^2y}{dx^2} = \dfrac{2}{(x - 2)^3}$

9. 10:00 A.M.

11. Minimal slope = $\frac{9}{2}$ at $(\frac{1}{2}, \frac{5}{2})$

13. The speed is decreasing at the rate of 6 meters per second per second.

15. (a) $A(t) = \frac{20}{3} - \frac{4}{3}t$
    (b) The speed is decreasing at the rate of $\frac{4}{3}$ kilometers per hour per hour.
    (c) 2 kilometers per hour
    (d) After 5 hours

17. $f''(x) > 0$ for $-4 < x < -2$ and $x > 1$; $f''(x) < 0$ for $x < -4$ and $-2 < x < 1$.

19.   21.

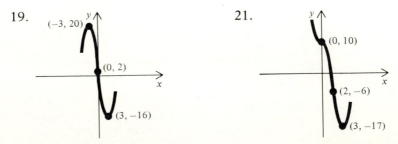

23.

25.

(−√5, 0)   (√5, 0)

(−1, −64)   (1, −64)

(0, −125)

27.

(1, 2)

(−1, −2)

29.

(6, 12)

(0, 0)   3

31.

(−1, 0)

33.

(−1, 0)

35.

(0, 1)

37.

2 4 5 7 x

39. (a) Increasing for $x < 0$ and $x > 4$; decreasing for $0 < x < 4$.
   (b) Concave upward for $x > 2$; concave downward for $x < 2$.
   (c) Relative maximum when $x = 0$; relative minimum when $x = 4$; inflection point when $x = 2$.

41. Increasing for $x > 2$; decreasing for $x < 2$; concave upward for all x; relative minimum when $x = 2$.

43. Increasing for $x > 2$; decreasing for $x < 2$; concave upward for $x < -3$ and $x > -1$; concave downward for $-3 < x < -1$; relative minimum when $x = 2$; inflection points when $x = -3$ and $x = -1$.

	Relative maxima	Relative minima
45.	$(-2, 5)$	$(0, 1)$
47.	$(0, 81)$	$(3, 0)$ and $(-3, 0)$
49.	$(-3, -11)$	$(3, 13)$

**Chapter 3,
Section 3
(page 151)**

1. $26,000 per year

3. 48π square inches per second

5. 10 feet per second　　　　　7. 50 miles per hour

9. $\dfrac{20}{9\pi}$ feet per minute　　　11. 200 feet per second

13. $\dfrac{dy}{dx} = -\dfrac{x}{y}$　　　　15. $\dfrac{dy}{dx} = \dfrac{y - 3x^2}{3y^2 - x}$

17. $\dfrac{dy}{dx} = \dfrac{3 - 2y^2}{2y(1 + 2x)}$　　19. $\dfrac{dy}{dx} = \dfrac{1}{3(2x + y)^2} - 2$

21. $\frac{1}{3}$　　23. $-\frac{1}{2}$　　25. $\frac{13}{12}$

27. $\dfrac{dy}{dx} = \dfrac{2x - y}{x + 2} = \dfrac{x(x + 4)}{(x + 2)^2}$　　29. $\dfrac{dy}{dx} = \dfrac{y - 1}{1 - x} = \dfrac{-3}{(x - 1)^2}$

31. $\dfrac{d^2y}{dx^2} = \dfrac{6y^3 - 8x^2}{9y^5} = -\dfrac{2}{9y^2}$

**Chapter 3,
Section 4
(page 157)**

1. 200　　　　　　　　　3. 0.05 parts per million

5. Cost will decrease by approximately $50.08.

7. Daily output will increase by approximately 8 units.

9. 2.16 centimeters per second

11. Accurate to within 3 percent.

13. Area will increase by approximately 2 percent.

15. $9.03\pi = 28.37$ cubic inches

**Chapter 3,
Proficiency test
(page 159)**

1. $75 per camera

2. Each plot should be 50 meters by 37.5 meters.

3. 12　　　　　　　　　4. 77 or 78

5. (a) 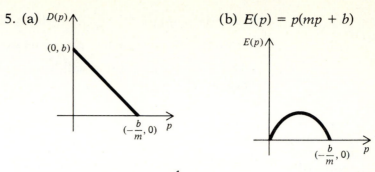   (b) $E(p) = p(mp + b)$

(c) Optimal price $= -\dfrac{b}{2m}$

6. After 2 hours and 20 minutes on the job.

7. (a) 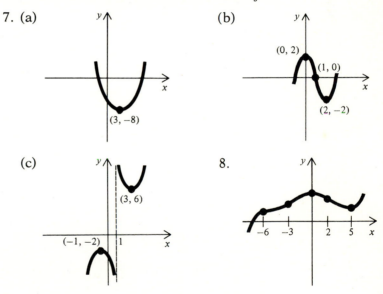   (b)

(c)   8.

9.

Relative maxima	**Relative minima**
(a) (2, 15)	(−1, −12)
(b) (−2, −4)	(0, 0)
(c) (−2, −6)	(2, 10)

10. $\dfrac{1}{\pi}$ feet per minute

11. Increasing at the rate of 6 kilometers per hour.

12. (a) $\dfrac{dy}{dx} = -\dfrac{5}{3}$ (b) $\dfrac{dy}{dx} = -\dfrac{2y}{x}$ (c) $\dfrac{dy}{dx} = \dfrac{1 - 10(2x + 3y)^4}{15(2x + 3y)^4}$

13. (a) $-\frac{2}{3}$ (b) -28

14. Output will decrease by approximately 5,000 units.

15. The level of air pollution will increase by approximately 10 percent.

**Chapter 4,
Section 1
(page 169)**

1.

n	1,000	10,000	25,000	50,000
$\left(1 + \frac{1}{n}\right)^n$	2.71692	2.71815	2.71823	2.71825

3. $e^2 = 7.389$, $e^{-2} = 0.135$, $e^{0.05} = 1.051$, $e^{-0.05} = 0.951$, $e^0 = 1$,

$e = 2.718$, $\sqrt{e} = 1.649$, $\dfrac{1}{\sqrt{e}} = 0.607$

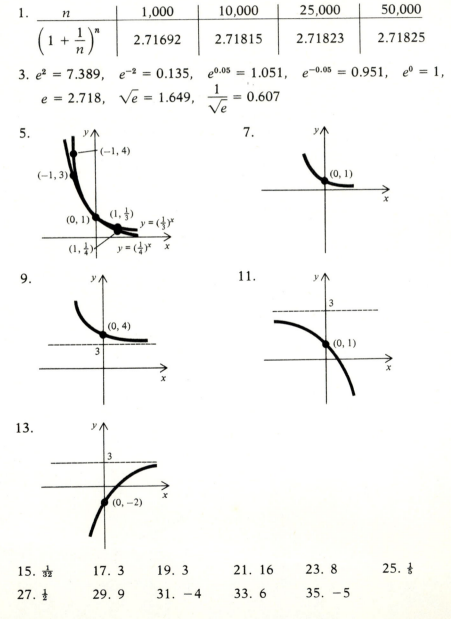

5.

7.

9.

11.

13.

15. $\frac{1}{32}$ 17. 3   19. 3   21. 16   23. 8   25. $\frac{1}{5}$

27. $\frac{1}{2}$ 29. 9   31. -4 33. 6   35. -5

37. $-\frac{3}{2}$ 39. $\frac{1}{6}$ 41. $\frac{13}{3}$ 43. 4   45. 8   47. $\frac{110}{3}$

49. (a) $1,967.15   (b) $2,001.60
    (c) $2,009.66   (d) $2,013.75

51. (a) $P = Be^{-rt}$ (b) $5,488.12

Chapter 4,
Section 2
(page 176)

1. (a) 50 million   (b) 91.11 million

3. 202.5 million

5. 324 billion dollars

7. (a) 12,000 people per square mile
   (b) 5,959 people per square mile

9. 204.8 grams

11. (a) $f(t)$ (b) 0.7408
                        (c) 0.0888

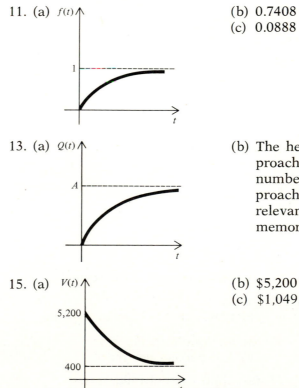

13. (a) $Q(t)$

(b) The height of the graph approaches A because the number of facts recalled approaches the total number of relevant facts in the person's memory.

15. (a) $V(t)$ (b) $5,200
                        (c) $1,049.61

17. 18.75° Celsius

19. (a) 4 million   (b) 9.31 million
    (c) The population will approach 10 million.

21. 4 hours

Chapter 4,
Section 3
(page 186)

1. $\ln 1 = 0$, $\ln 2 = 0.693$, $\ln e = 1$, $\ln 5 = 1.609$, $\ln \frac{1}{5} = -1.609$, $\ln e^2 = 2$, $\ln 0$ and $\ln -2$ are undefined.

3. $\frac{1}{2}$ 5. 9   7. $\frac{19}{6}$ 9. 0.58

11. $e^{-b/2}$ 13. $e^2 = 7.39$ 15. $\frac{9}{25}$ 17. $\ln a$

19. e or $\dfrac{1}{e}$

21. 5   23. 5.33 percent   25. In the year 2095

27. 5,614.06 years   31. $Q(t) = 6,000e^{0.02t}$

33. $Q(t) = 500 - 200e^{-0.133t}$ 35. 9,081.90 years old

39. $\log_a x = \dfrac{\ln x}{\ln a}$

Chapter 4,
Section 4
(page 196)

1. $f'(x) = 5e^{5x}$ 3. $f'(x) = 2(x + 1)e^{x^2+2x-1}$

5. $f'(x) = -0.5e^{-0.05x}$ 7. $f'(x) = (6x^2 + 20x + 33)e^{6x}$

9. $f'(x) = (1 - x)e^{-x}$ 11. $f'(x) = -6e^x(1 - 3e^x)$

13. $f'(x) = \dfrac{3}{x}$ 15. $f'(x) = \dfrac{2x + 5}{x^2 + 5x - 2}$

17. $f'(x) = 2x \ln x + x$ 19. $f'(x) = \dfrac{1}{x^2}(1 - \ln x)$

21. $f'(x) = \dfrac{-2}{(x + 1)(x - 1)}$ 23. $f'(x) = 2$

25. (a) 1.22 million per year
    (b) Constant rate of 2 percent per year

27. (a) $1,082.68 per year
    (b) Constant rate of 40 percent per year

31. (a) 0.7548 billion per year
    (b) 4.95 percent per year

33. (a) Approximately 406 copies
    (b) 368 copies

35. $15 per radio

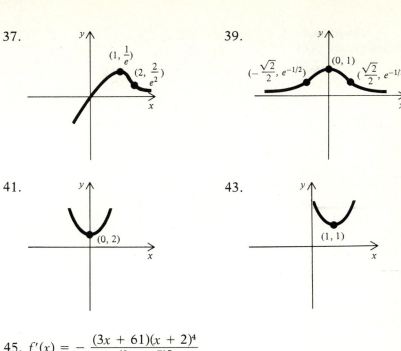

37. (1, $\frac{1}{e}$) (2, $\frac{2}{e^2}$)

39. $(-\frac{\sqrt{2}}{2}, e^{-1/2})$ (0, 1) $(\frac{\sqrt{2}}{2}, e^{-1/2})$

41. (0, 2)

43. (1, 1)

45. $f'(x) = -\dfrac{(3x + 61)(x + 2)^4}{(3x - 5)^7}$

47. $f'(x) = 2^x \ln 2$

49. $f'(x) = x^x(1 + \ln x)$

51. 207.94 years from now

55. 12.5 percent per year

**Chapter 4,
Section 5
(page 207)**

1. (a) \$10,000.00   (b) \$13,425.32   (c) \$13,591.41

3. (a) 8.75 years   (b) 8.66 years

5. Doubling time $= \dfrac{1}{r}$ 7. 18.58 years

9. Tripling time $= \dfrac{\ln 3}{k \ln (1 + r/k)}$

11. (a) 15.39 years   (b) 15.27 years

13. (a) 6.14 percent   (b) 6.18 percent

15. 10.20 percent compounded continuously

17. 10 percent   19. 5.83 percent

21. (a) \$6,095.65   (b) \$6,023.88

23. 69.44 years from now   25. 6.5 years from now

27. \$209.18   29. \$1,732.55

**Chapter 4,
Proficiency test
(page 211)**

1. (a)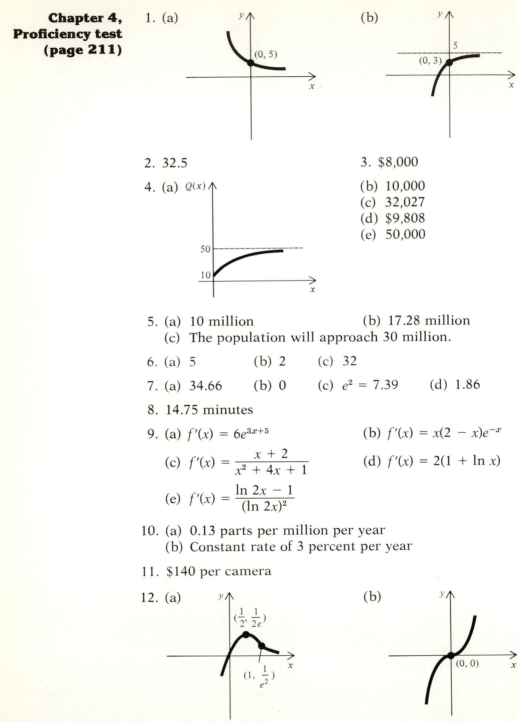

(b)

2. 32.5

3. $8,000

4. (a) $Q(x)$

(b) 10,000
(c) 32,027
(d) $9,808
(e) 50,000

5. (a) 10 million   (b) 17.28 million
(c) The population will approach 30 million.

6. (a) 5   (b) 2   (c) 32

7. (a) 34.66   (b) 0   (c) $e^2 = 7.39$ (d) 1.86

8. 14.75 minutes

9. (a) $f'(x) = 6e^{3x+5}$ (b) $f'(x) = x(2 - x)e^{-x}$

(c) $f'(x) = \dfrac{x + 2}{x^2 + 4x + 1}$ (d) $f'(x) = 2(1 + \ln x)$

(e) $f'(x) = \dfrac{\ln 2x - 1}{(\ln 2x)^2}$

10. (a) 0.13 parts per million per year
(b) Constant rate of 3 percent per year

11. $140 per camera

12. (a)   (b)

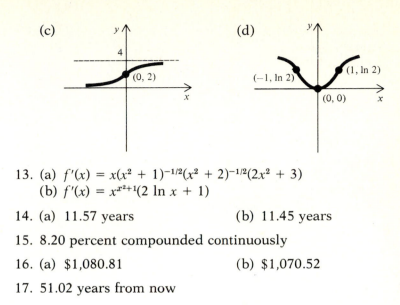

(c)   (0, 2)

(d)   (−1, ln 2)   (1, ln 2)   (0, 0)

13. (a) $f'(x) = x(x^2 + 1)^{-1/2}(x^2 + 2)^{-1/2}(2x^2 + 3)$
     (b) $f'(x) = x^{x^2+1}(2 \ln x + 1)$

14. (a) 11.57 years   (b) 11.45 years

15. 8.20 percent compounded continuously

16. (a) \$1,080.81   (b) \$1,070.52

17. 51.02 years from now

Chapter 5, Section 1 (page 221)

1. $\frac{1}{6}x^6 + C$

3. $-\dfrac{1}{x} + C$

5. $5x + C$

7. $x^3 - \frac{5}{2}x^2 + 2x + C$

9. $2x^{3/2} + \dfrac{1}{x^2} + \ln |x| + C$

11. $2e^x + \ln x^6 + C$

13. $x + \ln x^2 - \dfrac{1}{x} + C$

15. $\frac{2}{7}x^{7/2} - \frac{2}{3}x^{3/2} + C$

17. 10,128   19. \$1,000   21. \$436   23. \$2,300

25. $y = 2x^2 + x - 1$

27. $y = \dfrac{1}{4}x^4 + \dfrac{2}{x} + 2x - \dfrac{5}{4}$

29. $\frac{1}{3}e^{3x} + C$

31. $\frac{1}{12}(2x + 3)^6 + C$

33. $\frac{1}{2} \ln |2x + 1| + C$

Chapter 5, Section 2 (page 227)

1. $\frac{1}{12}(2x + 6)^6 + C$

3. $\frac{1}{6}(4x - 1)^{3/2} + C$

5. $-e^{1-x} + C$

7. $\frac{1}{2}e^{x^2} + C$

9. $\frac{1}{12}(x^2 + 1)^6 + C$

11. $\frac{4}{21}(x^3 + 1)^{7/4} + C$

13. $\frac{2}{5} \ln |x^5 + 1| + C$

15. $\frac{1}{26}(x^2 + 2x + 5)^{13} + C$

17. $\frac{3}{5} \ln |x^5 + 5x^4 + 10x + 12| + C$

19. $\frac{1}{2}(\ln 5x)^2 + C$

21. $\frac{1}{2}[\ln(x^2 + 1)]^2 + C$

23. $x + \ln |x - 1| + C$

25. $-\frac{1}{4}(x - 5)^{-4} - (x - 5)^{-5} + C$

27. $\ln |x - 4| - 7(x - 4)^{-1} + C$

29. $y = -\frac{1}{3} \ln |1 - 3x^2| + 5$

31. $849.61

33. $510.56 per acre

Chapter 5, Section 3 (page 233)

1. $\frac{dQ}{dt} = kQ$

3. $\frac{dQ}{dt} = 0.07Q$

5. $\frac{dP}{dt} = 500$

7. $\frac{dQ}{dt} = k(Q - M)$, where M is the temperature of the surrounding medium.

9. $\frac{dQ}{dt} = k(N - Q)$, where N is the total number of relevant facts in the person's memory.

11. $\frac{dQ}{dt} = kQ(N - Q)$, where N is the total number of people involved.

17. $y = x^3 + \frac{5}{2}x^2 - 6x + C$

19. $V = \ln (x + 1)^2 + C$

21. $P = 25t^2 + C_1t + C_2$

23. $y = \frac{1}{5}e^{5x} + \frac{4}{5}$

25. $V = 2(t^2 + 1)^4 - 1$

27. $A = 4e^{-t/2} + 3t - 2$

29. (a) $V(t) = 4,800e^{-t/5} - 4,800 + V_0$ (b) $1,049.61

31. (a) 30,000   (b) 45,000

33. (a) $Q(t) = 0.03(36 + 16t - t^2)^{1/2} + 0.07$
    (b) 0.37 parts per million at 3:00 P.M.

35. $75

37. (a) $D(t) = -14t^2 + S_0t$ (b) 138.29 feet

Chapter 5, Section 4 (page 244)

1. $y = Ce^{3x}$

3. $y = -\ln (C - x)$

5. $y = \pm\sqrt{x^2 + C}$

7. $y = Ae^x - 10$

9. $y = 500e^{0.05x}$

11. $y = \dfrac{2}{3 - 2x^4}$

13. $Q(t) = 1,000e^{0.07t}$

17. $Q(t) = B(1 - e^{-kt})$, where B is the total number of relevant facts in the person's memory.

19. $Q(t) = 80 - 40e^{-0.014t}$

21. $Q(t) = S - (S - Q_0)e^{-kAt}$, where S is the concentration of the solute outside the cell and A is the area of the cell wall.

23. $Q(t) = 400 + 200e^{-t/40}$ 25. $\frac{1}{2}$

Chapter 5, Section 5 (page 249)

1. $-(x + 1)e^{-x} + C$ 3. $(2 - x)e^x + C$

5. $\frac{1}{2}x^2(\ln x - \frac{1}{2}) + C$ 7. $\frac{2}{3}x(x - 6)^{3/2} - \frac{4}{15}(x - 6)^{5/2} + C$

9. $\frac{1}{9}x(x + 1)^9 - \frac{1}{90}(x + 1)^{10} + C$ 11. $2x(x + 2)^{1/2} - \frac{4}{3}(x + 2)^{3/2} + C$

13. $-(x^2 + 2x + 2)e^{-x} + C$ 15. $(x^3 - 3x^2 + 6x - 6)e^x + C$

17. $\frac{1}{3}x^3 \ln x - \frac{1}{9}x^3 + C$ 19. $\frac{1}{2}(x^2 - 1)e^{x^2} + C$

21. $\frac{1}{36}x^4(x^4 + 5)^9 - \frac{1}{360}(x^4 + 5)^{10} + C$

23. 176.87

25. (b) $(\frac{1}{5}x^3 - \frac{3}{25}x^2 + \frac{6}{125}x - \frac{6}{625})e^{5x} + C$

Chapter 5, Section 6 (page 254)

1. $-\frac{1}{3} \ln \left| \dfrac{x}{2x - 3} \right| + C$ 3. $\ln |x + \sqrt{x^2 + 25}| + C$

5. $\frac{1}{4} \ln \left| \dfrac{2 + x}{2 - x} \right| + C$ 7. $\frac{1}{2} \ln \left| \dfrac{x}{3x + 2} \right| + C$

9. $(\frac{1}{3}x^2 - \frac{2}{9}x + \frac{2}{27})e^{3x} + C$ 11. $-\frac{1}{2} \ln |2 - x^2| + C$

13. $x(\ln 2x)^2 - 2x \ln 2x + 2x + C$

15. $\dfrac{1}{3\sqrt{5}} \ln \left| \dfrac{\sqrt{2x + 5} - \sqrt{5}}{\sqrt{2x + 5} + \sqrt{5}} \right| + C$

17. $\frac{1}{2}x + \frac{1}{2} \ln |2 - 3e^{-x}| + C$

Chapter 5, Proficiency test (page 257)

1. $\dfrac{1}{6} x^6 - x^3 - \dfrac{1}{x} + C$ 2. $\frac{3}{5}x^{5/3} - \ln |x| + 5x + \frac{2}{3}x^{3/2} + C$

3. $\frac{2}{9}(3x + 1)^{3/2} + C$ 4. $\frac{1}{3}(3x^2 + 2x + 5)^{3/2} + C$

5. $\frac{1}{12}(x^2 + 4x + 2)^6 + C$ 6. $\frac{1}{2} \ln |x^2 + 4x + 2| + C$

7. $\frac{1}{13}(x - 5)^{13} + C$

8. $\dfrac{1}{14} (x - 5)^{14} + \dfrac{5}{13} (x - 5)^{13} + C = \dfrac{x}{13} (x - 5)^{13} - \dfrac{1}{182} (x - 5)^{14} + C$

9. $\frac{5}{3}e^{3x} + C$ 10. $(\frac{5}{3}x - \frac{5}{9})e^{3x} + C$

11. $\frac{1}{2}x^2 \ln 3x - \frac{1}{4}x^2 + C$ 12. $x \ln 3x - x + C$

13. $\frac{1}{2}(\ln 3x)^2 + C$

14. $\frac{1}{18}x^2(x^2 + 1)^9 - \frac{1}{180}(x^2 + 1)^{10} + C$

15. $y = \frac{1}{8}(x^2 + 1)^4 + 3$

16. 11,250   17. 10,945   18. $2,265.80

19. $y = \frac{1}{4}x^4 - x^3 + 5x + C$ 20. $y = Ce^{0.02x}$

21. $y = 80 - Ce^{-kx}$ 22. $y = \dfrac{1}{1 + Ce^{-x}}$

23. $y = x^5 - x^3 - 2x + 6$ 24. $y = 100e^{0.06x}$

25. $y = 3 - e^{-x}$ 26. $y = x^2 + 3x + 5$

27. $22,857   28. 18.75 pounds

Chapter 6,
Section 1
(page 263)

1. $\frac{9}{20}$ 3. 144   5. $\frac{8}{3} + \ln 3$ 7. $\frac{2}{9}$

9. $-\frac{16}{3}$ 11. $\frac{4}{3}$ 13. 0   15. e 17. $e^2 + 1$

19. $-3e^{-2} - e^2$ 21. 98 people

23. $1,870   25. $774

27. $75   29. $4,081,077.40

31. (b) 1   (c) $\frac{70}{3}$

Chapter 6,
Section 2
(page 275)

1. $\frac{8}{3}$ 3. 15   5. $\frac{38}{3}$ 7. $\frac{4}{3}$ 9. $\frac{1}{2}$

11. (a) 0.0577   (b) 0.4512   (c) 0.5488

13. (a) 0.6321   (b) 0.3012

15. 33   17. $\frac{1}{6}$ 19. $\frac{1}{3}$ 21. $\frac{625}{12}$ 23. $\frac{128}{3}$ 25. (a) $625

Chapter 6,
Section 3
(page 282)

1. 30 meters   3. $\displaystyle\int_0^5 r(t)\, dt$

5. $7,040,000   7. $\displaystyle\int_0^{12} n(x)p(x)\, dx$

9. $75   11. $480

Chapter 6,
Section 4
(page 292)

1. 2   3. $\frac{4}{3}$ 5. 18.7° Celsius

7. 492.83 letters per hour

9. (a) $\dfrac{1}{N} \displaystyle\int_0^N S(t)\, dt$ (b) $\displaystyle\int_0^N S(t)\, dt$

11. $13,994.35   13. $27,124.92

15. $10,367.27   17. $5,183.64

19. The spy should take the 35,000 pounds. The present value of the pension is only 31,606 pounds.

21. 4,207   23. $P_0 f(N) + \displaystyle\int_0^N r(t)f(N-t)\, dt$

25. $\displaystyle\int_0^R 2\pi r S(r)\, dr$ 27. 116,039   29. 19,567 pounds

Chapter 6, Proficiency test (page 297)

1. 0   2. $\frac{17}{3}$ 3. 1,710   4. $1 - e^{-1}$

5. $\frac{65}{8}$ 6. $2e^{-1}$ 7. 126 people   8. $76.80

9. 36   10. $\frac{3}{4}$ 11. $\frac{9}{2}$ 12. $\frac{3}{10}$

13. (a) 22.10 percent   (b) 55.07 percent   (c) 44.93 percent

14. (a) 15   (b) $33,750

15. $20,700   16. $\displaystyle\int_0^{12} D(x)P(x)\, dx$ 17. $1.32 per pound

18. $7,377.37   19. $7,191.64   20. 62   21. 73,186

Chapter 7, Section 1 (page 310)

1. ∞ 3. $-\infty$ 5. $\frac{1}{2}$ 7. 0   9. ∞

11. $-\infty$ 13. $\frac{3}{2}$ 15. ∞ 17. ∞ 19. 0

21. 0   23. 0   25. 0   27. 1   29. ∞

31. 0   33. 0   35. 1   37. e^2

Chapter 7, Section 2 (page 318)

1. $\frac{1}{2}$ 3. ∞ 5. ∞ 7. $\frac{1}{10}$ 9. $\frac{5}{2}$

11. $\frac{1}{9}$ 13. ∞ 15. $\dfrac{2}{e}$ 17. $\frac{2}{9}$ 19. $5e^{10}$

21. ∞ 23. 2   25. $20,000   27. $150,000

31. 200   33. 50

**Chapter 7,
Section 3
(page 332)**

1. (a) 1   (b) $\frac{1}{3}$ (c) $\frac{1}{3}$

3. (a) 1   (b) $\frac{3}{16}$ (c) $\frac{9}{16}$

5. (a) 1   (b) $\frac{7}{8}$ (c) $\frac{1}{8}$

7. (a) 1   (b) 0.3496   (c) 0.9817

9. $\frac{1}{3}$ 11. $\frac{1}{6}$ 13. 0.1353

15. $E(x) = \frac{7}{2}$; $\text{Var}(x) = \frac{3}{4}$

17. $E(x) = \frac{4}{3}$; $\text{Var}(x) = \frac{8}{9}$

19. $E(x) = 10$; $\text{Var}(x) = 100$

21. 10 minutes   23. 4 minutes

**Chapter 7,
Section 4
(page 343)**

1. $-0.1 \le x \le 0.1$; $-0.2 \le x \le 0.2$; $-0.3 \le x \le 0.3$

3. $350 \le x \le 608$; $221 \le x \le 737$; $92 \le x \le 866$

5. 0.5239   7. 0.3446   9. 0.0166   11. 0.9987

13. 0.8902   15. 0.7698

17. (a) 0.3811   (b) 0.8849

19. (a) 0.3050   (b) 0.0851

21. 10.56 percent   23. 0.2302

**Chapter 7,
Proficiency test
(page 347)**

1. $-\infty$ 2. $-\frac{3}{2}$ 3. $-\infty$ 4. $\frac{2}{3}$ 5. 0   6. ∞

7. ∞ 8. $\frac{1}{2}$ 9. ∞ 10. ∞ 11. 1   12. e^6

13. ∞ 14. 1   15. ∞ 16. $\frac{3}{5}$ 17. $\frac{1}{4}$ 18. $\frac{2}{3}$

19. $\frac{1}{4}$ 20. $\dfrac{1}{\ln 2}$ 21. $\frac{1}{3}$

22. 10,000   23. $120,000

24. The population will increase without bound.

25. (a) 1   (b) $\frac{1}{3}$ (c) $\frac{1}{3}$

26. (a) 1   (b) $\frac{1}{3}$

27. (a) 1   (b) 0.3694   (c) 0.3679

28. $E(x) = \frac{5}{2}$; $\text{Var}(x) = \frac{3}{4}$

29. $E(x) = 1$; $\text{Var}(x) = \frac{1}{2}$

30. $E(x) = 5$; $\text{Var}(x) = 25$

31. $\frac{2}{9}$

32. (a) 0.0498   (b) 2 minutes

33. $30 \le x \le 34$; $28 \le x \le 36$; $26 \le x \le 38$

34. (a) 0.7422   (b) 0.0222   (c) 0.5
    (d) 0.9251   (e) 0.6584   (f) 0.8664

35. (a) 0.5764   (b) 0.5   (c) 0.2620

36. 36.74 percent

Chapter 8, Section 1 (page 357)

1. (a) $R(x_1, x_2) = 200x_1 - 10x_1^2 + 25x_1x_2 + 100x_2 - 10x_2^2$
   (b) $1,840

3. (a) 160,000
   (b) The level will increase by 16,400 units.
   (c) The level will increase by 4,000 units.
   (d) The level will increase by 20,810 units.

5. $f_x = 15(3x + 2y)^4$, $f_y = 10(3x + 2y)^4$

7. $\dfrac{\partial z}{\partial x} = 3(1 + y)(x + xy + y)^2$, $\dfrac{\partial z}{\partial y} = 3(1 + x)(x + xy + y)^2$

9. $\dfrac{\partial w}{\partial x} = 9(3 + z^2)(3x + 2y + xz^2)^8$, $\dfrac{\partial w}{\partial y} = 18(3x + 2y + xz^2)^8$,

   $\dfrac{\partial w}{\partial z} = 18xz(3x + 2y + xz^2)^8$

11. $f_r = \dfrac{2r}{s^3}$, $f_s = \dfrac{-3r^2}{s^4}$

13. $\dfrac{\partial z}{\partial x} = (1 + xy)e^{xy}$, $\dfrac{\partial z}{\partial y} = x^2e^{xy}$

15. $f_x = -\dfrac{1}{x^2y^2}$, $f_y = -\dfrac{1}{y^2} - \dfrac{2}{xy^3}$

17. $\dfrac{\partial z}{\partial x} = 1 + \ln xy$, $\dfrac{\partial z}{\partial y} = \dfrac{x}{y}$

19. Daily output will increase by approximately 10 units.

21. Monthly profit will increase by approximately $3,360.

23. (a) Butter and margarine

    (b) $\dfrac{\partial Q_1}{\partial p_2} \ge 0$ and $\dfrac{\partial Q_2}{\partial p_1} \ge 0$ (c) No

25. $f_{xx} = 60x^2y^3$, $f_{yy} = 30x^4y$, $f_{xy} = f_{yx} = 60x^3y^2 + 2$

27. $f_{xx} = 2y(1 + 2x^2y)e^{x^2y}$, $f_{yy} = x^4e^{x^2y}$, $f_{xy} = f_{yx} = 2x(1 + x^2y)e^{x^2y}$

29. $f_{xx} = \dfrac{y^2}{(x^2 + y^2)^{3/2}}$, $f_{yy} = \dfrac{x^2}{(x^2 + y^2)^{3/2}}$, $f_{xy} = f_{yx} = \dfrac{-xy}{(x^2 + y^2)^{3/2}}$

31. $\dfrac{\partial^2 Q}{\partial K^2} \approx$ change in marginal product of capital generated by an increase in capital investment of $1,000.

33. (a) $\dfrac{\partial^2 Q}{\partial L^2} > 0$ for $L < L_0$ and $\dfrac{\partial^2 Q}{\partial L^2} < 0$ for $L > L_0$

**Chapter 8,
Section 2
(page 364)**

1. $\dfrac{dz}{dt} = 7$

3. $\dfrac{dz}{dt} = \dfrac{1}{3}$

5. $\dfrac{dz}{dt} = -3t^{-4}$

7. $\dfrac{dz}{dt} = -162t^5$

9. 23   11. $-\frac{1}{3}$ 13. 5

15. The monthly demand will be increasing at the rate of 7 per month.

17. Approximately 61.6 additional units will be produced.

19. Daily profit will increase by approximately 24 cents.

21. 112 square yards

23. $\dfrac{d^2z}{dt^2} = a^2f_{xx} + 2abf_{xy} + b^2f_{yy}$

**Chapter 8,
Section 3
(page 372)**

1.   3.

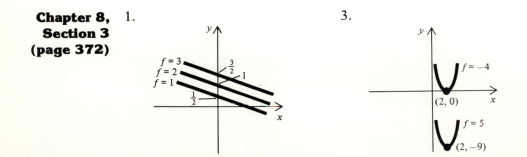

5.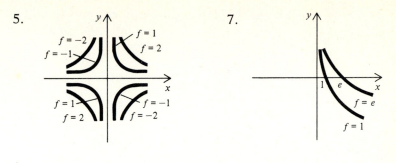

7.

9. $x^2 + y^2 = 4$ 11. $x \ln y = 0$ 13. $\dfrac{y}{x}$

15. $\dfrac{3 - 2xy}{x^2 + 6y^2 - 2}$ 17. $-\dfrac{y}{x} \ln y$ 19. $-\frac{1}{6}$ 21. 2

23. The level of unskilled labor should be reduced by approximately 2.4 hours.

25. Approximately 250 worker-hours

Chapter 8, Section 4 (page 381)

	Relative maxima	Relative minima	Saddle points
1.	(0, 0)	none	none
3.	none	none	(0, 0)
5.	(−2, −1)	(1, 1)	(−2, 1) and (1, −1)
7.	none	(4, $\frac{19}{2}$)	(2, $\frac{7}{2}$)

9. Price the first system at $3,000 and the second at $4,500.

11. $x = 200$, $y = 300$

13. (a) Relative minimum   (b) Saddle point

Chapter 8, Section 5 (page 391)

1. $f(\frac{1}{2}, \frac{1}{2}) = \frac{1}{4}$ 3. $f(1, 1) = f(-1, -1) = 2$

5. $f(0, 2) = f(0, -2) = -4$ 7. 40 meters by 80 meters

9. 3,456 cubic inches

11. Radius $= 1$ inch, height $= 4$ inches

13. $40,000 on labor; $80,000 on equipment.

15. Maximum output would increase by approximately 31.75 units.

17. (a) $4,000 on development and $6,000 on promotion.
    (b) $\lambda = 0$

Chapter 8,
Proficiency test
(page 394)

1. (a) $f_x = 6x^2y + 3y^2 - \dfrac{y}{x^2}, f_y = 2x^3 + 6xy + \dfrac{1}{x}$

    (b) $f_x = 5y^2(xy^2 + 1)^4, f_y = 10xy(xy^2 + 1)^4$
    (c) $f_x = y(1 + xy)e^{xy}, f_y = x(1 + xy)e^{xy}$

2. (a) $f_{xx} = 2, f_{yy} = 6y - 4x, f_{xy} = f_{yx} = -4y$
    (b) $f_{xx} = (2 + 4x^2)e^{x^2+y^2}, f_{yy} = (2 + 4y^2)e^{x^2+y^2},$
    $f_{xy} = f_{yx} = 4xye^{x^2+y^2}$

    (c) $f_{xx} = 0, f_{yy} = \dfrac{-x}{y^2}, f_{xy} = f_{yx} = \dfrac{1}{y}$

3. Daily output will increase by approximately 16 units.

4. (a) $\dfrac{dz}{dt} = -30t^4 + 24t^2$ (b) $\dfrac{dz}{dt} = 4t$

5. The demand will drop by approximately 46 cans per week.

6. (a)   (b)

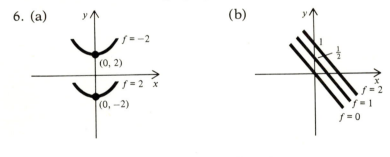

7. (a) $\frac{2}{3}$ (b) $-\frac{1}{2}$

8. The level of unskilled labor should be decreased by approximately 2 workers.

9. (a) Relative maximum at $(-2, 3)$; relative minimum at $(0, 9)$; saddle points at $(0, 3)$ and $(-2, 9)$.

    (b) Relative minimum at $(-\frac{23}{2}, 5)$; saddle point at $(\frac{1}{2}, 1)$.

10. Maximum value $= f(1, \sqrt{3}) = f(1, -\sqrt{3}) = 12$;
    minimum value $= f(-2, 0) = 3$.

12. $5,000 on development and $8,000 on promotion.

13. $4,000 on development and $7,000 on promotion.

14. Maximum profit will increase by approximately $235.

Chapter 9,
Section 1
(page 404)

1. $\displaystyle\iint_R 600{,}000\ dA$

3. $\displaystyle\iint_R \sqrt{x^2 + y^2}\ dA$

5. $\displaystyle\iint_R 1\ dA$

7. $\displaystyle\iint_R 1{,}400f(x, y)\ dA$

9. $\frac{7}{6}$ 11. -1 13. $4\ln 2$ 15. 32

17. $\frac{1}{3}$ 19. $\frac{3}{2}e^4 + \frac{1}{2}$ 21. $\frac{1}{3}\ln 2$

23. $\frac{1}{2}$ 25. $-\frac{1}{4}e^2 + e - \frac{1}{4}$

Chapter 9,
Section 2
(page 416)

1. $0 \le x \le 3$ and $x^2 \le y \le 3x$; $0 \le y \le 9$ and $\frac{y}{3} \le x \le \sqrt{y}$

3. $-1 \le x \le 2$ and $1 \le y \le 2$; $1 \le y \le 2$ and $-1 \le x \le 2$

5. $1 \le x \le e$ and $0 \le y \le \ln x$; $0 \le y \le 1$ and $e^y \le x \le e$

7. $\frac{3}{2}$ 9. $\frac{1}{2}$ 11. $\frac{44}{15}$ 13. $32\ln 2 - 12$

15. 3 17. 1 19. $\frac{3}{2}\ln 5$ 21. $2e - 4$

23. $\displaystyle\int_0^4 \int_0^{\sqrt{4-y}} f(x, y)\ dx\ dy$

25. $\displaystyle\int_0^1 \int_{y^2}^{\sqrt[3]{y}} f(x, y)\ dx\ dy$

27. $\displaystyle\int_0^2 \int_1^{e^y} f(x, y)\ dx\ dy$

29. $\displaystyle\int_1^2 \int_{-\sqrt{y-1}}^{\sqrt{y-1}} f(x, y)\ dx\ dy$

31. $\displaystyle\int_0^1 \int_0^y f(x, y)\ dx\ dy + \int_1^2 \int_0^{2-y} f(x, y)\ dx\ dy$

33. $\displaystyle\int_0^4 \int_{-\sqrt{y}}^{\sqrt{y}} f(x, y)\ dx\ dy + \int_4^9 \int_{-\sqrt{y}}^{6-y} f(x, y)\ dx\ dy$

Chapter 9,
Section 3
(page 429)

1. 18 3. $\frac{16}{3}$ 5. $\frac{4}{3}$ 7. 1 9. $\frac{19}{6}$

11. 8 13. $e - 1$ 15. $\frac{1}{2}e^2 - e + \frac{1}{2}$

17. $\displaystyle\iint_R f(x, y)\ dA \Big/ \iint_R 1\ dA$

19. \$188.61 per acre

21. 2 23. 0 25. $3(e - 2)$

27. 0.2285 29. 0.0803

31. 0.8452   33. 0.5269

35. $500 \iint\limits_{R} f(x, y)\, dA$

Chapter 9, Proficiency test (page 433)

1. -4 2. $\frac{1}{9}$ 3. $\frac{4}{5}\ln 2$ 4. $(e^{-2} - 1)(e^{-1} - 1) = 0.5466$

5. $\frac{1}{8}(e^2 - 1)$ 6. 81 7. $\frac{9}{2}$ 8. 256

9. $e - 2$ 10. $1{,}026$

11. $\displaystyle\int_0^2 \int_{y^2}^{\sqrt{8y}} f(x, y)\, dx\, dy$ 12. $\displaystyle\int_1^{e^2} \int_{\ln x}^2 f(x, y)\, dy\, dx$

13. $\displaystyle\int_0^4 \int_0^{\sqrt{y}} f(x, y)\, dx\, dy + \int_4^8 \int_0^{4-y/2} f(x, y)\, dx\, dy$

14. $\displaystyle\int_0^1 \int_{-\sqrt{y}}^{\sqrt{y}} f(x, y)\, dx\, dy + \int_1^5 \int_{-\sqrt{y}}^1 f(x, y)\, dx\, dy + \int_5^9 \int_{-\sqrt{y}}^{6-y} f(x, y)\, dx\, dy$

15. $\frac{9}{2}$ 16. $e - 1$ 17. $\frac{5}{6}$ 18. $\frac{1}{3}$ 19. $\frac{1}{2}(3 - e)$

20. $\frac{1}{2}$

21. (a) $10^6 \times \iint\limits_{R} f(x, y)\, dA$ (b) $\iint\limits_{R} f(x, y)\, dA \Big/ \iint\limits_{R} 1\, dA$

22. (a) 0.8625 (b) 0.9500

23. 0.4582

24. $1{,}200 \iint\limits_{R} g(x, y) f(x, y)\, dA$

Chapter 10, Section 1 (page 445)

1. $\displaystyle\sum_{n=1}^{\infty} \frac{1}{3^n}$ 3. $\displaystyle\sum_{n=1}^{\infty} \frac{n}{n + 1}$ 5. $\displaystyle\sum_{n=1}^{\infty} \frac{(-1)^{n+1} n^2}{n + 1}$

7. $\frac{15}{16}$ 9. $-\frac{7}{12}$

11. $\frac{1}{4}$ 13. $\frac{1}{2}$ 15. $\frac{2}{3}$ 17. 2

19. Diverges   21. -1 23. $-\frac{2}{5}$ 25. Diverges

Chapter 10, Section 2 (page 454)

1. 5 3. 3 5. Diverges   7. $\frac{3}{20}$

9. 45 11. $\frac{3}{16}$ 13. 100 15. $\frac{1}{3}$ 17. $\frac{25}{99}$

19. 575 billion dollars   21. $12,358.32

23. 27   25. 0.8202

Chapter 10, Section 3 (page 464)

1. Converges   3. Diverges   5. Converges   7. Converges

9. Converges   11. Converges   13. Diverges   15. Converges

17. Converges   19. Diverges   21. Converges   23. Converges

25. Diverges   27. Converges

Chapter 10, Section 4 (page 470)

1. Converges conditionally   3. Converges absolutely

5. Converges absolutely   7. Diverges

9. Diverges   11. Diverges

13. Diverges   15. Converges absolutely

17. Converges conditionally   19. Converges absolutely

21. Converges absolutely

Chapter 10, Section 5 (page 480)

1. $R = 1;\ -1 < x < 1$ 3. $R = 1;\ -1 < x < 1$

5. $R = 2;\ -2 \le x < 2$ 7. $R = \infty;\ -\infty < x < \infty$

9. $R = 1;\ -1 < x < 1$ 11. $R = \infty;\ -\infty < x < \infty$

13. $R = \frac{1}{3};\ -\frac{1}{3} < x < \frac{1}{3}$ 15. $R = 1;\ -3 < x < -1$

17. $R = 0;\ x = -5$

19. $\sum\limits_{n=0}^{\infty} 5^n x^{n+1};\ |x| < \frac{1}{5}$ 21. $\sum\limits_{n=1}^{\infty} \frac{n x^{n-1}}{(-2)^{n-1}};\ |x| < 2$

23. $\sum\limits_{n=2}^{\infty} \frac{n(n-1)x^{n-1}}{2};\ |x| < 1$ 25. $\sum\limits_{n=0}^{\infty} -\frac{3^{n+1} x^{n+1}}{n+1};\ |x| < \frac{1}{3}$

29. $\sum\limits_{n=0}^{\infty} \frac{x^n}{(-2)^n n!}$ 31. $\sum\limits_{n=0}^{\infty} \frac{x^{2n+1}}{2^{n-1} n!}$

33. 7.3810

**Chapter 10,
Section 6
(page 492)**

1. $\displaystyle\sum_{n=0}^{\infty} \frac{3^n}{n!} x^n$

3. $\displaystyle\sum_{n=1}^{\infty} \frac{(-1)^{n+1}}{n} x^n$

5. $\displaystyle\sum_{n=0}^{\infty} \frac{1}{(2n)!} x^{2n}$

7. $\displaystyle\sum_{n=0}^{\infty} \left(\frac{n+1}{n!}\right) x^n$

9. $\displaystyle\sum_{n=0}^{\infty} \frac{2^n e^2}{n!} (x-1)^n$

11. $\displaystyle\sum_{n=0}^{\infty} (-1)^n (x-1)^n$

13. $\displaystyle\sum_{n=0}^{\infty} (x-1)^n$

15. 1.94936   17. 0.09531   19. 1.34984

21. 0.46127   23. 0.09967

25. 0.26246; $|R_5| \leq 0.00012$ 27. 1.22133; $|R_3| \leq 0.00008$

29. 1.009950; $|R_2| \leq 0.0000005$ 31. 0.1823; $|R_5| \leq 0.00001$

33. 2.7183; $|R_8| \leq 0.000008$ 35. 1.0488; $|R_3| \leq 0.000004$

**Chapter 10,
Proficiency test
(page 495)**

1. $\frac{5}{6}$ 2. 1   3. $-\frac{1}{2}$

4. $-\frac{1}{2}$ 5. Diverges   6. 1.5415   7. 72

8. $\frac{17}{11}$ 9. 54 feet   10. \$3,921.67

11. 18.16 units   12. 0.3297

13. Diverges   14. Converges   15. Converges

16. Converges   17. Diverges   18. Converges

19. Converges   20. Diverges

21. Converges conditionally   22. Diverges

23. Converges absolutely

24. $R = \frac{1}{3}$; $-\frac{1}{3} \leq x < \frac{1}{3}$ 25. $R = \infty$; $-\infty < x < \infty$

26. $R = \frac{1}{6}$; $-\frac{1}{6} < x < \frac{1}{6}$ 27. $R = \frac{1}{2}$; $-\frac{3}{2} \leq x \leq -\frac{1}{2}$

28. $\displaystyle\sum_{n=0}^{\infty} (-1)^n \frac{(3x^2)^{n+1}}{n+1}$; $|x| < \frac{1}{\sqrt{3}}$ 29. $\displaystyle\sum_{n=1}^{\infty} (-2)^n n x^{n+1}$; $|x| < \frac{1}{2}$

30. $\displaystyle\sum_{n=0}^{\infty} (-1)^n \left(\frac{1}{3}\right)^{n+1} x^n$ 31. $\displaystyle\sum_{n=0}^{\infty} \frac{(-3)^n}{n!} x^n$

32. $\displaystyle\sum_{n=0}^{\infty} \left(\frac{2n+1}{n!}\right) x^n$ 33. $\displaystyle\sum_{n=0}^{\infty} (n+1)(x+2)^n$

34. 0.94869

35. 0.11919

36. 0.60651; $|R_5| < 0.00002$

37. -0.1053; $|R_3| < 0.00003$

Chapter 11,
Section 1
(page 507)

1. $\frac{25}{16} = 1.5625$

3. $-\frac{7}{16} = -0.4375$

5. $\frac{17}{16} = 1.0625$

7. $\frac{23}{16} = 1.4375$

9. $\frac{33}{16} = 2.0625$

11. 3.464

13. 0.755

15. -0.352

17. 4.791 and 0.209

19. 1.41421

21. 2.08008

Chapter 11,
Section 2
(page 517)

1. (a) 2.3438

(b) 2.3333

3. (a) 0.7828

(b) 0.7854

5. (a) 1.1515

(b) 1.1478

7. (a) 0.7430

(b) 0.7469

9. (a) 0.5090; $|E_4| \le 0.0313$

(b) 0.5004; $|E_4| \le 0.0026$

11. (a) 2.7967; $|E_{10}| \le 0.0017$

(b) 2.7974; $|E_{10}| \le 0.00002$

13. (a) 1.4907; $|E_4| \le 0.0849$

(b) 1.4637; $|E_4| \le 0.0045$

15. 0.7881

17. (a) $n = 164$

(b) $n = 18$

19. (a) $n = 36$

(b) $n = 6$

21. (a) $n = 179$

(b) $n = 8$

Chapter 11,
Section 3
(page 526)

1. $y = \frac{1}{4}x + \frac{3}{2}$

3. $y = 3$

5. $y = 0.78x + 1.06$

7. $y = -\frac{1}{2}x + 4$

9. (b) $y = 0.42x - 0.71$

(c) 1,306

11. (b) $y = 3.05x + 6.10$

(c) 42.7 percent

13. $y = 4.94e^{0.19x}$

15. (b) $y = 49.95e^{0.02x}$

(c) 85.71 million

Chapter 11,
Proficiency test
(page 529)

1. $\frac{23}{16} = 1.4375$

2. $-\frac{5}{16} = -0.3125$

3. $\frac{119}{16} = 7.4375$

4. 7.41620

5. -3.104

6. 0.258

7. (a) 1.1016; $|E_{10}| \leq 0.0133$
   (b) 1.0987; $|E_{10}| \leq 0.0004$

8. (a) 17.5651; $|E_8| \leq 10.2372$
   (b) 16.5386; $|E_8| \leq 1.0901$

9. (a) $n = 58$ (b) $n = 8$

10. (a) $n = 59$ (b) $n = 6$

11. $y = \frac{4}{9}x + 1$

12. (b) $y = 11.54x + 44.45$ (c) \$102,150

13. (b) $y = 4.08e^{0.03x}$ (c) 10.04 units

**Chapter 12,
Section 1
(page 544)**

1. 30°   3. 120°   5. −120°

7.   9.   11.

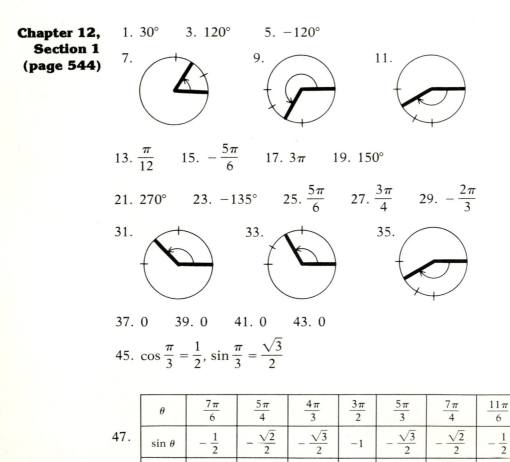

13. $\dfrac{\pi}{12}$ 15. $-\dfrac{5\pi}{6}$ 17. 3π 19. 150°

21. 270°   23. −135°   25. $\dfrac{5\pi}{6}$ 27. $\dfrac{3\pi}{4}$ 29. $-\dfrac{2\pi}{3}$

31.   33.   35.

37. 0   39. 0   41. 0   43. 0

45. $\cos\dfrac{\pi}{3} = \dfrac{1}{2}$, $\sin\dfrac{\pi}{3} = \dfrac{\sqrt{3}}{2}$

47.

θ	$\dfrac{7\pi}{6}$	$\dfrac{5\pi}{4}$	$\dfrac{4\pi}{3}$	$\dfrac{3\pi}{2}$	$\dfrac{5\pi}{3}$	$\dfrac{7\pi}{4}$	$\dfrac{11\pi}{6}$	2π
$\sin\theta$	$-\dfrac{1}{2}$	$-\dfrac{\sqrt{2}}{2}$	$-\dfrac{\sqrt{3}}{2}$	-1	$-\dfrac{\sqrt{3}}{2}$	$-\dfrac{\sqrt{2}}{2}$	$-\dfrac{1}{2}$	0
$\cos\theta$	$-\dfrac{\sqrt{3}}{2}$	$-\dfrac{\sqrt{2}}{2}$	$-\dfrac{1}{2}$	0	$\dfrac{1}{2}$	$\dfrac{\sqrt{2}}{2}$	$\dfrac{\sqrt{3}}{2}$	1

49. $-\frac{1}{2}$ 51. $\frac{1}{2}$ 53. $\frac{\sqrt{3}}{3}$ 55. 2   57. -1

59. $\frac{3}{4}$ 61. $\frac{3}{4}$ 63. $\frac{4}{5}$

65. $\theta = \frac{\pi}{2}$, $\theta = \frac{3\pi}{2}$, $\theta = \frac{\pi}{6}$, or $\theta = \frac{5\pi}{6}$

67. $\theta = \frac{\pi}{2}$, $\theta = \frac{\pi}{3}$, or $\theta = \frac{2\pi}{3}$

69. $\theta = \frac{\pi}{2}$ or $\theta = \frac{\pi}{4}$

71. $\theta = \frac{\pi}{2}$

73. $\theta = \frac{\pi}{6}$ or $\theta = \frac{5\pi}{6}$

Chapter 12,
Section 2
(page 554)

1. $f'(\theta) = 3 \cos 3\theta$

3. $f'(\theta) = -2 \cos (1 - 2\theta)$

5. $f'(\theta) = -3\theta^2 \sin (\theta^3 + 1)$

7. $f'(\theta) = 2 \cos \left(\frac{\pi}{2} - \theta\right) \sin \left(\frac{\pi}{2} - \theta\right) = \sin (\pi - 2\theta)$

9. $f'(\theta) = -6(1 + 3\theta) \sin (1 + 3\theta)^2$

11. $f'(\theta) = -e^{-\theta/2} (2\pi \sin 2\pi\theta + \frac{1}{2} \cos 2\pi\theta)$

13. $f'(\theta) = \dfrac{\cos \theta}{(1 + \sin \theta)^2}$

15. $f'(\theta) = -5\theta^4 \sec^2 (1 - \theta^5)$

17. $f'(\theta) = -4\pi \tan \left(\frac{\pi}{2} - 2\pi\theta\right) \sec^2 \left(\frac{\pi}{2} - 2\pi\theta\right)$

19. $f'(\theta) = 2 \dfrac{\cos \theta}{\sin \theta} = 2 \cot \theta$

Chapter 12,
Section 3
(page 564)

1. 60 radians per hour

3. 0.15 radian per minute

5. $\frac{\pi}{2}$ radians

7. $\frac{\pi}{3}$ radians

9. 8 feet

Chapter 12,
Section 4
(page 572)

5. 0.7074292; $|R_4| \le 0.002490$

7. 0.0872665; $|R_1| \le 0.0038077$

9. 0.6192282; $|R_2| \le 0.110762$

11. 0.2079117; $|R_5| \le 0.0000001$

13. $n = 6$; 0.9396926   15. $n = 9$; 0.9659263

17. $n = 4$; 0.9975641

Chapter 12,
Proficiency test
(page 574)

1. (a) $\frac{2\pi}{3}$ radians; 120°   (b) $-\frac{5\pi}{4}$ radians; −225°

2. 0.8727 radian   3. 14.3239°

4. (a) $\frac{\sqrt{3}}{2}$ (b) $\frac{\sqrt{2}}{2}$ (c) 2   (d) $-\frac{1}{\sqrt{3}}$

5. $\frac{4}{3}$ 6. $\frac{3}{2}$

7. $\theta = \frac{\pi}{2}$ or $\theta = \frac{3\pi}{2}$ 8. $\theta = \frac{\pi}{3}$ or $\theta = \frac{2\pi}{3}$

9. $\theta = \frac{\pi}{6}$ or $\theta = \frac{5\pi}{6}$ 10. $\theta = \frac{\pi}{6}$ or $\theta = \frac{5\pi}{6}$

14. $f'(\theta) = 6(3\theta + 1) \cos (3\theta + 1)^2$

15. $f'(\theta) = -6 \cos (3\theta + 1) \sin (3\theta + 1)$

16. $f'(\theta) = 6\theta \sec^2 (3\theta^2 + 1)$

17. $f'(\theta) = 12\theta \tan (3\theta^2 + 1) \sec^2 (3\theta^2 + 1)$

18. $f'(\theta) = \dfrac{\cos \theta - 1}{(1 - \cos \theta)^2}$

19. $f'(\theta) = -2 \tan \theta$

21. 0.012 radian per minute   22. $\frac{\pi}{3}$ radians   23. 27 feet

25. 0.3089916; $|R_3| \le 0.0004059$

26. 0.8091019; $|R_4| \leq 0.0008161$

27. $n = 4$; 0.9848078   28. $n = 8$; 0.7660436

**Appendix,
Section A
(page 584)**

1. $x^4(x - 4)$ 3. $(x + 2)(x - 2)$

5. $(x + 2)(x - 1)$ 7. $(x - 3)(x - 4)$

9. $(x - 1)^2$ 11. $(x - 1)(x^2 + x + 1)$

13. $x^5(x + 1)(x - 1)$ 15. $2x(x - 5)(x + 1)$

17. $x = -2, x = 4$ 19. $x = -5$ 21. $x = -4, x = 4$

23. $x = -1, x = -\frac{1}{2}$ 25. none   27. $x = -\frac{3}{2}$

29. 125   31. 4   33. 4   35. $\frac{1}{2}$ 37. 1

39. $\frac{1}{4}$ 41. 2   43. $\frac{1}{4}$ 45. $n = 3$ 47. $n = -3$

49. $n = -2$ 51. $n = -\frac{1}{6}$ 53. 34   55. 0

57. $\sum\limits_{i=1}^{10} 3i$ 59. $\sum\limits_{i=1}^{6} 2x_i$ 61. $\sum\limits_{i=1}^{8} i(-1)^{i+1}$

**Appendix,
Section B
(page 596)**

1. $\lim\limits_{x \to a} f(x) = b$ 3. $\lim\limits_{x \to a} f(x) = b$

5. Limit does not exist.   7. 4   9. 7   11. 16

13. $\frac{3}{4}$ 15. Limit does not exist.   17. 2

19. 7   21. $\frac{5}{3}$ 23. 5   25. $\frac{1}{4}$ 27. 2

29. Yes   31. Yes   33. No   35. No

37. No   39. No   41. Yes   43. None

45. $x = 3$ 47. $x = -1$ 49. $x = -5, x = 1$

51. $x = -1, x = 2$ 53. $x = 2$ 55. $f(1) = 4$

57. Impossible

INDEX

INDEX